How to Design Heating-Cooling Comfort Systems

Fourth Edition

Joseph B. Olivieri, P.E.

BNP Business News Publishing Company
Troy, Michigan

Library of Congress Cataloging in Publication Data

Olivieri, Joseph B.
  How to design heating-cooling comfort systems.

  Includes index.
  1. Heating. 2. Air conditioning. I. Title.

TH7222.04   1986   697   85-31387
ISBN 0-912524-36-7

Fourth Edition
Administrative Editor: Phillip R. Roman
Technical Editor: Matthew McCann

Copyright History
Third Edition Copyright © 1973
Second Edition Copyright © 1971
First Edition Copyright © 1970

Printed in the United States of America
10 9 8 7 6 5 4 3 2 1

Contents

Acknowledgement

The author wishes to express his gratitude to the many manufacturers, professional societies, associates, students and editors who have freely given of both their time and resources to this book possible. Foremost is the American Society of Heating, Refrigerating and Air-Conditioning Engineers whom, in addition to the materials expressly credited throughout this book, have granted permission to reprint and extract supplementary materials from the ASHRAE Handbook of Fundamentals.

Disclaimer

Foreword

At this time, I want to thank all those responsible for helping me get the fourth revision of this book in print. The original edition of this book was inspired by the lecture notes Mr. Eric F. Hyde passed on to me when I was asked to step in for him as lecturer on mechanical design at the Engineering Society of Detroit. This book is partially dedicated to the memory of Mr. Hyde.

Others I'd like to acknowledge for their assistance are Joseph Bobbio of Hyde and Bobbio Consulting Engineers, Professor John J. Uicher of the University of Detroit College of Engineering and David S. Falk, manufacturer's representative.

Finally, this book is dedicated to the most important people in my life: my late parents, my wife, children and grandchildren.

Note, I initially prepared the psychrometric chapter in partial fulfillment of the requirements for a Master of Science Degree from Wayne State University School of Medicine.

Other books available from
Business News Publishing Company

Hydronics
Good Piping Practice
Humidification Handbook
The Schematic Wiring Book Set
Refrigeration Licenses Unlimited
The Valve Selection and Service Guide
Marine Refrigeration and Fish Preservation
Systematic Commercial Refrigeration Service
How To Design Heating/Cooling Comfort Systems
How To Solve Your Refrigeration and A/C Service Problems
Getting Started in Heating and A/C Service
The (MSAC) Service Hot Line Handbooks
How To Make It In The Service Business
The A/C Cutter's Ready Reference
HVAC/R Reference Notebook Set
Industrial Refrigeration Systems

For more information, write to:
Business News Publishing Company
Book Division, P.O. Box 2600
Troy, MI 48007

CHAPTER 1

Introduction

IT IS SAID that a house without a tree "ain't" fit for a dog. In a similar fashion, a building without a good environmental system is nothing more than a mausoleum. At no other time in history has man had more control over his environment, yet, rarely do we take full advantage of this knowledge. We too often install heating systems or cooling systems instead of a thermal environmental control system. There is much ugliness in the world and the architect must do all he can to reduce the amount of ugliness by creating well designed buildings. However, a well designed building is more than a pretty facade. If not, it is like a Hollywood set—all front, with emptiness inside. A well designed building includes an environmental control system which will comfort the body as the beautiful building comforts the soul. To understand how to design a proper environmental control system, we must learn what constitutes a comfortable environment. Before we study comfort, we should study definitions and explanations of terms commonly used.

AIR CONDITIONING

True air conditioning is a complete control of environment. To have true air conditioning you must be able to heat or cool, humidify or de-humidify, filter the room air, and distribute the treated air to the space.

HEAT

Heat is that form of energy that exists by virtue of a temperature difference.

TEMPERATURE

The temperature of a substance is a measure of its molecular activity. There are two arbitrary scales in use for measuring temperature. The fahrenheit scale (which is used in the U.S.A.) sets a degree as $1/_{180}$ of the difference between the temperature of the formation of ice and the boiling point of water. The centigrade scale is based on $1/_{100}$ of the difference between ice and steam.

SENSIBLE HEAT

Sensible heat is that heat added or removed which can be measured by a change in temperature of the fluid or substance.

DRY-BULB TEMPERATURE

A thermometer measures dry-bulb temperature, that is, the sensible heat.

LATENT HEAT

Latent heat is the heat required to change the state of a substance. It takes approximately one Btu to heat one pound of water from 211° to 212°, but to change this pound of water from a liquid at 212° to a vapor (steam) at 212° takes about 1000 Btu. This 1000 Btu is the latent heat. Similarly, the heat required to change water to ice is the latent heat: with no change in temperature, 144 Btu/lb.

WET-BULB TEMPERATURE

Wet-bulb temperature is the temperature measured by wrapping a wet cloth about the bulb of a thermometer and forcing air across the bulb, either by whirling the thermometer through the air or using a fan. This air evaporates moisture from the wet cloth, which causes a depression in temperature. The drier the air, the more moisture is evaporated and the more the temperature is depressed. The dry-bulb temperature and wet-bulb temperature are used to determine relative humidity.

RELATIVE HUMIDITY

Relative humidity is an expression of the amount of moisture in the air compared to the maximum amount that can exist without condensation.

BTU

A Btu is a British thermal unit, which is the name given to the amount of heat required to raise one pound of water one degree Fahrenheit. (This is not completely accurate, but close enough for our purpose.)

HUMIDIFY

To add moisture to air.

DE-HUMIDIFY

To remove moisture from air.

DEW POINT

The temperature at which condensation of water vapor begins for a given state of humidity and pressure.

CONDENSATE

The liquid that is formed by the condensation of a vapor. When steam is condensed, the liquid formed (water) is the condensate.

AMBIENT AIR

The air surrounding an object.

TON OF REFRIGERATION

The amount of cooling from a ton of ice melting in 24 hours; 288,000 Btu/24 hours or 12,000 Btuh.

HEAT TRANSFER

Heat is transferred from a high level source to an object at a lower temperature in three ways: by conduction, convection, and radiation.

CONDUCTION

"The mechanism of heat transfer whereby the molecules of higher kinetic energy transmit part of their energy to adjacent molecules of lower kinetic energy by direct molecular action." *

The classic example of conduction of heat is that of holding a cold poker in a hot fire. As the end in the fire becomes hot due to the fire, the opposite end which is being held gradually becomes hot also. The transmission of heat from one end of the poker to the other is by conduction. The molecules in the end in the fire begin to move faster because they are heated. They jump up and down just as you do when you try to walk barefoot across the hot sand at the beach. The reaction set up is something like an old slapstick movie. Remember that scene where one person in a crowd has fleas and is scratching and jumping around? Soon, the two persons next to the scratcher begin to scratch. They pass it on to two each and start four scratching and soon the people way at the other end are scratching and jumping. This is conduction. The heated molecules hit adjacent molecules, passing some of their increased energy on to them, finally reaching the end you are holding, and it gets hot.

CONVECTION

"Transfer of heat by convection involves energy by movement of a fluid or gas from an object of high temperature to one of a lower temperature. Transfer of heat by convection is in part dependent upon conduction transfer wherein the liquid or gas molecules are in contact with the warm object and are heated by direct molecular action." *

Have you ever sat in your living room in the evening, quietly smoking and thinking great thoughts? Perhaps only one lamp is on. Notice how the smoke from your cigarette slowly drifts over to the lamp, entering the bottom of the shade. Now watch the velocity increase, and isn't it traveling when it comes out the top of the shade? This is convection. You are able to see it thanks to the smoke. The air near the heat source is warmed and becomes less dense, or lighter, and so rises. It is replaced by heavier air which, in turn, is heated. The heated air, when it gets away from the heat source, begins to cool and drop, returning to the heat source to be reheated.

RADIATION

"Transfer of heat by radiation takes place by a change in energy form from internal energy at the source to electro-magnetic energy for transmission, then back to internal energy at the receiver." *

Examples of radiation heat transfer are the sun and fireplaces. I am sure you have had the experience of driving in your car on a bright, sunshiny, cold, winter day and finding that you had to turn off your heater. The sun radiates heat directly and is not affected by the surrounding temperature. Radiation is like a machine gun. The sun takes packets of energy and hurls them rapidly like bullets. Nothing happens until these bullets of energy hit a target. Remember how, when you stand in front of a fireplace, the part facing the fire intercepts all these bullets of energy, but the opposite side does not. So, while your front may get too hot, your back is cold and you have to turn around.

STANDARD AIR

Standard air is air with a density of 0.075 lb per cu ft and an absolute viscosity of 0.0379×10^{-5} lb mass per (ft) (sec). This is equivalent to dry air at 70°F and 29.92 in. Hg barometric pressure.

*ASHRAE GUIDE AND DATA BOOK, Pg. 49, Chapter 4, Heat Transfer, 1961 Edition.

CHAPTER 2

Comfort

A BUILDING is a system designed to provide people with a comfortable environment. We are accustomed to thinking about heating and air conditioning systems when we should be thinking in terms of comfort systems. To discuss comfort systems, we must understand what creates a feeling of comfort. In order to do this, we must understand how our body works.

Every person's body is a machine. This machine's fuel is food. The body combines food and oxygen in a low temperature combustion process and converts the fuel into energy and heat.

A person at rest or engaged in light activity gives off 410 Btuh. 300 Btuh is given off by radiation and convection, 65 Btuh by evaporation, and 45 Btuh by respiration. If this heat loss is accelerated, we feel cold. If it is retarded, we feel hot.

The rate of heat loss depends upon our degree of activity. Our machine rejects much more heat when engaged in frenzied activity than when sitting at rest. Table 2-1 gives the amount of heat rejected by our bodies while engaged in various activities. The unit *met* is used. A *met* is 18.4 Btuh/sq ft of skin surface per hour. An average man will have 21.8 sq ft of surface.

Age is also a factor. The maximum heat loss per sq ft of skin area occurs at age 5. It is 25 Btuh/sq ft. At 40, the rate is 20 Btuh/sq ft and at 80 the rate is 13 Btuh/sq ft.

The amount of clothing that is worn is a factor. After all, clothing is an insulator and so more clothing means more insulation. The insulating value of clothing is given in *clothes,* often abbreviated as *clo*.

Table 2-2 gives the insulating value of various cloth-ing ensembles. A 0.1 *clo* is equal to a 1°F drop in temperature. So, you can see that a girl in a minidress requires a temperature as much as 10° warmer than a man in a midweight suit.

The ambient temperature and relative humidity also are factors, as is the temperature of surrounding surfaces. For most people, a winter temperature of 75°, along with 30% relative humidity, appears to be ideal for comfort. (In summer the conditions are 75° and 50%.) If the temperature is held at 75° and the relative humidity is reduced, we feel cold because our machine rejects more moisture and so more heat to the space. Conversely, if the relative humidity is raised, we cannot reject enough moisture and we feel hot and sweaty. However, the comfort range of relative humidity is quite large. The body notices no difference in comfort from 50% RH down to 20%. During the cooling season, the body will not feel uncomfortable until the relative humidity reaches a point where wet sweating occurs. This is roughly above 60% RH at 75°.

Surface temperatures are an important factor which are too often neglected or forgotten when considering comfort. Our machine radiates heat to the surroundings. If the enclosing surfaces are too cold, the heat lost by radiation is accelerated and our machine is cold. Conversely, if too hot, then the body will absorb heat and be too hot. The skin surface temperature of the average North American is about 90°F and will radiate heat to anything at a colder temperature and be heated by anything at a warmer temperature. This is why glass is a villain.

The temperature of the skin is the most important

TABLE 2—1   Met Rates of Various Activities

Activity	Metabolic Rate in met units*
RESTING	
Sleeping	0.7
Reclining	0.8
Seated, quiet	1.0
Standing, relaxed	1.2
WALKING	
On the level mph	
2	2.0
3	2.6
4	3.8
MISCELLANEOUS OCCUPATIONS	
Bakery (e.g., cleaning tins, packing boxes)	1.4–2.0
Brewery (e.g., filling bottles, loading beer boxes onto belt)	1.2–2.4
Carpentry	
Machine sawing, table	1.8–2.2
Sawing by hand	4.0–4.8
Planning by hand	5.6–6.4
Foundry Work	
Using a pneumatic hammer	3.0–3.4
Tending furnaces	5.0–7.0
Garage Work (e.g., replacing tires, raising cars by jack)	2.2–3.0
General Laboratory Work	1.4–1.8
Machine Work	
Light (e.g., electrical industry)	2.0–2.4
Heavy (e.g., steel work)	3.5–4.5
Shop Assistant	2.0
Teacher	1.6
Watch repairer, seated	1.1
Vehicle driving	
Car	1.5
Motorcycle	2.0
Heavy vehicle	3.2
Aircraft flying, routine	1.4
Instrument landing	1.8
Combat flying	2.4
DOMESTIC WORK, WOMEN	
House cleaning	2.0–3.4
Cooking	1.6–2.0
Washing by hand and ironing	2.0–3.6
Shopping	1.4–1.8
OFFICE WORK	
Typing	1.2–1.4
Miscellaneous office work	1.1–1.3
Drafting	1.1–1.3
LEISURE ACTIVITIES	
Stream fishing	1.2–2.0
Calisthenics exercise	3.0–4.0
Dancing, social	2.4–4.4
Tennis, singles	3.6–4.6
Squash, singles	5.0–7.2
Basketball, half court, intramural	5.0–7.6
Wrestling-competitive or intensive	7.0–8.7

* Ranges are given for those activities which may vary considerably from one place of work or leisure to another or when performed by different people. 1 met = 58.2 W/m²; 50 kcal/hr·m²; 18.4 Btu/hr·ft². Some occupational and leisure time activities are difficult to evaluate because of differences in exercise intensity and body position.

TABLE 2—2   Insulating Value of Clothing

Type of Clothing	clo
Heavy suit, men's	1.0
Midweight suit, men's	0.7
Dress	0.5
Minidress	0.1
Bikini	0.01
Arctic clothing	5.0

factor in comfort. A drop in skin temperature of 1°F is equal to a 4°F drop in air temperature. A 1°F drop in mean radiant temperature is the same as a 1°F drop in air temperature.

Mean radiant temperature (MRT) is the surface temperature of a large, uniformly heated *surround* in which the occupant would experience a net radiant loss equal to that which occurs in the actual room. So, when we say that we have a MRT of 50° we mean that the occupant is losing as much heat as he would if he were surrounded by 50° surfaces. Remember that we defined radiant heat transfer as the shooting of little bullets of energy from a hot body to a cold one. The sun, which is hot, fires off bullets which strike and heat us but do not heat the space through which they pass. When we are in a room that has a MRT which is lower than our skin temperature, we act like the sun and try to heat these cold surfaces.

A few years ago, I did some research on just this problem. Thanks to my understanding wife, I turned our living room into a calorimeter or test chamber. Our living room has an open beam or studio ceiling. The wall adjacent to the patio has two 6-ft x 7-ft sliding glass doors. The chamber was built around one of these doors. In the living room, I built a room 6 ft x 7 ft x 16 ft deep. On the patio, I constructed a cold room so I could maintain any outside temperature I wished. I maintained the cold side at temperatures ranging from −10°F to 30°F. while testing a variety of heating systems. I used the following:

1. Air blowing up the glass.
2. Air blowing down the glass.
3. Baseboard radiator under the glass.
4. Radiant ceiling panel.
5. Radiant floor panel.
6. Radiant floor and ceiling panel.

At 0°, the closest one could sit to this glass and be comfortable was 4 feet. At 30° one could sit about 2 feet from the glass and be reasonable comfortable. The type of heating system had little effect on comfort. Although single glass was not tested, it is obvious that the results would have been horrendous. This problem is really very simple to solve: be judicious in the use of glass. If large glass areas are desirable or necessary to fulfill an architectural concept, use large glass areas in a large room so that furniture, and therefore people, can be kept away from the wall. It is ridiculous to use floor-to-ceiling glass in a 10-ft x 15-ft living room. If the amount of glass in relationship to the wall area is small, then blanketing the glass with heat will provide sufficient comfort.

Table 2-3 gives some examples of surface temperatures for various wall constructions:

TABLE 2—3 Surface Temperatures of Exterior Walls

Wall	"U"	Surface Temp. @0° Outside	Surface Temp. @20° Outside
8-in. Cinder block	.41	54	60
8-in. Cavity wall	.25	62	66
8-in. Cavity wall with 2-in. rigid insulation	.125	69	70
Brick veneer, stud wall, m.l. and plaster, plywood sheathing	.36	57	62
Brick veneer, stud wall, with 2-in. blanket insulation	.10	70	71
Single glass	1.13	17	33
Double glass ½-in. air space	.55	47	54

To maintain optimum comfort conditions, a 75°F mean radiant temperature is necessary along with a 75°F space temperature. Space temperature and the mean radiant temperature are interrelated. The values for this interrelation are given different values by different researchers. McNall reports approximately a 1-to-1 relationship. That is, a one degree rise in space temperature is offset by a one degree drop in MRT and vice versa.

If the MRT is equal to 75°F, then a 75°F space temperature will provide comfort. As the MRT drops, the space temperature must rise. Let us examine a 20-ft x 20-ft x 10-ft room with a slab-on-grade floor, insulated roof, insulated wall and a 58 sq ft window with single glass. The MRT, when the subject stands in the center of the room facing the window, is 65.3°. So this 9.7° drop in MRT must be offset by a 9.7° rise in air temperature to 84.7°. At 2 feet from the window, the MRT is 57°. Therefore the air temperature must rise to 93°. So, which air temperature do we maintain, 84.7° or 93°? The problem is better solved by raising the MRT at the outside wall. If double glass is used, the MRT rises to 62°. Adding panel heat raises it even more. However, studies by McNall and his colleagues at Kansas State University showed that the surface temperature could not be more than 20° lower than the room temperature, even with panel heating.

Franger states that, for a sedentary person wearing light clothing, the MRT should be 90°, at 75° room temperature and approximately 40 fpm velocity. Light clothing would be the kind a young female office worker would wear. For typing and related activities, the MRT should be 80°. With the usual men's office

dress (middleweight business suit), the MRT should be 75° for a sedentary person and about 60° for a person who is typing. You can see how difficult it is to keep people comfortable when they vary in age, in type of clothing, in types of activities and are at varying distances from cold surfaces.

Another cause of thermal discomfort is air motion. Under certain conditions, we call air motion a pleasant, cooling breeze, but under others it is called a draft. A draft is defined by ASHRAE as: *any localized feeling of coolness or warmth of any portion of the body due to both air movement and air temperature.* A draft equation is also presented by ASHRAE.

$$t_{ed} = (t_x - 76) - 0.07 (V_x - 30)$$
t_{ed} = equivalent draft temperature
t_x = temperature of local air stream
V_x = velocity of local air stream

As you can see, every 15 fpm increase in velocity over 30 fpm is equal to a one degree drop in temperature.

This equation was used to construct the graphs shown in Figures 2-1 and 2-2. Figure 2-1 is for the neck and Figure 2-2 for the ankle region. The reaction of people to these various conditions was then tested and, from these tests, the *percent uncomfortable* lines were drawn.

As you can see, when we either raise the local air velocity or drop the temperature of the air stream, we will feel a draft. At 50 fpm and at a temperature 3° lower than room temperature, 40% of the people felt discomfort at the neck region.

The same conditions caused only 10% of the people to feel uncomfortable in the ankle region. You could

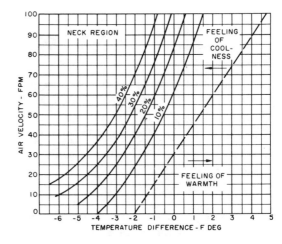

FIG. 2—1 Percent of Occupants Objecting to Drafts at the Neck

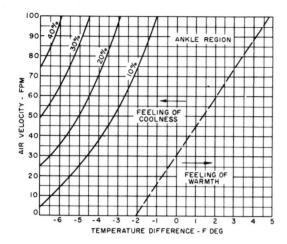

FIG. 2—2 Percent of Occupants Objecting to Drafts at the Ankles

conclude from this that we don't have to be too concerned about drafts at the ankle region. This is fine for adults, but crawling babies spend all day near their parents' ankles.

The ADPI (Air Distribution Performance Index), Figure 2-3, proposed by Miller and Nevins* makes use of this formula. If the draft temperature is kept between −3° and +2° at 30 fpm and air velocity does not exceed 70 fpm, 80% of the population will be comfortable.

To determine the ADPI, the room temperature and velocity are measured at 216 points within a space. The percentage of these 216 measurements that meet the −3° to +2° and less than 70 fpm velocity standards is the ADPI. A comfortable heating and air conditioning system must have a high ADPI, at least 90.

*Room Air Distribution with an Air Distributing Ceiling, Part II, ASHRAE Transactions, Vol. 75, Part I, 1969.

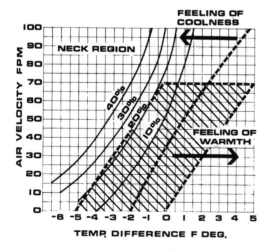

FIG. 2—3 Air Distribution Performance Index

The ASHRAE *Handbook of Fundamentals* presents the work done by Straub, Gillman and others on the effect of diffuser and grille location on room air distribution.

Figure 2-4A shows heating from a floor or sill location. Notice how the warm air blankets the outside wall preventing drafts. The stagnant area is at the inside wall.

Figure 2-4B shows cooling from the same location. The cold air pretty much hangs along the outside wall causing little if any turnover of air in the rest of the room.

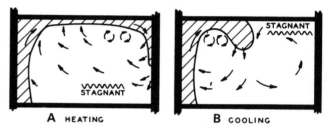

FIG. 2—4 Heating and Cooling from a Floor or Sill Outlet

FIG. 2—5 Heating and Cooling from an Overhead Outlet

When heated air is introduced from either high on the inside wall or at the ceiling, Figure 2-5A, the stagnant zone is at the outside wall and cold drafts wash across the floor. In fact, we would be cold from the hips down.

Figure 2-5B shows cooling from the same position. Notice the fine distribution of cooling with this arrangement. Incidently, Straub and Gillman showed that return air location had little effect on room air distribution.

We now have some parameters for comfort. A comfortable environment consists of:

1. Room temperature of 75°.
2. Room relative humidity of 30% to 50%.
3. Surface temperature of 50° minimum with 75° mean radiant temperature.
4. No perceptible air motion.
5. No drafts.
6. An ADPI above 90.

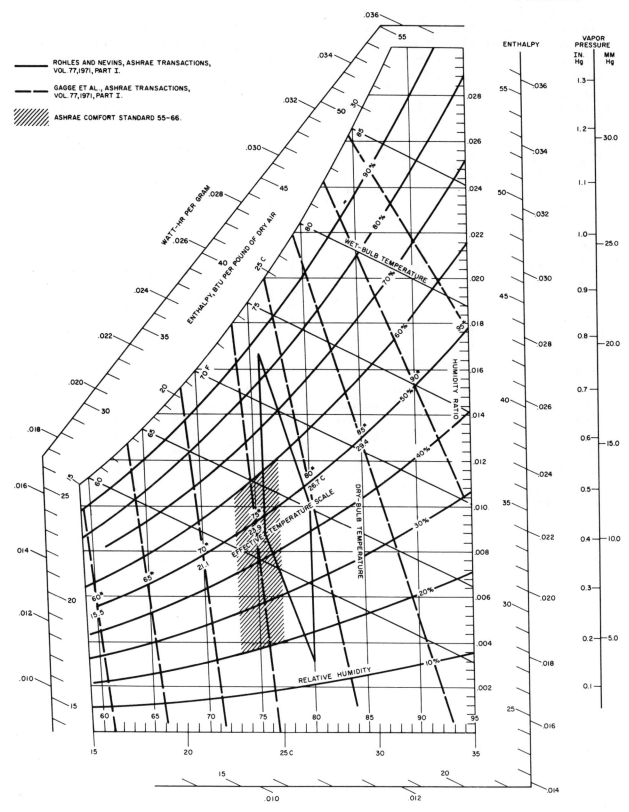

FIG. 2—6 The New ASHRAE Comfort Chart

But is this all we need to be comfortable? In later chapters, ventilation and sound will be discussed in detail. Keep in mind that a fresh, quiet room is as important to total comfort as the items previously mentioned are to thermal comfort.

Two other items must be mentioned: temperature control and reliability. If thermal comfort is influenced by all the variables mentioned, then we must install a temperature control system that will keep the temperature and relative humidity within the limits set. But remember individual variation. If you have four private offices occupied by:

1. An older man wearing a traditional suit,
2. A young girl in a brief dress,
3. A young fireball executive who works in his shirtsleeves,
4. A mature woman in a tweed suit with a mid-calf-length skirt,

who gets the thermostat? You had better select a system with a thermostat in each room.

ASHRAE and Kansas State University have produced the new ASHRAE comfort chart shown in Figure 2-6. Formerly, we had a table in which we listed recommended dry bulb temperatures for various occupants. Now we recommend the use of the ASHRAE comfort chart, with adjustments for clothing, physical activity and MRT. The chart was constructed from tests using persons wearing 0.5 to 0.7 *clo* clothing. If the clothing of the occupants is less, raise the temperature; if more than this, lower the temperature.

Adjust the temperature 1°F for every 0.1 *clo* for sedentary or light work (less than 1.6 *met*). For medium to heavy work, adjust the temperature 2°F for every 0.1 *clo*.

If we use 400 Btuh as our base, then we should decrease the dry bulb temperature by 3°F for every 100 Btuh increase in activity. Be sure that the space relative humidity remains below 60% to prevent wet sweating.

Now for MRT. Calculating MRT is difficult and time consuming. For the purposes of adjusting the comfort chart, the area-weighted average surface temperature may be used.

Watch your room air velocity. If a room has a high rate of air change, such as a *clean room* or operating room, which causes room velocities to be more than 30 fpm, then the room air temperature must be increased. Remember the draft formula. For every 15 fpm over 30 fpm we must increase the air temperature 1°F. If the room air velocity is 120 fpm then the temperature increase must be 6°, (120 - 30)/15 = 6.

Finally notice the slanted lines called ET. These are lines of *effective temperature*. The old ASHRAE *effective temperature* chart overemphasized the effect of relative humidity on comfort. These new effective temperature lines incorporate the latest knowledge on the effect of temperature and relative humidity on comfort. What these lines tell us is that 75° at 45% RH and 76° at 20% RH have the same effective temperature of 75°. That is, your feeling of comfort or discomfort would be the same at these two conditions.

CHAPTER 3

Heating and Cooling Loads

To properly size the heating-cooling comfort system we are are going to design, we must first determine the heating and cooling load.

Heat loss calculations are usually done on a room-by-room basis. The mathematics is really quite simple and involves only the use of arithmetic. In Chapter 1, we defined heat transfer by conduction. Heat loss calculations involve primarily conduction heat transfer. If we wish to determine the rate of heat transfer through one square foot of material, we could do it by experimentation. We could take the material under test and place it in an insulated chamber, so that all the heat is transferred through the material. Then, we could apply heat to one side of the chamber and maintain a fixed temperature on this side of the material. If we measured how much heat was needed to keep the chamber at the selected temperature and measured the temperature on each side of one square foot of the material, we would know the rate of heat transfer through the material. If it took 10 Btu/hr to keep one side of the material at 100°, while the other side stayed at 50°, we would say that 10 Btu/hr was transferred per square foot per 50° (100° − 50°) temperature difference. We could state it mathematically as:

$$\frac{10 \text{ Btu}}{(\text{hr})(\text{sq ft})(50°)}$$

which can be reduced to:

$$\frac{10 \text{ Btu}}{50 \ (\text{hr})(\text{sq ft})(°F)} = \frac{.20 \text{ Btu}}{(\text{hr})(\text{sq ft})(°F)}$$

It would be most inconvenient to have to set up your apparatus each time you wanted to calculate the heat loss in a room. Fortunately, most common materials have been measured and the results recorded. These rates of heat transfer, per hour per square foot per °F temperature difference, are called U factors. In the example above, U = .20. The amount of heat transferred per hour is calculated by multiplying the area of the surface, through which the heat is being transferred, by its U factor and the difference in temperature from one side of the surface to the other. We can write a simple equation as follows:

$$Q = A(U)(t_i - t_o)$$

Q = Heat transferred in Btuh (Btu/hr)
A = Area in sq ft
U = Heat transmission coefficient
  $(\text{Btu}/(\text{hr})(\text{sq ft})(°F))$
t_i = Inside air temperature (°F)
t_o = Outside air temperature (°F)

Let's try a simple problem.

Example #1

How much heat is lost through a 100 sq ft surface, whose U factor is 0.40 (Btuh) (sq ft) (°F), if the outside temperature is 30° and the inside temperature is 70°?

$$Q = A(U)(t_i - t_o)$$

A = 100 sq ft
U = 0.40(Btuh)(sq ft)(°F)
t_i = 70°
t_o = 30°

Q = 100(0.40)(70 − 30)
Q = 100(0.40)(40)
Q = 1600 Btuh

As you can see, the heat transferred is directly related to the temperature difference between outdoors and indoors. If, in Example #1, the outdoor temperature is −10° (10 below zero) then the heat loss is doubled.

Example #2

How much heat is transferred if the outdoor temperature, in Example #1, is −10°?

$Q = A(U)(t_i - t_o)$

$Q = 100(0.40)(70 - -10)$

$Q = 100(0.40)(80)$

$Q = 3200$ Btuh

However, we can conserve heat by adding insulation to the wall to lower its heat transmission coefficient

Example #3

How much heat is transferred if the U factor, in Example #2, is reduced to 0.20 Btu/(hr)(sq ft)(°F)?

$Q = A(U)(t_i - t_o)$

$Q = 100(0.20)(70 - -10)$

$Q = 100(0.20)(80)$

$Q = 1600$ Btuh

You can see that insulation can do much to reduce heat loss in winter and heat gain in summer.

Determining the amount of heat being transferred is simple enough as long as you know the area, the heat transmission coefficient, and the temperature difference. The area is determined by measuring either the space itself, in the case of an existing building, or by using scale drawings prepared by an architect or designer.

At one time, as many wall and roof combinations as possible were listed in tables, in order to reduce the burden of arithmetic. With the universal use of hand-held electronic calculators, only values for representative wall and roof sections are now shown in most handbooks. These can be easily converted to your own roof or wall by deletion and substitution of the appropriate values.

However, the process of calculating U factors is so simple that it is almost as easy as the substitution method. First, some definitions:

Heat transfer coefficients are usually given in terms of heat transferred, per hour and per degree of temperature difference, through a square foot of surface that is perpendicular to the heat flow.

When the heat is transferred through one inch of an homogeneous material, we call the heat transfer coefficient *thermal conductivity* or the *k factor*. The temperature difference is between the hotter and the colder surface. With an homogeneous material, we only need to know the heat transfer coefficient for one inch because it is the same for every inch. The first inch of a poured concrete wall is the same as the third and fourth inches and so on.

When we have a nonhomogeneous material, such as a concrete block, the first inch is not the same as the second inch. The third inch may be the same as the second, but the fourth inch may again be different. Therefore, we need a heat transfer coefficient for the whole unit.

This coefficient is called *thermal conductance or the C factor*. The temperature difference is again measured from surface to surface.

Heat is transferred through a solid by conduction. At the surface of a material, heat is transferred to the air by a combination of conduction, convection, and radiation. The heat transmission coefficient for this process is called the *film* or *surface conductance*. It is often called the film coefficient. This book refers to the film coefficient as the *f factor* although, in some books, it is called the h factor. The value of the f factor varies with wind velocity and is a maximum with still air conditions.

When dealing with a composite unit (wall, roof, floor, etc.), we call the overall heat transmission coefficient the *U factor*. While the temperature difference for C or k factors is taken from surface to opposite surface, the U factor includes the air film on both surfaces. For example, the temperature difference for a wall would be the difference in temperature between the room air and the outside air.

Resistance or resistivity to heat transfer is called simply *R*. When we speak of resistance, R, we mean the reciprocal of the conductance or C factor. When we speak of resistivity, R, we mean the reciprocal of the conductivity or k. To convert resistivity to resistance, simply multiply the resistivity by the thickness of the material in inches.

To recap the coefficients, remember that

$$U = \frac{Btu}{(hr)(sq\ ft)(°F)}$$

$$C = \frac{Btu}{(hr)(sq\ ft)(°F)}$$

$$k = \frac{Btu\ per\ inch}{(hr)(sq\ ft)(°F)}$$

$$f = \frac{Btu}{(hr)(sq\ ft)(°F)}$$

and $$U = \frac{1}{R}$$

Isn't that simple? All we need to do is find the resistance of a material and take its reciprocal to convert it to a U factor. How do we find the resistance? Well the total resistance is the sum of the resistances of the individual components that make up a wall, roof or floor. Or

$$R = R_1 + R_2 + R_3 + R_4 .... + R_n$$

If we were dealing with a material that has a C factor, we need only take the reciprocal of the C factor to get the resistance R.

$$R = \frac{1}{C}$$

When the material is homogeneous, and the k factor is used, we must multiply the thickness of the material, x, by the reciprocal of its k factor to determine the resistance R.

$$R = x\frac{1}{k}$$

This leaves only the film conductance to worry about. The film conductance is treated in the same fashion as the C factor. The resistance is the reciprocal of the film conductance.

$$R = \frac{1}{f}$$

Now that we know how to determine the resistances we can proceed to calculate the U factor.

Example #4

Calculate the U factor for a four inch exterior wall constructed of common brick. Wind velocity is 15 mph.

We have only three items to worry about, the four inch common brick and the two air film conductances. One film conductance is for still air on the room side, the other is for outside air.

Table	Item	C, f or k	R
3-1	Outside air film	6.00	1/6 = 0.17
3-3	4″ brick	5.0	4/5 = 0.80
3-1	Inside air film	1.46	1/1.46 = 0.68
	Total resistance		1.65

$$U = \frac{1}{R} = \frac{1}{1.65} = 0.606$$

As you can see, we calculate the U factor in much the same way that we build a hero sandwich. Example #4 was simple because we had only three layers. Would six layers be more difficult? Not really.

Example #5

Calculate the U factor for an exterior wall constructed of four inch face brick, backed by four inch common brick, one inch of insulation (smooth skin, expanded polystyrene, 1.8 lb/ft³ density), and ½ inch sand aggregate, gypsum plaster.

Table	Item	C, f or k	R
3-1	Outside air film	6.00	1/6 = 0.17
3-3	4″ face brick	9.0	4/9 = 0.44
3-3	4″ common brick	5.0	4/5 = 0.80
	1″ insulation	0.25	1/.25 = 4.00
	½ in. plaster	11.1	1/11.1 = 0.09
3-1	Inside air film	1.46	1/1.46 = 0.68
	Total resistance		6.18

$$U = \frac{1}{R} = \frac{1}{6.18} = 0.162$$

Now that you have had a chance to examine Tables 3-1 and 3-3, you will notice that not only are the thermal conductivity (k factor), thermal conductance (C factor) and film conductance (f factor) given, but the resistance (R) is also listed.

Be careful. While the resistance (R) for the conductance (C) is given directly, resistance for conductivity (k) is given for one inch of the material and must be multiplied by the thickness of the material.

Table 3-2 lists the thermal resistance of air spaces. Values are given for various thicknesses (0.5, 0.75, 1.5, and 3.5 inches), for materials of varying emittance, and on the basis of air space position: whether it is vertical, horizontal or sloping.

What is emittance? You probably learned all about this in physics but, in case you have forgotten, let's review a bit. The brighter the material, the more of the radiant heat transfer it will reflect. The more radiant energy is reflected, the better the insulating value of an air space. How do we determine the emittance of an air space? Turn to Table 3-1. Section B lists reflectivity and emittance of various materials along with the effective emittance of an air space.

Example #6

What is the thermal resistance of the following vertical air spaces:
a. 0.5 inch bounded by face brick
b. 3.5 inch bounded by face brick
c. 0.5 inch bounded by bright aluminum foil
d. 3.5 inch bounded by bright aluminum foil

First we must find the effective emittance in section B of Table 3-1. For ordinary materials, E = 0.82 and for bright aluminum, E = 0.03. We still need to know two other things: the mean air space temperature and the temperature difference across the air space. Let's assume a 50° air space and a 10° temperature difference. With these facts at hand, we can enter Table 3-2 and read the following:

a. 1.5 inch bounded by ordinary materials, R = 1.02
b. 3.5 inch bounded by ordinary materials, R = 1.01
c. 0.5 inch bounded by aluminum foil, R = 2.66
d. 3.5 inch bounded by aluminum foil, R = 3.63

Deciding what is the mean temperature and the temperature difference across an air space are the most difficult. To do it properly requires repeated trial and error solutions. To ease the burden, it is usually acceptably accurate to use a 50° mean temperature and a 10° temperature difference for heat loss calculations. For heat gain calculations, use a 90° mean temperature and a 10° temperature difference.

Example #7

Calculate the U factor for a frame wall constructed of 4″ face brick, ²⁵⁄₃₂″ insulating sheathing, 2 × 4 studs, and ⅜″ dry wall.

Let's find the resistance of these materials in Table 3-3. The face brick had a resistance per inch of 0.11. When multiplied by the 4″ thickness, the total resistance is 0.44. The insulating sheathing is listed under building boards. Generally, a vegetable fiberboard is used. All dimensions are given as decimals, so you must convert ²⁵⁄₃₂ to 0.78125. Now you should have no trouble finding a resistance of 2.06. The 2 × 4 stud is listed under Wood near the end of Table 3-3. While we speak of 2 × 4's, we are really giving the nominal dimension. Once upon a time, 2 × 4's were 2″ by 4″. Today, they are 1½″ by 3½″. Under Woods, find *fir, pine, etc., 3.5 in.* and read R = 4.35.

The air space is bounded by ordinary materials and we will use a 50° mean and a 10° temperature difference. The air space resistance is 1.01. This leaves the drywall. You won't find drywall listed, but you will find gypsum board, which is the proper name. Converting ⅜″ to 0.375, we find that R = 0.32. Now we need the film or surface resistance. We have been using 15 mph, which is the standard wind velocity. However, when dealing with real world problems, the wind velocity at the building site must be used.

When you think about a frame wall, you can visualize two heat transfer paths: one through the studs and the other through the space between the studs. We will calculate the U factor for both paths.

Item	Resistance Path 1	Path 2
Outside air film	0.17	0.17
4″ face brick	0.44	0.44
0.78125 sheathing	2.06	2.06
2″ × 4″ wood	4.35	
Air space		1.01
0.375 gypsum	0.32	0.32
Inside air film	0.68	0.68
Total resistance	8.02	4.68

To arrive at the U factor, we must combine these two total resistances according to their proportions. When studs are spaced 16″ on center, the framing will occupy 20% of the wall. So the average resistance is:

$$U = .20U_1 + .80U_2$$

$$U = .20\frac{1}{R_1} + .80\frac{1}{R_2}$$

$$U = .20\frac{1}{8.02} + .80\frac{1}{4.68}$$

$$U = 0.025 + 0.171$$

$$U = 0.196$$

In the past, the path with the least resistance was used. Had we done this, the U factor would have been 1/4.68 or 0.214. The difference between 0.196 and 0.214 is 9%, which isn't much. Had we filled the space between the studs with insulation, our resistance for path 2 would have been 15.68 and the U factor would be 0.068. Now, the combined U factor becomes 0.0795. Had we used 0.068 instead of 0.0795, the error would have been 16.9%. There is one big difference between the two cases: the 9% error gave a safety factor, but the 16.9% difference creates a significant error.

While we have only considered walls, the same technique is used for roofs. Roofs may be constructed of wood, metal or masonry. The roof assembly can be a roof alone or a combination of roof and suspended ceiling. A suspended ceiling improves the U factor due to both its resistance and the resistance of the air space between the ceiling and the roof.

The outside film resistance is the same as the one used for walls, but the inside or still air resistance is different. You will remember that the air space resistance is different, as well.

Example #8

What is the U factor of a flat roof, constructed of built-up roofing, 2″ of roof insulation, 4″ of concrete and a suspended, acoustical ceiling that is ³⁄₄″ thick. The air space is 3½″ deep.

The roofing resistance, appropriately enough, is found under Roofing in Table 3-3. Its resistance is 0.33. Roof insulation is found under the heading Insulating Materials in Table 3-3. There are a number of subheads and you will find roof insulation under the fourth subhead. If possible, get the k factor from the manufacturer. If you cannot, the ASHRAE tables can be safely used. The Table is a bit confusing, so just use R = 2.77/in. For two inches, R = 5.54.

Concrete is found under Masonry Materials. You will notice that there are several different concretes. Unless you know otherwise, always use *Sand and gravel or stone aggregate (not dried)* which has a resistance of 0.08/in. or a total resistance of 0.32.

The air space resistance is found in Table 3-2. The air space is horizontal and bounded by ordinary materials

(E = 0.82) so the resistance is 0.93.

Just as with the wall, use a 50° mean temperature and a 10° temperature difference.

The acoustical tile is given under Building Board and has a resistance of 1.89.

Item	Resistance
Outside air film	0.17
$^3/_8$ " built up roofing	0.33
2" roof insulation	5.54
4" concrete	0.32
$3^1/_2$ " air space	0.93
$^3/_4$ " acoustical tile	1.89
Inside air film	0.61
Total resistance	9.79

$$U = \frac{1}{R} = \frac{1}{9.79} = 0.102$$

Example #9

What is the U factor for a single sheet of vertical glass, for two sheets of glass separated by a $^1/_2$ " air space, and for three vertical sheets of glass separated by $^1/_2$ " air spaces?

Turn to Table 3-4 which lists the U factors for vertical glass, horizontal glass, and solid wood doors. The U factor for single vertical glass is 1.10 for an outdoor exposure and 0.73 for an indoor exposure. The indoor U factor is lower because of the higher, insulating film coefficient of still air. The U factor for two sheets of glass, with a $^1/_2$ " air space, are 0.49 and 0.46. When three sheets of glass are divided by $^1/_2$ " air spaces, the U factors are 0.31 and 0.30.

As you can see, multiple sheets of glass dramatically reduce the U factor. A single sheet of glass has virtually no resistance to heat transfer, what U factor it has is due to the resistance of the film coefficients.

Table 3-5 shows the U factor for a typical frame wall. Two types of construction are shown. Construction 1 has no insulation between the framing while construction 2 has 3.5 in. of R-11 blanket insulation.

Table 3-6 is a typical masonry wall, a combination of concrete blocks and brick. Table 3-7 is a roof built on a metal deck. Table 3-8 gives the U factor for a masonry roof with and without insulation. Table 3-9 is for a roof constructed of wood, with and without insulation. Table 3-10 is used for floor and ceiling combinations.

If you study these Tables, you will notice that they are really demonstrations of the substitution and deletion technique. For example, column 1 in Table 3-10 is for a floor and ceiling combination without insulation. This is a typical intermediate floor. If the ceiling were below an attic, you would delete items 5 through 8 and replace item 4, an air space, with insulation. You may use this

technique if you wish, but I find it easier to start from scratch each time.

Speaking of attics, they do present a special problem. When dealing with building components that have an outdoor exposure, you use the temperature difference between indoors and outdoors. But what is the temperature difference in an attic that is ventilated with outdoor air to prevent condensation? This means that the temperature will be close to the outdoor temperature, but how close? We can use this equation:

$$t_a = \frac{A_c U_c t_i + t_o (2.16\, A_c V_c + A_r U_r + A_w U_w + A_g U_g)}{A_c (U_c + 2.16 V_c + A_r U_r + A_w U_w + A_g U_g)}$$

where: t_a = temperature in unheated space, °F
t_i = indoor design temperature of heated room, °F
t_o = outdoor design temperature, °F
A_c = area, ceiling
A_w = area, wall
A_g = area, glass
A_r = area, roof
U_c = overall heat transfer coefficient, ceiling
U_w = overall heat transfer coefficient, wall
U_g = overall heat transfer coefficient, glass
U_r = overall heat transfer coefficient, roof
V_c = rate of introduction of outside air into the unheated space by infiltration and/or ventilation, cfm/ft²

The areas the walls, glass, ceiling, and roof are simply measured from the architect's plans. But what do you do about ventilation? The recommended rate of attic ventilation is 0.5 cfm/ft² for power ventilation and 0.1 cfm/ft² for natural ventilation. Unless you have reason to believe that the ventilation rate is different, use 0.1 cfm/ft².

Example #10

Calculate the temperature in an unheated attic if the outside temperature is 5 °F and the room temperature is 70 °F. The areas are as follows: ceiling, A_c = 1500 ft², roof, A_r F 1800 ft², end walls, A_w = 290 ft², glass in the end wall, A_g = 10 ft². The U factors are: ceiling U_c = 0.05, roof U_r = 0.50, walls U_w = 0.08, glass U_g = 1.10. Assume natural ventilation, V_c = 0.1 cfm/ft²

$$t_a = \frac{A_c U_c t_i + t_o (2.16\, A_c V_c + A_r U_r + A_w U_w + A_g U_g)}{A_c (U_c + 2.16\, V_c) + A_r U_r + A_w U_w + A_g U_g}$$

$$t_a = \frac{1500(0.05)(70) + 5[(2.16)(1500)(0.1) + (1800)(0.5) + (290)(0.08) + (10)(1.10)]}{1500[0.05 + 2.16(0.1)] + 1800(0.5) + 290(0.08) + 10(1.10)}$$

$$t_a = 8.66$$

Basement and floor heat losses present a problem. At one time, arbitrary values were used for soil temperatures adjacent to basement walls. Now we can use a technique, developed by ASHRAE, that assumes that the heat transfer path is circular as in Figure 3-1. The first step is to calculate the length of the path from the basement wall to the surface. Next, calculate the U factor, using a k for soil of 9.6 Btuh/in./ft²/°F.

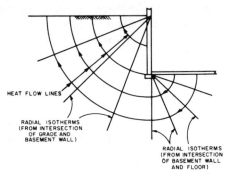

FIG. 3-1 Heat Flow from Basement

To determine the length of the path, arcs are drawn at one foot intervals. The center of each arc is on the inside of the basement wall, at ground level. The arc length through the wall must be subtracted from the total arc length. If the wall thickness is used, instead of the wall arc length, an error of only 0.5% results. Therefore, we will use the wall thickness.

Example #11

Calculate the average U factor for a basement wall eight feet below grade. The wall is 10 in. thick, poured concrete with an R value of 0.08 per inch.

Depth	Total arc length, in.	Wall thickness in.	Length through soil, in.	Resistance
1	18.85	10	8.85	.92
2	37.70	10	27.70	2.89
3	56.55	10	46.55	4.84
4	75.40	10	65.40	6.81
5	94.25	10	85.25	8.78
6	113.10	10	103.10	10.74
7	131.95	10	121.95	12.70
8	150.80	10	140.80	14.67
				62.35

$$R_{avg} = \frac{62.35}{8}$$
$$R_{avg} = 7.79$$

To this average resistance, add the resistance of the wall and the inside and outside air films. The outside film is at ground level since the heat flow path terminates there. Use a 15 mph wind velocity.

$$U = \frac{1}{R}$$
$$U = \frac{1}{0.17 + 7.79 + 10(0.08) + 0.68}$$
$$U = \frac{1}{9.44} = 0.106$$

This newer method was shown to give you an understanding of how these computations are made. With increasing use of underground structures and earth berms, it is good to know how to calculate the U factor for underground walls.

However, a shorter method is available for standard basements, using Tables 3-11 and 3-12. The beauty of Table 3-11 is that it can be used for insulated or uninsulated basement walls, even partially insulated walls. As you can see, the U factor for uninsulated walls decreases as we go down into the earth. Therefore, it may be economically feasible to insulate only the top three feet.

Let's compare the heat loss through an uninsulated, an insulated and a partially insulated basement wall. To use Table 3-11, add the U factors for each foot of wall, then multiply this composite U by the perimeter of the wall and the temperature difference between indoors and outdoors.

Example #12

Use Table 3-11 to calculate the heat loss through the below grade portion of a basement wall. The basement is 30 × 50 ft, with 7 ft of wall below grade. Outside temperature is 0°F, indoors is 70°F.

Determine the heat loss through an uninsulated wall, a wall that is fully insulated with one in. of R = 4.17 material, and a wall whose top three feet are insulated with R = 4.17 insulation.

Depth	Uninsulated	Insulated	Three feet insulated
1	0.410	0.152	0.152
2	0.222	0.116	0.116
3	0.155	0.094	0.094
4	0.119	0.079	0.119
5	0.096	0.069	0.096
6	0.079	0.060	0.079
7	0.069	0.054	0.069
	1.150	0.624	0.725

$$Q = PU_a(t_i - t_o)$$
P = Basement perimeter
= 50 + 30 + 50 + 30
U_a = Composite U
t_i = Inside temperature
t_o = Outside temperature

Uninsulated wall Q = 160(1.150)(70 − 0)
= 12,880 Btuh
Insulated wall Q = 160(0.624)(70] 0)
= 6989 Btuh
Partially insulated wall Q = 160(0.725)(70 − 0)
= 8120 Btuh

As you can see, there is a big savings if the wall is insulated and almost as big a savings if only the top three feet are insulated.

Calculating the floor loss is even easier. There are two things to be considered: the basement depth and the building width. For example, if the basement is 7 ft below grade and the building is 32 ft wide, the average U factor is 0.021 Btuh/ft²/°F. Multiplying the basement area by this factor and the temperature difference between outdoors and indoors gives the heat loss.

Example #13

What is the heat loss through a basement floor that is 40 ft long by 24 ft wide, if the basement is 6 ft below grade and the outside temperature is 0°F?

From Table 3-12, the U factor for this floor is 0.027.

$$Q = (40 \times 24)(0.027)(70 - 0)$$
$$Q = 1814 \text{ Btuh}$$

The heat loss from slab floors on grade is expressed in Btuh/ft of perimeter. Table 3-13 give the heat loss, for various types of construction, for slabs on grade.

The floor over a crawl space is handled in two ways.

If the crawl space is ventilated, the heat loss is figured for the floor above. Even though it is ventilated, the temperature of the crawl space is less than outdoors. In general, assume that the temperature difference between the room and the crawl space is ⅔ of the difference between the room and outdoors. If the room is 70° and outdoors is 10°, the temperature difference between the room and the crawl space is ⅔ (70 − 10) = 40°.

When the crawl space is not ventilated, treat it as another room. Assume that the temperature in the space is 70° and calculate the heat loss through the crawl space walls in the normal manner. The floor loss is treated just like a slab on grade.

INFILTRATION

The wind blows and the heat goes. This is called infiltration.

All buildings have intentional openings, such as windows and doors. In addition, there are the unintentional openings, such as cracks in walls, cracks around window and door frames, the crack around the door itself and around the ventilating section of windows. As the wind blows against a building, it forces itself in through all these cracks and crevices. If the entering air is at space temperature, it will not affect the space temperature. But, if it is warmer or colder than the space, it will increase or decrease the space temperature. Infiltration becomes as much of a load as heat transmission by conduction.

Another type of infiltration is caused by the *stack effect* or chimney effect. Stack effect occurs whenever the inside temperature is higher than the outside temperature. Because the density of warm air is less than cold air, there is a difference in pressure that causes cold air to enter the building at lower levels and warm air to flow out at higher levels. This can be a very serious problem in tall buildings. The only way to offset this type of infiltration is by pressurizing the building with outside air. For high rise buildings, consult the *ASHRAE Handbook–Fundamentals*.

Two common methods of calculating infiltration are the *volume air change method* and the *crack method*.

With the volume air change method, it is assumed that a room with one outside wall will experience enough infiltration to change the air in the room once every hour. With two outside walls, there will be 1½ air changes an hour and, with three outside walls, there will be two air changes every hour.

Example #14

What is the infiltration in a 10 ft by 10 ft by 8 ft room with one outside wall?

$$\text{Volume} = 10 \times 10 \times 8 = 800 \text{ ft}^3$$
$$\text{ft}^3/\text{hr} = 800 \times 1 \text{ AC/hr}$$
$$\text{ft}^3/\text{hr} = 800$$

Example #15

If the same room has two exposed walls, what is the infiltration?

$$\text{ft}^3/\text{hr} = 800 \times 1.5 \text{ AC/hr}$$
$$\text{ft}^3/\text{hr} = 1200$$

Example #16

Now, consider the same room with three exposed walls.

$$\text{ft}^3/\text{hr} = 800 \times 2 \text{ AC/hr}$$
$$\text{ft}^3/\text{hr} = 1600$$

While this method is simple, care must be used in its application. It was developed for use with traditional, residential construction and cannot and should not be used with buildings having large glass areas and lightweight construction. Nor should it be used with nonresidential construction.

You will find, through experience, that some engineers have developed their own factors and use them rather than the ASHRAE values.

The *crack* method is more accurate and rational. For

windows, the *crack* length is the perimeter of the ventilating section. The only exception is the double-hung window, where crack is the perimeter of the window plus the seam where the two sections meet. For a door, the crack length is the perimeter of the door.

Tables 3-15 and 3-17 give the infiltration rates for various windows in cubic feet per hour, per foot of crack, at typical wind velocities. Table 3-17 lists the window types found in existing buildings while Table 3-15 lists those used in new construction.

If possible, obtain the the actual infiltration rate from the window manufacturer. When working with an architect, ask him to specify a maximum infiltration rate and use that value.

Multiplying the given factor by the feet of crack equals the infiltration rate in cubic feet per hour.

Because the wind blows in a single direction, the air entering the building on the windward side will leave on the lee side. If the wind hits the building diagonally, not as much of it will infiltrate as when blowing directly at the windows. Therefore, when a room has more than one wall, the infiltration is adjusted as follows:

With two exposed walls, use the wall with the greatest infiltration.

With three exposed walls, use the wall with the most infiltration, but not less than ½ the total.

Infiltration through doors is calculated in a variety of ways. Table 3-15 can be used for sliding doors, but be careful. The factors shown are for feet of crack with windows and overall area with sliding doors.

For residential swinging doors, use the door perimeter multiplied by the Table 3-16 values. Notice that two values are given: one for weather-stripped and one for plain doors.

Commercial doors also have two values. In commercial buildings, there are doors that are rarely opened and those that are frequently opened. Doors that are opened, at most, two times an hour are considered infrequently opened. With all actively opened doors, use the *per opening* factor. Estimating how often a door is opened is a real challenge. However, the building owner or architect should have some idea how often a door will be opened. Multiply the number of openings per hour, by the factor from Table 3-16, to obtain the cubic feet per hour (cfh) of infiltration.

Infiltration through walls is normally neglected because, with good construction, it is minimal. However, masonry walls that have not been plastered can be quite porous, particularly if the workmanship is poor. With good workmanship, the infiltration will be 9 cfh/ft² while, with poor workmanship, it will be 25 cfh/ft². These values are with a 15 mph wind. Infiltration

through a plastered masonry wall is only 0.08 cfh/ft².

Another wall to watch is a curtain wall where the windows and solid, insulated wall panels are installed in a lightweight frame. Depending on the workmanship, leakage can vary from minimal to unbelievable.

Now that we know how much air is infiltrating, how is this converted to Btuh? Multiply the weight of the infiltrating air by its specific heat and the temperature rise and you have your answer. The specific heat of air is 0.24 Btu/lb/°F. But we don't know the weight of air, only the cfh. Multiplying the air volume, in cfh, by the density of air (0.075 lb/ft³) will convert air volume to air weight. Multiplying density by specific heat (0.075 × 0.24), we obtain a simple, useful factor of 0.018 Btu/ft³/°F.

Example #17

What is the heat loss due to the infiltration of 1000 cfh of 20°F air?

$$Q = 1000 \text{ cfh} \times 0.018 \text{ Btu/ft}^3/°F \times (70 - 20)$$
$$Q = 900 \text{ Btuh}$$

Example #18

Calculate the heat loss from a 10 ft by 10 ft by 10 ft room with two outside walls. There are two double-hung, 3 ft by 4 ft, wood sash windows in the south wall and another in the east wall. Each window has double glass with a ¼" air space.

The walls are constructed of 4 in. face brick veneer, 0.5 in. vegetable fiber board sheathing (regular density), 2 × 4 studs, 3½ in. batt insulation between the studs, and 0.5 in. gypsum wallboard.

The roof is constructed of 2 in. wood, 2 in. rigid roof insulation, and built-up roofing. There is no suspended ceiling.

The floor is slab on grade with 2 in., L-type insulation.

The indoor temperature will be 70°F and the outdoor design temperature is 0°.

Now, let us assemble the U factors.

First, the glass, from Table 3-4. In the second line, under insulating glass, double, we find that the winter U factor is 0.58.

Next, the wall. When we consult Table 3-5a, we find a frame wall with 3½ in. insulation. The veneer is wood siding rather than brick, but we can use the substitution technique. Deduct the resistance of the wood siding (0.81) from the total (14.43) and add the resistance of the face brick.

In Table 3-3, under Masonry Units, face brick is shown to have an R of 0.11 per inch. Therefore, the R for a 4 in. brick is 0.44. Substituting, the resistance between the framing is $R_f = 14.43 - 0.81 + 0.44 = 14.06$. Resistance at the framing is $R_s = 7.81 - 0.81 +$

0.44 = 7.44. Now combine these.

$$U = \frac{1}{R} = \frac{1}{14.06} = 0.071$$

$$U = \frac{1}{R} = \frac{1}{7.44} = 0.134$$

$$U = 0.8(0.071) + 0.2(0.134) = 0.0836$$
Round to 0.084

Now figure the roof from scratch, using Table 3-1 for film coefficients and Table 3-3 for the materials.

Item	Resistance
Outside air film	0.17
built up roofing	0.33
2″ roof insulation	5.54
2″ wood, (1.25 × 2)	2.50
Inside air film	0.61
Total resistance	9.15

$$U = \frac{1}{R} = \frac{1}{9.15} = 0.109$$

If you had trouble finding the roof insulation, remember that we said earlier that we could use R = 2.77/in. when we didn't know the type of insulation. And, we assumed softwood which is generally used for roof decks.

The floor slab is easy. Just pluck the first value for a floor slab out of Table 3-13. Use the coefficient on the last line because the outdoor design temperature falls into to the 0° to −10° range. Our coefficient is 40 Btuh per foot of exposed edge. By the way, if the outdoor design temperature were above 0°, we would still use 40 Btuh/ft because there is no other value to use.

Let's not forget about infiltration. We will use the crack method and a value, from Table 3-15, of 24 cfh/ft of crack. Each window has 3 × 3′ + 2 × 4′ = 17′ of crack. When you encounter double-hung windows, remember that crack is equal to 3 times the width plus 2 times the height.

You don't use all three windows to calculate infiltration, just the wall with the most infiltration which will be the one with two windows. Let's put it all together.

G$_e$ 3′ × 4′	= 12 ft² × 0.58 ×70°	=	487.20
W$_e$ (10′ × 10′) − G$_e$	= 88 ft² × 0.084 ×70°	=	517.44
G$_s$ 2(3′ × 4′)	= 24 ft² × 0.58 ×70°	=	974.40
W$_e$ (10′ × 10′) − G$_s$	= 76 ft² × 0.084 ×70°	=	446.88
R 10′ × 10′	= 100 ft² × 0.109 ×70°	=	763.00
F 20′	= 20 ft × 40 Btuh	=	800.00
Inf 2(17′ × 24)	= 816 cfh × 0.018 ×70°	=	1028.16
	Total		5017.08

Voila! as the French say, it is complete.

Perhaps you are wondering why we calculated the heat loss for both the East and South walls. Yes, we

could have lumped all the glass and both the walls together and simplified the arithmetic. However, at a later time, we will calculate the heat gain for this same room and will need to treat each exposure separately. So, I did it now to avoid repetition.

Our calculations were carried to two decimal places which is easy enough to do with an electronic calculator. But it's a bit unnecessary. Had we rounded the decimals, the resulting error would have been 1/5017.08 or 0.02%. Therefore, from now on, we will round off all the decimals.

Example #19

We have calculated the heat loss from one room, now let's try a whole building. Figures 3-2 through 3-5 are architectural plans for a clinic. You will notice that they are neither final nor complete. Because the heating design must be completed at the same time as the architectural plans, we must start with the preliminary drawings. So, if dimensions are not shown, the drawing is scaled.

The building is to be located in Pontiac, Michigan. To find the outdoor design conditions, consult Table 3-33. Column 5 lists the probabilities of typical low temperatures occuring. The temperature found in the 99% column is 0°, in the 97½% column, it is 4°.

What do these probabilities mean? Winter weather data is based on the months of December, January and February, a total of 2160 hours. 99% of the time, the temperature in Pontiac will be zero or above. 1% of the time, or for 21.6 hours, the temperature will be below 0°. 2½% of the time, or for 54 hours, the temperature will be below 4°. Which design temperature should you use? The choice is yours except in states or municipalities that have adopted ASHRAE Standard 90-80 which specifies the use of the 97½% temperature. Michigan has adopted the standard, so we are compelled to use 4° as the design temperature. The inside design temperature is 72°, so our temperature difference is 72 − 4 = 68°.

The weather table does not list wind velocities so what wind velocity should we use for the infiltration calculations? This example is located in a place I know very well so I will tell you that the wind velocity in Pontiac, Michigan is 12 mph. As you will see later, we will adjust our calculations accordingly. No adjustment will be made to the film coefficients since the difference is so slight that only a negligible error will occur.

Let's assemble the heat transfer coefficients. From Table 3-4, the winter U factor for glass is 1.10 Btuh/ft². The 1½ in., solid wood doors have a U of 0.49. The wall is calculated from scratch, using Tables 3-1, 3-2, and 3-3.

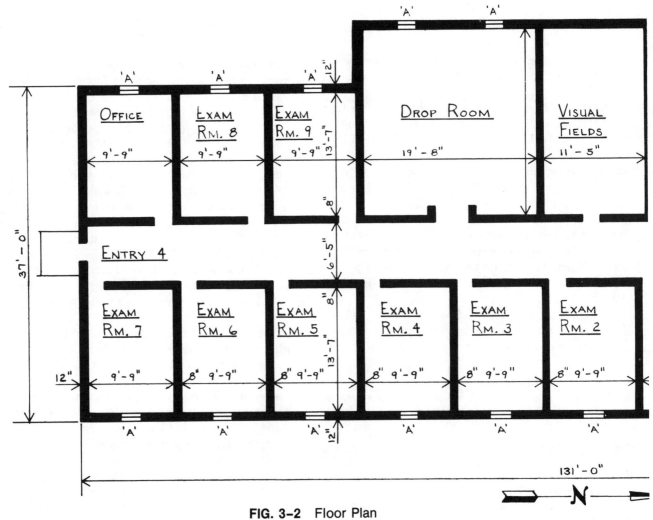

FIG. 3-2   Floor Plan

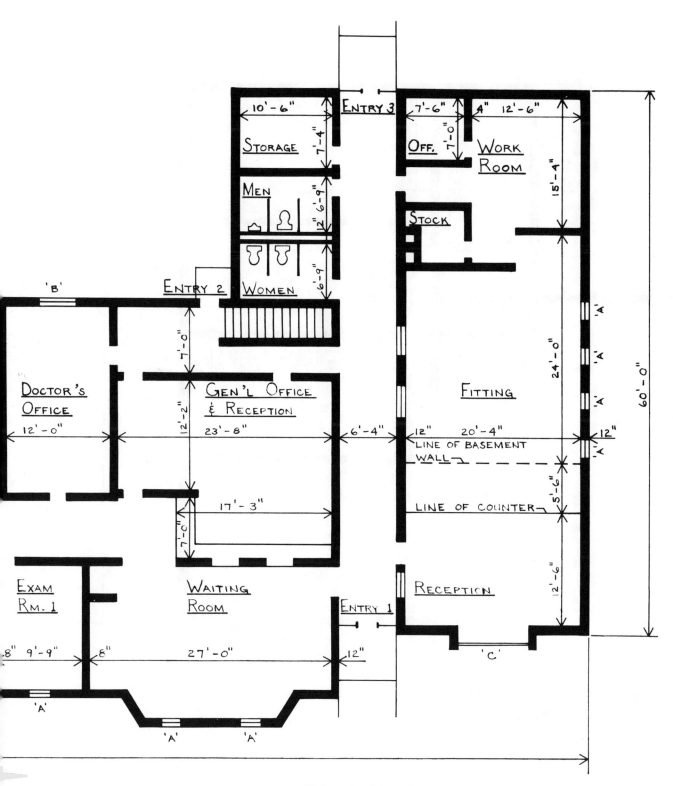

FIG. 3-2 Floor Plan

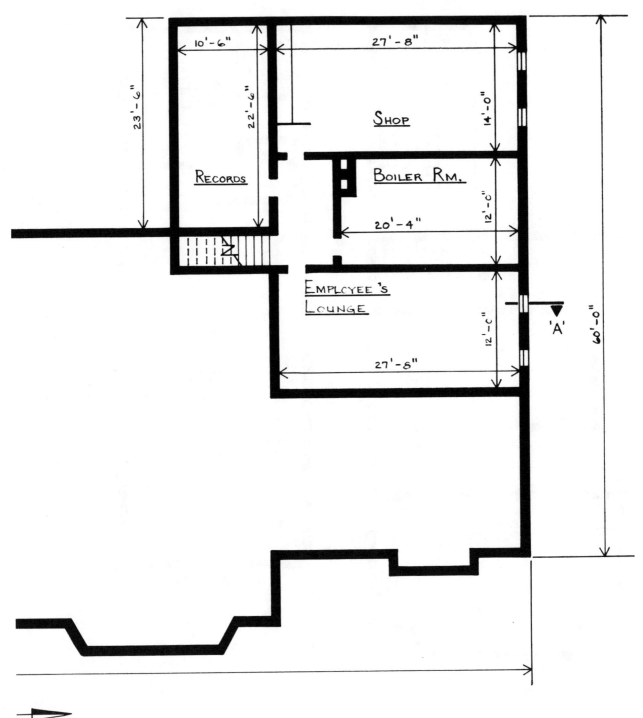

FIG. 3-3 Foundation and Basement Plan

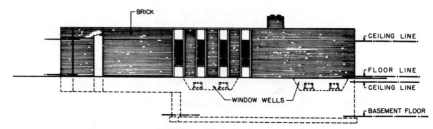

NORTH ELEVATION

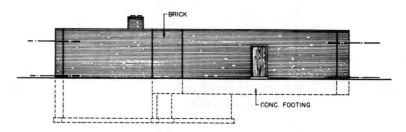

SOUTH ELEVATION

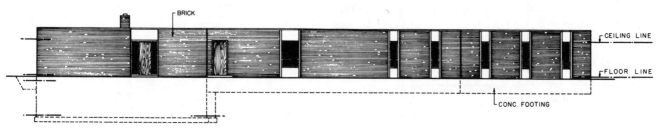

REAR ELEVATION - WEST

FRONT ELEVATION - EAST

FIG. 3-4 Four Elevations

Item	Resistance
Inside air film (still air)	0.68
5/8″ vinyl gypsum board	0.56
1″ insulation, cellular glass	2.63
8″ cinder block	1.72
2″ air space (50° mean temperature)	1.02
4″ face brick	0.44
Outside air film (15 mph)	0.17
Total resistance	7.22

$$U = \frac{1}{R} = \frac{1}{7.22} = 0.139 \quad \text{(use 0.14)}$$

We will use the substitution method is calculating the roof. From the wall section, we find that the roof consists of built-up roofing over one inch of roof insulation on 1½ in. metal deck with ¾ in. acoustic tile on metal furring. The architect says that he is specifying roof insulation with an R of 2.78/in.

The roof in column 1 of Table 3-7 has a total resistance of 6.34. The only difference between it and our roof is the insulation and the ceiling. The remaining elements have an R of 2.04 (6.34 − 4.17 − 0.13 = 2.04). Now add the resistance of our insulation and the ¾″ acoustic tile. R = 2.04 + 2.78 + 1.89 = 6.71.

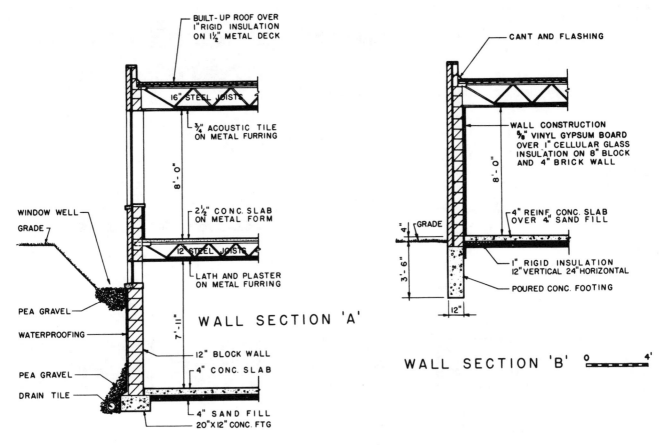

FIG. 3-5   Two Wall Sections

$$U = \frac{1}{R} = \frac{1}{6.71} = 0.149 \qquad \text{(use 0.15)}$$

The floor loss of the unexcavated area is 40 Btuh/ft. The basement wall and floor are calculated from Tables 3-11 and 3-12. The basement wall is uninsulated and is 7'8" deep. Table 3-11 shows a depth of 7 ft and is for a 10 in. block wall. Therefore, we will calculate the U factor using the same technique as in Example 11. The resistance for 12" concrete block is 1.28.

Depth	Total arc length, in.	Wall thickness in.	Length through soil, in.	Resistance
1	18.85	12	0.00	0.00
2	37.70	12	25.70	2.68
3	56.55	12	44.55	4.64
4	75.40	12	63.40	6.60
5	94.25	12	82.25	8.57
6	113.10	12	101.10	10.53
7	131.95	12	119.95	12.49
8	144.95	12	132.51	13.80
				59.31

$$\frac{59.31}{7'8''} = 7.74$$

$$U = \frac{1}{R}$$
$$U = \frac{1}{0.17 + 7.74 + 1.28 + 0.68}$$
$$U = \frac{1}{9.87} = 0.10$$

Table 3-12 only goes to 7 ft deep and 32 ft wide. To calculate the exact factor for the basement floor isn't worth the effort because, at a 7 ft depth and at 32 ft wide, we have reached the point of diminishing returns. Therefore, we will use a U factor of 0.021. Let's recap, using 68° as the temperature difference.

Windows	1.10 × 68° =	74.80
Walls	0.14 × 68° =	9.52
Roof	0.15 × 68° =	10.20
Basement wall	0.10 × 68° =	6.80
Basement floor	0.021 × 68° =	1.43
Floor edge loss	=	40 Btuh/ft
Door	0.49 × 68° =	33.32

The basement plans show 2'0" × 1'5" windows but do not tell what type they are. In a case like this, call the architect and ask him what window he is using and, if available, his infiltration specification. Let's assume that I am the architect and I tell you that these will be projected type windows, but that I haven't yet written my infiltration specification. So, you must use Table 3-15. A projected window's crack is equal to its perimeter. The perimeter of this window is $2' + 2' + 1'5'' + 1'5''$ or $7'$. From Table 3-15, we find that, at 12 mph, a projected window has an infiltration of 9 cfh/ft.

The first floor plan shows that there are three windows: type A $2'0'' \times 5'11''$, type B $4'0'' \times 5'11''$, type C $8'0'' \times 8'0''$. Only types A and B have ventilating sections. The ventilating section for type A is $2'0'' \times 2'0''$ projected, so its perimeter is $8'0''$. Type B has a ventilating section equal to $4'0'' \times 2'0''$ and a perimeter of $12'0''$. The infiltration rate is the same as the basement windows, 9 cfh/ft.

Door infiltration will be 1,956 cfh for infrequently opened doors, such as entries 2 and 4. For entries 1 and 3, figure 704 cfh each time the door opens. You must estimate how often the doors open. Myself, I would guess 10 times an hour. My thinking goes like this: there are two doctors and two opticians. If each one sees four patients an hour, there will be 16 people entering and 16 people leaving every hour. Assuming that one third of the people will arrive and leave simultaneously, then $32 \times \frac{2}{3} = 22$ openings. Then I round off to 20 and assume half of the openings for each door.

To convert cfh to Btuh, multiply cfh by $0.018 \times (72° - 4°)$ or 1.22.

Records

W$_{bg}$	$33'0'' \times 7'8''$	=	253×6.80	= 1720
F	$22'6'' \times 10'6''$	=	236×1.43	= <u>337</u>
				2057 Btuh

Shop

G$_N$	$2(2'0'' \times 1'5'')$	=	6×74.8	= 449
W$_{bg}$	$(41'8'' \times 7'8'') - G$	=	313×6.80	= 2128
F	$27'8'' \times 14'0''$	=	387×1.43	= 553
I	$2(7'0'' \times 9)$	=	$126 \frac{3}{4} 1.22$	= <u>154</u>
				3284 Btuh

Employees' Lounge

G$_N$	$2(2'0'' \times 1'5'')$	=	6×74.8	= 449
W$_{bg}$	$(12'0'' \times 7'8'') - G$	=	86×6.80	= 585
F	$27'8'' \times 12'0''$	=	332×1.43	= 475
I	$2(7'0'' \times 9)$	=	126×1.22	= <u>154</u>
				1663 Btuh

Exam Rooms #1 through #6

G$_E$	$2'0'' \times 5'11''$	=	12×74.8	= 898
W$_E$	$(9'9'' \times 8'0'') - G$	=	66×9.52	= 628
R	$9'9'' \times 13'7''$	=	132×10.2	= 1346
F	$9'9''$	=	10×40	= 400
I	$8'0'' \times 9$	=	72×1.22	= <u>88</u>
				3360 Btuh

Exam Room #7

G$_E$	$2'0'' \times 5'11''$	=	12×74.8	= 898
W$_E$	$(9'9'' \times 8'0'') - G$	=	66×9.52	= 628
W$_S$	$13'7'' \times 8'0''$	=	109×9.52	= 1038
R	$9'9'' \times 13'7''$	=	132×10.2	= 1346
F	$23'4''$	=	23×40	= 920
I	$8'0'' \times 9$	=	72×1.22	= <u>88</u>
				4918 Btuh

Entry #4

D$_S$	$3'0'' \times 7'0''$	=	21×33.3	= 700
W$_S$	$(6'5'' \times 8'0'') - D$	=	30×9.52	= 286
R	$6'5'' \times 20'0''$	=	128×10.2	= 1306
F	$6'5''$	=	7×40	= 280
I	1956	=	1956×1.22	= <u>2386</u>
				4958 Btuh

Hall

R	$62'0'' \times 6'5''$	=	398×10.2	= 4060 Btuh

Office

G$_W$	$2'0'' \times 5'11''$	=	12×74.8	= 898
W$_W$	$(9'9'' \times 8'0'') - G$	=	66×9.52	= 628
W$_S$	$13'7'' \times 8'0''$	=	109×9.52	= 1038
R	$9'9'' \times 13'7''$	=	132×10.2	= 1346
F	$23'4''$	=	23×40	= 920
I	$8'0'' \times 9$	=	72×1.22	= <u>88</u>
				4918 Btuh

Exam Rooms #8, #9

G$_W$	$2'0'' \times 5'11''$	=	12×74.8	= 898
W$_W$	$(9'9'' \times 8'0'') - G$	=	66×9.52	= 628
R	$9'9'' \times 13'7''$	=	132×10.2	= 1346
F	$9'9$	=	10×40	= 400
I	$8'0'' \times 9$	=	72×1.22	= <u>88</u>
				3360 Btuh

Drop room

G$_W$	$2(2'0'' \times 5'11'')$	=	24×74.8	= 1795
W$_W$	$(19'8'' \times 8'0'') - G$	=	133×9.52	= 1266
W$_S$	$6'10'' \times 8'0''$	=	55×9.52	= 524
R	$19'8'' \times 20'5''$	=	402×10.2	= 4100
F	$26'6''$	=	27×40	= 1080
I	$2(8'0'' \times 9)$	=	144×1.22	= <u>176</u>
				8901 Btuh

Visual fields

W$_W$	$11'5'' \times 8'0''$	=	91×9.52	= 866
R	$11'5'' \times 20'5''$	=	233×10.2	= 2377
F	$11'5''$	=	11×40	= <u>440</u>
				3683 Btuh

Doctor's Office

G$_W$	$4'0'' \times 5'11''$	=	24×74.8	= 1795
W$_W$	$(12'0'' \times 8'0'') - G$	=	72×9.52	= 685
R	$12'0'' \times 20'5''$	=	245×10.2	= 2499
F	$12'0''$	=	12×40	= 480
I	$12'0'' \times 9$	=	108×1.22	= <u>132</u>
				5591 Btuh

Entry #2

D$_S$	$3'0'' \times 7'0''$	=	21×33.3	= 700
W$_W$	$(12'2'' \times 8'0'') - D$	=	76×9.52	= 724
R	$23'8'' \times 7'0''$	=	166×10.2	= 1693
F	$11'6''$	=	12×40	= 480
I	1956	=	1956×1.22	= <u>2386</u>
				5983 Btuh

Women's toilet, men's toilet

W_S	$6'9'' \times 8'0''$	=	54×9.52	=	514
R	$6'9'' \times 10'6''$	=	71×10.2	=	724
F	$6'9''$	=	7×40	=	280
					1518 Btuh

Storage

W_S	$7'4'' \times 8'0''$	=	59×9.52	=	562
W_W	$10'6'' \times 8'0''$	=	84×9.52	=	800
R	$7'4'' \times 10'6''$	=	77×10.2	=	785
F	$17'10''$	=	18×40	=	720
					2867 Btuh

Entry #3

D_W	$3'0'' \times 7'0''$	=	21×33.3	=	700
G_W	$2(1'8'' \times 7'0'')$	=	23×74.8	=	1720
W_W	$(6'4'' \times 8') - G - D$	=	6×9.52	=	57
R	$20'0'' \times 6'4''$	=	127×10.2	=	1295
F	$6'4''$	=	6×40	=	240
I	704×10	=	7040×1.22	=	8589
					12601 Btuh

Hall

R	$19'0'' \times 6'4''$	=	120×10.2	= 1224 Btuh

Office

W_W	$7'6'' \times 8'0''$	=	60×9.52	=	571
R	$7'6'' \times 7'0''$	=	53×10.2	=	541
					1112 Btuh

Work Room

W_W	$12'6'' \times 8'0''$	=	100×9.52	=	952
W_N	$15'4'' \times 8'0''$	=	123×9.52	=	1171
R	$(12'6'' \times 15'4'') +$				
	$(7'10'' \times 4'0'')$	=	223×10.2	=	2275
					4398 Btuh

Stock

R	$7'6'' \times 7'0''$	=	53×10.2	= 541 Btuh

Fitting

G_N	$4(2'0'' \times 5'11'')$	=	48×74.8	=	3590
W_N	$(29'6'' \times 8'0'') - G$	=	188×9.52	=	1790
R	$29'6'' \times 20'4''$	=	600×10.2	=	6120
F	$5'6''$	=	6×40	=	240
I	$4(8'0'' \times 9)$	=	288×1.22	=	351
					12091 Btuh

Reception

G_E	$8'0'' \times 8'0''$	=	64×74.8	=	4787
W_E	$(20'4'' \times 8'0'') - G$	=	99×9.52	=	942
W_N	$12'6'' \times 8'0''$	=	100×9.52	=	952
R	$12'6'' \times 20'4''$	=	254×10.2	=	2591
F	$32'10''$	=	33×40	=	1320
					10592 Btuh

Entry #1

D_E	$3'0'' \times 7'0''$	=	21×33.3	=	700
G_E	$2(1'8'' \times 7'0'')$	=	23×74.8	=	1720
W_E	$(6'4'' \times 8') - G - D$	=	6×9.52	=	57
R	$6'4'' \times 20'0''$	=	127×10.2	=	1295
F	$6'4''$	=	6×40	=	240
I	704×10	=	7040×1.22	=	8589
					12601 Btuh

Waiting room

G_E	$2(2'0'' \times 5'11'')$	=	24×74.8	=	1795
W_E	$(33'6'' \times 8'0'') - G$	=	244×9.52	=	2323
W_N	$7'0'' \times 8'0''$	=	56×9.52	=	533
R	$(27'0'' \times 13'7'') +$				
	$(2'10'' \times 17'9'')$	=	417×10.2	=	4253
F	$40'6''$	=	41×40	=	1640
I	$2(8'0'' \times 9)$	=	144×1.22	=	176
					10720 Btuh

General office and reception

R	$(7'0'' \times 17'3'') +$			
	$(12'9'' \times 23'8'')$	=	422×10.2	= 4304 Btuh

HEAT GAIN CALCULATIONS (TETD METHOD)

In calculating the heat loss, we ignored the effect of the sun and internal loads, such as lights, because they produce heat and reduce the heat loss. These items are very important in heat gain calculations.

In making heat gain calculations, we must consider not only the heat gain due to transmission, but also the effect of the sun's direct radiation on glass, walls, and roof.

In addition, internal loads caused by lights, people, cooking, motors, anything that produces heat, must be considered. Latent heat, the heat required to dehumidify or humidify the space, is another load factor.

Solar heat gains through glass have been tabulated in Btuh/ft² for different orientations, times of day, and months of the year. Tables 3-18 through 3-22 give these values. For example, in Table 3-20, which is for 40° north latitude, we find that on July 21, at 12:00 noon, a window facing east has a heat gain of 41 Btuh/ft²; south glass gains 109 Btuh/ft²; and north glass gains 38 Btuh/ft².

Be careful when you use these tables. For morning values (A.M.), read across the top headings for compass orientation and along the left margin for time. For the afternoon hours (P.M.), read along the bottom headings for orientation and along the right margin for time. Notice that the morning values for east become the afternoon values for west. Thus, the maximum solar heat gain is on east glass at 8:00 A.M., on west glass at 4:00 P.M., and on south glass at 12 noon.

What month do you use? For most cooling load calculations, I use July 21 for all orientations except south, southeast, and southwest which are actually highest in December. If you are have a building whose cooling load is largely solar gain, use September or even October for southern orientations.

Solar heat gains are truly formidable, so we must do all we can to reduce them. Tables 3-23 through 3-25 give shading coefficients.

Solar heat entering the room varies with solar transmittance, reflectance, and absorptance. Transmittance is the amount of solar radiation that passes through the glass, reflectance is the amount reflected from the glass, and absorptance is the amount absorbed by the glass. The values give in Tables 3-18 through 3-22 are for double strength glass with a transmittance of 0.87, a reflectance of 0.08, and an absorptance of 0.05. Glass thickness, and any variation in its physical properties, can alter the transmittance, reflectance and absorptance.

Table 3-24 gives the shading coefficients of various glasses. Notice that 1½″ plate glass with a solar transmittance of 0.71 has a shading coefficient of 0.88. This means that, if the solar heat gain is 109 Btuh/ft², the net gain through our plate glass is 109 × 0.88 = 96 Btuh/ft². If ½ in. grey plate is used, the 109 becomes 109 × 0.50 = 54.5 Btuh/ft².

Tables 3-23 and 3-24 and 3-25 are shading coefficients for single or insulating glass with venetian blinds, roller shades, and draperies. Notice that with draperies, you must consider both the color and the weave of the material. A closed weave is a weave through which no objects can be seen, but light and dark areas may show. Details cannot be seen through a medium weave, but larger objects are clearly defined. Details can be seen through an open weave and the view is relatively clear, with no confusion of vision.

Using an overhang to shade glass is an excellent way to reduce solar heat gain. What value do you use for a shaded window? I use the values for north glass. North glass does not have the sun shining on it, so the loads listed are the diffuse portion of the solar heat gain. If the win window is partially shaded, I use the north value for the shaded portion and the regular (east, south, or west) value for the rest of the window. Table 3-27 gives the length of overhang required to shade 10 ft down. As you can see, east and west require some pretty long overhangs. Don't eliminate overhangs on these exposures for this reason because some overhang is better than none.

The total heat gain through glass must include the transmission load, as shown in the following formula:

Total Window Heat Gain =
$$(SC \times SHGF)(A) + (U)(A)(t_o - t_i)$$

where SC = shade coefficient, SHGF = sensible heat gain factor, t_i = inside temperature and t_o = outside temperature. The outside temperature varies during the day as shown in column 2, Table 3-28. For example, for 10 ft² of east plate glass at 40°N latitude, at 8:00 a.m.:

Total Window Heat Gain =
$$(0.88 \times 216)(10) + (1.10)(10)(77 - 75) = 1003 + 22 = 1025$$

The solar load on walls and roofs is treated in a different manner. To compensate for the flywheel effect, caused by the mass of the walls and roof, the total equivalent temperature differential (TETD) is used. ASHRAE has tabulated these equivalent temperature differences in Tables 3-29 and 3-30. The area of the wall or roof is multiplied by the U factor and the total equivalent temperature difference to calculate the heat gain due to solar radiation and transmission. It is not necessary to do a separate calculation for transmission because the TETD includes this.

The second column in Table 3-28 lists air temperatures for each hour of July 21. The remaining columns show the effect of orientation on the basic temperature. There are two tables: one for $\alpha/h_o = .15$ and one for $\alpha/h_o = .30$. α is the absorptance of solar radiation by the surface and h_o is the coefficient of heat transfer by radiation and convection. $\alpha/h_o = .15$ is used for light surfaces while $\alpha/h_o = .30$ is used for dark surfaces. The sol-air temperatures are used to calculate the equivalent temperature differences given in Tables 3-29 and 3-30.

The sol-air temperature for walls can be estimated from the following equation:

$$T_{SA} = T_{OA} + \alpha/h_o \, [1.15(SHGF)]$$
where
$$T_{SA} = \text{sol-air temperature}$$
$$T_{OA} = \text{outside air temperature}$$
$$SHGF = \text{solar heat gain from Tables 3-18 to 3-22}$$
$$\alpha/h_o = 0.15 \text{ for light walls, } 0.30 \text{ for dark walls}$$

When the sol-air temperature for a roof is desired, the above equation is used with a −7° correction factor. The horizontal SHGF is used for roofs.

Example #20

Calculate the sol-air temperature for a light colored, south facing wall on September 21 at noon and 4:00 p.m. The outside temperature is 75° at noon and 70° at 4:00 p.m.

From Table 3-20, the SHGF is 71 Btuh/ft² at 4:00 p.m. and 200 Btuh/ft² at noon.

noon
$$T_{SA} = T_{OA} + \alpha/h_o \, [1.15(SHGF)]$$
$$= 75° + 0.15[1.15(200)]$$
$$= 75° + 34.5°$$
$$= 109.5°$$
4:00 p.m.
$$T_{SA} = T_{OA} + \alpha/h_o \, [1.15(SHGF)]$$
$$= 70° + 0.15[1.15(71)]$$
$$= 70° + 12.25°$$
$$= 82.25°$$

Once you have the sol-air temperature, the TETD can be calculated with the following equation:

$$TETD = T_{OA} - T_i + \lambda(T_{SA\delta} - T_{OA})$$

T_i = space temperature
λ = effect decrement factor
δ = hour reduction factor
$T_{SA\delta}$ = sol-air temperature at δ hours before the time at which the TETD is desired

The factors λ and δ are listed in Table 3-29B. If you wish the TETD for noon, the $T_{SA\delta}$ will be the sol-air temperature at noon minus δ. For example, if δ is 3, then the sol-air temperature at noon minus 3, or 9:00 a.m. is used for $T_{SA\delta}$.

Example #21

Calculate the TETD for a south facing, light colored, Group E wall at 4:00 p.m. on September 21.

From Table 3-29B, we find that $\lambda = 0.48$ and $\delta = 4$ hours. From the previous example, the sol-air temperature at noon is 109.5° and 82° at 4:00 p.m. Assume a space temperature of 78°. Because $\delta = 4$ hours, we must calculate the sol-air temperature at 4 hours before 4:00 p.m. or noon. In Example 20, we found that $T_{SA\delta}$ (sol-air temperature at noon) = 109.5°.

$$\begin{aligned} TETD &= T_{OA} - T_i + \lambda(T_{SA\delta} - T_{OA}) \\ &= 70° - 78° + 0.48(109.5 - 70) \\ &= -8 + 18.96 \\ &= 10.96 \end{aligned}$$

The lighting load, in watts, is multiplied by 3.4 Btu/watt to determine the heat gain from lights. Table 3-31 lists the heat gain from people at different levels of activity. Tables 3-32A and B give the heat gains from equipment and appliances.

The heat gain from infiltration is calculated in the usual fashion. Sensible heat gain is equal to the infiltration, in cfh, multiplied by 0.018 and the temperature difference. To determine latent heat gain, we must first determine the amount of water in the outside air and the desired amount of water in the room air. Moisture content is given as lb of water/lb of dry. Obviously, we must determine how many pounds of air are being handled. Multiply the cfh by the density (0.075 lb/ft³) and we have lb/hr.

To remove moisture from the air, we must condense the water vapor to liquid water. A pound of water releases approximately 1076 Btu as it condenses. So, if multiply the infiltration, in cfh, by 0.075 lb/ft³ and by 1076 Btu/lb and by the moisture difference, we will have the latent heat gain from infiltration. While 0.075 _ 1076 = 80.7, it is customary to round off to 80.

The following heat gain is for the room used for Example #18. The wall with the window faces east, the other wall faces south, and the heat gain will peak at noon. No shading devices are used. Assume 800 watts of lighting and five moderately active people. Use July 21 in selecting glass factors.

A word about selecting outdoor design conditions. Table 3-33 lists summer and winter design conditions. But these are conditions that only occur for short periods. Notice, in column 6, that there are three choices: 1%, 2½%, and 5%. These percentages refer to the number of hours when the temperature will occur in a typical 2928 hour summer. If the room is located in Baltimore, Maryland, we can expect 1% of 2928 or approximately 30 hours when the temperature equals or exceeds 94°F, dry bulb (db), and 79°F wet bulb (wb). Column 7 gives the daily range of temperature as 21°.

The 2½% column lists 91° db, 78° wb, as the temperatures occuring for 73 hours. Finally, the 5% column shows that the dry bulb, and coincident wet bulb, temperatures of 89° and 77° will occur during 146 hours. Which one to use? For most of my professional life, I have used the 1% column but now, in the interest of energy conservation, I use the 2½% column.

You have probably noticed that the table lists two Baltimores. There is really only one, but it has two weather stations: one at the airport, designated AP, and the other in the city, designated CO for city office. Our room is located in the suburbs, so I have used the AP values.

From a psychrometric chart, we find that, at 91° db and 78° wb, the humidity ratio is 0.0178 lb of water per pound of dry air. From the daily range of 21°, we know that the db temperature will vary from 70° to 91°. However, the humidity ratio will remain fairly constant at the design point.

Table 3-28 lists the temperatures, at hours 1 to 24, for a 95° db design temperature and a 21° range. These temperatures are in the second column under *Air Temperature, F.* Since our daily range is also 21°, we merely subtract 4° (95° − 91°) from each temperature. So, our temperature will be 90 − 4 = 86° at noon.

If the daily range had been other than 21°, we would have used Table 3-28A to determine the hourly air temperature. It gives, as a % of the daily range, the number of degrees to be subtracted from the design temperature to arrive at the hourly temperature. If the daily range is 40° and the design temperature is 100°, the temperature at 5:00 p.m. (hour 17) is:

$$\begin{aligned} \text{Hourly}° &= \text{design}° - (\text{daily range})(\% \text{ of range}) \\ \text{17th hour} &= 100° - (40°)(.10) \\ &= 100° - 4° \\ &= 96° \end{aligned}$$

Interior design temperatures present another problem. In Chapter 2, we learned how many different factors affect our feeling of comfort. At one time, systems were designed to maintain 80° and 50% R.H. These conditions were fine for short term occupancy but, when people began to spend their working hours in an air-conditioned environment, they demanded lower temperatures. So, we saw inside design temperatures drop to 75° and some owners ordering their consultants to design for 72°. Now the cycle is being reversed. Many codes and standards, including ASHRAE 90-80, set the design dry bulb temperature at 78°. The design relative humidity is specified as the relative humidity, within the comfort envelope, that requires the minimum energy consumption. We will use 45%.

Example #22

In this example, we will do a heat gain for the room used in Example –18. The building is located in suburban Baltimore, Maryland, so we will use the airport (AP) values given in Table 3-33. The 2½% design dry bulb is 91° and the coincident wet bulb is 75°. Why don't we use the 2½% design wet bulb listed in column 8? Because this is normally used only when selecting cooling towers, evaporative condensers, evaporative coolers, and other equipment whose capacity is a function of wet bulb temperature.

The solar factors we select are from Table 3-20 for 40° N latitude. This is close enough to the latitude of Baltimore (39°1′ N) and no correction is needed.

When selecting the TETD for the walls and the roof, we will have to make some corrections. The TETD tables are based on a 75° inside temperature, a 95° outdoor temperature, and an average daily temperature of 85°. We will subtract 3° from the TETD because we are designing for 78° indoors. Next, correct for the average outdoor temperature. The table average is 85° and ours is $91 - 21/2 = 80.5°$. The difference is 4.5° which we will subtact from the TETD. Here are the corrected noon TETD's.

South wall	$18° - 3° - 4.5° = 10.5°$
East wall	$22° - 3° - 4.5° = 14.5°$
Roof	$20° - 3° - 4.5° = 12.5°$

The wall TETD's were selected from Table 3-29. A 2×4 stud wall is a light wall and falls into Group A. It is also assumed, in the interest of energy conservation, that light colors are used. Table 3-30 does not show a roof with with 2″ wood and 2″ insulation. So, we come as close as we can, using the values for 2½″ wood and 2″ insulation.

The activity of the people is assumed to be typing in an office. From Table 3-31, they produce 255 Btuh of sensible heatand 255 Btuh of latent heat.

Why do we separate the heat into sensible and latent? Two things must be accomplished: cooling and dehumidification. By selecting the proper supply air

temperature, it is possible to do both at once. (This is shown in Chapter 7, Psychrometrics.) However, it is necessary to know the ratio of room sensible heat to the combined sensible heat plus latent heat. If the latent heat load becomes too great, then simultaneous cooling and dehumidification is not possible. If the supply air temperature is selected to satisfy the sensible load, not enough moisture is removed and the relative humidity remains high. Conversely, if the supply air temperature is selected to dehumidify, the room temperature becomes too low. There are two solutions: chemically dehumidify or cool the air down to its dew point, to remove the required moisture, and then reheat it so that it won't be too cool. This is called a reheat system and is the method generally used. You are likely to have this problem in auditoriums, churches, stores, beauty parlors, or any place with a large number of people.

Design conditions

Peak outside	91° db	78.0° wb	0.0178 lb water/lb air
Peak room	86° db	73.5° wb	0.0178 lb water/lb air
Room	78° db	63.5° wb	0.0093 lb water/lb air
Difference	8° db		0.0085 lb water/lb air

Solar (Table 3-20)

				Sensible	Latent
G_S	$2 (3 \times 4)$	$= 24 \times 149$	$=$	3576	
G_E	(3×4)	$= 12 \times 38$	$=$	456	

Transmission

G	$(24 + 12)$	$= 36 \times 0.61 \times 8°$	$=$	176

Combined solar and transmission
(Wall, Table 3-29B. Roof, Table 3-30.)

W_S	$(10 \times 10) - G$	$= 76 \times 0.084 \times 10.5°$	$=$	67
W_E	$(10 \times 10) - G$	$= 88 \times 0.084 \times 14.5°$	$=$	107
R	10×10	$= 100 \times 0.109 \times 12.5°$	$=$	136

Internal

Lights		$= 800 \text{ W} \times 3.4$	$=$	2720	
People		$= 5 \times 255$	$=$	1275	
		$= 5 \times 255$	$=$		1275

Infiltration

Sensible	$= 816 \times 0.018 \times 8°$	$=$	118	
Latent	$= 816 \times 80 \times 0.0085$	$=$		555
	Total		8631	1830

Total sensible plus latent   10461

Now, let's do a heat gain for a complete building. We will use the same building as Example #18. First, let's get the factors assembled.

Example #23

The city of Pontiac, Michigan has a 2½% summer outdoor design condition of 87° db, 72° wb, and a 21° daily range. This means that the average outdoor dry bulb temperature will be $87° - 21°/2 = 76.5°$. All of the TETD charts are based on an 85° daily average, so we must subtract 8.5°. In addition, we must subtract 3° because we will design for indoor conditions of 78° and 45% RH instead of 75°. This gives a total correction of the TETD tables of $- 8.5° - 3° = - 11.5°$.

We must apply a correction to Table 3-28 for hourly temperatures since the table assumes a 95° peak and a 21° range. We have the proper range, so we have only to subtract 95 − 87 or 8° from each hourly temperature. This gives an 8:00 a.m. temperature of 77 − 8 = 69°, a noon temperature of 90 − 8 = 82°, and a 4:00 p.m. temperature of 94 − 8 = 86°.

Glass

Glass is single glass covered by dark, closed weave draperies. The shade factor is 0.63. Notice that we only apply the shade factor to orientations that are in the sun: 8:00 a.m. east, noon south, 4:00 p.m. west.

Time	8:00 a.m.		noon		4:00 p.m.	
N Solar		= 28.00		= 38.00		= 28.00
Trans	1.04 × (69 − 78) =	−9.36	1.04 × (82 − 78) =	4.16	1.04 × (86 − 78) =	8.32
Total		18.64		42.16		36.32
E Solar	216 × 0.63 =	136.08		= 41.00		= 26.00
Trans	1.04 × (69 − 78) =	−9.36	1.04 × (82 − 78) =	4.16	1.04 × (86 − 78) =	8.32
Total		126.72		45.16		34.32
S Solar		= 29.00	109 × 0.63 =	68.67		= 29.00
Trans	1.04 × (69 − 78) =	−9.36	1.04 × (82 − 78) =	4.16	1.04 × (86 − 78) =	8.32
Total		19.64		72.83		37.32
W Solar		= 26.00		= 41.00	216 × 0.63 =	136.08
Trans	1.04 × (69 − 78) =	−9.36	1.04 × (82 − 78) =	4.16	1.04 × (86 − 78) =	8.32
Total		16.64		45.16		144.40

Lights

Whenever possible, get a lighting plan from the lighting engineer. Usually, he is starting his design at the same time that you are designing the air conditioning, so you will probably only have preliminary lighting loads. We will assume that the lighting loads in the calculations to follow came from the lighting engineer. Be sure they include the ballast load since a 40-watt fluorescent lamp requires a 10-watt ballast for a total of 50 watts.

Motors

When a motor and the machine being driven are located in an air conditioned space, all of the electrical energy supplied to it become part of the air conditioning load. The equation for energy consumed by a motor is:

$$Q = \frac{motor\ horsepower}{motor\ efficiency} \times load\ factor \times 2545$$

If only the motor is in the air conditioned space, but the driven machine is not, then:

$$Q = motor\ horsepower \times load\ factor \times 2545$$

Appliances

Tables 3-32A and B give the sensible and latent heat gain from various appliances on the right side of the tables. Note that, when the appliance is hooded, the heat gain is all sensible. When there is no hood, both latent and sensible heat must be considered.

As you can see, from the first line of Table 3-32A, a broiler griddle, when hooded, gives off 3,600 Btuh, all

sensible. Without a hood, the sensible load is 11,700 Btuh plus a latent load of 6,300 Btuh.

Wall

Our wall is not listed in Table 29A, but wall D in Table 29B is a very close match, so we will use it. Remember, we must make corrections to the table values because of our differing outdoor and indoor conditions. The correction factor is −11.5°.

	8:00 a.m.	noon	4:00 p.m.
N	(4 − 11.5) × 0.14 = −1.05	(6 − 11.5) × 0.14 = −0.77	(14 − 11.5) × 0.14 = 0.35
E	(6 − 11.5) × 0.14 = −0.77	(20 − 11.5) × 0.14 = 1.19	(24 − 11.5) × 0.14 = 1.75
S	(4 − 11.5) × 0.14 = −1.05	(7 − 11.5) × 0.14 = −0.63	(21 − 11.5) × 0.14 = 1.33
W	(6 − 11.5) × 0.14 = −0.77	(6 − 11.5) × 0.14 = −0.77	(16 − 11.5) × 0.14 = 0.63

Notice that some of these factors are negative. This means that heat is being lost even though it is summer.

Roof

Our roof is the first one listed in Table 3-30. Notice that it has a U factor of 0.213. Don't let that confuse you. That value is for the roof alone while our 0.15 is for the roof and a suspended ceiling. The TETD for a dark colored roof were used.

8:00 a.m.	noon	4:00 p.m.
(28 − 11.5) × 0.15 = 2.48	(90 − 11.5) × 0.15 = 11.78	(78 − 11.5) × 0.15 = 9.98

Doors

The doors are wood, with a U factor of 0.47. A door falls into the category of light construction so the TETD for Wall A, Table 29A, are used.

	8:00 a.m.	noon	4:00 p.m.
E	(18 − 11.5) × 0.47 = 3.06	(22 − 11.5) × 0.47 = 4.94	(20 − 11.5) × 0.47 = 4.00
S	(9 − 11.5) × 0.47 = −1.18	(18 − 11.5) × 0.47 = 3.06	(21 − 11.5) × 0.47 = 4.47
W	(11 − 11.5) × 0.47 = −0.24	(16 − 11.5) × 0.47 = 2.12	(27 × 11.5) × 0.47 = 7.29

Humidity

As stated in Example #22, the humidity ratio can be considered to be constant during the day. At 87° db and 72° wb, the humidity ratio is 0.0134 lb per lb of dry air. The room humidity ratio is 0.0093 lb/lb.

Lights

The lighting loads we use were given us by the lighting engineer. An additional 740 watts were added for equipment in the shop.

People

The architect or building owner will give you the number of people per room. Most often, you will receive a furniture layout that will show you the number of desks. For light activity, allow:

255 Btuh sensible   255 Btuh latent

Infiltration

An air conditioning system, unlike a heating only system, often introduces tempered outside air into the space. This pressurizes the building, preventing infiltration. Therefore, infiltration is not calculated as part of the room load, except at entries.

The sensible and latent heat gains per cfh of infiltration are:

$$\text{sensible} = .018(t_o - t_i)$$
$$\text{latent} = 80(w_o - w_i)$$

w being the humidity ratio.

	8:00 a.m.		
Sensible	$(69° - 78°) \times 0.018$	=	-0.162 Btu/cfh
Latent	$(0.0134 - 0.0093) \times 80$	=	0.328 Btu/cfh
	Noon		
Sensible	$(82° - 78°) \times 0.018$	=	0.072 Btu/cfh
Latent	$(0.0134 - 0.0093) \times 80$	=	0.328 Btu/cfh
	4:00 p.m.		
Sensible	$(86° - 78°) \times 0.018$	=	0.144 Btu/cfh
Latent	$(0.0134 - 0.0093) \times 80$	=	0.328 Btu/cfh

Floor and wall below grade

There is no heat gain when floors are on earth because the earth temperature is equal to, or less than, the room. The same is true for below-grade walls.

Peak periods

In calculating heat gain, it is necessary to select the time of day at which the heat gain will peak. Most of the time this is obvious. For instance, a room with a large amount of glass facing west will peak at 4:00 p.m. Sometimes the peak is not obvious, so it is necessary to do alternate heat gains for each room at 8:00 a.m., noon, and 4:00 p.m. The areas of each building component are taken from Example #19 and the preliminary calculations need not be repeated.

Heat gain elements are designated as follows:

G_N = Glass (north) R = Roof   L = Lights
W_W = Wall (west) P = People I = Infiltration

		8:00 a.m.		noon		4:00 p.m.	
		Sensible	Latent	Sensible	Latent	Sensible	Latent
Shop							
G_N	6×18.64	112		253		218	
L	1440×3.4	4896		4896		4896	
P	2×255	510	510	510	510	510	510
		5518	510	5659	510	5624	510
Employees' Lounge							
G_N	6×18.64	112		253		218	
L	800×3.4	2720		2720		2720	
P	4×255	1020	1020	1020	1020	1020	1020
		3852	1020	3993	1020	3958	1020
Exam rooms #1 through #6							
G_E	12×126.72	1521		542		412	
W_E	66×-0.77	-51		79		116	
R	132×2.48	327		1555		1317	
L	300×3.4	1020		1020		1020	
P	1×255	255	255	255	255	255	255
		3072	255	3451	255	3120	255

Exam room #7							
G_E	12×126.72	1521		542		412	
W_E	66×-0.77	-51		79		116	
W_S	109×-1.05	-114		-69		145	
R	132×2.48	327		1555		1317	
L	300×3.4	1020		1020		1020	
P	1×255	255	255	255	255	255	255
		2958	255	3382	255	3265	255
Entry #4							
D_S	21×-1.18	-25		64		94	
W_S	30×-1.05	-32		-19		40	
R	$128 \& 2.48$	317		1508		1277	
L	200×3.4	680		680		680	
P	0×255	0	0	0	0	0	0
I	1956×-0.162	-317		141		282	
	$\times 0.328$		642		642		642
		623	642	2374	642	2373	642
Hall							
R	398×2.48	987		4688		3972	
L	300×3.4	1020		1020		1020	
P	0×255	0	0	0	0	0	0
		2007	0	5708	0	4992	0
Office							
G_W	12×16.64	200		542		1728	
W_W	66×-0.77	-51		-51		42	
W_S	109×-1.05	-114		-69		145	
R	132×2.48	327		1555		1317	
L	300×3.4	1020		1020		1020	
P	2×255	510	510	510	510	510	510
		1892	510	3507	510	4762	510
Exam rooms #8 and #9							
G_W	12×16.64	200		542		1728	
W_W	66×-0.77	-51		-51		42	
R	132×2.48	327		1554		1317	
L	300×3.4	1020		1020		1020	
P	1×255	255	255	255	255	255	255
		1751	255	3320	255	4362	255
Drop room							
G_W	24×16.64	399		1084		3456	
W_W	133×-0.77	-102		-102		84	
W_S	55×-1.05	-58		-35		73	
R	402×2.48	997		4736		4012	
L	800×3.4	2720		2720		2720	
P	20×255	5100	5100	5100	5100	5100	5100
		9056	5100	13503	5100	15445	5100
Visual fields							
W_W	91×-0.77	-70		-70		57	
R	233×2.48	578		2745		2325	
L	0×3.4	0		0		0	
P	2×255	510	510	510	510	510	510
		1018	510	3185	510	2892	510
Doctor's office							
G_W	24×16.64	399		1084		3466	
W_W	72×-0.77	-55		-55		45	
R	245×2.48	608		2886		2445	
L	400×3.4	1360		1360		1360	
P	2×255	510	510	510	510	510	510
		2822	510	5785	510	7826	510
Entry #2							
D_W	21×-0.24	-5		45		154	
W_W	76×-0.77	-59		-59		48	
R	166×2.48	412		1956		1657	
L	200×3.4	680		680		680	
P	0×255	0		0		0	
I	1956×-0.162	-317		141		282	
	$\times 0.328$		642		642		642
		711	642	2763	642	2821	642

		8:00 a.m.		noon		4:00 p.m.	
		Sensible	Latent	Sensible	Latent	Sensible	Latent
General office & reception							
R	422 × 2.48	1047		4971		4212	
L	800 × 3.4	2720		2720		2720	
P	5 × 255	1275	1275	1275	1275	1275	1275
		5042	1275	8966	1275	8207	1275
Women's toilet, men's toilet							
W_S	54 × −1.05	−57		−34		72	
R	71 × 2.48	176		836		709	
L	100 × 3.4	340		340		340	
P	0 × 255	0		0		0	
		459		1142		1121	
Storage							
W_S	59 × −1.05	−62		−37		78	
W_W	84 × −0.77	−65		−65		53	
R	77 × 2.48	191		907		768	
L	0 × 3.4	0		0		0	
P	0 × 0	0		0		0	
		64		805		899	
Entry #3							
D_W	21 × −0.24	−5		45		154	
G_W	23 × 16.64	383		1038		3221	
W_W	6 × −0.77	−5		−5		4	
R	127 × 2.48	315		1496		1267	
L	200 × 3.4	680		680		680	
P	0 × 255	0		0		0	
I	7040 × −0.162 × 0.328	−1140	2309	507	2309	1014	2309
		228	2309	3761	2309	6340	2309
Hall							
R	120 × 2.48	298		1414		1198	
L	100 × 3.4	340		340		340	
P	0 × 255	0		0		0	
		638	0	1754	0	1538	0
Office							
W_W	60 × −0.77	−46		−20		43	
R	53 × 2.48	131		624		529	
L	100 × 3.4	340		340		340	
P	1 × 255	255	255	255	255	255	255
		680	255	1199	255	1167	255
Work room							
W_W	100 × −0.77	−77		−77		63	
W_N	123 × −1.05	−129		−95		43	
R	223 × 2.48	553		2627		2226	
L	800 × 3.4	2720		2720		2720	
P	3 × 255	765	765	765	765	765	765
		3832	765	5940	765	5817	765
Stock							
R	53 × 2.48	131		624		529	
Fitting							
G_N	48 × 18.64	895		2021		1743	
W_N	188 × −1.05	−197		−145		66	
R	600 × 2.48	1488		7068		5988	
L	1400 × 3.4	4760		4760		4760	
P	5 × 255	1275	1275	1275	1275	1275	1275
		8221	1275	14979	1275	13832	1275
Reception							
G_E	64 × 126.72	8110		2890		2196	
W_E	99 × −0.77	−76		118		173	
W_N	100 × −1.05	−105		−77		35	
R	254 × 2.48	630		2992		2534	
L	400 × 3.4	1360		1360		1360	
P	1 × 255	255	255	255	255	255	255
		10174	255	7538	255	6553	255
Entry #1							
D_E	21 × 3.06	64		104		84	
G_E	23 × 126.72	2915		1039		789	
W_E	6 × −0.77	−5		7		11	
R	127 × 2.48	315		1496		1267	
L	200 × 3.4	680		680		680	
P	0 × 255	0		0		0	
I	7040 × −0.162 × 0.328	−1140	2309	507	2309	1014	2309
		2829	2309	3833	2309	3845	2309
Waiting room							
G_E	24 × 126.72	3041		1083		824	
W_E	244 × −0.77	−188		290		427	
W_N	56 × −1.05	−59		−43		20	
R	417 × 2.48	1034		4912		4167	
L	800 × 3.4	2720		2720		2720	
P	4 × 255	1020	1020	1020	1020	1020	1020
		7568	1020	9982	1020	9178	1020

CLTD METHOD

Calculating heat gain with the TETD method gives what amounts to an instaneous heat gain. With the exception of the time lag used in wall and roof calculations, all heat gains are assumed to happen *right now*.

There has been much evidence that field-monitored cooling loads do not match the calculated heat gains. This difference was attributed to what old-timers called *storage*. We theorized that, when the sun shined through a window, some of the heat was stored by the structure and by the building's furniture and furnishings. Many designers developed storage factors that they applied to heat gain calculations to arrive at the cooling load. The CLTD method (total equivalent cooling load temperature difference) was developed by ASHRAE-sponsored research to calculate the cooling load directly.

The technique is similar to the TETD method when working with roofs and walls. Merely substitute the CLTD for the TETD. These are given in Tables 3-35 and 3-36. Correction factors are applied in the same manner as with TETD. While TETD assumes a 75° space temperature, the CLTD tables assume 78°, in line with ASHRAE 90-80 recommendations. The outside temperatures are the same for both tables: 95° db, 21° range, and 84.5° average temperature.

Let's try a roof first. Looking at Table 3-36, note that the second column describes the type of roof, the third column the roof weight, the fourth gives the U factor, and finally we see the CLTD for each of the 24 hours in a day. the table is divided into two halves; the top is for roofs without a suspended ceiling, the bottom is for roofs with a suspended ceiling.

Example #24

What is the maximum cooling load through 100 ft² of

roof that is constructed of metal deck with one inch of insulation? This roof is the first one listed. As you follow it across the page, you find that the maximum CLTD occurs in the 14th hour (2:00 p.m.) and is 79°. The U factor is 0.213.

$$Q = U \times A \times CLTD$$
$$= 0.213(100)(79)$$
$$= 1682.7 \text{ Btuh}$$

Walls are handled in a similar fashion. First, consult Table 3-34, which describes various walls, to select the wall grouping.

There is an interesting difference between the CLTD wall table and TETD wall table. Where the TETD table ranked the walls from the lightest, A, to the heaviest, M, the CLTD table does just the opposite. It ranks the walls from the heaviest, A, to the lightest, G.

Example #25

What is the cooling load, at noon, from a west-facing wall that is constructed of 4″ face brick, 1_0 insulation, and 8″ cinder block. The wall's U factor is 0.14.

First, consult Table 3-34 to determine the wall category. The first four major headings are all headed by 4″ face brick. The third group is the one that describes our wall, "(4-in.) Face Brick + *(L.W. or H.W. Concrete Block)*." (L.W. and H.W. are contractions of lightweight and heavyweight.) In this group, the fourth line down reads, "Air Space or 1-in. Insulation + 6-in. or 8-in. Block." This is the wall we want and it belongs in Group C. From Table 3-35, the CLTD of a Group C, west-facing wall is 12° at noon.

$$Q = U \times A \times CLTD$$
$$= 0.14(100)(12)$$
$$= 168 \text{ Btuh}$$

Notice that, at the far right of Table 3-34, there is a column headed *Code Numbers of Layers*. This column lists the data used to arrive at the wall classification.

Up to this point, there is little difference between the two methods, only the numbers differ. However, with glass, there is a significant difference. To arrive at the cooling load caused by solar heat gain, we use the maximum solar heat gain, the shading coefficient, and a cooling load factor for glass. The cooling load factors are given in Tables 3-38 and 3-39.

Why this big difference from the TETD method? When the sun shines through a window, it first warms the glass. As it strikes the floor, walls, and furnishings, it heats each of these which delays the warming of the air. With the TETD method, it was assumed that the solar heat was absorbed at once or instantaneously.

The conduction portion of the heat gain is treated in

the same way as the TETD method. Let's try an example.

Example #26

What is the cooling load, at noon on July 21, for a 10 ft² south-facing window at 40°N latitude?

Compare this to the instantaneous heat gain. Assume interior drapes with a shade coefficient of 0.63. The space temperature is 78°, the outdoor temperature is 90°, and the U factor is 0.65. From Table 3-37, the maximum south heat gain is 109. The cooling load factor (CLF), given in Table 3-38, is 0.82 at noon with heavy construction.

$$Q =$$
$$(A \times HG \times SC \times CLF) + (U \times A \times TD)$$
$$= (10 \times 109 \times 0.63 \times 0.82) +$$
$$(0.65 \times 10 \times (90 - 78))$$
$$= 563 + 78$$
$$= 641 \text{ Btuh/ft}^2$$

The instantaneous heat gain is:

$$Q =$$
$$(10 \times 109 \times 0.63) + (0.65 \times 10 \times (90 - 78))$$
$$= 687 + 78$$
$$= 765$$

Example #27

What is the cooling load for the glass in Example #26, if there is exterior shading with a SC of 0.63? The room construction is considered to be heavy (H).

Again, we use a HG of 109 and a SC of 0.63, but now our noon CLF, from Table 3-38, is 0.51.

$$Q =$$
$$(A \times HG \times SC \times CLF) + (U \times A \times TD)$$
$$= (10 \times 109 \times 0.63 \times 0.51) +$$
$$(0.65 \times 10 \times (90 - 78))$$
$$= 350 + 78$$
$$= 428 \text{ Btuh/ft}^2$$

Obviously, it pays to use outside, rather than inside, shading.

Interior loads

When we get to interior loads, such as lights, people and equipment, additional factors are introduced. For lights, we must consider the construction of the space, the type of furnishings, the location of the supply and return air terminals, the type of light fixture, the hours that the lights are on, and the hours of cooling equipment operation.

Table 3-40 lists the first factor, the *a* coefficient. As you can see, the first decision is to evaluate the type of furniture and furnishings from the descriptions in the second column. The third column lists types of supply and return systems and air supply rates (V). The fourth column lists types of lighting fixtures. The term *vented* refers to whether or not the return air is drawn across or through the fixture.

Next, consult Table 3-41 to determine the *b* classification. The table list four air circulation rates: low, medium, high, and very high. The problem with this table is in deciding what is high and what is low. The footnotes are some help. While they don't give cfm/ft², they do give an *h* value which is the surface coefficient used by the researchers. A *h* value of 4.0 requires a velocity, across the ceiling, of 350 fpm. A value of 6.00 requires 1300 fpm. These are difficult to determine! It is probably easier to use the Low column for all projects that utilize floor, wall or ceiling diffusers and use the High column if fan coil or induction units are used.

After selecting the *a* coefficient and the *b* classification, use Table 4-42 to determine the CLF. A word of caution: these tables only apply if the cooling system operates 24 hours a day. If the cooling system operates only when the lights are on, the CLF is 1. If the lights are on 24 hours a day, the CLF is 1.

People

The sensible heat portion of the total heat given off by people does not immediately become part of the cooling load. Seventy percent of this sensible heat is given off in the form of radiation and is absorbed by the structure and furnishings before it is eventually released into the space. *All of the latent heat becomes an immediate part of the cooling load.* The CLF for occupants is given in Table 3-43.

As with lighting loads, the occupant CLF is 1 if the the air conditioning system operates only when the space is occupied. In churches, theaters, and classrooms, where the occupant density is high, the CLF is 1. If the air conditioning system operates for 24 hours or the space has a low occupant density, an hourly CLF can be selected. Remember to apply the CLF to the sensible load only.

Appliances

Appliances have a CLF that applies if the air conditioning system operates 24 hours a day. These are listed in Tables 3-44 and 3-45.

Example #28

What is the sensible cooling load, at 8:00 a.m., noon, 4:00 p.m., and 6:00 p.m., in an interior room that has 1000 watts of lighting and 4 people engaged in moderately active office work such as typing? The room is occupied from 8:00 a.m to 4:00 p.m. and the air conditioning operates 24 hours a day. The lights have an *a* co-

efficient of 0.45 and a *b* classification of B. First, select the CLF for lights and people from Tables 3-42 and 3-43 and the sensible load for people from Table 3-31.

	CLF Lights	CLF People	People Sensible
8:00 a.m.	0.07	0.51	255
Noon	0.65	0.72	255
4:00 p.m.	0.77	0.84	255
6:00 p.m.	0.31	0.30	255

Now, the cooling load.

$Q = CLF_L(watts)(3.4) + CLF_P(people)(sensible heat)$

8:00 a.m.	$0.07(1000)(3.4) + 0.51(4)(255) =$ 758 Btuh
Noon	$0.65(1000)(3.4) + 0.72(4)(255) =$ 2944 Btuh
4:00 p.m.	$0.77(1000)(3.4) + 0.84(4)(255) =$ 3475 Btuh
6:00 a.m.	$0.31(1000)(3.4) + 0.30(4)(255) =$ 1360 Btuh
Instantaneous	$1.0(1000)(3.4) + 1.0(4)(255) =$ 4420 Btuh

Let's recompute Example #22 and see how the loads compare. We will first assume that there is air conditioning only when the space is occupied and alternately that the air condition system operates 24 hours a day. We will only calculate the sensible load because the latent load will not change.

The room construction is considered to be medium and there is interior shading.

Notice that we use a CLF of 1.0 for lights and people in the 8 hour example because the air conditioning system operates only while the space is occupied. If the system operates for 24 hours, the CLF at noon will be 0.65 for lights and 0.72 for people. The lights have an *a* coefficient of 0.45 and a *B* classification. They are turned on at 8:00 a.m., when the people arrive, and stay on for 8 hours.

Solar			8 Hour Operation	24 Hour Operation
G_S	$24 \times 109 \times 0.83$	=	2171	2171
G_E	$12 \times 216 \times 0.27$	=	700	700
Transmission				
G	$36 \times 0.61 \times 8°$	=	176	176
Combined solar and transmission				
W_S	$76 \times 0.084 \times 39°$	=	249	249
W_E	$88 \times 0.084 \times 40°$	=	296	296
R	$100 \times 0.109 \times 17°$	=	185	185
Internal				
Lights	$800 W \times 3.4 \times 1.0$	=	2720	
	$800 W \times 3.4 \times 0.65$	=		1768
People	$5 \times 255 \times 1.0$	=	1275	
	$5 \times 255 \times 0.72$	=		918

Infiltration

Sensible	816 × 0.018 × 8°	=	118	118
	Total		7890	6581

How do the they methods compare? The sensible load is reduced from 8627 to 7890, if the CLTD is used and the system operates only when the space is occupied. If we assume 24 hour operation, then the load is 6581.

Let's see how our complete building (Example #21) does with the CLTD method. We will assume that the doctors have office hours from 8:00 a.m. to 6:00 p.m. All rooms are used continuously, because the doctors' hours overlap, and the air conditioning system will operate 24 hours a day.

We will calculate the cooling loads for 8:00 a.m., noon, 4:00 p.m., and 6:00 p.m. with a 8° correction for the outdoor temperature. Only the 8° will be subtracted from the CLTD wall, door, and roof tables because these are based on a 78° space temperature and no further correction is necessary. The lights have an *a* coefficient of 0.55 and a *B* classification. Only the sensible heat calculations are shown because the latent heat load is the same as with the TETD method.

Glass (solar) Maximum Solar (Table 3-37)

		CLF (Table 3-38)			
		8:00 a.m.	Noon	4:00 p.m.	6:00 p.m.
N	38	0.65	0.89	0.75	0.91
E	216	0.80	0.27	0.17	0.11
S	109	0.22	0.83	0.35	0.19
W	216	0.11	0.17	0.82	0.61

Outdoor temperature		69°	82°	86°	83°
Wall		CLTD (Table 3-35) Group B			
N		2	0	2	4
E		7	11	17	18
S		5	3	7	11
W		11	7	7	11
Roof		(Roof #1, Table 3-36)			
		1°	54°	66°	48°
		CLF			
Lights (Table 3-42)		0.12	0.72	0.78	0.81
People (Table 3-43)		0.53	0.74	0.85	0.89
Door		CLTD (Table 3-35) Group G			
E		39	32	21	16
S		−3	31	29	17
W		−3	11	59	59

Shop

			8:00 a.m.	Noon	4:00 p.m.	6:00 p.m.
G_N	6 × 38 × 0.65	=	148	203	171	207
*G_T	6 × 1.04 × (69 − 78)	=	−56	25	50	31
L	1440 × 3.4 × 0.12	=	588	3525	3819	3966
P	2 × 255 × 0.53	=	270	377	434	454
			950	4130	4474	4658

Employees' Lounge

G_N	6 × 38 × 0.65	=	148	203	171	207
G_T	6 × 1.04 × (69 − 78)	=	−56	25	50	31
L	800 × 3.4 × 0.12	=	326	1958	2122	2203
P	4 × 255 × 0.53	=	541	755	867	908
			959	2941	3210	3349

*Glass transmittance

Exam Rooms #1 through #6

G_E	12 × 216 × 0.8 × 0.63	=	1306	700	441	285
G_T	12 × 1.04 × −9°	=	−112	50	100	62
W_E	66 × 0.14 × 7°	=	65	102	157	166
R	132 × .15 × 1°	=	20	1069	1307	950
L	300 × 3.4 × 0.12	=	122	734	796	826
P	1 × 255 × 0.53	=	135	189	217	227
			1536	2844	3018	2516

Exam Room #7

G_E	12 × 216 × 0.8 × 0.63	=	1306	700	441	285
G_T	12 × 1.04 × −9°	=	−112	50	100	62
W_E	66 × 0.14 × 7°	=	65	102	157	166
W_S	109 × 0.14 × 5°	=	76	46	107	168
R	132 × .15 × 1°	=	20	1069	1307	950
L	300 × 3.4 × 0.12	=	122	734	796	826
P	1 × 255 × 0.53	=	135	189	217	227
			1612	2890	3125	2684

Entry #4

D_S	21 × 0.47 × −3°	=	−30	306	286	168
W_S	30 × 0.14 × 5	=	21	13	29	46
R	128 × .15 × 1°	=	19	1037	1267	922
I	1956 × − 0.162	=	−317	141	282	211
L	200 × 3.4 × 0.12	=	82	490	530	551
P	0 × 255 × 0.53	=	0	0	0	0
			−225	1987	2394	1898

Hall

R	398 × .15 × 1°	=	60	3224	3940	2866
L	300 × 3.4 × 0.12	=	122	734	796	826
P	0 × 255 × 0.53	=	0	0	0	0
			182	3958	4736	3692

Office

G_W	12 × 216 × 0.11	=	285	441	1339	996
G_T	12 × 1.04 × −9°	=	−112	50	100	62
W_W	66 × 0.14 × 11°	=	102	65	65	102
W_S	109 × 0.14 × 5°	=	76	46	107	168
R	132 × .15 × 1°	=	20	1069	1307	950
L	300 × 3.4 × 0.12	=	122	734	796	826
P	2 × 255 × 0.53	=	270	377	434	454
			763	2782	4148	3558

Exam Rooms #8, #9

G_W	12 × 216 × 0.11	=	285	441	1339	996
G_T	12 × 1.04 × −9°	=	−112	50	100	62
W_W	66 × 0.14 × 11°	=	102	65	65	102
R	132 × .15 × 1°	=	20	1069	1307	950
L	300 × 3.4 × 0.12	=	122	734	796	826
P	1 × 255 × 0.53	=	135	189	217	227
			552	2548	3824	3163

Drop room

G_W	24 × 216 × 0.11	=	570	881	4251	3162
G_T	22 × 1.04 × −9°	=	−225	100	200	125
W_W	145 × 0.14 × 11°	=	223	142	142	223
W_S	55 × 0.14 × 5°	=	39	23	54	85
R	402 × 0.15 × 1°	=	60	3256	3980	3894
L	800 × 3.4 × 0.12	=	326	1958	2122	2203
P	20 × 255 × 0.53	=	2703	3774	4335	4539
			3696	10134	15084	14231

Visual fields

W_W	91 × 0.14 × 11°	=	140	89	89	140
R	233 × .15 × 1°	=	35	1887	2307	2771
L	0 × 3.4 × 0.12	=	0	0	0	0
P	2 × 255 × 0.53	=	270	377	434	454
			445	2353	2830	3365

Doctor's Office

			8:00 a.m.	Noon	4:00 p.m.	6:00 p.m.
G_W	$24 \times 216 \times 0.13$	=	570	881	4251	3162
G_T	$24 \times 1.04 \times -9°$	=	−225	100	200	125
W_W	$72 \times 0.14 \times 11°$	=	111	71	71	111
R	$245 \times 0.15 \times 1°$	=	37	1985	2426	1764
L	$400 \times 3.4 \times 0.12$	=	163	979	1061	1102
P	$2 \times 255 \times 0.53$	=	270	378	434	454
			926	4394	8443	6718

Entry #2

D_W	$21 \times 0.47 \times -3°$	=	−30	109	582	582
W_W	$76 \times 0.14 \times 11°$	=	117	74	74	117
R	$166 \times .15 \times 1°$	=	25	1345	1643	1195
L	$200 \times 3.4 \times 0.12$	=	82	490	530	551
P	$0 \times 255 \times 0.53$	=	0	0	0	0
I	1956×-0.162	=	−317	141	282	211
			−123	2159	3111	2656

General office & reception

R	$422 \times 0.15 \times 1°$	=	63	3418	4178	3038
L	$800 \times 3.4 \times 0.12$	=	326	1958	2122	2203
P	$5 \times 255 \times 0.53$	=	676	944	1084	1135
			1065	6320	7384	6376

Women's toilet, men's toilet

W_S	$54 \times 0.14 \times 5°$	=	38	23	53	83
R	$71 \times 0.15 \times 1°$	=	11	575	703	511
L	$100 \times 3.4 \times 0.12$	=	41	245	265	275
P	$0 \times 255 \times 0.53$	=	0	0	0	0
			90	843	1021	869

Storage

W_S	$59 \times 0.14 \times 5°$	=	41	25	58	91
W_W	$84 \times 0.14 \times 11°$	=	129	82	82	129
R	$77 \times 0.15 \times 1°$	=	12	624	762	554
L	$0 \times 3.4 \times 0.12$	=	0	0	0	0
P	$0 \times 255 \times 0.53$	=	0	0	0	0
			182	731	902	774

Entry #3

D_W	$21 \times 0.47 \times -3°$	=	−30	109	582	582
G_W	$23 \times 216 \times 0.13$	=	546	845	2566	1909
G_T	$23 \times 1.04 \times -9°$	=	−215	96	191	120
W_W	$6 \times 0.14 \times 11°$	=	9	6	6	120
R	$127 \times 0.15 \times 1°$	=	19	1029	1257	914
L	$200 \times 3.4 \times 0.12$	=	82	490	530	551
P	$0 \times 255 \times 0.53$	=	0	0	0	0
I	7040×-0.162	=	−1140	506	1014	760
			−729	3081	6146	4956

Hall

R	$120 \times 0.15 \times 1°$	=	18	972	1188	. 864
L	$100 \times 3.4 \times 0.12$	=	41	245	265	275
P	$0 \times 255 \times 0.53$	=	0	0	0	0
			59	1217	1453	1139

Office

W_W	$60 \times 0.14 \times 11°$	=	92	59	59	92
R	$53 \times 0.15 \times 1°$	=	8	429	525	382
L	$100 \times 3.4 \times 0.12$	=	41	245	265	275
P	$1 \times 255 \times 0.53$	=	135	189	217	227
			276	922	1066	976

Work Room

W_N	$123 \times 0.14 \times 2°$	=	34	0	34	69
W_W	$100 \times 0.14 \times 11°$	=	154	98	98	154
R	$223 \times 0.15 \times 1°$	=	33	1806	2208	1606
L	$800 \times 3.4 \times 0.12$	=	326	1958	2122	2203
P	$3 \times 255 \times 0.53$	=	405	566	658	681
			952	4428	5120	4713

Stock

			8:00 a.m.	Noon	4:00 p.m.	6:00 p.m.
R	$53 \times 0.15 \times 1°$	=	8	429	525	382
L	$0 \times 3.4 \times 0.12$	=	0	0	0	0
P	$0 \times 255 \times 0.53$	=	0	0	0	0
			8	429	525	382

Fitting

G_N	$48 \times 38 \times 0.65$	=	1186	1623	1368	1660
G_T	$48 \times 1.04 \times -9°$	=	−449	200	400	250
W_N	$189 \times 0.14 \times 2°$	=	53	0	53	106
R	$600 \times 0.15 \times 1°$	=	90	4860	5940	4320
L	$1400 \times 3.4 \times 0.12$	=	571	3427	3713	3856
P	$5 \times 255 \times 0.53$	=	676	944	1084	1135
			2127	11054	12558	11327

Reception

G_E	$64 \times 216 \times 0.8 \times 0.63$	=	6967	3732	2350	1521
G_T	$64 \times 1.04 \times -9°$	=	−599	266	532	323
W_E	$99 \times 0.14 \times 7°$	=	97	152	236	249
W_N	$100 \times 0.14 \times 2°$	=	28	0	28	56
R	$254 \times 0.15 \times 1°$	=	38	2057	2515	1829
L	$400 \times 3.4 \times 0.12$	=	163	979	1060	1102
P	$1 \times 255 \times 0.53$	=	135	189	217	227
			6829	7375	6938	5307

Entry #1

D_E	$21 \times 0.47 \times 39°$	=	−385	316	207	158
G_E	$23 \times 216 \times 0.8 \times 0.63$	=	2504	1341	845	546
G_T	$23 \times 1.04 \times -9°$	=	−215	96	191	120
W_E	$6 \times 0.14 \times 7°$	=	6	9	14	15
R	$127 \times 0.15 \times 1°$	=	19	1029	1257	914
L	$200 \times 3.4 \times 0.12$	=	82	490	530	551
P	$0 \times 255 \times 0.53$	=	0	0	0	0
I	7040×-0.162	=	−1140	506	1014	760
			871	3787	4058	3064

Waiting room

G_E	$24 \times 216 \times 0.8 \times 0.63$	=	2613	1400	881	570
G_T	$24 \times 1.04 \times -9°$	=	−225	100	200	125
W_E	$244 \times 0.14 \times 7°$	=	239	376	581	615
W_N	$56 \times 0.14 \times 2°$	=	16	0	16	31
R	$417 \times 0.15 \times 1°$	=	63	3378	4128	3002
L	$800 \times 3.4 \times 0.12$	=	326	1958	2122	2203
P	$4 \times 255 \times 0.53$	=	541	755	867	908
			3573	7967	8795	7454

Totals	*8:00 a.m.*	26,576	*Noon*	91,274
	4:00 p.m.	114,363	*6:00 p.m.*	99,825

REDUCING COOLING LOADS

Once I overheard three architects discussing the cost of air conditioning and they chorused, "Can't something be done about air conditioning costs?" Yes, something can be done! The architect and engineer, working together, can reduce air conditioning loads and, in this way, reduce air conditioning costs. As you have just seen, air conditioning loads are caused by direct solar radiation, transmission and internal loads, such as lights, equipment, and people. The architect can incorporate elements into his design that will reduce the load caused by these heat producers.

Many of our modern buildings are built with large amounts of glass and about 75% of the air conditioning load is due to direct solar radiation on the glass. For example, the peak load on unshaded glass facing east or west is 225 Btuh/ft². If heat-absorbing, insulating glass is used, the load is 65 Btuh/ft² and, if this glass is shaded, the load is only 40 Btuh/ft². Here we have dramatic evidence of what shading can do to reduce air conditioning loads.

Not as dramatic, but also significant, is orientation. If the glass in the building is oriented north and south, the peak load on the unshaded, south-facing glass is 110 Btuh/ft² and 30 Btuh/ft² on the north-facing glass. Without spending money for heat-absorbing double glazing or for shading devices, we have reduced the sun load on the glass by a significant amount.

Even the configuration of a building can make a difference. A 15,000 ft² building that is 300 ft by 50 ft has 700 lineal feet of wall. A square building, with the same area, will have only 490 lineal feet of exposed wall.

Lighting is another item that deserves attention. Of course, we must provide good illumination. However, substituting fluorescent lamps for incandescents will reduce the wattage consumed by half while maintaining the same illumination. Consequently, the air conditioning load, due to lighting, is reduced by half.

Another item that can be reduced is the sun load on the roof. Flooding the roof with one inch of water will cut the heat load by 60%, spraying it with water will reduce the load by 80%. Of course, the use of insulation is as important in cooling as in heating. I would recommend that the minimum U factor for roofs be 0.10 Btuh/ft².

To summarize, the air conditioning load can be reduced by:

1. Shading the glass or using heat-absorbing double glazing
2. Orientation of the building
3. Shape of the building
4. Fluorescent lighting
5. Flooding or spraying the roof
6. Insulation

Let's apply some of these strategies to a specific building to see how much air conditioning loads can be reduced. Assume a 15,000 ft² building with three possible designs: a rectangular, 300 ft by 50 ft, building with the long side running north and south, the same building with the long side running east and west, and a square building. Here are the heat gains of the buildings as various strategies are applied:

	Rectangular, long axis N-S	*Rectanglar, long axis E-W*	*Square Building*
Single glass and incandescent lights	82 Tons	57 Tons	52 Tons
Single glass and fluorescent lighting	72 Tons	47 Tons	42 Tons
Shaded single glass, fluorescent lighting	35 Tons	33 Tons	31 Tons
Heat-absorbing double glazing, fluorescent lighting	48 Tons	38 Tons	35 Tons
Shaded single glass, fluorescent lighting, flooded roof	33 Tons	31 Tons	29 Tons

You will notice that the basic building has a load of 82 tons. Ten tons of that can be saved by the use of fluorescent lighting alone.

Probably the two most significant savings are in orientation and shading the glass. Orienting the building, so that the long axis runs east and west, saves 25 tons, shading the glass saves 37 tons. Notice that, if the glass is not shaded, orientation and configuration create significant savings. If the the glass is shaded, the load reductions are minimal.

TABLE 3-1 Surface Conductances and Resistances for Air

All conductance values expressed in Btu/(hr · ft^2 · F).

A surface cannot take credit for both an air space resistance value and a surface resistance value. No credit for an air space value can be taken for any surface facing an air space of less than 0.5 in.

SECTION A. Surface Conductances and Resistances[a,b,d]							SECTION B. Reflectivity and Emittance Values of Various Surfaces[c] and Effective Emittances of Air Spaces				
		Surface Emittance								**Effective Emittance E of Air Space**	
Position of Surface	**Direction of Heat Flow**	**Non-reflective $\varepsilon = 0.90$**		**Reflective $\varepsilon = 0.20$**		**Reflective $\varepsilon = 0.05$**	**Surface**	**Reflectivity in Percent**	**Average Emittance ε**	**One surface emittance ε; the other 0.90**	**Both surfaces emittances ε**
		h_i	R	h_i	R	h_i R					
STILL AIR											
Horizontal	Upward	1.63	0.61	0.91	1.10	0.76 1.32	Aluminum foil, bright	92 to 97	0.05	0.05	0.03
Sloping—45 deg	Upward	1.60	0.62	0.88	1.14	0.73 1.37	Aluminum sheet	80 to 95	0.12	0.12	0.06
Vertical	Horizontal	1.46	0.68	0.74	1.35	0.59 1.70	Aluminum coated paper, polished	75 to 84	0.20	0.20	0.11
Sloping—45 deg	Downward	1.32	0.76	0.60	1.67	0.45 2.22	Steel, galvanized, bright. . .	70 to 80	0.25	0.24	0.15
Horizontal	Downward	1.08	0.92	0.37	2.70	0.22 4.55	Aluminum paint	30 to 70	0.50	0.47	0.35
MOVING AIR (Any Position)		h_0	R	h_0	R	h_0 R	Building materials: wood, paper, masonry, nonmetallic paints	5 to 15	0.90	0.82	0.82
15-mph Wind (for winter)	Any	6.00	0.17				Regular glass	5 to 15	0.84	0.77	0.72
7.5-mph Wind (for summer)	Any	4.00	0.25								

TABLE 3-2 Thermal Resistances of Plane[a] Air Spaces[d,e]

All resistance values expressed in (hour)(square foot)(degree Fahrenheit temperature difference) per Btu

Values apply only to air spaces of uniform thickness bounded by plane, smooth, parallel surfaces with no leakage of air to or from the space.

Thermal resistance values for multiple air spaces must be based on careful estimates of mean temperature differences for each air space.

See the Caution section, under Overall Coefficients and Their Practical Use.

Position of Air Space	Direction of Heat Flow	Mean Temp,[b] (F)	Temp Diff,[b] (deg F)	0.5-in. Air Space[d] Value of E[b,c]					0.75-in. Air Space[d] Value of E[b,c]				
				0.03	0.05	0.2	0.5	0.82	0.03	0.05	0.2	0.5	0.82
Horiz.	Up	90	10	2.13	2.03	1.51	0.99	0.73	2.34	2.22	1.61	1.04	0.75
		50	30	1.62	1.57	1.29	0.96	0.75	1.71	1.66	1.35	0.99	0.77
		50	10	2.13	2.05	1.60	1.11	0.84	2.30	2.21	1.70	1.16	0.87
		0	20	1.73	1.70	1.45	1.12	0.91	1.83	1.79	1.52	1.16	0.93
		0	10	2.10	2.04	1.70	1.27	1.00	2.23	2.16	1.78	1.31	1.02
		−50	20	1.69	1.66	1.49	1.23	1.04	1.77	1.74	1.55	1.27	1.07
		−50	10	2.04	2.00	1.75	1.40	1.16	2.16	2.11	1.84	1.46	1.20
45° Slope	Up	90	10	2.44	2.31	1.65	1.06	0.76	2.96	2.78	1.88	1.15	0.81
		50	30	2.06	1.98	1.56	1.10	0.83	1.99	1.92	1.52	1.08	0.82
		50	10	2.55	2.44	1.83	1.22	0.90	2.90	2.75	2.00	1.29	0.94
		0	20	2.20	2.14	1.76	1.30	1.02	2.13	2.07	1.72	1.28	1.00
		0	10	2.63	2.54	2.03	1.44	1.10	2.72	2.62	2.08	1.47	1.12
		−50	20	2.08	2.04	1.78	1.42	1.17	2.05	2.01	1.76	1.41	1.16
		−50	10	2.62	2.56	2.17	1.66	1.33	2.53	2.47	2.10	1.62	1.30
Vertical	Horiz.	90	10	2.47	2.34	1.67	1.06	0.77	3.50	3.24	2.08	1.22	0.84
		50	30	2.57	2.46	1.84	1.23	0.90	2.91	2.77	2.01	1.30	0.94
		50	10	2.66	2.54	1.88	1.24	0.91	3.70	3.46	2.35	1.43	1.01
		0	20	2.82	2.72	2.14	1.50	1.13	3.14	3.02	2.32	1.58	1.18
		0	10	2.93	2.82	2.20	1.53	1.15	3.77	3.59	2.64	1.73	1.26
		−50	20	2.90	2.82	2.35	1.76	1.39	2.90	2.83	2.36	1.77	1.39
		−50	10	3.20	3.10	2.54	1.87	1.46	3.72	3.60	2.87	2.04	1.56
45° Slope	Down	90	10	2.48	2.34	1.67	1.06	0.77	3.53	3.27	2.10	1.22	0.84
		50	30	2.64	2.52	1.87	1.24	0.91	3.43	3.23	2.24	1.39	0.99
		50	10	2.67	2.55	1.89	1.25	0.92	3.81	3.57	2.40	1.45	1.02
		0	20	2.91	2.80	2.19	1.52	1.15	3.75	3.57	2.63	1.72	1.26
		0	10	2.94	2.83	2.21	1.53	1.15	4.12	3.91	2.81	1.80	1.30
		−50	20	3.16	3.07	2.52	1.86	1.45	3.78	3.65	2.90	2.05	1.57
		−50	10	3.26	3.16	2.58	1.89	1.47	4.35	4.18	3.22	2.21	1.66
Horiz.	Down	90	10	2.48	2.34	1.67	1.06	0.77	3.55	3.29	2.10	1.22	0.85
		50	30	2.66	2.54	1.88	1.24	0.91	3.77	3.52	2.38	1.44	1.02
		50	10	2.67	2.55	1.89	1.25	0.92	3.84	3.59	2.41	1.45	1.02
		0	20	2.94	2.83	2.20	1.53	1.15	4.18	3.96	2.83	1.81	1.30
		0	10	2.96	2.85	2.22	1.53	1.16	4.25	4.02	2.87	1.82	1.31
		−50	20	3.25	3.15	2.58	1.89	1.47	4.60	4.41	3.36	2.28	1.69
		−50	10	3.28	3.18	2.60	1.90	1.47	4.71	4.51	3.42	2.30	1.71

TABLE 3-2 (Cont.) Thermal Resistances of Plane[a] Air Spaces[d,e]

Position of Air Space	Direction of Heat Flow	Air Space Mean Temp,[b] (F)	Air Space Temp Diff,[b] (deg F)	1.5-in. Air Space[d] Value of E[b,c]					3.5-in. Air Space[d] Value of E[b,c]				
				0.03	0.05	0.2	0.5	0.82	0.03	0.05	0.2	0.5	0.82
Horiz	Up	90	10	2.55	2.41	1.71	1.08	0.77	2.84	2.66	1.83	1.13	0.80
		50	30	1.87	1.81	1.45	1.04	0.80	2.09	2.01	1.58	1.10	0.84
		50	10	2.50	2.40	1.81	1.21	0.89	2.80	2.66	1.95	1.28	0.93
		0	20	2.01	1.95	1.63	1.23	0.97	2.25	2.18	1.79	1.32	1.03
		0	10	2.43	2.35	1.90	1.38	1.06	2.71	2.62	2.07	1.47	1.12
		−50	20	1.94	1.91	1.68	1.36	1.13	2.19	2.14	1.86	1.47	1.20
		−50	10	2.37	2.31	1.99	1.55	1.26	2.65	2.58	2.18	1.67	1.33
45° Slope	Up	90	10	2.92	2.73	1.86	1.14	0.80	3.18	2.96	1.97	1.18	0.82
		50	30	2.14	2.06	1.61	1.12	0.84	2.26	2.17	1.67	1.15	0.86
		50	10	2.88	2.74	1.99	1.29	0.94	3.12	2.95	2.10	1.34	0.96
		0	20	2.30	2.23	1.82	1.34	1.04	2.42	2.35	1.90	1.38	1.06
		0	10	2.79	2.69	2.12	1.49	1.13	2.98	2.87	2.23	1.54	1.16
		−50	20	2.22	2.17	1.88	1.49	1.21	2.34	2.29	1.97	1.54	1.25
		−50	10	2.71	2.64	2.23	1.69	1.35	2.87	2.79	2.33	1.75	1.39
Vertical	Horiz.	90	10	3.99	3.66	2.25	1.27	0.87	3.69	3.40	2.15	1.24	0.85
		50	30	2.58	2.46	1.84	1.23	0.90	2.67	2.55	1.89	1.25	0.91
		50	10	3.79	3.55	2.39	1.45	1.02	3.63	3.40	2.32	1.42	1.01
		0	20	2.76	2.66	2.10	1.48	1.12	2.88	2.78	2.17	1.51	1.14
		0	10	3.51	3.35	2.51	1.67	1.23	3.49	3.33	2.50	1.67	1.23
		−50	20	2.64	2.58	2.18	1.66	1.33	2.82	2.75	2.30	1.73	1.37
		−50	10	3.31	3.21	2.62	1.91	1.48	3.40	3.30	2.67	1.94	1.50
45° Slope	Down	90	10	5.07	4.55	2.56	1.36	0.91	4.81	4.33	2.49	1.34	0.90
		50	30	3.58	3.36	2.31	1.42	1.00	3.51	3.30	2.28	1.40	1.00
		50	10	5.10	4.66	2.85	1.60	1.09	4.74	4.36	2.73	1.57	1.08
		0	20	3.85	3.66	2.68	1.74	1.27	3.81	3.63	2.66	1.74	1.27
		0	10	4.92	4.62	3.16	1.94	1.37	4.59	4.32	3.02	1.88	1.34
		−50	20	3.62	3.50	2.80	2.01	1.54	3.77	3.64	2.90	2.05	1.57
		−50	10	4.67	4.47	3.40	2.29	1.70	4.50	4.32	3.31	2.25	1.68
Horiz.	Down	90	10	6.09	5.35	2.79	1.43	0.94	10.07	8.19	3.41	1.57	1.00
		50	30	6.27	5.63	3.18	1.70	1.14	9.60	8.17	3.86	1.88	1.22
		50	10	6.61	5.90	3.27	1.73	1.15	11.15	9.27	4.09	1.93	1.24
		0	20	7.03	6.43	3.91	2.19	1.49	10.90	9.52	4.87	2.47	1.62
		0	10	7.31	6.66	4.00	2.22	1.51	11.97	10.32	5.08	2.52	1.64
		−50	20	7.73	7.20	4.77	2.85	1.99	11.64	10.49	6.02	3.25	2.18
		−50	10	8.09	7.52	4.91	2.89	2.01	12.98	11.56	6.36	3.34	2.22

TABLE 3-3 Thermal Properties of Typical Building and Insulating Materials (Design Values)[a]

(For Industrial Insulation Design Values, see Table 3B). These constants are expressed in Btu per (hour) (square foot) (degree Fahrenheit temperature difference). Conductivities *(k)* are per inch thickness, and conductances *(C)* are for thickness or construction stated, not per inch thickness. **All values are for a mean temperature of 75 F, except as noted by an asterisk (*) which have been reported at 45 F.** The SI units for Resistance (last two columns) were calculated by taking the the values from the two Resistance columns under Customary Unit, and multiplying by the factor 1/k (r/in.) and 1/C (R) for the appropriate conversion factor in Table 18.

Description	Density (lb/ft³)	Conductivity (k)	Conductance (C)	Customary Unit Resistance[b] (R) Per inch thickness (1/k)	Customary Unit Resistance[b] (R) For thickness listed (1/C)	Specific Heat, Btu/(lb) (deg F)	SI Unit Resistance[b] (R) (m·K) W	SI Unit Resistance[b] (R) (m²·K) W
BUILDING BOARD								
Boards, Panels, Subflooring, Sheathing								
Woodboard Panel Products								
Asbestos-cement board	120	4.0	—	0.25	—	0.24	1.73	
Asbestos-cement board.............0.125 in.	120	—	33.00	—	0.03			0.005
Asbestos-cement board.............0.25 in.	120	—	16.50	—	0.06			0.01
Gypsum or plaster board ....0.375 in.	50	—	3.10	—	0.32	0.26		0.06
Gypsum or plaster board ....0.5 in.	50	—	2.22	—	0.45			0.08
Gypsum or plaster board ....0.625 in.	50	—	1.78	—	0.56			0.10
Plywood (Douglas Fir).................	34	0.80	—	1.25	—	0.29	8.66	
Plywood (Douglas Fir).............0.25 in.	34	—	3.20	—	0.31			0.05
Plywood (Douglas Fir).............0.375 in.	34	—	2.13	—	0.47			0.08
Plywood (Douglas Fir).............0.5 in.	34	—	1.60	—	0.62			0.11
Plywood (Douglas Fir).............0.625 in.	34	—	1.29	—	0.77			0.19

TABLE 3-3 (Cont.) Thermal Properties of Typical Building and Insulating Materials (Design Values)[u/a/]

Description	Density (lb/ft³)	Conductivity (k)	Conductance (C)	Resistance[b] (R) Per inch thickness (1/k)	Resistance[b] (R) For thickness listed (1/C)	Specific Heat, Btu/(lb) (deg F)	Resistance[b] (R) (m·K) W	Resistance[b] (R) (m²·K) W
Mineral fiber with resin binder...................	15	0.29	—	3.45	—	0.17	23.91	
Mineral fiberboard, wet felted								
Core or roof insulation......................	16–17	0.34	—	2.94	—		20.38	
Acoustical tile............................	18	0.35	—	2.86	—	0.19	19.82	
Acoustical tile............................	21	0.37	—	2.70	—		18.71	
Mineral fiberboard, wet molded								
Acoustical tile[g].........................	23	0.42	—	2.38	—	0.14	16.49	
Wood or cane fiberboard								
Acoustical tile[g].....................0.5 in.	—	—	0.80	—	1.25	0.31		0.22
Acoustical tile[g]....................0.75 in.	—	—	0.53	—	1.89			0.33
Interior finish (plank, tile).............	15	0.35	—	2.86	—	0.32	19.82	
Wood shredded (cemented in								
preformed slabs)......................	22	0.60	—	1.67	—	0.31	11.57	
LOOSE FILL								
Cellulosic insulation (milled paper or								
wood pulp)...........................	2.3–3.2	0.27–0.32	—	3.13–3.70	—	0.33	21.69–25.64	
Sawdust or shavings....................	8.0–15.0	0.45	—	2.22	—	0.33	15.39	
Wood fiber, softwoods	2.0–3.5	0.30	—	3.33	—	0.33	23.08	
Perlite, expanded.....................	5.0–8.0	0.37	—	2.70	—	0.26	18.71	
Mineral fiber (rock, slag or glass)								
approx.[e] 3.75–5 in.................	0.6–2.0	—	—		11	0.17		1.94
approx.[e] 6.5–8.75 in...............	0.6–2.0	—	—		19			3.35
approx.[e] 7.5–10 in.................	0.6–2.0	—	—		22			3.87
approx.[e] 10.25–13.75 in.	0.6–2.0	—	—		30			5.28
Vermiculite, exfoliated.................	7.0–8.2	0.47	—	2.13	—	3.20	14.76	
	4.0–6.0	0.44	—	2.27	—		15.73	
ROOF INSULATION[h]								
Preformed, for use above deck								
Different roof insulations are available in different			0.72		1.39		—	0.24
thicknesses to provide the design C values listed.[h]			to		to			to
Consult individual manufacturers for actual			0.12		8.33		—	1.47
thickness of their material...................								
MASONRY MATERIALS								
CONCRETES								
Cement mortar............................	116	5.0	—	0.20	—		1.39	
Gypsum-fiber concrete 87.5% gypsum,								
12.5% wood chips	51	1.66	—	0.60	—	0.21	4.16	
Lightweight aggregates including ex-	120	5.2	—	0.19	—		1.32	
panded shale, clay or slate; expanded	100	3.6	—	0.28	—		1.94	
slags; cinders; pumice; vermiculite;	80	2.5	—	0.40	—		2.77	
also cellular concretes	60	1.7	—	0.59	—		4.09	
	40	1.15	—	0.86	—		5.96	
	30	0.90	—	1.11	—		7.69	
	20	0.70		1.43			9.91	
Perlite, expanded.........................	40	0.93		1.08			7.48	
	30	0.71		1.41			9.77	
	20	0.50		2.00		0.32	13.86	
Sand and gravel or stone aggregate								
(oven dried)	140	9.0	—	0.11	—	0.22	0.76	
Sand and gravel or stone aggregate								
(not dried)	140	12.0	—	0.08	—		0.55	
Stucco	116	5.0	—	0.20	—		1.39	
MASONRY UNITS								
Brick, common[i]...........................	120	5.0	—	0.20	—	0.19	1.39	
Brick, face[i]	130	9.0	—	0.11	—		0.76	
Clay tile, hollow:								
1 cell deep ....3 in.	—	—	1.25	—	0.80	0.21		0.14
1 cell deep ....4 in.	—	—	0.90	—	1.11			0.20
2 cells deep.......................6 in.	—	—	0.66	—	1.52			0.27
2 cells deep.......................8 in.	—	—	0.54	—	1.85			0.33
2 cells deep......................10 in.	—	—	0.45	—	2.22			0.39
3 cells deep......................12 in.	—	—	0.40	—	2.50			0.44

TABLE 3-3 (Cont.) Thermal Properties of Typical Building and Insulating Materials (Design Values)[u/a/]

Description	Density (lb/ft³)	Conductivity (k)	Conductance (C)	Resistance[b] (R) Per inch thickness (1/k)	Resistance[b] (R) For thickness listed (1/C)	Specific Heat, Btu/(lb) (deg F)	SI Unit Resistance[b] (R) (m·K) W	SI Unit Resistance[b] (R) (m²·K) W
Plywood or wood panels.................0.75 in.	34	—	1.07	—	0.93	0.29		0.16
Vegetable Fiber Board								
Sheathing, regular density...............0.5 in.	18	—	0.76	—	1.32	0.31		0.23
..............0.78125 in.	18	—	0.49	—	2.06			0.36
Sheathing intermediate density ....0.5 in.	22	—	0.82	—	1.22	0.31		0.21
Nail-base sheathing.....................0.5 in.	25	—	0.88	—	1.14	0.31		0.20
Shingle backer......................0.375 in.	18	—	1.06	—	0.94	0.31		0.17
Shingle backer......................0.3125 in.	18	—	1.28	—	0.78			0.14
Sound deadening board ....0.5 in.	15	—	0.74	—	1.35	0.30		0.24
Tile and lay-in panels, plain or								
acoustic	18	0.40	—	2.50	—	0.14	17.33	
..........................0.5 in.	18	—	0.80	—	1.25			0.22
..........................0.75 in.	18	—	0.53	—	1.89			0.33
Laminated paperboard.....................	30	0.50	—	2.00	—	0.33	13.86	
Homogeneous board from								
repulped paper	30	0.50	—	2.00	—	0.28	13.86	
Hardboard								
Medium density	50	0.73	—	1.37	—	0.31	9.49	
High density, service temp. service								
underlay	55	0.82	—	1.22	—	0.32	8.46	
High density, std. tempered	63	1.00	—	1.00	—	0.32	6.93	
Particleboard								
Low density	37	0.54	—	1.85	—	0.31	12.82	
Medium density	50	0.94	—	1.06	—	0.31	7.35	
High density	62.5	1.18	—	0.85	—	0.31	5.89	
Underlayment.....................0.625 in.	40	—	1.22	—	0.82	0.29		0.14
Wood subfloor.......................0.75 in.		—	1.06	—	0.94	0.33		0.17
BUILDING MEMBRANE								
Vapor—permeable felt......................	—	—	16.70	—	0.06			0.01
Vapor—seal, 2 layers of mopped								
15-lb felt..........................	—	—	8.35	—	0.12			0.02
Vapor—seal, plastic film	—	—	—	—	Negl.			
FINISH FLOORING MATERIALS								
Carpet and fibrous pad	—	—	0.48	—	2.08	0.34		0.37
Carpet and rubber pad	—	—	0.81	—	1.23	0.33		0.22
Cork tile...........................0.125 in.	—	—	3.60	—	0.28	0.48		0.05
Terrazzo...............................1 in.	—	—	12.50	—	0.08	0.19		0.01
Tile—asphalt, linoleum, vinyl, rubber	—	—	20.00	—	0.05	0.30		0.01
vinyl asbestos						0.24		
ceramic............................						0.19		
Wood, hardwood finish................0.75 in.			1.47		0.68			0.12
INSULATING MATERIALS								
BLANKET AND BATT								
Mineral Fiber, fibrous form processed								
from rock, slag, or glass								
approx.[e] 2–2.75 in.....................	0.3–2.0	—	0.143	—	7[d]	0.17–0.23		1.23
approx.[e] 3–3.5 in.....................	0.3–2.0	—	0.091	—	11[d]			1.94
approx.[e] 3.50–6.5	0.3–2.0	—	0.053	—	19[d]			3.35
approx.[e] 6–7 in.	0.3–2.0		0.045		22[d]			3.87
approx.[d] 8.5 in.	0.3–2.0		0.033		30[d]			5.28
BOARD AND SLABS								
Cellular glass	8.5	0.38	—	2.63	—	0.24	18.23	
Glass fiber, organic bonded	4–9	0.25	—	4.00	—	0.23	27.72	
Expanded rubber (rigid).....................	4.5	0.22	—	4.55	—	0.40	31.53	
Expanded polystyrene extruded								
Cut cell surface.......................	1.8	0.25	—	4.00	—	0.29	27.72	
Expanded polystyrene extruded								
Smooth skin surface....................	2.2	0.20	—	5.00	—	0.29	34.65	
Expanded polystyrene extruded								
Smooth skin surface....................	3.5	0.19	—	5.26	—		36.45	
Expanded polystyrene, molded beads............	1.0	0.28	—	3.57	—	0.29	24.74	
Expanded polyurethane[f] (R-11 exp.)	1.5	0.16	—	6.25	—	0.38	43.82	
(Thickness 1 in. or greater)	2.5							

TABLE 3-3 (Cont.) Thermal Properties of Typical Building and Insulating Materials (Design Values)[u/a/]

Description	Density (lb/ft³)	Conductivity (k)	Conductance (C)	Resistance[b] (R) Per inch thickness (1/k)	Resistance[b] (R) For thickness listed (1/C)	Specific Heat, Btu/(lb) (deg F)	SI Unit Resistance[b] (R) (m·K)/W	SI Unit Resistance[b] (R) (m²·K)/W
Concrete blocks, three oval core:								
Sand and gravel aggregate .... 4 in.	—	—	1.40	—	0.71	0.22		0.13
.... 8 in.	—	—	0.90	—	1.11			0.20
.... 12 in.	—	—	0.78	—	1.28			0.23
Cinder aggregate .... 3 in.	—	—	1.16	—	0.86	0.21		0.15
.... 4 in.	—	—	0.90	—	1.11			0.20
.... 8 in.	—	—	0.58	—	1.72			0.30
.... 12 in.	—	—	0.53	—	1.89			0.33
Lightweight aggregate .... 3 in.	—	—	0.79	—	1.27	0.21		0.22
(expanded shale, clay, slate .... 4 in.	—	—	0.67	—	1.50			0.26
or slag; pumice) .... 8 in.	—	—	0.50	—	2.00			0.35
.... 12 in.	—	—	0.44	—	2.27			0.40
Concrete blocks, rectangular core.*ʲ								
Sand and gravel aggregate								
2 core, 8 in. 36 lb.ᵏ*	—	—	0.96	—	1.04	0.22		0.18
Same with filled coresˡ*	—	—	0.52	—	1.93	0.22		0.34
Lightweight aggregate (expanded shale,								
clay, slate or slag, pumice):								
3 core, 6 in. 19 lb.ᵏ*	—	—	0.61	—	1.65	0.21		0.29
Same with filled coresˡ*	—	—	0.33	—	2.99			0.53
2 core, 8 in. 24 lb.ᵏ*	—	—	0.46	—	2.18			0.38
Same with filled coresˡ*	—	—	0.20	—	5.03			0.89
3 core, 12 in. 38 lb.ᵏ*	—	—	0.40	—	2.48			0.44
Same with filled coresˡ*	—	—	0.17	—	5.82			1.02
Stone, lime or sand.	—	12.50	—	0.08	—	0.19	0.55	
Gypsum partition tile:								
3 × 12 × 30 in. solid	—	—	0.79	—	1.26	0.19		0.22
3 × 12 × 30 in. 4-cell	—	—	0.74	—	1.35			0.24
4 × 12 × 30 in. 3-cell	—	—	0.60	—	1.67			0.29
METALS **(See Chapter 37, Table 3)**								
PLASTERING MATERIALS								
Cement plaster, sand aggregate	116	5.0	—	0.20	—	0.20	1.39	
Sand aggregate .... 0.375 in.	—	—	13.3	—	0.08	0.20		0.01
Sand aggregate .... 0.75 in.	—	—	6.66	—	0.15	0.20		0.03
Gypsum plaster:								
Lightweight aggregate.................. 0.5 in.	45	—	3.12	—	0.32			0.06
Lightweight aggregate .... 0.625 in.	45	—	2.67	—	0.39			0.07
Lightweight agg. on metal lath .... 0.75 in.	—	—	2.13	—	0.47			0.08
Perlite aggregate......................	45	1.5	—	0.67	—	0.32	4.64	
Sand aggregate	105	5.6	—	0.18	—	0.20	1.25	
Sand aggregate .... 0.5 in.	105	—	11.10	—	0.09			0.02
Sand aggregate .... 0.625 in.	105	—	9.10	—	0.11			0.02
Sand aggregate on metal lath .... 0.75 in.	—	—	7.70	—	0.13			0.02
Vermiculite aggregate.................	45	1.7	—	0.59	—		4.09	
ROOFING								
Asbestos-cement shingles	120	—	4.76	—	0.21	0.24		0.04
Asphalt roll roofing	70	—	6.50	—	0.15	0.36		0.03
Asphalt shingles..............................	70	—	2.27	—	0.44	0.30		0.08
Built-up roofing .... 0.375 in.	70	—	3.00	—	0.33	0.35		0.06
Slate................................ 0.5 in.	—	—	20.00	—	0.05	0.30		0.01
Wood shingles, plain and plastic film faced	—	—	1.06	—	0.94	0.31		0.17
SIDING MATERIALS (ON FLAT SURFACE)								
Shingles								
Asbestos-cement............................	120	—	4.75	—	0.21			0.04
Wood, 16 in., 7.5 exposure	—	—	1.15	—	0.87	0.31		0.15
Wood, double, 16-in., 12-in. exposure	—	—	0.84	—	1.19	0.28		0.21
Wood, plus insul. backer board, 0.3125 in........	—	—	0.71	—	1.40	0.31		0.25
Siding								
Asbestos-cement, 0.25 in., lapped..............	—	—	4.76	—	0.21	0.24		0.04
Asphalt roll siding	—	—	6.50	—	0.15	0.35		0.03
Asphalt insulating siding (0.5 in. bed.)	—	—	0.69	—	1.46	0.35		0.26
Wood, drop, 1 × 8 in........................	—	—	1.27	—	0.79	0.28		0.14
Wood, bevel, 0.5 × 8 in., lapped..............	—	—	1.23	—	0.81	0.28		0.14
Wood, bevel, 0.75 × 10 in., lapped	—	—	0.95	—	1.05	0.28		0.18
Wood, plywood, 0.375 in., lapped	—	—	1.59	—	0.59	0.29		0.10
Wood, medium density siding, 0.4375 in.	40	1.49	—	0.67	—	0.28	4.65	

TABLE 3-3 (Cont.) Thermal Properties of Typical Building and Insulating Materials (Design Values)[u/a/]

Description	Density (lb/ft³)	Conductivity (k)	Conductance (C)	Resistance[b](R) Per inch thickness (1/k)	Resistance[b](R) For thickness listed (1/C)	Specific Heat, Btu/(lb) (deg F)	SI Unit Resistance[b] (R) (m·K) W	SI Unit Resistance[b] (R) (m²·K) W
Aluminum or Steel[m], over sheathing								
Hollow-backed	—	—	1.61	—	0.61	0.29		0.11
Insulating-board backed nominal 0.375 in.	—	—	0.55	—	1.82	0.32		0.32
Insulating-board backed nominal 0.375 in., foil backed			0.34		2.96			0.52
Architectural glass	—	—	10.00	—	0.10	0.20		0.02
WOODS								
Maple, oak, and similar hardwoods	45	1.10	—	0.91	—	0.30	6.31	
Fir, pine, and similar softwoods	32	0.80	—	1.25	—	0.33	8.66	
Fir, pine, and similar softwoods . . . . 0.75 in.	32	—	1.06	—	0.94	0.33		0.17
. . . . 1.5 in.		—	0.53	—	1.89			0.33
. . . . 2.5 in.		—	0.32	—	3.12			0.60
. . . . 3.5 in.		—	0.23	—	4.35			0.75

TABLE 3-4 Coefficients of Transmission (U) of Windows, Skylights and Light Transmitting Partitions

These values are for heat transfer from air to air, Btu/(hr · ft² · F). To calculate total heat gain including solar transmission, see Chapter 28.

PART A—VERTICAL PANELS (EXTERIOR WINDOWS, SLIDING PATIO DOORS, AND PARTITIONS)— FLAT GLASS, GLASS BLOCK, AND PLASTIC SHEET

Description	Winter	Summer	Interior
Flat Glass[b]			
single glass	1.10	1.04	0.73
insulating glass—double[c]			
0.1875-in. air space[d]	0.62	0.65	0.51
0.25-in. air space[d]	0.58	0.61	0.49
0.5-in. air space[e]	0.49	0.56	0.46
0.5-in. air space, low emittance coating[f]			
e = 0.20	0.32	0.38	0.32
e = 0.40	0.38	0.45	0.38
e = 0.60	0.43	0.51	0.42
insulating glass—triple[c]			
0.25-in. air spaces[d]	0.39	0.44	0.38
0.5-in. air spaces[g]	0.31	0.39	0.30
storm windows			
1-in. to 4-in. air space[d]	0.50	0.50	0.44
Plastic Sheet			
single glazed			
0.125-in. thick	1.06	0.98	—
0.25-in. thick	0.96	0.89	—
0.5-in. thick	0.81	0.76	—
insulating unit—double[c]			
0.25-in. air space[d]	0.55	0.56	—
0.5-in. air space[c]	0.43	0.45	—
Glass Block[h]			
6 × 6 × 4 in. thick	0.60	0.57	0.46
8 × 8 × 4 in. thick	0.56	0.54	0.44
—with cavity divider	0.48	0.46	0.38
12 × 12 × 4 in. thick	0.52	0.50	0.41
—with cavity divider	0.44	0.42	0.36
12 × 12 × 2 in. thick	0.60	0.57	0.46

PART B—HORIZONTAL PANELS (SKYLIGHTS)— FLAT GLASS, GLASS BLOCK, AND PLASTIC DOMES

Description	Winter[i]	Summer[j]	Interior[f]
Flat Glass[e]			
single glass	1.23	0.83	0.96
insulating glass—double[c]			
0.1875-in. air space[d]	0.70	0.57	0.62
0.25-in. air space[d]	0.65	0.54	0.59
0.5-in. air space[c]	0.59	0.49	0.56
0.5-in. air space, low emittance coating[f]			
e = 0.20	0.48	0.36	0.39
e = 0.40	0.52	0.42	0.45
e = 0.60	0.56	0.46	0.50

TABLE 3-4 Coefficients of Transmission (U) for Slab Doors

Btu per (hr·ft²·F)

Thickness[a]	Winter Solid Wood, No Storm Door	Winter Storm Door[b] Wood	Winter Storm Door[b] Metal	Summer No Storm Door
1-in.	0.64	0.30	0.39	0.61
1.25-in.	0.55	0.28	0.34	0.53
1.5-in.	0.49	0.27	0.33	0.47
2-in.	0.43	0.24	0.29	0.42
Steel Door[14]				
1.75 in.				
A[c]	0.59	—	—	0.58
B[d]	0.19	—	—	0.18
C[e]	0.47	—	—	0.46

[a]Nominal thickness.
[b]Values for wood storm doors are for approximately 50% glass; for metal storm door values apply for any percent of glass.
[c]A = Mineral fiber core (2 lb/ft³).
[d]B = Solid urethane foam core with thermal break.
[e]C = Solid polystyrene core with thermal break.

TABLE 3-5 Coefficients of Transmission (*U*) of Frame Walls[a]

These coefficients are expressed in Btu per (hour) (square foot) (degree Fahrenheit) difference in temperature between the air on the two sides), and are based on an outside wind velocity of 15 mph

Replace Air Space with 3.5-in. R-11 Blanket Insulation (New Item 4)

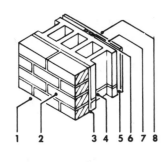

Construction	1 Resistance (R)		2 Resistance (R)	
	Between Framing	At Framing	Between Framing	At Framing
1. Outside surface (15 mph wind)	0.17	0.17	0.17	0.17
2. Siding, wood, 0.5 in.× 8 in. lapped (average)	0.81	0.81	0.81	0.81
3. Sheathing, 0.5-in. asphalt impregnated	1.32	1.32	1.32	1.32
4. Nonreflective air space, 3.5 in. (50 Fmean; 10 deg F temperature difference)	1.01	—	11.00	—
5. Nominal 2-in. × 4-in. wood stud	—	4.38	—	4.38
6. Gypsum wallboard, 0.5 in.	0.45	0.45	0.45	0.45
7. Inside surface (still air)	0.68	0.68	0.68	0.68
Total Thermal Resistance (R)	R_i=4.44	R_s=7.81	R_i=14.43	R_s=7.81

Construction No. 1: U_i = 1/4.44=0.225; U_s=1/7.81 =0.128. With 20% framing (typical of 2-in. × 4-in. studs @ 16-in. o.c.), U_{av} = 0.8 (0.225) + 0.2 (0.128) = 0.206 (See Eq 9)

Construction No. 2: U_i = 1/14.43 = 0.069; U_s = 0.128. With framing unchanged, U_{av} = 0.8(0.069) + 0.2(0.128) = 0.081

TABLE 3-6 Coefficients of Transmission (*U*) of Masonry Walls[a]

Coefficients are expressed in Btu per (hour) (square foot) (degree Fahrenheit difference in temperature between the air on the two sides), and are based on an outside wind velocity of 15 mph

Replace Cinder Aggregate Block with 6-in. Light-weight Aggregate Block with Cores Filled (New Item 4)

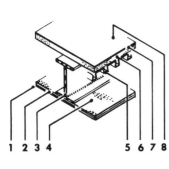

Construction	1 Resistance (R)		2 Resistance (R)	
	Between Furring	At Furring	Between Furring	At Furring
1. Outside surface (15 mph wind)	0.17	0.17	0.17	0.17
2. Face brick, 4 in.	0.44	0.44	0.44	0.44
3. Cement mortar, 0.5 in.	0.10	0.10	0.10	0.10
4. Concrete block, cinder aggregate, 8 in.	1.72	1.72	2.99	2.99
5. Reflective air space, 0.75 in. (50 F mean; 30 deg F temperature difference)	2.77	—	2.77	—
6. Nominal 1-in. × 3-in. vertical furring	—	0.94	—	0.94
7. Gypsum wallboard, 0.5 in., foil backed	0.45	0.45	0.45	0.45
8. Inside surface (still air)	0.68	0.68	0.68	0.68
Total Thermal Resistance (R)	R_i= 6.33	R_s= 4.50	R_i= 7.60	R_s= 5.77

Construction No. 1: U_i = 1/6.33 = 0.158; U_s = 1/4.50 = 0.222. With 20% framing (typical of 1-in. × 3-in. vertical furring on masonry @ 16-in. o.c.), U_{av} = 0.8 (0.158) + 0.2 (0.222) = 0.171

Construction No. 2: U_i = 1/7.60 = 0.132, U_s = 1/5.77 = 0.173. With framing unchanged, U_{av} = 0.8(0.132) + 0.2(0.173) = 1.40

TABLE 3-7 Coefficients of Transmission (*U*) of Metal Construction Flat Roofs and Ceilings[a]
(Winter Conditions, Upward Flow)

Coefficients are expressed in Btu per (hour) (square foot) (degree Fahrenheit) difference in temperature between the air on the two sides), and are based on upon outside wind velocity of 15 mph

Replace Rigid Roof Deck Insulation (*C* = 0.24) and Sand Aggregate Plaster with Rigid Roof Deck Insulation, *C* = 0.36 and Lightweight Aggregate Plaster (New Items 2 and 6)		
Construction (Heat Flow Up)	1	2
1. Inside surface (still air)	0.61	0.61
2. Metal lath and sand aggregate plaster, 0.75 in	0.13	0.47
3. Structural beam	0.00*	0.00*
4. Nonreflective air space (50 F mean; 10 deg F temperature difference)	0.93**	0.93**
5. Metal deck	0.00*	0.00*
6. Rigid roof deck insulation, *C* = 0.24(*R* = 1/*c*)	4.17	2.78
7. Built-up roofing, 0.375 in.	0.33	0.33
8. Outside surface (15 mph wind)	0.17	0.17
Total Thermal Resistance (R)	6.34	5.29

Construction No. 1: *U* = 1/6.34 = 0.158
Construction No. 2: *U* = 1/5.29 = 0.189

TABLE 3-8 Coefficients of Transmission (U) of Flat Masonary Roofs w/Built-up Roofing, with and without Suspended Ceilings[a,b] (Winter Conditions, Upward Flow)

These Coefficients are expressed in Btu per (hour) (square foot) (degree Fahrenheit difference in temperature between the air on the two sides), and are based upon an outside wind velocity of 15 mph

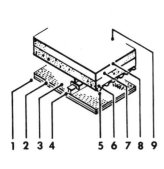

Add Rigid Roof Deck Insulation, $C = 0.24$ ($R = 1/C$) (New Item 7) Construction (Heat Flow Up)	1	2
1. Inside surface (still air)	0.61	0.61
1. Metal lath and lightweight aggregate plaster, 0.75 in.	0.47	0.47
3. Nonreflective air space, greater than 3.5 in. (50 F mean; 10 deg F temperature difference)	0.93*	0.93*
4. Metal ceiling suspension system with metal hanger rods	0**	0**
5. Corrugated metal deck	0	0
6. Concrete slab, lightweight aggregate, 2 in.	2.22	2.22
7. Rigid roof deck insulation (none)	—	4.17
8. Built-up roofing, 0.375 in.	0.33	0.33
9. Outside surface (15 mph wind)	0.17	0.17
Total Thermal Resistance (R)	4.73	8.90

Construction No. 1: $U_{av} = 1/4.73 = 0.211$
Construction No. 2: $U_{av} = 1/8.90 = 0.112$

TABLE 3-9 Coefficients of Transmission (U) of Wood Construction Flat Roofs and Ceilings[a] (Winter Conditions, Upward Flow)

Coefficients are expressed in Btu per (hour) (square foot) (degree Fahrenheit difference in temperature between the air on the two sides), and are based upon an outside wind velocity of 15 mph

Replace Roof Deck Insulation and 7.25-in. Air Space with 6-in. R-19 Blanket Insulation and 1.25-in. Air Space (New Items 5 and 7)

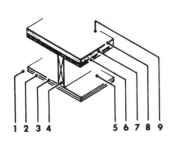

Construction (Heat Flow Up)	1 Resistance (R) Between Joists	At Joists	2 Between Joists	At Joists
1. Inside surface (still air)	0.61	0.61	0.61	0.61
2. Acoustical tile, fiberboard, glued, 0.5 in.	1.25	1.25	1.25	1.25
3. Gypsum wallboard, 0.5 in.	0.45	0.45	0.45	0.45
4. Nominal 2-in. × 8-in. ceiling joists	—	9.06	—	9.06
5. Nonreflective air space, 7.25 in. (50 F mean; 10 deg F temperature difference)	0.93*	—	1.05**	—
6. Plywood deck, 0.625 in.	0.78	0.78	0.78	0.78
7. Rigid roof deck insulation, c = 0.72, ($R = 1/C$)	1.39	1.39	19.00	—
8. Built-up roof	0.33	0.33	0.33	0.33
9. Outside surface (15 mph wind)	0.17	0.17	0.17	0.17
Total Thermal Resistance (R)	R_i=5.91	R_s=14.04	R_i=23.64	R_s=12.65

Construction No. 1 $U_i = 1/5.91 = 0.169$; $U_s = 1/14.04 = 0.071$. With 10% framing (typical of 2-in. joists @ 16-in. o.c.), $U_{av} = 0.9 (0.169) + 0.1 (0.071) = 0.159$
Construction No. 2 $U_i = 1/23.64 = 0.042$; $U_s = 1/12.65 = 0.079$. With framing unchanged, $U_{av} = 0.9 (0.042) + 0.1 (0.079) = 0.046$

TABLE 3-10 Coefficients of Transmission (U) of Frame Construction Ceilings and Floors[a]

Coefficients are expressed in Btu per (hour) (square foot) (degree Fahrenheit difference between the air on the two sides), and are based on still air (no wind) on both sides

Assume Unheated Attic Space above Heated Room with Heat Flow Up—Remove Tile, Felt, Plywood, Subfloor and Air Space—Replace with R-19 Blanket Insulation (New Item 4)

Heated Room Below Unheated Space Construction (Heat Flow Up)	1 Resistance (R) Between Floor Joists	At Floor Joist	2 Between Floor Joists	At Floor Joists
1. Bottom surface (still air)	0.61	0.61	0.61	0.61
2. Metal lath and lightweight aggregate, plaster, 0.75 in.	0.47	0.47	0.47	0.47
3. Nominal 2-in. × 8-in. floor joist	—	9.06	—	9.06
4. Nonreflective airspace, 7.25-in.	0.93*	—	19.00	—
5. Wood subfloor, 0.75 in.	0.94	0.94	—	—
6. Plywood, 0.625 in.	0.78	0.78	—	—
7. Felt building membrane	0.06	0.06	—	—
8. Resilient tile	0.05	0.05	—	—
9. Top surface (still air)	0.61	0.61	0.61	0.61
Total Thermal Resistance (R)	R_i= 4.45	R_s= 12.58	R_i= 20.69	R_s=10.75

Construction No. 1 $U_i = 1/4.45 = 0.225$; $U_s = 1/12.58 = 0.079$. With 10% framing (typical of 2-in. joists @ 16-in. o.c.), $U_{av} = 0.9 (0.225) + 0.1 (0.079) = 0.210$
Construction No. 2 $U_i = 1/20.69 = 0.048$; $U_s = 1/10.75 = 0.093$. With framing unchanged, $U_{av} = 0.9 (0.048) + 0.1 (0.093) = 0.053$

TABLE 3-11 Heat Loss below Grade in Basement Walls[a] [Btuh/(ft²)(F)]

Depth (ft)	Path Length through Soil (ft)	Heat Loss			
			Insulation		
		Uninsulated	1-in.	2-in.	3-in.
0–1 (1st)	0.68	0.410	0.152	0.093	0.067
1–2 (2nd)	2.27	0.222	0.116	0.079	0.059
2–3 (3rd)	3.88	0.155	0.094	0.068	0.053
3–4 (4th)	5.52	0.119	0.079	0.060	0.048
4–5 (5th)	7.05	0.096	0.069	0.053	0.044
5–6 (6th)	8.65	0.079	0.060	0.048	0.040
6–7 (7th)	10.28	0.069	0.054	0.044	0.037

[a] $k_{soil} = 9.6 (Btuh)(in.)/(ft^2)(F)$; $k_{insulation} = 0.24 (Btuh)(in.)/(ft^2)(F)$.

TABLE 3-12 Heat Loss through Basement Floors [Btuh/(h)(ft²)(F)]

Depth of Foundation Wall below Grade (ft)	Width of House			
	20 (ft)	24 (ft)	28 (ft)	32 (ft)
5	0.032	0.029	0.026	0.023
6	0.030	0.027	0.025	0.022
7	0.029	0.026	0.023	0.021

TABLE 3-13 Heat Loss of Concrete Floors at or near Grade Level per Foot of Exposed Edge

Outdoor Design Temperature F	Heat Loss per Foot of Exposed Edge, BTU/Hr			
	Recommended 2-In. Edge Insulation	1-In. Edge Insulation	1-In. Edge Insulation	No Edge Insulation*
−20 to −30	50	55	60	75
−10 to −20	45	50	55	65
0 to −10	40	45	50	60

*This construction not recommended; shown for comparison only.

TABLE 3-14 Floor Heat Loss Used When Warm Air Perimeter Heating Ducts are Embedded in Slabs[a]

Btuh per (linear foot of heated edge)

Outdoor Design Temperature, F	Edge Insulation		
	1-in. Vertical Extending Down 18 in. Below Floor Surface	1-in. L-Type Extending at Least 12 in. Deep and 12 in. Under	2-in. L-Type Extending at Least 12 in. Down and 12 in. Under
−20	105	100	85
−10	95	90	75
0	85	80	65
10	75	70	55
20	62	57	45

[a] Factors include loss downward through inner area of slab.

TABLE 3-15 Infiltration Through Windows

Type of Window	Wind Velocity					
	5	8	12	15	20	25
Aluminum						
Awning, double-hung, sliding	3	7	14	20	31	45
Casement, projected, pivoted	2	5	9	13	21	30
Wood						
Double-hung, sliding	2	5	9	13	21	30
Doors, sliding*						
Wood	2	5	9	13	21	30
Aluminum (best)	2	5	9	13	21	30
Aluminum (avg.)	5	9	18	26	42	60

*These values are cfh/ft²

TABLE 3-16 Infiltration Through Doors

Type of Door or Use	Wind Velocity					
	5	8	12	15	20	25
Residential						
Non-weatherstripped	27	69	111	154	199	299
Weatherstripped	13	34	55	77	98	122
Commercial						
*Infrequent use	747	1252	1956	2500	3430	4385
**Per opening	269	450	704	900	1235	1579

*These values are cfh/ 3' × 7' door

*These values are cfh/door opening

TABLE 3-17 Infiltration Through Windows
Expressed in ft³/ft of crack/hr

Type of Window	Remarks	Wind Velocity, Miles per Hour					
		5	10	15	20	25	30
Double-Hung Wood Sash Windows (Unlocked)	Around frame in masonry wall—not calked[b]	3	8	14	20	27	35
	Around frame in masonry wall—calked[b]	1	2	3	4	5	6
	Around frame in wood frame construction[b]	2	6	11	17	23	30
	Total for average window, non-weatherstripped, $1/16$-in. crack and $3/64$-in. clearance.[c] Includes wood frame leakage[d]	7	21	39	59	80	104
	Ditto, weatherstripped[d]	4	13	24	36	49	63
	Total for poorly fitted window, non-weatherstripped, $3/32$-in. crack and $3/32$-in. clearance.[e] Includes wood frame leakage[d]	27	69	111	154	199	249
	Ditto, weatherstripped[d]	6	19	34	51	71	92
Double-Hung Metal Windows[f]	Non-weatherstripped, locked	20	45	70	96	125	154
	Non-weatherstripped, unlocked	20	47	74	104	137	170
	Weatherstripped, unlocked	6	19	32	46	60	76
Rolled Section Steel Sash Windows[k]	Industrial pivoted, $1/16$-in. crack[g]	52	108	176	244	304	372
	Architectural projected, $1/32$-in. crack[h]	15	36	62	86	112	139
	Architectural projected, $3/64$-in. crack[h]	20	52	88	116	152	182
	Residential casement, $1/64$-in. crack[i]	6	18	33	47	60	74
	Residential casement, $1/32$-in. crack[i]	14	32	52	76	100	128
	Heavy casement section, projected, $1/64$-in. crack[j]	3	10	18	26	36	48
	Heavy casement section, projected, $1/32$-in. crack[j]	8	24	38	54	72	92
Hollow Metal, Vertically Pivoted Window[f]		30	88	145	186	221	242

[a] The values given in this table, with the exception of those for double-hung and hollow metal windows are 20 percent less than test values to allow for building up of pressure in rooms, and are based on test data, reported in the papers listed in chapter footnotes.

[b] The values given for frame leakage are per foot of sash perimeter, as determined for double-hung wood windows. Some of the frame leakage in masonry walls originates in the brick wall itself, and cannot be prevented by calking. For the additional reason that calking is not done perfectly and deteriorates with time, it is considered advisable to choose the masonry frame leakage values for calked frames as the average determined by the calked and non-calked tests.

[c] The fit of the average double-hung wood window was determined as $1/16$-in. crack and $3/64$-in. clearance by measurements on approximately 600 windows under heating season conditions.

[d] The values given are the totals for the window opening per foot of sash perimeter, and include frame leakage and so-called *elsewhere leakage*. The frame leakage values included are for wood frame construction, but apply as well to masonry construction assuming a 50 percent efficiency of frame calking.

[e] A $3/32$-in. crack and clearance represent a poorly fitted window, much poorer than average.

[f] Windows tested in place in building, so that no reduction from test values is necessary, as mentioned in footnote a.

[g] Industrial pivoted window generally used in industrial buildings. Ventilators horizontally pivoted at center or slightly above, lower part swinging out.

[h] Architecturally projected made of same sections as industrial pivoted, except that outside framing member is heavier, and it has refinements in weathering and hardware. Used in semi-monumental buildings such as schools. Ventilators swing in or out and are balanced on side arms. $1/32$-in. crack is obtainable in the best practice of manufacture and installation, $3/64$-in. crack considered to represent average practice.

[i] Of same design and section shapes as so-called *heavy section casement*, but of lighter weight. $1/64$-in. crack is obtainable in the best practice of manufacture and installation, $1/32$-in. crack considered to represent average practice.

[j] Made of heavy sections. Ventilators swing in or out and stay set at any degree of opening. $1/64$-in. crack is obtainable in the best practice of manufacture and installation, $1/32$-in. crack considered to represent average practice. Known as Intermediate Windows by steel window manufacturers.

[k] With reasonable care in installation, leakage at contacts where windows are attached to steel framework and at mullions, is negligible. With $1/64$-in. crack, representing poor installation, leakage at contact with steel framework is about one-third, and at mullions, about one-sixth of that given for industrial pivoted windows in the table.

TABLE 3–18   Solar Position and Intensity; Solar Heat Gain Factors for 24 Deg North Latitude

Date	Time A.M.	Solar Position Alt.	Solar Position Azimuth	Direct Normal Irradiation, Btu	N	NE	E	SE	S	SW	W	NW	Hor.	Time P.M.
Jan 21	7	4.8	65.6	70	2	20	61	63	25	2	2	2	4	5
	8	16.9	58.3	239	11	41	190	218	114	11	11	11	55	4
	9	27.9	48.8	287	18	22	190	253	166	19	18	18	120	3
	10	37.2	36.1	308	23	23	144	245	200	37	23	23	172	2
	11	43.6	19.6	317	26	26	72	211	220	94	26	26	204	1
	12	46.0	0.0	320	27	27	28	160	227	160	28	27	215	12
	Half Day Totals				93	142	662	1064	833	238	93	93	660	
Feb 21	7	9.0	73.9	153	6	67	141	128	33	6	6	6	16	5
	8	21.9	66.4	261	14	80	220	224	89	14	14	14	83	4
	9	33.9	56.8	297	21	45	208	243	133	22	21	21	153	3
	10	44.5	43.5	313	26	27	157	229	165	28	26	26	205	2
	11	52.2	24.5	321	29	29	80	191	185	67	29	29	238	1
	12	55.2	0.0	323	30	30	31	133	192	133	31	30	249	12
	Half Day Totals				111	269	833	1095	701	199	111	111	817	
Mar 21	7	13.7	83.8	194	10	115	186	145	17	9	9	9	36	5
	8	27.2	76.8	267	18	124	234	204	48	18	18	18	111	4
	9	40.2	67.9	295	24	85	215	214	82	24	24	24	180	3
	10	52.3	54.8	308	29	41	162	195	111	30	29	29	232	2
	11	61.9	33.4	315	32	33	84	154	130	42	32	32	264	1
	12	66.0	0.0	317	33	33	35	95	137	95	35	33	275	12
	Half Day Totals				130	431	922	981	457	163	129	129	960	
Apr 21	6	4.7	100.6	40	5	33	39	22	2	2	2	2	3	6
	7	18.3	94.9	203	19	151	198	127	15	14	14	14	58	5
	8	32.0	89.0	257	24	159	229	165	24	22	22	22	132	4
	9	45.6	81.9	281	29	126	209	169	39	28	28	28	196	3
	10	59.0	71.8	293	34	75	158	148	56	33	33	33	245	2
	11	71.1	51.6	298	36	39	85	107	70	38	36	36	275	1
	12	77.6	0.0	300	37	37	39	58	75	58	39	37	284	12
	Half Day Totals				166	611	953	780	245	164	155	154	1053	
May 21	6	8.0	108.4	85	25	79	83	38	5	5	5	5	12	6
	7	21.2	103.2	203	43	171	196	105	17	17	17	17	72	5
	8	34.6	98.5	248	38	178	218	132	26	24	24	24	142	4
	9	48.3	93.6	269	35	150	198	132	33	31	31	31	201	3
	10	62.0	87.7	280	37	102	150	111	38	35	35	35	247	2
	11	75.5	76.9	286	40	55	83	74	44	39	38	38	274	1
	12	86.0	0.0	287	41	41	41	44	46	44	41	41	282	12
	Half Day Totals				237	754	950	616	186	173	171	171	1090	
June 21	6	9.3	111.6	97	35	93	94	38	7	7	7	7	17	6
	7	22.3	106.8	200	55	176	192	94	18	18	18	18	77	5
	8	35.5	102.6	242	49	184	212	117	26	26	26	26	144	4
	9	49.0	98.7	262	43	158	192	116	33	31	31	31	201	3
	10	62.6	95.0	273	41	113	146	95	38	36	36	36	245	2
	11	76.3	90.7	279	41	64	82	63	41	39	38	39	271	1
	12	89.5	0.0	280	42	42	42	42	43	42	42	42	279	12
	Half Day Totals				282	804	936	544	184	177	176	177	1095	
July 21	6	8.2	109.0	81	25	76	80	36	5	5	5	5	13	6
	7	21.4	103.8	195	45	168	190	101	17	17	17	17	72	5
	8	34.8	99.2	239	40	176	214	128	27	25	25	25	141	4
	9	48.4	94.5	261	37	150	195	128	34	31	31	31	199	3
	10	62.1	89.0	272	39	104	149	108	39	36	36	36	243	2
	11	75.7	79.2	278	41	57	83	73	44	40	38	38	270	1
	12	86.6	0.0	279	42	42	42	44	46	44	42	42	278	12
	Half Day Totals				247	751	933	598	189	176	172	172	1078	
Aug 21	6	5.0	101.3	34	5	29	34	19	2	2	2	2	4	6
	7	18.5	95.6	186	21	144	186	118	16	15	15	15	58	5
	8	32.2	89.7	240	25	155	220	157	26	23	23	23	129	4
	9	45.9	82.9	265	31	126	202	162	39	30	30	30	191	3
	10	59.3	73.0	277	35	77	154	143	55	34	34	34	238	2
	11	71.6	53.2	283	38	42	85	103	67	39	37	37	267	1
	12	78.3	0.0	285	38	39	41	58	72	58	41	39	276	12
	Half Day Totals				176	603	916	742	242	170	162	161	1027	
Sept 21	7	13.7	83.8	172	11	105	169	132	17	10	10	10	35	5
	8	27.2	76.8	248	19	119	222	194	48	19	19	19	108	4
	9	40.2	67.9	277	26	84	207	206	81	26	26	26	174	3
	10	52.3	54.8	292	30	42	158	190	110	32	30	30	224	2
	11	61.9	33.4	298	33	35	84	151	128	44	33	33	256	1
	12	66.0	0.0	301	34	34	37	94	134	94	37	34	267	12
	Half Day Totals				137	417	879	939	451	172	137	136	930	
Oct 21	7	9.1	74.1	137	6	62	129	117	30	6	6	6	16	5
	8	22.0	66.7	246	15	78	211	213	85	15	15	15	82	4
	9	34.1	57.1	284	22	46	202	235	128	22	22	22	150	3
	10	44.7	43.8	300	27	28	153	222	160	29	27	27	201	2
	11	52.5	24.7	308	30	30	79	186	180	66	30	30	233	1
	12	55.5	0.0	311	31	31	32	130	186	130	32	31	244	12
	Half Day Totals				115	265	800	1050	676	198	116	115	802	
Nov 21	7	4.9	65.8	66	2	20	58	60	24	2	2	2	4	5
	8	17.0	58.4	232	12	41	186	213	111	12	12	12	54	4
	9	28.0	48.9	281	18	23	187	249	162	19	18	18	120	3
	10	37.3	36.3	302	23	24	142	241	197	37	23	23	171	2
	11	43.8	19.7	311	26	26	72	209	217	93	26	26	202	1
	12	46.2	0.0	314	27	27	29	158	224	158	29	27	213	12
	Half Day Totals				93	144	651	1046	817	237	94	93	655	
Dec 21	7	3.2	62.6	29	0	7	25	27	12	0	0	0	1	5
	8	14.9	55.3	225	10	29	174	209	118	10	10	10	44	4
	9	25.5	46.0	281	17	18	180	252	174	18	17	17	106	3
	10	34.3	33.7	304	21	22	137	247	209	44	21	21	157	2
	11	40.4	18.2	314	24	24	69	216	230	104	24	24	188	1
	12	42.6	0.0	317	25	25	27	167	237	167	27	25	199	12
	Half Day Totals				83	107	581	1019	851	254	84	83	593	
					N	NW	W	SW	S	SE	E	NE	HOR.	←P.M.

TABLE 3–19 Solar Position and Intensity; Solar Heat Gain Factors for 32 Deg North Latitude

Date	Time A.M.	Solar Position Alt.	Azimuth	Direct Normal Irradiation, Btu	N	NE	E	SE	S	SW	W	NW	Hor.	Time P.M.
Jan 21	7	1.4	65.2	1	0	0	1	1	0	0	0	0	0	5
	8	12.5	56.5	202	8	29	160	189	103	9	8	8	32	4
	9	22.5	46.0	269	15	16	175	246	169	16	15	15	88	3
	10	30.6	33.1	295	19	20	135	249	212	45	19	19	136	2
	11	36.1	17.5	306	22	22	67	221	238	110	22	22	166	1
	12	38.0	0.0	309	23	23	25	174	246	174	25	23	176	12
	Half Day Totals				75	91	529	974	834	262	75	75	509	
Feb 21	7	6.7	72.8	111	4	47	102	95	26	4	4	4	9	5
	8	18.5	63.8	244	12	64	205	217	95	12	12	12	63	4
	9	29.3	52.8	287	19	32	199	248	149	19	19	19	127	3
	10	38.5	38.9	305	23	24	151	241	189	31	23	23	176	2
	11	44.9	21.0	314	26	26	76	208	213	87	26	26	207	1
	12	47.2	0.0	316	27	27	29	155	221	155	29	27	217	12
	Half Day Totals				97	207	749	1091	780	227	98	97	689	
Mar 21	7	12.7	81.9	184	9	105	176	142	19	9	9	9	31	5
	8	25.1	73.0	260	17	107	227	209	62	17	17	17	99	4
	9	36.8	62.1	289	23	64	210	227	107	23	23	23	163	3
	10	47.3	47.5	304	27	30	158	215	144	29	27	27	211	2
	11	55.0	26.8	310	30	31	82	179	168	58	30	30	242	1
	12	58.0	0.0	312	31	31	33	122	176	122	33	31	252	12
	Half Day Totals				122	368	891	1054	588	191	123	122	872	
Apr 21	6	6.1	99.9	66	9	54	65	37	3	3	3	3	7	6
	7	18.8	92.2	206	17	147	201	136	15	14	14	14	61	5
	8	31.5	84.0	256	23	144	228	178	30	22	22	22	130	4
	9	43.9	74.2	278	28	103	206	188	58	27	27	27	189	3
	10	55.7	60.3	290	32	52	156	173	87	33	32	32	234	2
	11	65.4	37.5	296	34	36	83	135	108	40	34	34	263	1
	12	69.6	0.0	298	35	35	35	82	115	82	35	35	272	12
	Half Day Totals				159	55ᵃ	965	898	359	174	150	149	1022	
May 21	6	10.4	107.2	118	32	108	116	55	8	8	8	8	21	6
	7	22.8	100.1	211	35	170	204	118	18	18	18	18	81	5
	8	35.4	92.9	249	29	165	220	149	27	25	25	25	146	4
	9	48.1	84.7	269	32	128	198	155	37	30	30	30	201	3
	10	60.6	73.3	279	36	76	150	138	54	35	35	35	243	2
	11	72.0	51.9	285	38	41	82	102	68	39	37	37	269	1
	12	78.0	0.0	286	38	39	41	59	74	59	41	39	277	12
	Half Day Totals				217	697	983	747	248	181	172	171	1100	
June 21	6	12.2	110.2	130	44	123	127	55	10	10	10	10	28	6
	7	24.3	103.4	209	46	176	201	108	19	19	19	19	88	5
	8	36.9	96.8	244	36	171	214	135	28	26	26	26	151	4
	9	49.6	89.4	263	34	136	193	139	35	32	32	32	203	3
	10	62.2	79.7	273	38	86	146	122	45	36	36	36	244	2
	11	74.2	60.9	278	40	46	81	88	56	40	38	38	268	1
	12	81.5	0.0	280	40	41	42	52	60	52	42	41	276	12
	Half Day Totals				252	744	972	672	222	186	180	180	1119	
July 21	6	10.7	107.7	113	33	105	112	53	8	8	8	8	22	6
	7	23.1	100.6	203	37	167	198	114	19	18	18	18	81	5
	8	35.7	93.6	241	31	163	216	145	28	26	26	26	145	4
	9	48.4	85.5	261	34	128	195	150	37	31	31	31	199	3
	10	60.9	74.3	271	37	78	148	134	53	35	35	35	240	2
	11	72.4	53.3	277	39	43	82	99	66	40	38	38	265	1
	12	78.6	0.0	278	40	40	42	58	71	58	42	40	273	12
	Half Day Totals				227	694	964	724	245	184	176	175	1089	
Aug 21	6	6.5	100.5	59	9	50	59	34	3	3	3	3	7	6
	7	19.1	92.8	189	18	140	189	127	16	15	15	15	61	5
	8	31.8	84.7	239	25	141	219	170	30	23	23	23	127	4
	9	44.3	75.0	263	30	104	200	180	56	29	29	29	185	3
	10	56.1	61.3	275	33	55	152	167	84	34	33	33	229	2
	11	66.0	38.4	281	36	38	83	131	104	41	36	36	256	1
	12	70.3	0.0	283	37	37	40	80	111	80	40	37	265	12
	Half Day Totals				169	552	929	858	349	180	159	158	999	
Sep 21	7	12.7	81.9	163	10	95	159	128	19	9	9	9	30	5
	8	25.1	73.0	240	18	103	215	199	60	18	18	18	96	4
	9	36.8	62.1	272	24	64	202	218	105	24	24	24	158	3
	10	47.3	47.5	287	29	32	154	208	141	30	29	29	204	2
	11	55.0	26.8	294	31	32	81	174	164	59	31	31	234	1
	12	58.0	0.0	296	32	32	34	120	171	120	34	32	244	12
	Half Day Totals				128	355	846	1004	575	194	128	127	844	
Oct 21	7	6.8	73.1	98	4	43	92	85	23	4	4	4	9	5
	8	18.7	64.0	229	13	63	195	205	90	13	13	13	62	4
	9	29.5	53.0	273	19	33	193	239	144	20	19	19	125	3
	10	38.7	39.1	292	24	25	147	234	183	32	24	24	173	2
	11	45.1	21.1	301	27	27	75	202	206	85	27	27	203	1
	12	47.5	0.0	304	28	28	30	151	214	151	30	28	213	12
	Half Day Totals				100	205	718	1044	750	225	101	100	677	
Nov 21	7	1.5	65.4	1	0	0	1	1	0	0	0	0	0	5
	8	12.7	56.6	196	9	29	156	183	100	9	9	9	32	4
	9	22.6	46.1	262	15	17	172	241	166	16	15	15	87	3
	10	30.8	33.2	288	20	20	134	244	209	45	20	20	135	2
	11	36.2	17.6	300	22	22	67	218	234	108	22	22	165	1
	12	38.2	0.0	303	23	23	25	171	243	171	25	23	175	12
	Half Day Totals				76	92	521	955	820	258	77	76	505	
Dec 21	8	10.3	53.8	176	7	18	135	166	96	7	7	7	22	4
	9	19.8	43.6	257	13	14	162	238	171	15	13	13	72	3
	10	27.6	31.2	287	18	18	127	246	217	52	18	18	119	2
	11	32.7	16.4	300	20	20	63	222	243	116	20	20	148	1
	12	34.6	0.0	304	21	21	23	177	252	177	23	21	158	12
	Half Day Totals				67	76	482	947	844	273	68	67	440	
					N	NW	W	SW	S	SE	E	NE	HOR.	←P.M.

TABLE 3–20   Solar Position and Intensity; Solar Heat Gain Factors for 40 Deg North Latitude

Date	Time A.M.	Alt.	Azimuth	Direct Normal Irradiation, Btu	N	NE	E	SE	S	SW	W	NW	Hor.	Time P.M.
Jan 21	8	8.1	55.3	141	5	17	111	133	75	5	5	5	13	4
	9	16.8	44.0	238	11	12	154	224	160	13	11	11	54	3
	10	23.8	30.9	274	16	16	123	241	213	51	16	16	96	2
	11	28.4	16.0	289	18	18	61	222	244	118	18	18	123	1
	12	30.0	0.0	293	19	19	20	179	254	179	20	19	133	12
	Half Day Totals				59	68	449	903	815	271	59	59	353	
Feb 21	7	4.3	72.1	55	1	22	50	47	13	1	1	1	3	5
	8	14.8	61.6	219	10	50	183	199	94	10	10	10	43	4
	9	24.3	49.7	271	16	22	186	245	157	17	16	16	98	3
	10	32.1	35.4	293	20	21	142	247	203	38	20	20	143	2
	11	37.3	18.6	303	23	23	71	219	231	103	23	23	171	1
	12	39.2	0.0	306	24	24	25	170	241	170	25	24	180	12
	Half Day Totals				81	144	634	1035	813	250	81	81	546	
Mar 21	7	11.4	80.2	171	8	93	163	135	21	8	8	8	26	5
	8	22.5	69.6	250	15	91	218	211	73	15	15	15	85	4
	9	32.8	57.3	281	21	46	203	236	128	21	21	21	143	3
	10	41.6	41.9	297	25	26	153	229	171	28	25	25	186	2
	11	47.7	22.6	304	28	28	78	198	197	77	28	28	213	1
	12	50.0	0.0	306	28	28	30	145	206	145	30	28	223	12
	Half Day Totals				112	310	849	1100	692	218	112	112	764	
Apr 21	6	7.4	98.9	89	11	72	88	52	5	4	4	4	11	6
	7	18.9	89.5	207	16	141	201	143	16	14	14	14	61	5
	8	30.3	79.3	253	22	128	225	189	41	21	21	21	124	4
	9	41.3	67.2	275	26	80	203	204	83	26	26	26	177	3
	10	51.2	51.4	286	30	37	153	194	121	32	30	30	218	2
	11	58.7	29.2	292	33	34	81	161	146	52	33	33	244	1
	12	61.6	0.0	294	33	33	36	108	155	108	36	33	253	12
	Half Day Totals				153	509	969	1003	489	196	146	145	962	
May 21	5	1.9	114.7	1	0	0	0	0	0	0	0	0	0	7
	6	12.7	105.6	143	35	128	141	71	10	10	10	10	30	6
	7	24.0	96.6	216	28	165	209	131	20	18	18	18	87	5
	8	35.4	87.2	249	27	149	220	164	29	25	25	25	146	4
	9	46.8	76.0	267	31	105	197	175	53	30	30	30	196	3
	10	57.5	60.9	277	34	54	148	163	83	35	34	34	234	2
	11	66.2	37.1	282	36	38	81	130	105	42	36	36	258	1
	12	70.0	0.0	284	37	37	40	82	112	82	40	37	265	12
	Half Day Totals				203	643	1002	874	356	194	171	170	1083	
June 21	5	4.2	117.3	21	10	21	20	6	1	1	1	1	2	7
	6	14.8	108.4	154	47	142	151	70	12	12	12	12	39	6
	7	26.0	99.7	215	37	172	207	122	21	20	20	20	97	5
	8	37.4	90.7	246	29	156	215	152	29	26	26	26	153	4
	9	48.8	80.2	262	33	113	192	161	45	31	31	31	201	3
	10	59.8	65.8	272	35	62	145	148	69	36	35	35	237	2
	11	69.2	41.9	276	37	40	80	116	88	41	37	37	260	1
	12	73.5	0.0	278	38	38	41	71	95	71	41	38	267	12
	Half Day Totals				242	714	1019	810	311	197	181	180	1121	
July 21	5	2.3	115.2	2	0	2	1	0	0	0	0	0	0	7
	6	13.1	106.1	137	37	125	137	68	10	10	10	10	31	6
	7	24.3	97.2	208	30	163	204	127	20	19	19	19	88	5
	8	35.8	87.8	241	28	148	216	160	29	26	26	26	145	4
	9	47.2	76.7	259	32	106	194	170	52	31	31	31	194	3
	10	57.9	61.7	269	35	56	146	159	80	36	35	35	231	2
	11	66.7	37.9	274	37	39	81	127	102	42	37	37	255	1
	12	70.6	0.0	276	38	38	41	80	109	80	41	38	262	12
	Half Day Totals				211	645	986	850	347	197	177	176	1074	
Aug 21	6	7.9	99.5	80	12	67	82	48	5	5	5	5	11	6
	7	19.3	90.0	191	17	135	191	135	17	15	15	15	62	5
	8	30.7	79.9	236	23	126	216	180	40	22	22	22	122	4
	9	41.8	67.9	259	28	82	197	196	79	28	28	28	174	3
	10	51.7	52.1	271	32	40	149	187	116	34	32	32	213	2
	11	59.3	29.7	277	34	35	81	156	140	52	34	34	238	1
	12	62.3	0.0	279	35	35	38	105	149	105	38	35	247	12
	Half Day Totals				161	503	936	961	471	202	154	153	945	
Sep 21	7	11.4	80.2	149	8	84	146	121	21	8	8	8	25	5
	8	22.5	69.6	230	16	87	205	199	71	16	16	16	82	4
	9	32.8	57.3	263	22	47	195	226	124	23	22	22	138	3
	10	41.6	41.9	279	26	28	148	221	165	30	26	26	180	2
	11	47.7	22.6	287	29	29	77	192	191	77	29	29	206	1
	12	50.0	0.0	290	30	30	32	141	200	141	32	30	215	12
	Half Day Totals				116	300	803	1045	672	221	117	116	738	
Oct 21	7	4.5	72.3	48	1	20	45	41	12	1	1	1	3	5
	8	15.0	61.9	203	10	49	173	187	88	10	10	10	43	4
	9	24.5	49.8	257	17	23	180	235	151	18	17	17	96	3
	10	32.4	35.6	280	21	22	139	238	196	38	21	21	140	2
	11	37.6	18.7	290	23	23	70	212	224	100	23	23	167	1
	12	39.5	0.0	293	24	24	26	165	234	165	26	24	177	12
	Half Day Totals				83	143	610	989	783	245	84	83	535	
Nov 21	8	8.2	55.4	136	5	17	107	128	72	5	5	5	14	4
	9	17.0	44.1	232	12	13	151	219	156	13	12	12	54	3
	10	24.0	31.0	267	16	16	122	237	209	50	16	16	96	2
	11	28.6	16.1	283	19	19	61	218	240	116	19	19	123	1
	12	30.2	0.0	287	19	19	21	176	250	176	21	19	132	12
	Half Day Totals				61	71	442	884	798	267	62	61	353	
Dec 21	8	5.5	53.0	88	2	7	67	83	49	3	2	2	6	4
	9	14.0	41.9	217	9	10	135	205	151	12	9	9	39	3
	10	20.7	29.4	261	14	14	113	232	210	55	14	14	77	2
	11	25.0	15.2	279	16	16	56	217	242	120	16	16	103	1
	12	26.6	0.0	284	17	17	18	177	253	177	18	17	113	12
	Half Day Totals				49	54	380	831	781	273	50	49	282	
					N	NW	W	SW	S	SE	E	NE	HOR.	←P.M.

TABLE 3–21 Solar Position and Intensity; Solar Heat Gain Factors for 48 Deg North Latitude

Date	Time A.M.	Solar Position Alt.	Solar Position Azimuth	Direct Normal Irradiation, Btu	N	NE	E	SE	S	SW	W	NW	Hor.	Time P.M.
Jan 21	8	3.5	54.6	36	1	4	28	34	19	1	1	1	2	4
	9	11.0	42.6	185	7	8	117	176	128	9	7	7	25	3
	10	16.9	29.4	239	11	11	105	216	195	50	11	11	55	2
	11	20.7	15.1	260	14	14	52	208	233	115	14	14	77	1
	12	22.0	0.0	267	15	15	16	171	245	171	16	15	85	12
			Half Day Totals		41	44	319	735	705	256	41	41	202	
Feb 21	7	1.8	71.7	3	0	1	3	3	0	0	0	0	0	5
	8	10.9	60.0	180	7	36	149	166	82	7	7	7	24	4
	9	19.0	47.3	247	13	15	168	230	155	14	13	13	66	3
	10	25.5	33.0	275	17	17	131	242	207	43	17	17	105	2
	11	29.7	17.0	288	19	19	65	221	240	112	19	19	129	1
	12	31.2	0.0	291	20	20	21	176	251	176	21	20	138	12
			Half Day Totals		65	88	508	936	803	260	65	65	392	
Mar 21	7	10.0	78.7	152	7	80	145	123	22	7	7	7	20	5
	8	19.5	66.8	235	13	75	204	206	81	13	13	13	67	4
	9	28.2	53.4	270	19	33	193	239	142	19	19	19	117	3
	10	35.4	37.8	287	22	23	146	237	189	33	22	22	156	2
	11	40.3	19.8	295	24	24	73	210	218	94	24	24	180	1
	12	42.0	0.0	297	25	25	27	161	228	161	27	25	188	12
			Half Day Totals		98	256	790	1111	765	244	99	98	634	
Apr 21	6	8.6	97.8	108	12	86	107	64	6	6	6	6	14	6
	7	18.6	86.7	205	15	133	200	149	17	14	14	14	60	5
	8	28.5	74.9	247	20	111	219	197	55	20	20	20	114	4
	9	37.8	61.2	269	25	60	198	216	107	25	25	25	161	3
	10	45.8	44.6	281	28	30	148	210	150	30	28	28	197	2
	11	51.5	24.0	287	30	30	78	181	177	69	30	30	219	1
	12	53.6	0.0	289	31	31	33	132	187	132	33	31	227	12
			Half Day Totals		143	459	961	1086	604	225	139	138	879	
May 21	5	5.2	114.3	41	16	39	38	13	2	2	2	2	4	7
	6	14.7	103.7	162	35	140	160	84	12	12	12	12	39	6
	7	24.6	93.0	218	23	158	212	142	21	19	19	19	91	5
	8	34.6	81.6	248	26	132	218	178	38	24	24	24	142	4
	9	44.3	68.3	264	29	82	194	192	77	29	29	29	186	3
	10	53.0	51.3	274	32	39	145	184	116	34	32	32	219	2
	11	59.5	28.6	279	34	35	79	155	142	54	34	34	240	1
	12	62.0	0.0	280	35	35	38	106	150	106	38	35	247	12
			Half Day Totals		209	637	1058	1002	483	220	170	169	1043	
June 21	5	7.9	116.5	77	35	76	72	23	5	5	5	5	12	7
	6	17.2	106.2	172	46	154	169	84	14	14	14	14	51	6
	7	27.0	95.8	219	29	165	211	135	22	21	21	21	102	5
	8	37.1	84.6	245	28	139	215	167	34	26	26	26	152	4
	9	46.9	71.6	260	31	90	190	179	66	31	31	31	193	3
	10	55.8	54.8	269	34	45	142	171	101	36	34	34	225	2
	11	62.7	31.2	273	36	37	78	142	125	49	36	36	245	1
	12	65.4	0.0	275	36	36	39	96	134	96	39	36	252	12
			Half Day Totals		258	728	1098	952	434	224	186	185	1105	
July 21	5	5.7	114.7	42	18	42	40	14	2	2	2	2	5	7
	6	15.2	104.1	155	36	138	156	82	12	12	12	12	41	6
	7	25.1	93.5	211	24	156	207	138	21	20	20	20	92	5
	8	35.1	82.1	240	27	132	214	174	38	25	25	25	142	4
	9	44.8	68.8	256	30	83	191	187	75	30	30	30	184	3
	10	53.5	51.9	266	33	41	143	180	113	35	33	33	217	2
	11	60.1	29.0	271	35	37	79	151	138	53	35	35	238	1
	12	62.6	0.0	272	36	36	39	104	146	104	39	36	245	12
			Half Day Totals		218	643	1044	978	472	222	175	174	1040	
Aug 21	6	9.1	98.3	98	13	81	100	60	7	6	6	6	16	6
	7	19.1	87.2	189	16	127	189	140	18	15	15	15	61	5
	8	29.0	75.4	231	21	110	211	188	53	21	21	21	113	4
	9	38.4	61.8	253	26	62	192	208	102	26	26	26	159	3
	10	46.4	45.1	265	30	32	145	202	144	32	30	30	193	2
	11	52.2	24.3	271	32	32	78	175	171	68	32	32	215	1
	12	54.3	0.0	273	33	33	35	128	180	128	35	33	222	12
			Half Day Totals		152	454	927	1040	584	227	147	146	868	
Sep 21	7	10.0	78.7	131	7	71	128	108	21	7	7	7	19	5
	8	19.5	66.8	215	14	72	191	193	77	14	14	14	65	4
	9	28.2	53.4	251	20	33	184	227	136	20	20	20	113	3
	10	35.4	37.8	269	23	25	141	228	182	34	23	23	151	2
	11	40.3	19.8	277	26	26	73	203	211	92	26	26	174	1
	12	42.0	0.0	280	26	26	28	156	220	156	28	26	182	12
			Half Day Totals		104	246	744	1050	736	242	104	104	612	
Oct 21	7	2.0	71.9	3	0	1	3	3	0	0	0	0	0	5
	8	11.2	60.2	165	8	35	139	155	76	8	8	8	24	4
	9	19.3	47.4	232	13	16	161	219	147	15	13	13	66	3
	10	25.7	33.1	261	17	18	127	233	199	43	17	17	103	2
	11	30.0	17.1	274	20	20	64	213	231	109	20	20	127	1
	12	31.5	0.0	278	20	20	22	170	242	170	22	20	136	12
			Half Day Totals		67	91	488	895	768	256	68	67	387	
Nov 21	8	3.6	54.7	36	1	4	28	34	19	1	1	1	2	4
	9	11.2	42.7	178	7	8	115	171	125	9	7	7	25	3
	10	17.1	29.5	232	12	12	104	211	191	49	12	12	55	2
	11	20.9	15.1	254	14	14	52	204	228	113	14	14	77	1
	12	22.2	0.0	260	15	15	16	168	240	168	16	15	85	12
			Half Day Totals		41	45	316	719	690	252	42	41	202	
Dec 21	9	8.0	40.9	140	5	5	86	133	100	7	5	5	13	3
	10	13.6	28.2	214	9	9	91	194	179	49	9	9	37	2
	11	17.3	14.4	242	12	12	46	195	220	111	12	12	57	1
	12	18.6	0.0	250	12	12	14	163	233	163	14	12	64	12
			Half Day Totals		32	32	241	621	623	244	33	32	139	
					N	NW	W	SW	S	SE	E	NE	HOR.	←P.M.

TABLE 3–22　Solar Position and Intensity; Solar Heat Gain Factors for 56 Deg North Latitude

Date	Time A.M.	Solar Position Alt.	Solar Position Azimuth	Direct Normal Irradiation, Btu	N	NE	E	SE	S	SW	W	NW	Hor.	Time P.M.
Jan 21	9	5.0	41.8	77	2	2	48	74	54	3	2	2	5	3
	10	9.9	28.5	170	6	6	73	156	143	38	6	6	20	2
	11	12.9	14.5	206	9	9	39	169	190	96	9	9	34	1
	12	14.0	0.0	216	9	9	10	143	205	143	10	9	39	12
			Half Day Totals		21	21	168	475	489	205	22	21	78	
Feb 21	8	6.9	59.0	115	4	20	94	107	54	4	4	4	9	4
	9	13.5	45.6	207	9	10	139	197	136	10	9	9	36	3
	10	18.7	31.2	245	13	13	115	223	196	45	13	13	64	2
	11	22.0	15.9	262	15	15	56	210	232	112	15	15	84	1
	12	23.2	0.0	267	15	15	17	171	244	171	17	15	91	12
			Half Day Totals		48	60	405	819	738	252	49	48	239	
Mar 21	7	8.3	77.5	127	5	64	121	104	21	5	5	5	14	5
	8	16.2	64.4	215	11	61	185	194	83	11	11	11	49	4
	9	23.3	50.3	253	16	23	179	233	148	17	16	16	89	3
	10	29.0	34.9	272	19	20	136	238	198	38	19	19	122	2
	11	32.7	17.9	281	21	21	68	215	230	106	21	21	142	1
	12	34.0	0.0	284	22	22	23	170	241	170	23	22	149	12
			Half Day Totals		83	205	712	1080	799	260	83	83	490	
Apr 21	5	1.4	108.8	0	0	0	0	0	0	0	0	0	0	7
	6	9.6	96.5	122	12	96	121	75	7	7	7	7	18	6
	7	18.0	84.1	201	14	123	196	152	21	13	13	13	56	5
	8	26.1	70.9	240	19	95	212	202	68	19	19	19	101	4
	9	33.6	56.3	261	23	44	190	224	126	23	23	23	141	3
	10	39.9	39.7	273	26	27	142	221	172	32	26	26	171	2
	11	44.1	20.7	279	27	27	74	196	201	86	27	27	189	1
	12	45.6	0.0	280	28	28	30	149	211	149	30	28	196	12
			Half Day Totals		132	413	940	1144	699	251	128	128	773	
May 21	4	1.2	125.5	0	0	0	0	0	0	0	0	0	0	8
	5	8.5	113.4	92	36	89	87	33	6	6	6	6	14	7
	6	16.5	101.5	175	32	148	173	97	14	13	13	13	48	6
	7	24.8	89.3	219	21	149	212	152	21	19	19	19	92	5
	8	33.1	76.3	244	24	115	215	189	52	24	24	24	135	4
	9	40.9	61.6	259	27	62	189	206	102	27	27	27	171	3
	10	47.6	44.2	268	30	33	141	200	145	33	30	30	199	2
	11	52.3	23.4	273	32	32	75	174	172	70	32	32	216	1
	12	54.0	0.0	275	33	33	35	129	181	129	35	33	222	12
			Half Day Totals		223	651	1115	1120	602	252	168	167	986	
June 21	4	4.2	127.2	21	13	21	17	2	1	1	1	1	2	8
	5	11.4	115.3	121	52	119	114	40	9	9	9	9	24	7
	6	19.3	103.6	185	42	160	182	97	16	16	16	16	61	6
	7	27.6	91.7	221	24	156	213	147	23	21	21	21	105	5
	8	35.9	78.8	243	27	121	212	181	46	26	26	26	146	4
	9	43.8	64.1	256	29	69	186	195	91	29	29	29	181	3
	10	50.7	46.4	264	32	35	139	189	132	35	32	32	208	2
	11	55.6	24.9	268	34	35	76	164	158	64	34	34	225	1
	12	57.4	0.0	270	34	34	37	119	167	119	37	34	230	12
			Half Day Totals		268	734	1158	1079	560	255	186	185	1067	
July 21	4	1.7	125.8	0	0	0	0	0	0	0	0	0	0	8
	5	9.0	113.7	91	37	89	87	33	6	6	6	6	15	7
	6	17.0	101.9	169	34	145	170	94	14	14	14	14	50	6
	7	25.3	89.7	212	22	147	208	148	22	20	20	20	93	5
	8	33.6	76.7	236	25	115	211	185	51	25	25	25	135	4
	9	41.4	62.0	251	28	63	186	201	99	28	28	28	171	3
	10	48.2	44.6	260	31	34	139	196	141	34	31	31	198	2
	11	52.9	23.7	265	33	33	76	171	168	70	33	33	215	1
	12	54.6	0.0	267	34	34	37	126	177	126	37	34	221	12
			Half Day Totals		231	650	1101	1096	589	256	175	174	987	
Aug 21	5	2.0	109.2	1	0	1	1	0	0	0	0	0	0	7
	6	10.2	97.0	112	13	91	114	70	8	7	7	7	19	6
	7	18.5	84.5	186	16	119	186	144	21	15	15	15	58	5
	8	26.7	71.3	224	20	94	203	192	66	20	20	20	101	4
	9	34.3	56.7	245	24	46	184	215	121	25	24	24	139	3
	10	40.5	40.0	257	27	29	139	213	165	34	27	27	168	2
	11	44.8	20.9	263	29	29	74	189	193	84	29	29	187	1
	12	46.3	0.0	265	30	30	32	145	203	145	32	30	193	12
			Half Day Totals		142	412	908	1095	673	254	137	136	768	
Sep 21	7	8.3	77.5	107	5	56	104	90	19	5	5	5	13	5
	8	16.2	64.4	194	12	57	171	179	78	12	12	12	48	4
	9	23.3	50.3	233	17	24	170	220	140	18	17	17	86	3
	10	29.0	34.9	253	20	21	131	227	189	39	20	20	117	2
	11	32.7	17.9	263	22	22	67	206	220	102	22	22	137	1
	12	34.0	0.0	266	23	23	25	163	231	163	25	23	144	12
			Half Day Totals		87	195	664	1013	760	255	88	87	472	
Oct 21	8	7.1	59.1	103	4	19	86	98	50	4	4	4	10	4
	9	13.8	45.7	192	10	11	132	185	128	11	10	10	36	3
	10	19.0	31.3	230	13	13	110	212	186	43	13	13	64	2
	11	22.3	16.0	247	15	15	55	201	222	108	15	15	83	1
	12	23.5	0.0	252	16	16	18	164	234	164	18	16	90	12
			Half Day Totals		50	61	386	776	702	244	50	50	238	
Nov 21	9	5.2	41.9	75	2	2	47	72	53	3	2	2	5	3
	10	10.1	28.5	164	7	7	72	152	139	37	7	7	21	2
	11	13.1	14.5	200	9	9	39	165	185	93	9	9	34	1
	12	14.2	0.0	210	10	10	11	140	200	140	11	10	40	12
			Half Day Totals		22	22	166	464	476	199	23	22	79	
Dec 21	9	1.9	40.5	5	0	0	3	4	3	0	0	0	0	3
	10	6.6	27.5	113	3	3	46	103	96	27	3	3	9	2
	11	9.5	13.9	165	6	6	29	135	154	78	6	6	19	1
	12	10.6	0.0	180	7	7	8	120	171	120	8	7	23	12
			Half Day Totals		12	12	76	294	330	162	12	12	39	↑
					N	NW	W	SW	S	SE	E	NE	HOR.	←P.M.

TABLE 3-23

Shading Coefficients — Single Glass with Indoor Shading by Venetian Blinds, Roller Shades and Draperies

Type of Glass	Nominal Thickness[a]	Solar Trans.[a]	Venetian Blinds		Roller Shade			Draperies								
					Opaque		Translucent	Open Weave			Medium Weave			Closed Weave		
			Medium	Light	Dark	White	Light	D	M	L	D	M	L	D	M	L
Regular Sheet	$\frac{3}{32}$ to $\frac{1}{4}$	0.87–0.80														
Regular Plate	$\frac{1}{4}$ to $\frac{1}{2}$	0.80–0.71														
Regular Pattern	$\frac{1}{8}$ to $\frac{9}{32}$	0.87–0.79	0.64	0.55	0.59	0.25	0.39	0.73	0.68	0.50	0.68	0.62	0.55	0.63	0.54	0.45
Heat-Absorbing Pattern	$\frac{1}{8}$	—														
Grey Sheet	$\frac{3}{16}, \frac{7}{32}$	0.74, 0.71														
Heat-Absorbing Sheet	$\frac{7}{32}$	0.51														
Heat-Absorbing Plate	$\frac{1}{4}$	0.46														
Heat-Absorbing Pattern	$\frac{3}{16}, \frac{1}{4}$	—	0.57	0.53	0.45	0.30	0.36	0.57	0.55	0.53	0.53	0.48	0.46	0.51	0.45	0.38
Grey Sheet	$\frac{1}{8}, \frac{7}{32}$	0.59, 0.45														
Grey Plate	$\frac{13}{64}, \frac{1}{4}$	0.52, 0.45														
Heat-Absorbing Sheet, Plate or Pattern	—	0.44–0.30														
Heat-Absorbing Plate	$\frac{3}{8}$	0.34	0.54	0.52	0.40	0.28	0.32	0.46	0.43	0.41	0.42	0.40	0.38	0.41	0.37	0.33
Grey Plate	$\frac{3}{8}$	0.33														
Heat-Absorbing Sheet, Plate or Pattern	—	0.29–0.15	0.51	0.50	0.36	0.28	0.31	0.36	0.35	0.34	0.35	0.33	0.32	0.33	0.32	0.30
Grey Plate	$\frac{1}{2}$	0.24														

[a] Refer to manufacturer's literature for values.

[b] For vertical blinds with opaque white and beige louvers in the tightly closed position, SC is 0.25 and 0.29 when used with glass of 0.71 to 0.80 transmittance.

TABLE 3-24

Shading Coefficients for Insulating Glass[a] with Indoor Shading by Venetian Blinds, Roller Shades and Draperies

Type of Glass	Nominal Thickness, each light	Solar Trans.[b]		Venetian Blinds		Roller Shade			Draperies								
		Outer Pane	Inner Pane			Opaque		Translucent	Open Weave			Medium Weave			Closed Weave		
				Medium	Light	Dark	White	Light	D	M	L	D	M	L	D	M	L
Regular Sheet Out	$\frac{3}{32}, \frac{1}{8}$	0.87	0.87														
Regular Sheet In				0.57	0.51	0.60	0.25	0.37	0.62	0.57	0.55	0.57	0.53	0.50	0.56	0.48	0.42
Regular Plate Out	$\frac{1}{4}$	0.80	0.80														
Regular Plate In																	
Heat-Absorbing Plate Out	$\frac{1}{4}$	0.46	0.80														
Regular Plate In				0.39	0.36	0.40	0.22	0.30	0.46	0.44	0.42	0.44	0.42	0.38	0.42	0.38	0.33
Grey Plate Out	$\frac{1}{4}$	0.46	0.80														
Regular Plate In																	

[a] Refers to factory-fabricated units with $\frac{3}{16}, \frac{1}{4}$, or $\frac{1}{2}$ in. air space, or to prime windows plus storm windows.

[b] Refer to manufacturer's literature for exact values.

[c] For vertical blinds with opaque white or beige louvers, tightly closed, SC is approximately the same as for opaque white roller shades.

TABLE 3-24 (Cont.) Shading Coeficients for Single Glass and Insulating Glass[a]

A. Single Glass

Type of Glass	Nominal Thickness[b]	Solar Trans.[b]	Shading Coefficient
Regular Sheet	$\frac{3}{32}$, $\frac{1}{8}$	0.86	1.00
Regular Plate	$\frac{1}{4}$	0.80	0.95
	$\frac{3}{8}$	0.75	0.91
	$\frac{1}{2}$	0.71	0.88
Heat-Abs Sheet	$\frac{7}{32}$	0.51	0.71
Heat-Abs Plate	$\frac{1}{4}$	0.46	0.67
	$\frac{3}{8}$	0.34	0.57
Grey Sheet	$\frac{1}{8}$	0.59	0.78
	$\frac{3}{16}$	0.74	0.90
	$\frac{7}{32}$	0.45	0.66
	$\frac{7}{32}$	0.71	0.88
	$\frac{1}{4}$	0.67	0.86
Grey Plate	$\frac{13}{64}$	0.52	0.72
	$\frac{1}{4}$	0.47	0.70
	$\frac{3}{8}$	0.33	0.56
	$\frac{1}{2}$	0.24	0.50

B. Insulating Glass[a]

Type of Glass	Nominal Thickness[c]	Solar Trans.[b] Outer Pane	Solar Trans.[b] Inner Pane	Shading Coefficient
Regular Sheet Out, Regular Sheet In	$\frac{3}{32}$, $\frac{1}{8}$	0.87	0.87	0.90
Regular Plate Out, Regular Plate In	$\frac{1}{4}$	0.80	0.80	0.83
Heat-Abs Plate Out, Regular Plate In	$\frac{1}{4}$	0.46	0.80	0.56
Grey Plate Out, Regular Plate In	$\frac{1}{4}$	0.46	0.80	0.56
Grey Plate with Sun Control Film Out, Regular Plate In	$\frac{1}{4}$	0.23	0.80	0.42
Regular Plate Out, Regular Plate with Sun Control Film In	$\frac{1}{4}$	0.80	0.12	0.29

[a] Refers to factory-fabricated units with $\frac{3}{16}$, $\frac{1}{4}$, or $\frac{1}{2}$ in. air space or to prime windows plus storm windows.
[b] Refer to manufacturer's literature for values.
[c] Thickness of each pane of glass, not thickness of assembled unit.

TABLE 3-25   Shading Coefficients—Double Glazing with Between-Glass Shading

Type of Glass	Nominal Thickness, each pane	Solar Trans.[a] Outer Pane	Solar Trans.[a] Inner Pane	Description of Air Space	Venetian Blinds Light	Venetian Blinds Medium	Louvered Sun Screen
Regular Sheet Out, Regular Sheet In	$\frac{3}{32}$, $\frac{1}{8}$	0.87	0.87	Shade in contact with glass or shade separated from glass by air space.	0.33	0.36	0.43
Regular Plate Out, Regular Plate In	$\frac{1}{4}$	0.80	0.80	Shade in contact with glass-voids filled with plastic.	—	—	0.49
Heat-Abs. Plate Out, Regular Plate In	$\frac{1}{4}$	0.46	0.80	Shade in contact with glass or shade separated from glass by air space.	0.28	0.30	0.37
Grey Plate Out, Regular Plate In	$\frac{1}{4}$	0.46	0.80	Shade in contact with glass-voids filled with plastic.	—	—	0.41

[a] Refer to manufacturer's literature for exact values.

TABLE 3—26 Shading Coefficients for Domed Skylights

Dome	Light Diffuser (Translucent)	Curb		Shading Coefficient	U-Value
		Height, in.	Width to Height Ratio		
Clear r = 0.86	yes r = 0.58	0	∞	0.61	0.46
		9	5	0.58	0.43
		18	2.5	0.50	0.40
Clear r = 0.86	None	0	∞	0.99	0.80
		9	5	0.88	0.75
		18	2.5	0.80	0.70
Translucent r = 0.52	None	0	∞	0.57	0.80
		9	5	0.51	0.75
		18	2.5	0.46	0.70
Translucent r = 0.27	None	0	∞	0.34	0.80
		9	5	0.30	0.75
		18	2.5	0.28	0.70

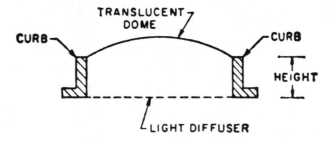

Terminology for Domed Skylights

TABLE 3–27   Length of Horizontal Projection Required for Shading Windows and Walls

(For shading 10 ft down from projection for April 11 through September 1)

Latitude	Sun Time AM→ ↓	Projection in Feet						Sun Time
		N	NE	E	SE	S	SW	
24 Deg North	6 a.m.	17.3	[a]	[a]	[a]	—	—	6 p.m.
	7	—	[a]	[a]	[a]	1.6	—	5
	8	—	10.8	16.5	15.8	2.3	—	4
	9	—	5.3	10.0	10.7	2.7	—	3
	10	—	2.3	5.8	7.5	2.7	—	2
	11	—	—	2.8	3.8	2.7	—	1
	12 N	—	—	—	2.2	2.8	2.2	12 N
32 Deg North	6 a.m.	15.8	[a]	[a]	[a]	—	—	6 p.m.
	7	—	[a]	[a]	[a]	1.7	—	5
	8	—	10.0	17.3	14.2	3.0	—	4
	9	—	4.6	10.3	10.0	3.8	—	3
	10	—	1.4	6.0	7.3	4.2	—	2
	11	—	—	2.8	5.1	4.2	1.2	1
	12 N	—	—	—	3.0	4.2	3.0	12 N
40 Deg North	6 a.m.	12.0	[a]	[a]	[a]	—	—	6 p.m.
	7	—	[a]	[a]	[a]	1.8	—	5
	8	—	9.7	18.9	16.1	4.3	—	4
	9	—	4.7	11.2	11.6	5.4	—	3
	10	—	1.3	6.5	9.1	5.8	—	2
	11	—	—	3.1	6.5	6.1	2.2	1
	12 N	—	—	—	4.3	6.3	4.3	12 N
48 Deg North	6 a.m.	7.3	[a]	[a]	[a]	—	—	6 p.m.
	7	—	[a]	[a]	[a]	3.6	—	5
	8	—	9.3	19.6	18.7	6.5	—	4
	9	—	3.3	12.0	13.7	7.5	—	3
	10	—	—	7.3	10.8	8.4	—	2
	11	—	—	3.2	8.4	8.4	3.2	1
	12 N	—	—	—	6.0	8.4	6.0	12 N
56 Deg North	6 a.m.	6.2	[a]	[a]	[a]	—	—	6 p.m.
	7	—	[a]	[a]	[a]	4.9	—	5
	8	—	9.6	[a]	[a]	7.8	—	4
	9	—	2.9	13.5	16.3	9.3	—	3
	10	—	—	7.9	13.0	10.4	1.4	2
	11	—	—	3.9	10.2	10.6	4.7	1
	12 N	—	—	—	7.5	10.7	7.5	12 N
		N	NW	W	SW	S	SE	←↑PM

[a] Projection greater than 20 ft required.

TABLE 3–28 Sol-Air Temperatures for July 21, 40 deg North Latitude

Time	Air Temp, F	N	NE	E	SE	S	SW	W	NW	HOR
					$\alpha/h_o = 0.15$					
1	76	76	76	76	76	76	76	76	76	69
2	76	76	76	76	76	76	76	76	76	69
3	75	75	75	75	75	75	75	75	75	68
4	74	74	74	74	74	74	74	74	74	67
5	74	74	74	74	74	74	74	74	74	67
6	74	82	95	97	86	75	75	75	75	74
7	75	82	103	109	97	78	78	78	78	85
8	77	82	103	114	105	83	81	81	81	96
9	80	85	101	114	110	92	85	85	85	106
10	83	89	96	110	112	100	89	89	89	115
11	87	93	94	104	111	108	96	93	93	123
12	90	96	96	97	107	112	107	97	96	127
13	93	99	99	99	102	114	117	110	100	129
14	94	100	100	100	100	111	123	121	107	126
15	95	100	100	100	100	107	125	129	116	121
16	94	99	98	98	98	100	122	131	120	113
17	93	100	96	96	96	96	115	127	121	103
18	91	99	92	92	92	92	103	114	112	91
19	87	87	87	87	87	87	87	87	87	80
20	85	85	85	85	85	85	85	85	85	78
21	83	83	83	83	83	83	83	83	83	76
22	81	81	81	81	81	81	81	81	81	74
23	79	79	79	79	79	79	79	79	79	72
24	77	77	77	77	77	77	77	77	77	70
	83	86	89	91	90	89	90	91	89	91
					$\alpha/h_o = 0.30$					
1	76	76	76	76	76	76	76	76	76	69
2	76	76	76	76	76	76	76	76	76	69
3	75	75	75	75	75	75	75	75	75	68
4	74	74	74	74	74	74	74	74	74	67
5	74	74	74	74	74	74	74	74	74	67
6	74	90	117	121	99	77	77	77	77	81
7	75	90	131	144	120	82	82	82	82	102
8	77	87	130	151	134	89	86	86	86	122
9	80	91	122	148	141	105	91	91	91	140
10	83	95	109	137	141	118	96	95	95	155
11	87	100	101	122	136	129	105	100	100	166
12	90	103	103	104	125	134	125	104	103	172
13	93	106	106	106	111	135	142	128	107	172
14	94	106	106	106	107	129	152	148	120	166
15	95	106	106	106	106	120	156	163	137	155
16	94	104	103	103	103	106	151	168	147	139
17	93	108	100	100	100	100	138	162	149	120
18	91	107	94	94	94	94	116	138	134	98
19	87	87	87	87	87	87	87	87	87	80
20	85	85	85	85	85	85	85	85	85	78
21	83	83	83	83	83	83	83	83	83	76
22	81	81	81	81	81	81	81	81	81	74
23	79	79	79	79	79	79	79	79	79	72
24	77	77	77	77	77	77	77	77	77	70
	83	89	95	100	99	95	99	100	95	107

TABLE 3-28A Percentage of the Daily Range

Time, hr	%	Time, hr	%	Time, hr	%	Time, hr	%
1	87	7	93	13	11	19	34
2	92	8	84	14	3	20	47
3	96	9	71	15	0	21	58
4	99	10	56	16	3	22	68
5	100	11	39	17	10	23	76
6	98	12	23	18	21	24	82

TABLE 3-29A Descriptions of Wall Constructions[a]

Group	Components	Wt, lb per sq ft	U Value
A	1" stucco +4" l.w. concrete block +air space	28.6	0.267
	1" stucco +air space +2" insulation	16.3	0.106
B	1" stucco +4" common brick	55.9	0.393
	1" stucco +4" h.w. concrete	62.5	0.481
C	4" face brick +4" l.w. concrete block +1" insulation	62.5	0.158
	1" stucco +4" h.w. concrete +2" insulation	62.9	0.114
D	1" stucco +8" l.w. concrete block +1" insulation	41.4	0.141
	1" stucco +2" insulation +4" h.w. concrete block	36.6	0.111
E	4" face brick +4" l.w. concrete block	62.2	0.333
	1" stucco +8" h.w. concrete block	56.6	0.349
F	4" face brick +4" common brick	89.5	0.360
	4" face brick +2" insulation +4" l.w. concrete block	62.5	0.103
G	1" stucco +8" clay tile +1" insulation	62.8	0.141
	1" stucco +2" insulation +4" common brick	56.2	0.108
H	4" face brick +8" clay tile +1" insulation	96.4	0.137
	4" face brick +8" common brick	129.6	0.280
	1" stucco +12" h.w. concrete	155.9	0.365
	4" face brick +2" insulation +4" common brick	89.8	0.106
	4" face brick +2" insulation +4" h.w. concrete	96.5	0.111
	4" face brick +2" insulation +8" h.w. concrete block	90.6	0.102
I	1" stucco +8" clay tile +air space	62.6	0.209
	4" face brick +air space +4" h.w. concrete block	69.9	0.282
J	face brick +8" common brick +1" insulation	129.8	0.145
	4" face brick +2" insulation +8" clay tile	96.5	0.094
	1" stucco +2" insulation +8" common brick	96.3	0.100
K	4" face brick +air space +8" clay tile	96.2	0.200
	4" face brick +2" insulation +8" common brick	129.9	0.098
	4" face brick +2" insulation +8" h.w. concrete	143.3	0.107
L	4" face brick +8" clay tile +air space	96.2	0.200
	4" face brick +air space +4" common brick	89.5	0.265
	4" face brick +air space +4" h.w. concrete	96.2	0.301
	4" face brick +air space +8" h.w. concrete block	90.2	0.246
	1" stucco +2" insulation +12" h.w. concrete	156.3	0.106
M	4" face brick +air space +8" common brick	129.6	0.218
	4" face brick +air space +12" h.w. concrete	189.5	0.251
	4" face brick +2" insulation +12" h.w. concrete	189.9	0.104

[a] In addition to the structure components listed above, all walls had an outside surface resistance (code number AO of table 41) and on the inside: a ¾ in. layer of plaster, gypsum or other similar finishing material and an inside surface resistance (code numbers E1 and E0 respectively of Table 41).

TABLE 3-29B Total Equivalent Temp. Differentials for Calculating Heat Gain Through Sunlit Walls

North Latitude Wall Facing	8 D	8 L	10 D	10 L	12 D	12 L	2 D	2 L	4 D	4 L	6 D	6 L	8 D	8 L	10 D	10 L	12 D	12 L	λ	δ	South Latitude Wall Facing
Group A[a]																					
NE	27	16	31	18	26	17	24	17	24	18	23	17	20	15	17	13	15	11			SE
E	32	18	41	24	37	22	29	20	28	20	26	19	23	16	20	14	18	13			E
SE	25	15	36	21	38	23	33	21	28	20	26	18	22	16	19	14	18	12			NE
S	14	9	20	13	28	18	33	22	31	21	25	18	20	15	17	13	15	11	0.34	2	N
SW	17	11	20	13	24	16	34	22	42	27	41	26	28	19	20	14	18	12			NW
W	17	11	20	13	24	16	30	20	42	27	48	30	33	22	22	15	19	13			W
NW	14	9	17	11	21	14	23	17	31	21	38	25	28	19	18	13	16	11			SW
N	14	9	15	10	17	12	20	15	21	16	21	16	18	14	14	11	12	9			S
Group B[a]																					
NE	12	7	27	14	31	17	30	19	31	21	30	22	27	20	21	17	16	13			SE
E	14	8	34	18	45	24	43	25	39	25	35	24	30	22	23	18	17	14			E
SE	9	5	25	13	39	21	44	26	41	26	37	25	31	23	24	18	17	14			NE
S	4	3	7	4	18	11	32	19	41	26	39	27	33	24	25	19	18	15	0.51	3	N
SW	5	3	7	4	11	7	23	15	41	26	54	34	51	33	38	25	26	19			NW
W	6	4	7	4	11	7	18	12	35	23	55	34	59	37	43	28	30	20			W
NW	5	3	6	4	11	7	17	12	26	18	41	27	47	31	36	24	25	18			SW
N	6	4	9	5	12	8	18	12	22	17	25	20	27	21	22	17	16	14			S
Group C[a]																					
NE	9	6	19	10	26	15	28	17	29	18	29	20	28	20	24	19	20	16			SE
E	10	7	22	12	36	19	40	23	39	23	36	24	33	23	28	20	22	17			E
SE	8	6	16	9	29	16	38	21	39	24	37	24	34	23	28	21	23	17			NE
S	7	5	7	4	12	7	22	14	32	20	36	24	34	24	29	21	23	17	0.40	4	N
SW	9	6	8	5	10	6	16	10	28	18	42	26	48	30	42	28	33	22			NW
W	10	7	9	5	10	6	14	9	24	16	40	25	52	32	47	30	37	24			W
NW	8	6	8	5	9	6	13	9	19	14	30	20	40	27	38	26	30	21			SW
N	7	5	8	5	10	7	14	9	18	13	22	16	25	19	23	18	19	16			S

Sun Time — A.M.: 8, 10, 12; P.M.: 2, 4, 6, 8, 10, 12. Exterior color of wall—D = dark, L = light. Amplitude Decrement Factor, λ; Time Lag, δ hr.

TABLE 3-29B (Cont.) Total Equivalent Temp. Differentials for Calculating Heat Gain Through Sunlit Walls

North Latitude Wall Facing	A.M. 8 D	L	10 D	L	12 D	L	P.M. 2 D	L	4 D	L	6 D	L	8 D	L	10 D	L	12 D	L	λ	δ	South Latitude Wall Facing
Group Iᵃ																					
NE	16	11	18	12	20	13	22	14	23	15	24	16	24	16	23	16	22	16			SE
E	19	13	21	14	25	16	29	17	30	18	30	19	29	19	28	19	26	18			E
SE	19	13	19	13	22	14	26	16	28	18	29	18	29	19	28	18	26	18			NE
S	16	12	15	11	16	11	18	12	21	14	24	16	25	17	25	17	23	16	0.13	6	N
SW	20	14	19	13	18	12	19	13	22	14	27	17	31	20	32	20	30	20			NW
W	22	14	20	13	19	13	20	13	22	14	26	17	31	20	33	21	32	21			W
NW	18	12	16	11	16	11	17	11	18	12	21	14	25	17	27	18	26	18			SW
N	13	10	12	9	13	9	13	10	15	11	16	12	18	13	18	14	18	14			S
Group Jᵃ																					
NE	18	13	17	12	18	12	19	13	21	13	22	14	23	15	23	16	23	16			SE
E	22	15	20	14	21	14	24	15	26	16	28	17	29	18	29	19	29	19			E
SE	21	15	20	14	20	13	21	14	24	15	26	16	28	17	28	18	28	18			NE
S	19	14	17	12	16	11	16	11	17	12	20	13	22	15	24	16	24	16	0.10	9	N
SW	24	16	22	15	20	13	19	13	19	13	21	14	24	16	28	18	30	19			NW
W	26	17	24	16	22	14	20	13	20	13	21	14	24	16	28	18	31	20			W
NW	21	15	19	13	18	12	17	11	17	11	17	12	19	13	22	15	25	17			SW
N	15	11	14	11	13	10	13	9	13	10	14	10	15	11	17	12	17	13			S
Group Kᵃ																					
NE	19	14	19	13	19	13	20	13	20	14	21	14	22	15	22	15	22	15			SE
E	23	16	22	15	23	15	24	16	26	16	27	17	27	17	28	18	27	18			E
SE	23	15	22	15	22	14	22	15	24	15	25	16	26	17	27	17	27	17			NE
S	20	14	19	13	18	12	18	12	18	13	20	13	21	14	22	15	23	15	0.08	11	N
SW	25	16	23	15	22	14	21	14	21	14	22	15	24	16	26	17	27	18			NW
W	26	17	24	16	23	15	22	15	22	14	23	15	24	16	27	17	28	18			W
NW	21	15	20	14	19	13	18	13	18	12	19	13	20	14	22	15	23	16			SW
N	15	11	14	11	14	10	14	10	14	10	14	11	15	11	16	12	16	12			S
Group Lᵃ																					
NE	18	13	18	13	19	13	20	13	21	14	22	15	23	15	23	16	22	15			SE
E	22	15	22	14	23	15	25	16	27	17	28	18	28	18	28	18	27	18			E
SE	21	14	21	14	22	14	23	15	25	16	27	17	27	17	27	18	26	18			NE
S	19	13	17	12	17	12	18	12	19	13	21	14	23	15	23	16	23	16	0.08	8	N
SW	23	15	22	14	21	14	20	13	21	14	23	15	26	17	28	18	28	18			NW
W	25	16	23	15	22	14	21	14	22	14	24	15	26	17	29	19	30	19			W
NW	20	14	19	13	18	12	18	12	18	12	19	13	21	15	23	16	24	16			SW
N	14	11	14	10	13	10	13	10	14	10	15	11	16	12	17	13	17	13			S
Group Mᵃ																					
NE	20	14	20	14	19	13	20	13	20	14	20	14	21	14	21	14	22	15			SE
E	25	16	24	16	24	16	24	16	25	16	25	16	26	17	27	17	27	17			E
SE	24	16	23	15	23	15	23	15	23	15	24	16	25	16	25	16	26	17			NE
S	21	14	20	14	19	13	19	13	19	13	19	13	20	14	21	14	21	15	0.05	12	N
SW	25	17	25	16	24	16	23	15	22	15	22	15	23	15	24	16	25	17			NW
W	27	17	26	17	25	16	24	16	23	15	23	15	24	15	25	16	26	17			W
NW	22	15	21	14	20	14	20	13	19	13	19	13	19	13	20	14	21	15			SW
N	15	12	15	11	14	11	14	11	14	11	14	11	15	11	15	11	16	12			S

Exterior color of wall—D = dark, L = light

Amplitude Decrement Factor, λ; Time Lag, δ hr

TABLE 3-29B (Cont.) Total Equivalent Temp. Differentials for Calculating Heat Gain Through Sunlit Walls

North Latitude Wall Facing	8 D	8 L	10 D	10 L	12 D	12 L	2 D	2 L	4 D	4 L	6 D	6 L	8 D	8 L	10 D	10 L	12 D	12 L	λ	δ	South Latitude Wall Facing
	A.M.						P.M.														
	Exterior color of wall—D = dark, L = light																				
Group Dᵃ																					
NE	8	5	19	10	28	15	29	17	30	19	30	21	28	21	24	19	19	16			SE
E	9	6	23	12	38	20	42	24	40	24	37	24	33	23	27	20	21	17			E
SE	7	5	16	9	30	16	40	22	41	25	38	25	34	24	28	21	22	17			NE
S	5	4	6	4	12	7	23	14	34	21	38	25	35	24	29	21	23	17	0.45	4	N
SW	8	5	7	4	9	6	16	10	30	19	44	28	51	32	43	28	33	22			NW
W	8	6	7	5	9	6	14	9	25	16	42	27	55	34	49	31	37	25			W
NW	7	5	7	4	9	6	13	9	20	14	31	21	42	28	40	27	31	21			SW
N	6	4	8	5	10	6	14	10	19	14	23	17	25	19	24	19	19	16			S
Group Eᵃ																					
NE	10	6	23	12	30	16	30	18	30	20	30	21	28	21	23	18	18	14			SE
E	11	6	28	15	42	22	43	24	39	24	36	24	32	23	25	19	19	15			E
SE	8	5	20	11	35	19	42	24	41	25	38	25	33	23	26	20	20	16			NE
S	4	3	6	4	15	9	28	17	38	24	39	26	34	24	27	20	20	16	0.48	4	N
SW	6	4	7	4	10	6	19	12	35	22	49	31	52	33	41	27	30	21			NW
W	7	5	7	4	10	6	16	11	30	20	48	31	57	36	47	30	34	23			W
NW	6	4	6	4	10	6	15	10	23	16	36	24	45	30	38	26	28	20			SW
N	6	4	8	5	11	7	16	11	21	15	24	18	26	20	23	18	18	15			S
Group Fᵃ																					
NE	9	7	14	9	21	12	25	15	27	17	29	19	28	20	26	19	23	17			SE
E	10	8	17	10	28	15	35	19	37	22	37	23	35	23	31	22	26	19			E
SE	10	7	13	8	22	12	31	17	36	21	37	23	35	23	32	22	27	19			NE
S	9	7	7	5	10	6	17	10	26	16	32	20	33	22	31	22	27	19	0.32	6	N
SW	12	9	10	6	9	6	13	8	22	14	33	21	42	27	42	27	37	25			NW
W	14	9	11	7	10	6	12	8	19	12	31	20	43	27	46	29	41	27			W
NW	12	8	9	6	9	6	11	8	16	11	24	16	33	22	36	24	33	23			SW
N	8	7	8	6	9	6	12	8	15	11	19	14	22	17	23	18	21	17			S
Group Gᵃ																					
NE	11	9	10	15	20	12	24	14	25	16	26	17	27	18	26	18	23	17			SE
E	13	9	17	11	26	15	32	18	34	20	34	21	33	22	31	21	27	19			E
SE	13	9	14	9	21	12	28	16	33	19	34	21	33	22	31	21	27	19			NE
S	12	9	10	7	11	8	16	10	23	15	29	18	30	20	29	20	26	19	0.25	6	N
SW	16	11	13	9	13	8	14	9	20	13	29	19	37	24	39	25	35	23			NW
W	18	12	15	10	14	9	14	9	18	12	27	18	38	24	42	26	38	25			W
NW	14	10	12	8	12	8	13	9	16	11	21	15	29	20	33	22	31	21			SW
N	10	8	10	7	10	7	12	8	15	10	18	13	20	15	21	16	20	16			S
Group Hᵃ																					
NE	15	11	16	11	18	12	20	13	22	14	24	15	25	16	25	17	24	17			SE
E	18	13	18	12	22	14	26	16	29	17	30	19	31	20	30	20	29	19			E
SE	18	13	17	12	19	12	23	14	27	16	29	18	30	19	30	20	28	19			NE
S	16	12	14	10	14	10	15	10	19	12	23	15	25	17	26	18	26	18	0.14	8	N
SW	22	14	19	12	17	11	16	11	18	12	23	15	29	18	32	21	32	21			NW
W	23	15	20	13	18	12	17	11	18	12	22	15	29	18	33	21	34	22			W
NW	19	13	17	11	15	10	15	10	15	11	18	12	23	15	26	18	27	19			SW
N	13	10	12	9	11	9	12	9	13	10	15	11	17	13	18	14	19	14			S

TABLE 3-30 Total Equivalent Temp Differentials for Calculating Heat Gain Through Flat Roofs

Description of Roof Construction[a,b]	Wt, lb per sq ft	U value Btu/(hr)(ft²)(F°)	8 A.M. D	8 A.M. L	10 A.M. D	10 A.M. L	12 A.M. D	12 A.M. L	2 P.M. D	2 P.M. L	4 P.M. D	4 P.M. L	6 P.M. D	6 P.M. L	8 P.M. D	8 P.M. L	10 P.M. D	10 P.M. L	12 P.M. D	12 P.M. L	λ	δ
Light Construction Roofs—Exposed to Sun																						
1″ insulation +steel siding	7.4	0.213	28	11	65	31	90	48	95	53	78	45	43	27	8	6	1	1	-3	-3	1.0	0
2″ insulation +steel siding	7.8	0.125	24	8	61	29	88	46	96	53	81	46	48	30	10	8	2	2	-3	-3	0.99	1
1″ insulation +1″ wood[c]	8.4	0.206	12	2	47	21	77	39	92	50	86	48	61	36	25	16	7	5	0	-1	0.93	2
2″ insulation +1″ wood[c]	8.5	0.122	8	0	41	18	72	36	90	48	88	49	65	38	30	19	9	7	1	0	0.93	2
1″ insulation +2.5″ wood[c]	12.7	0.193	2	-2	23	8	48	23	70	36	79	42	71	40	50	29	29	17	15	9	0.73	3
2″ insulation +2.5″ wood[c]	13.1	0.117	1	-2	19	6	43	20	65	33	76	41	72	40	53	31	33	20	18	11	0.68	4
Medium Construction Roofs—Exposed to Sun																						
1″ insulation +4″ wood[c]	17.3	0.183	5	0	14	5	31	14	49	24	62	32	65	35	56	31	41	24	29	17	0.51	5
2″ insulation +4″ wood[c]	17.8	0.113	6	1	13	4	28	12	45	22	58	30	63	34	56	31	43	25	32	18	0.48	5
1″ insulation +2″ h.w. concrete	28.3	0.206	4	-1	27	11	54	26	74	39	81	44	70	40	45	27	24	15	12	7	0.75	3
2″ insulation +2″ h.w. concrete	28.8	0.122	2	-2	23	9	49	23	70	36	79	43	71	40	49	29	28	17	15	9	0.73	3
4″ l.w. concrete	17.8	0.213	1	-3	28	11	59	28	82	43	88	48	74	42	44	27	19	12	6	4	0.87	3
6″ l.w. concrete	24.5	0.157	-2	-4	9	2	31	13	55	27	72	38	76	41	64	36	42	25	25	15	0.67	5
8″ l.w. concrete	31.2	0.125	6	2	6	1	16	6	32	14	49	24	61	32	63	34	55	31	41	24	0.50	6
Heavy Construction Roofs—Exposed to Sun																						
1″ insulation +4″ h.w. concrete	51.6	0.199	7	1	17	6	33	15	50	25	61	32	63	34	53	30	40	23	28	16	0.48	5
2″ insulation +4″ h.w. concrete	52.1	0.120	7	2	15	6	30	13	46	23	58	30	61	33	54	30	41	23	31	17	0.45	5
1″ insulation +6″ h.w. concrete	75.0	0.193	13	6	17	7	26	12	38	18	48	25	53	28	51	27	43	24	35	19	0.33	6
2″ insulation +6″ h.w. concrete	75.4	0.117	15	7	17	7	25	11	36	17	46	23	51	27	50	27	43	24	36	20	0.30	6

Roofs Covered with Water—Exposed to Sun

	Outside Air Dew Point (F)	Water Layer Thickness (in.)	8	10	12	2	4	6	8	10	12
Light Construction	60	6	-6	-6	-1	6	13	17	17	13	7
		1	-12	-6	4	15	21	22	17	8	0
		0	-12	-4	7	17	23	22	16	5	-3
	70	6	-1	0	4	11	18	21	21	17	12
		1	-5	0	10	19	25	26	21	12	5
		0	-5	2	12	21	26	26	19	9	2
Heavy Construction	60	6	-3	-4	-1	4	9	13	15	13	10
		1	-8	-6	1	8	15	18	17	11	6
		0	-9	-5	2	10	16	19	16	10	4
	70	6	-2	2	4	9	14	18	20	18	15
		1	-2	0	6	14	20	23	21	16	11
		0	-2	1	8	16	21	23	21	15	9

TABLE 3-31 Rates of Heat Gain from Occupants of Conditioned Spaces[a]

Degree of Activity	Typical Application	Total Heat Adults, Male Watts	Btuh	kcal/hr	Total Heat Adjusted[b] Watts	Btuh	kcal/hr	Sensible Heat Watts	Btuh	kcal/hr	Latent Heat Watts	Btuh	kcal/hr
Seated at rest	Theater, movie	115	400	100	100	350	90	60	210	55	40	140	30
Seated, very light work writing	Offices, hotels, apts	140	480	120	120	420	105	65	230	55	55	190	50
Seated, eating	Restaurant[c]	150	520	130	170	580[c]	145	75	255	60	95	325	80
Seated, light work, typing	Offices, hotels, apts	185	640	160	150	510	130	75	255	60	75	255	65
Standing, light work or walking slowly	Retail Store, bank	235	800	200	185	640	160	90	315	80	95	325	80
Light bench work	Factory	255	880	220	230	780	195	100	345	90	130	435	110
Walking, 3 mph, light machine work	Factory	305	1040	260	305	1040	260	100	345	90	205	695	170
Bowling[d]	Bowling alley	350	1200	300	280	960	240	100	345	90	180	615	150
Moderate dancing	Dance hall	400	1360	340	375	1280	320	120	405	100	255	875	220
Heavy work, heavy machine work, lifting	Factory	470	1600	400	470	1600	400	165	565	140	300	1035	260
Heavy work, athletics	Gymnasium	585	2000	500	525	1800	450	185	635	160	340	1165	290

TABLE 3-32A   Recommended Rate of Heat Gain from Commercial Cooking Appliances Located in the Air-Conditioned Area[a]

Appliance	Capacity	Overall Dim., Inches Width×Depth ×Height	Miscellaneous Data (Dimensions in Inches)	Manufacturer's Input Rating — Boiler hp or Watts	Manufacturer's Input Rating — Btu/Hr	Probable Max. Hourly Input Btu/Hr	Without Hood — Sensible	Without Hood — Latent	Without Hood — Total	With Hood[b] — All Sensible
Gas-Burning, Counter Type										
Broiler-griddle		31×20×18			36,000	18,000	11,700	6,300	18,000	3,600
Coffee brewer per burner			With *warm* position		5,500	2,500	1,750	750	2,500	500
Water heater burner			With storage tank		11,000	5,000	3,850	1,650	5,500	1,100
Coffee urn	3 gal.	12-inch dia.			10,000	5,000	3,500	1,500	5,000	1,000
	5 gal.	14-inch dia.			15,000	7,500	5,250	2,250	7,500	1,500
	8 gal. twin	25-inch wide			20,000	10,000	7,000	3,000	10,000	2,000
Deep fat fryer	15 lb fat	14×21×15			30,000	15,000	7,500	7,500	15,000	3,000
Dry food warmer per sq ft of top					1,400	700	560	140	700	140
Griddle, frying per sq ft of top					15,000	7,500	4,900	2,600	7,500	1,500
Short order stove, per burner			Open grates		10,000	5,000	3,200	1,800	5,000	1,000
Steam table per sq ft of top					2,500	1,250	750	500	1,250	250
Toaster, continuous	360 slices/hr	19×16×30	2 slices wide		12,000	6,000	3,600	2,400	6,000	1,200
	720 slices/hr	24×16×30	4 slices wide		20,000	10,000	6,000	4,000	10,000	2,000
Gas-Burning, Floor Mounted Type										
Broiler, unit		24×26 grid	Same burner heats oven		70,000	35,000	Exhaust hood required	Exhaust hood required	Exhaust hood required	7,000
Deep fat fryer	32 lb fat		14-inch kettle		65,000	32,500				6,500
	56 lb fat		18-inch kettle		100,000	50,000				10,000
Oven, deck, per sq ft of hearth area			Same for 7 and 12 high decks		4,000	2,000				400
Oven, roasting		32×32×60	Two ovens—24×28×15		80,000	40,000				8,000
Range, heavy duty		32×42×33								
Top section			32 wide×39 deep		64,000	32,000				6,400
Oven			25×28×15		40,000	20,000				4,000
Range, Jr., heavy duty		31×35×33								
Top section			31 wide×32 deep		45,000	22,500				4,500
Oven			24×28×15		35,000	17,500				3,500
Range, restaurant type										
Per 2-burner sect.			12 wide×28 deep		24,000	12,000				2,400
Per oven			24×22×14		30,000	15,000				3,000
Per broiler-griddle			24 wide×26 deep		35,000	17,500				3,500
Electric, Counter Type										
Coffee brewer										
Per burner				625	2,130	1,000	770	230	1,000	340
Per warmer				160	545	300	230	70	300	90
Automatic	240 cups per hr	27×21×22	4-burner+water htr.	5,000	17,000	8,500	6,500	2,000	8,500	1,700
Coffee urn	3 gal.			2,000	6,800	3,400	2,550	850	3,400	1,000
	5 gal.			3,000	10,200	5,100	3,850	1,250	5,100	1,600
	8 gal. twin			4,000	13,600	6,800	5,200	1,600	6,800	2,100
Deep fat fryer	14 lb fat	13×22×10		5,500	18,750	9,400	2,800	6,600	9,400	3,000
	21 lb fat	16×22×10		8,000	27,300	13,700	4,100	9,600	13,700	4,300
Dry food warmer, per sq ft of top				240	820	400	320	80	400	130
Egg boiler	2 cups	10×13×25		1,100	3,750	1,900	1,140	760	1,900	600
Griddle, frying, per sq ft of top				2,700	9,200	4,600	3,000	1,600	4,600	1,500
Griddle-Grill		18×20×13	Grid, 200 sq in.	6,000	20,400	10,200	6,600	3,600	10,200	3,200
Hotplate		18×20×13	2 heating units	5,200	17,700	8,900	5,300	3,600	8,900	2,800
Roaster		18×20×13		1,650	5,620	2,800	1,700	1,100	2,800	900
Roll warmer		18×20×13		1,650	5,620	2,800	2,600	200	2,800	900
Toaster, continuous	360 slices/hr	15×15×28	2 slices wide	2,200	7,500	3,700	1,960	1,740	3,700	1,200
	720 slices/hr	20×15×28	4 slices wide	3,000	10,200	5,100	2,700	2,400	5,100	1,600
Toaster, pop-up	4 slice	12×11×9		2,540	8,350	4,200	2,230	1,970	4,200	1,300
Waffle iron		18×20×13	2 grids	1,650	5,620	2,800	1,680	1,120	2,800	900

TABLE 3–32A Recommended Rate of Heat Gain from Commercial Cooking Appliances Located in the Air-Conditioned Area[a] (Continued)

Appliance	Capacity	Overall Dim., Inches Width×Depth×Height	Miscellaneous Data (Dimensions in Inches)	Manufacturer's Input Rating Boiler hp or Watts	Btu/Hr	Probable Max. Hourly Input Btu/Hr	Without Hood Sensible	Latent	Total	With Hood[b] All Sensible
Electric, Floor Mounted Type										
Broiler, no oven			23 wide×25 deep grid	12,000	40,900	20,500	Exhaust hood required	Exhaust hood required	Exhaust hood required	6,500
With oven			23×27×12 oven	18,000	61,400	30,700				9,800
Deep fat fryer	28 lb fat	20×38×36	14 wide×15 deep kettle	12,000	40,900	20,500				6,500
	60 lb fat	24×36×36	20 wide×20 deep kettle	18,000	61,400	30,700				9,800
Oven, baking, per sq ft of hearth			Compartment 8-in. high	500	1,700	850				270
Oven, roasting, per sq ft of hearth			Compartment 12-in. high	900	3,070	1,500				490
Range, heavy duty Top section		36×36×36		15,000	51,100	25,600				8,200
Oven				6,000	20,400	10,200				3,200
Range, medium duty Top section		30×32×36		8,000	27,300	13,600				4,300
Oven				3,600	12,300	6,200				1,900
Range, light duty Top section		30×29×36		6,600	22,500	11,200				3,600
Oven				3,000	10,200	5,100				1,600
Steam Heated										
Coffee urn	3 gal.			0.2	6,600	3,300	2,180	1,120	3,300	1,000
	5 gal.			0.3	10,000	5,000	3,300	1,700	5,000	1,600
	8 gal. twin			0.4	13,200	6,600	4,350	2,250	6,600	2,100
Steam table per sq ft of top			With insets	0.05	1,650	825	500	325	825	260
Bain marie per sq ft of top			Open tank	0.10	3,300	1,650	825	825	1,650	520
Oyster steamer				0.5	16,500	8,250	5,000	3,250	8,250	2,600
Steam kettles per gal. capacity			Jacketed type	0.06	2,000	1,000	600	400	1,000	320
Compartment steamer per compartment		24×25×12 compartment	Floor mounted	1.2	40,000	20,000	12,000	8,000	20,000	6,400
Compartment steamer	3 pans 12×20×2½		Single counter unit	0.5	16,500	8,250	5,000	3,250	8,250	2,600
Plate warmer per cu ft				0.05	1,650	825	550	275	825	260

[a] Heat gain from cooking appliances located in the conditioned area (but not included in the table) should be estimated as follows:
1. Obtain *probable maximum hourly input* by multiplying the manufacturer's hourly input rating by the usage factor of 0.50.
2. If appliances are installed without an exhaust hood, the estimated latent heat gain is 34 percent of the *probable maximum hourly input* and the sensible heat gain in 66 percent.
3. If appliances are installed under an effective exhaust hood, the estimated heat gain is all sensible heat and can be calculated from equations (18) and (20) in the text.

[b] For poorly designed or undersized exhaust systems the heat gains in this column should be doubled and half of the increase assumed as latent heat.

TABLE 3–32B Rate of Heat Gain from Miscellaneous Appliances

Appliance	Miscellaneous Data	Manufacturer's Rating Watts	Btu/Hr	Recommended Rate of Heat Gain, Btu/Hr Sensible	Latent	Total
Electrical Appliances						
Hair dryer	Blower type	1580	5400	2300	400	2700
Hair dryer	Helmet type	705	2400	1870	330	2200
Permanent wave machine	60 heaters @25 W, 36 in normal use	1500	5000	850	150	1000
Neon sign, per linear ft of tube	½ in., dia			30		30
	⅜ in. dia			60		60
Sterilizer, instrument		1100	3750	650	1200	1850

Appliance	Miscellaneous Data	Manufacturer's Rating Watts	Btu/Hr	Recommended Rate of Heat Gain, Btu/Hr Sensible	Latent	Total
Gas-Burning Appliances						
Lab burners Bunsen	7/16 in. barrel		3000	1680	420	2100
Fishtail	1½ in. wide		5000	2800	700	3500
Meeker	1 in. diameter		6000	3360	840	4200
Gas light, per burner	Mantle type		2000	1800	200	2000
Cigar lighter	Continuous flame		2500	900	100	1000

62

TABLE 3–33 Climatic Conditions for United States and Canada

Col. 1 State and Station[b]	Col. 2 Latitude[c] ° ′		Col. 3 Elev.[d] Ft	Winter Col. 4 Median of Annual Extremes	99%	97½%	Col. 5 Coincident Wind Velocity[e]	Summer Col. 6 Design Dry-Bulb 1%	2½%	5%	Col. 7 Outdoor Daily Range[f]	Col. 8 Design Wet-Bulb 1%	2½%	5%
ALABAMA														
Alexander City	33	0	660	12	16	20	L	96	94	93	21	79	78	77
Anniston AP	33	4	599	12	17	19	L	96	94	93	21	79	78	77
Auburn	32	4	730	17	21	25	L	98	96	95	21	80	79	78
Birmingham AP	33	3	610	14	19	22	L	97	94	93	21	79	78	77
Decatur	34	4	580	10	15	19	L	97	95	94	22	79	78	77
Dothan AP	31	2	321	19	23	27	L	97	95	94	20	81	80	79
Florence AP	34	5	528	8	13	17	L	97	95	94	22	79	78	77
Gadsden	34	0	570	11	16	20	L	96	94	93	22	78	77	76
Huntsville AP	34	4	619	8	13	17	L	97	95	94	23	78	77	76
Mobile AP	30	4	211	21	26	29	M	95	93	91	18	80	79	79
Mobile CO	30	4	119	24	28	32	M	96	94	93	16	80	79	79
Montgomery AP	32	2	195	18	22	26	L	98	95	93	21	80	79	78
Selma-Craig AFB	32	2	207	18	23	27	L	98	96	94	21	81	80	79
Talladega	33	3	565	11	15	19	L	97	95	94	21	79	78	77
Tuscaloosa AP	33	1	170r	14	19	23	L	98	96	95	22	81	80	79
ALASKA														
Anchorage AP	61	1	90	−29	−25	−20	VL	73	70	67	15	63	61	59
Barrow	71	2	22	−49	−45	−42	M	58	54	50	12	54	51	48
Fairbanks AP	64	5	436	−59	−53	−50	VL	82	78	75	24	64	63	61
Juneau AP	58	2	17	−11	−7	−4	L	75	71	68	15	66	64	62
Kodiak	57	3	21	4	8	12	M	71	66	63	10	62	60	58
Nome AP	64	3	13	−37	−32	−28	L	66	62	59	10	58	56	54
ARIZONA†														
Douglas AP	31	3	4098	13	18	22	VL	100	98	96	31	70	69	68
Flagstaff AP	35	1	6973	−10	0	5	VL	84	82	80	31	61	60	59
Fort Huachuca AP	31	3	4664	18	25	28	VL	95	93	91	27	69	68	67
Kingman AP	35	2	3446	18	25	29	VL	103	100	97	30	70	69	69
Nogales	31	2	3800	15	20	24	VL	100	98	96	31	72	71	70
Phoenix AP	33	3	1117	25	31	34	VL	108	106	104	27	77	76	75
Prescott AP	34	4	5014	7	15	19	VL	96	94	91	30	67	66	65
Tucson AP	33	1	2584	23	29	32	VL	105	102	100	26	74	73	72
Winslow AP	35	0	4880	2	9	13	VL	97	95	92	32	66	65	64
Yuma AP	32	4	199	32	37	40	VL	111	109	107	27	79	78	77
ARKANSAS														
Blytheville AFB	36	0	264	6	12	17	L	98	96	93	21	80	79	78
Camden	33	4	116	13	19	23	L	99	97	96	21	81	80	79
El Dorado AP	33	1	252	13	19	23	L	98	96	95	21	81	80	79
Fayetteville AP	36	0	1253	3	9	13	M	97	95	93	23	77	76	75
Fort Smith AP	35	2	449	9	15	19	M	101	99	96	24	79	78	77
Hot Springs Nat. Pk.	34	3	710	12	18	22	M	99	97	96	22	79	78	77
Jonesboro	35	5	345	8	14	18	M	98	96	95	21	80	79	78
Little Rock AP	34	4	257	13	19	23	M	99	96	94	22	80	79	78
Pine Bluff AP	34	1	204	14	20	24	L	99	96	95	22	81	80	79
Texarkana AP	33	3	361	16	22	26	M	99	97	96	21	80	79	78
CALIFORNIA†														
Bakersfield AP	35	2	495	26	31	33	VL	103	101	99	32	72	71	70
Barstow AP	34	5	2142	18	24	28	VL	104	102	99	37	73	72	71
Blythe AP	33	4	390	26	31	35	VL	111	109	106	28	78	77	76
Burbank AP	34	1	699	30	36	38	VL	97	94	91	25	72	70	69
Chico	39	5	205	23	29	33	VL	102	100	97	36	71	70	69
Concord	38	0	195	27	32	36	VL	96	92	88	32	69	67	66

* Data for U. S. stations extracted from *Evaluated Weather Data for Cooling Equipment Design, Addendum No. 1, Winter and Summer Data*, with the permission of the publisher, Fluor Products Company, Inc., Box 1267, Santa Rosa, California.

ª Data compiled from official weather stations, where hourly weather observations are made by trained observers, and from other sources. Table 1 prepared by ASHRAE Technical Committee 2.2, Weather Data and Design Conditions. Percentage of *winter* design data show the percent of 3-month period, December through February. Canadian data are based on January only. Percentage of *summer* design data show the percent of 4-month period, June through September. Canadian data are based on July only. Also see References 1 to 7.

ᵇ When airport temperature observations were used to develop design data, "AP" follows station name, and "AFB" follows Air Force Bases. Data for stations followed by "CO" came from office locations within an urban area and generally reflect an influence of the surrounding area. Stations without designation can be considered semirural and may be directly compared with most airport data.

ᶜ Latitude is given to the nearest 10 minutes, for use in calculating solar loads. For example, the latitude for Anniston, Alabama is given as 33 4, or 33°40′.

ᵈ Elevations are ground elevations for each station as of 1964. Temperature readings are generally made at an elevation of 5 ft above ground, except for locations marked r, indicating roof exposure of thermometer.

ᵉ Coincident wind velocities derived from approximately coldest 600 hours out of 20,000 hours of December through February data per station. Also see References 5 and 6. The four classifications are:

VL = Very Light, 70 percent or more of cold extreme hours ≤7 mph. M = Moderate, 50 to 74 percent cold extreme hours >7 mph.
L = Light, 50 to 69 percent cold extreme hours ≤7 mph. H = High, 75 percent or more cold extreme hours >7 mph, and 50 percent are >12 mph.

ᶠ The difference between the average maximum and average minimum temperatures during the warmest month.

† More detailed data on Arizona, California, and Nevada may be found in *Recommended Design Temperatures, Northern California*, published by the Golden Gate Chapter; and *Recommended Design Temperatures, Southern California, Arizona, Nevada*, published by the Southern California Chapter.

Reprinted by permission from ASHRAE Handbook of Fundamentals (1967).

TABLE 3–33 Climatic Conditions for United States and Canada

Col. 1 State and Station[b]	Col. 2 Latitude[c] ° '	Col. 3 Elev.[d] Ft	Winter Col. 4 Median of Annual Extremes	Winter Col. 4 99%	Winter Col. 4 97½%	Col. 5 Coincident Wind Velocity[e]	Summer Col. 6 Design Dry-Bulb 1%	Summer Col. 6 2½%	Summer Col. 6 5%	Col. 7 Outdoor Daily Range[f]	Summer Col. 8 Design Wet-Bulb 1%	Summer Col. 8 2½%	Summer Col. 8 5%
CALIFORNIA† (continued)													
Covina	34 0	575	32	38	41	VL	100	97	94	31	73	72	71
Crescent City AP	41 5	50	28	33	36	L	72	69	65	18	61	60	59
Downey	34 0	116	30	35	38	VL	93	90	87	22	72	71	70
El Cajon	32 4	525	26	31	34	VL	98	95	92	30	74	73	72
El Centro AP	32 5	−30	26	31	35	VL	111	109	106	34	81	80	79
Escondido	33 0	660	28	33	36	VL	95	92	89	30	73	72	71
Eureka/Arcata AP	41 0	217	27	32	35	L	67	65	63	11	60	59	58
Fairfield-Travis AFB	38 2	72	26	32	34	VL	98	94	90	34	71	69	67
Fresno AP	36 5	326	25	28	31	VL	101	99	97	34	73	72	71
Hamilton AFB	38 0	3	28	33	35	VL	89	85	81	28	71	68	66
Laguna Beach	33 3	35	32	37	39	VL	83	80	77	18	69	68	67
Livermore	37 4	545	23	28	30	VL	99	97	94	24	70	69	68
Lompoc, Vandenburg AFB	34 4	552	32	36	38	VL	82	79	76	20	65	63	61
Long Beach AP	33 5	34	31	36	38	VL	87	84	81	22	72	70	69
Los Angeles AP	34 0	99	36	41	43	VL	86	83	80	15	69	68	67
Los Angeles CO	34 0	312	38	42	44	VL	94	90	87	20	72	70	69
Merced-Castle AFB	37 2	178	24	30	32	VL	102	99	96	36	73	72	70
Modesto	37 4	91	26	32	36	VL	101	98	96	36	72	71	70
Monterey	36 4	38	29	34	37	VL	82	79	76	20	64	63	61
Napa	38 2	16	26	31	34	VL	94	92	89	30	69	68	67
Needles AP	34 5	913	27	33	37	VL	112	110	107	27	76	75	74
Oakland AP	37 4	3	30	35	37	VL	85	81	77	19	65	63	62
Oceanside	33 1	30	33	38	40	VL	84	81	78	13	69	68	67
Ontario	34 0	995	26	32	34	VL	100	97	94	36	72	71	70
Oxnard AFB	34 1	43	32	35	37	VL	84	80	78	19	70	69	67
Palmdale AP	34 4	2517	18	24	27	VL	103	101	98	35	70	68	67
Palm Springs	33 5	411	27	32	36	VL	110	108	105	35	79	78	77
Pasadena	34 1	864	31	36	39	VL	96	93	90	29	72	70	69
Petaluma	38 1	27	24	29	32	VL	94	90	87	31	70	68	67
Pomona CO	34 0	871	26	31	34	VL	99	96	93	36	73	72	71
Redding AP	40 3	495	25	31	35	VL	103	101	98	32	70	69	67
Redlands	34 0	1318	28	34	37	VL	99	96	93	33	72	71	70
Richmond	38 0	55	28	35	38	VL	85	81	77	17	66	64	63
Riverside-March AFB	33 5	1511	26	32	34	VL	99	96	94	37	72	71	69
Sacramento AP	38 3	17	24	30	32	VL	100	97	94	36	72	70	69
Salinas AP	36 4	74	27	32	35	VL	87	85	82	24	67	65	64
San Bernardino, Norton AFB	34 1	1125	26	31	33	VL	101	98	96	38	75	73	71
San Diego AP	32 4	19	38	42	44	VL	86	83	80	12	71	70	68
San Fernando	34 1	977	29	34	37	VL	100	97	94	38	73	72	71
San Francisco AP	37 4	8	32	35	37	L	83	79	75	20	65	63	62
San Francisco CO	37 5	52	38	42	44	VL	80	77	73	14	64	62	61
San Jose AP	37 2	70r	30	34	36	VL	90	88	85	26	69	67	65
San Luis Obispo	35 2	315	30	35	37	VL	89	85	82	26	65	64	63
Santa Ana AP	33 4	115r	28	33	36	VL	92	89	86	28	72	71	70
Santa Barbara CO	34 3	100	30	34	36	VL	87	84	81	24	67	66	65
Santa Cruz	37 0	125	28	32	34	VL	87	84	80	28	66	65	63
Santa Maria AP	34 5	238	28	32	34	VL	85	82	79	23	65	64	63
Santa Monica CO	34 0	57	38	43	45	VL	80	77	74	16	69	68	67
Santa Paula	34 2	263	28	33	36	VL	91	89	86	36	72	71	70
Santa Rosa	38 3	167	24	29	32	VL	95	93	90	34	70	68	67
Stockton AP	37 5	28	25	30	34	VL	101	98	96	37	72	70	69
Ukiah	39 1	620	22	27	30	VL	98	96	93	40	70	69	67
Visalia	36 2	354	26	32	36	VL	102	100	97	38	73	72	70
Yreka	41 4	2625	7	13	17	VL	96	94	91	38	68	66	65
Yuba City	39 1	70	24	30	34	VL	102	100	97	36	71	70	69
COLORADO													
Alamosa AP	37 3	7536	−26	−17	−13	VL	84	82	79	35	62	61	60
Boulder	40 0	5385	− 5	4	8	L	92	90	87	27	64	63	62
Colorado Springs AP	38 5	6173	− 9	− 1	4	L	90	88	86	30	63	62	61
Denver AP	39 5	5283	− 9	− 2	3	L	92	90	89	28	65	64	63
Durango	37 1	6550	−10	0	4	VL	88	86	83	30	64	63	62

Reprinted by permission from ASHRAE Handbook of Fundamentals (1967).

TABLE 3–33 Climatic Conditions for United States and Canada

Col. 1 State and Station[b]	Col. 2 Latitude[c] °	'	Col. 3 Elev[d] Ft	Winter Col. 4 Median of Annual Extremes	99%	97½%	Col. 5 Coincident Wind Velocity[e]	Summer Col. 6 Design Dry-Bulb 1%	2½%	5%	Col. 7 Outdoor Daily Range[f]	Col. 8 Design Wet-Bulb 1%	2½%	5%
COLORADO (continued)														
Fort Collins	40	4	5001	−18	− 9	− 5	L	91	89	86	28	63	62	61
Grand Junction AP	39	1	4849	− 2	8	11	VL	96	94	92	29	64	63	62
Greeley	40	3	4648	−18	− 9	− 5	L	94	92	89	29	65	64	63
La Junta AP	38	0	4188	−14	− 6	− 2	M	97	95	93	31	72	71	69
Leadville	39	2	10177	−18	− 9	− 4	VL	76	73	70	30	56	55	54
Pueblo AP	38	2	4639	−14	− 5	− 1	L	96	94	92	31	68	67	66
Sterling	40	4	3939	−15	− 6	− 2	M	95	93	90	30	67	66	65
Trinidad AP	37	2	5746	− 9	1	5	L	93	91	89	32	66	65	64
CONNECTICUT														
Bridgeport AP	41	1	7	− 1	4	8	M	90	88	85	18	77	76	75
Hartford, Brainard Field	41	5	15	− 4	1	5	M	90	88	85	22	77	76	74
New Haven AP	41	2	6	0	5	9	H	88	86	83	17	77	76	75
New London	41	2	60	0	4	8	H	89	86	83	16	77	75	74
Norwalk	41	1	37	− 5	0	4	M	91	89	86	19	77	76	75
Norwich	41	3	20	− 7	− 2	2	M	88	86	83	18	77	76	75
Waterbury	41	3	605	− 5	0	4	M	90	88	85	21	77	76	75
Windsor Locks, Bradley Field	42	0	169	− 7	− 2	2	M	90	88	85	22	76	75	73
DELAWARE														
Dover AFB	39	0	38	8	13	15	M	93	90	88	18	79	78	77
Wilmington AP	39	4	78	6	12	15	M	93	90	87	20	79	77	76
DISTRICT OF COLUMBIA														
Andrews AFB	38	5	279	9	13	16	M	94	91	88	18	79	77	76
Washington National AP	38	5	14	12	16	19	M	94	92	90	18	78	77	76
FLORIDA														
Belle Glade	26	4	16	31	35	39	M	93	91	90	16	80	79	79
Cape Kennedy AP	28	1	16	33	37	40	L	90	89	88	15	81	80	79
Daytona Beach AP	29	1	31	28	32	36	L	94	92	91	15	81	80	79
Fort Lauderdale	26	0	13	37	41	45	M	91	90	89	15	81	80	79
Fort Myers AP	26	4	13	34	38	42	M	94	92	91	18	80	80	79
Fort Pierce	27	3	10	33	37	41	M	93	91	90	15	81	80	79
Gainesville AP	29	4	155	24	28	32	L	96	94	93	18	80	79	79
Jacksonville AP	30	3	24	26	29	32	L	96	94	92	19	80	79	79
Key West AP	24	3	6	50	55	58	M	90	89	88	9	80	79	79
Lakeland CO	28	0	214	31	35	39	M	95	93	91	17	80	79	78
Miami AP	25	5	7	39	44	47	M	92	90	89	15	80	79	79
Miami Beach CO	25	5	9	40	45	48	M	91	89	88	10	80	79	79
Ocala	29	1	86	25	29	33	L	96	94	93	18	80	79	79
Orlando AP	28	3	106r	29	33	37	L	96	94	93	17	80	79	78
Panama City, Tyndall AFB	30	0	22	28	32	35	M	92	91	90	14	81	80	80
Pensacola CO	30	3	13	25	29	32	M	92	90	89	14	82	81	80
St. Augustine	29	5	15	27	31	35	L	94	92	90	16	81	80	79
St. Petersburg	28	0	35	35	39	42	M	93	91	90	16	81	80	79
Sanford	28	5	14	29	33	37	L	95	93	92	17	80	79	79
Sarasota	27	2	30	31	35	39	M	93	91	90	17	80	80	79
Tallahassee AP	30	2	58	21	25	29	L	96	94	93	19	80	79	79
Tampa AP	28	0	19	32	36	39	M	92	91	90	17	81	80	79
West Palm Beach AP	26	4	15	36	40	44	M	92	91	89	16	81	80	80
GEORGIA														
Albany, Turner AFB	31	3	224	21	26	30	L	98	96	94	20	80	79	78
Americus	32	0	476	18	22	25	L	98	96	93	20	80	79	78
Athens	34	0	700	12	17	21	L	96	94	91	21	78	77	76
Atlanta AP	33	4	1005	14	18	23	H	95	92	90	19	78	77	76
Augusta AP	33	2	143	17	20	23	L	98	95	93	19	80	79	78
Brunswick	31	1	14	24	27	31	L	97	95	92	18	81	80	79
Columbus, Lawson AFB	32	3	242	19	23	26	L	98	96	94	21	80	79	78
Dalton	34	5	720	10	15	19	L	97	95	92	22	78	77	76
Dublin	32	3	215	17	21	25	L	98	96	93	20	80	79	78
Gainesville	34	2	1254	11	16	20	L	94	92	89	21	78	77	76
Griffin	33	1	980	13	17	22	L	95	93	90	21	79	78	77
La Grange	33	0	715	12	16	20	L	96	94	92	21	79	78	77
Macon AP	32	4	356	18	23	27	L	98	96	94	22	80	79	78
Marietta, Dobbins AFB	34	0	1016	12	17	21	L	95	93	91	21	78	77	76

Reprinted by permission from ASHRAE Handbook of Fundamentals (1967).

TABLE 3–33 Climatic Conditions for United States and Canada

Col. 1 State and Station[b]	Col. 2 Latitude[c] ° ′	Col. 3 Elev,[d] Ft	Col. 4 Median of Annual Extremes	99%	97½%	Col. 5 Coincident Wind Velocity[e]	Col. 6 Design Dry-Bulb 1%	2½%	5%	Col. 7 Outdoor Daily Range[f]	Col. 8 Design Wet-Bulb 1%	2½%	5%
GEORGIA (continued)													
Moultrie	31 1	340	22	26	30	L	97	95	93	20	80	79	78
Rome AP	34 2	637	11	16	20	L	97	95	93	23	78	77	76
Savannah-Travis AP	32 1	52	21	24	27	L	96	94	92	20	81	80	79
Valdosta-Moody AFB	31 0	239	24	28	31	L	96	94	92	20	80	79	78
Waycross	31 2	140	20	24	28	L	97	95	93	20	80	79	78
HAWAII													
Hilo AP	19 4	31	56	59	61	L	85	83	82	15	74	73	72
Honolulu AP	21 2	7	58	60	62	L	87	85	84	12	75	74	73
Kaneohe	21 2	198	58	60	61	L	85	83	82	12	74	73	73
Wahiawa	21 3	215	57	59	61	L	86	84	83	14	75	74	73
IDAHO													
Boise AP	43 3	2842	0	4	10	L	96	93	91	31	68	66	65
Burley	42 3	4180	− 5	4	8	VL	95	93	89	35	68	66	64
Coeur d'Alene AP	47 5	2973	− 4	2	7	VL	94	91	88	31	66	65	63
Idaho Falls AP	43 3	4730r	−17	−12	− 6	VL	91	88	85	38	65	64	62
Lewiston AP	46 2	1413	1	6	12	VL	98	96	93	32	67	66	65
Moscow	46 4	2660	−11	− 3	1	VL	91	89	86	32	64	63	61
Mountain Home AFB	43 0	2992	− 3	2	9	L	99	96	93	36	68	66	64
Pocatello AP	43 0	4444	−12	− 8	− 2	VL	94	91	88	35	65	63	62
Twin Falls AP	42 3	4148	− 5	4	8	L	96	94	91	34	66	64	63
ILLINOIS													
Aurora	41 5	744	−13	− 7	− 3	M	93	91	88	20	78	77	75
Belleville, Scott AFB	38 3	447	0	6	10	M	97	95	92	21	79	78	77
Bloomington	40 3	775	− 7	− 1	3	M	94	92	89	21	79	78	77
Carbondale	37 5	380	1	7	11	M	98	96	94	21	80	79	78
Champaign/Urbana	40 0	743	− 6	0	4	M	96	94	91	21	79	78	77
Chicago, Midway AP	41 5	610	− 7	− 4	1	M	95	92	89	20	78	76	75
Chicago, O'Hare AP	42 0	658	− 9	− 4	0	M	93	90	87	20	77	75	74
Chicago, CO	41 5	594	− 5	− 3	1	M	94	91	88	15	78	76	75
Danville	40 1	558	− 6	− 1	4	M	96	94	91	21	79	78	76
Decatur	39 5	670	− 6	0	4	M	96	93	91	21	79	78	77
Dixon	41 5	696	−13	− 7	− 3	M	93	91	89	23	78	77	76
Elgin	42 0	820	−14	− 8	− 4	M	92	90	87	21	78	76	75
Freeport	42 2	780	−16	−10	− 6	M	92	90	87	24	78	77	75
Galesburg	41 0	771	−10	− 4	0	M	95	92	89	22	79	78	76
Greenville	39 0	563	− 3	3	7	M	96	94	92	21	79	78	77
Joliet AP	41 3	588	−11	− 5	− 1	M	94	92	89	20	78	77	75
Kankakee	41 1	625	−10	− 4	1	M	94	92	89	21	78	77	76
La Salle/Peru	41 2	520	− 9	− 3	1	M	94	93	90	22	78	77	76
Macomb	40 3	702	− 5	− 3	1	M	95	93	90	22	79	78	77
Moline AP	41 3	582	−12	− 7	− 3	M	94	91	88	23	79	77	76
Mt. Vernon	38 2	500	0	6	10	M	97	95	92	21	79	78	77
Peoria AP	40 4	652	− 8	− 2	2	M	94	92	89	22	78	77	76
Quincy AP	40 0	762	− 8	− 2	2	M	97	95	92	22	80	79	77
Rantoul, Chanute AFB	40 2	740	− 7	− 1	3	M	94	92	89	21	78	77	76
Rockford	42 1	724	−13	− 7	− 3	M	92	90	87	24	77	76	75
Springfield AP	39 5	587	− 7	− 1	4	M	95	92	90	21	79	78	77
Waukegan	42 2	680	−11	− 5	− 1	M	92	90	87	21	77	76	75
INDIANA													
Anderson	40 0	847	− 5	0	5	M	93	91	88	22	78	77	76
Bedford	38 5	670	− 3	3	7	M	95	93	90	22	79	78	77
Bloomington	39 1	820	− 3	3	7	M	95	92	90	22	79	78	76
Columbus, Bakalar AFB	39 2	661	− 3	3	7	M	95	92	90	22	79	78	76
Crawfordsville	40 0	752	− 8	− 2	2	M	95	93	90	22	79	77	76
Evansville AP	38 0	381	1	6	10	M	96	94	91	22	79	78	77
Fort Wayne AP	41 0	791	− 5	0	5	M	93	91	88	24	77	76	75
Goshen AP	41 3	823	−10	− 4	0	M	92	90	87	23	77	76	74
Hobart	41 3	600	−10	− 4	0	M	93	91	88	21	78	76	75
Huntington	40 4	802	− 8	− 2	2	M	94	92	89	23	78	76	75
Indianapolis AP	39 4	793	− 5	0	4	M	93	91	88	22	78	77	76
Jeffersonville	38 2	455	3	9	13	M	96	94	91	23	79	78	77
Kokomo	40 3	790	− 6	0	4	M	94	92	89	22	78	76	75
Lafayette	40 2	600	− 7	− 1	3	M	94	92	89	22	78	77	76

Reprinted by permission from ASHRAE Handbook of Fundamentals (1967).

TABLE 3–33 Climatic Conditions for United States and Canada

Col. 1 State and Station[b]	Col. 2 Latitude[c] ° '		Col. 3 Elev,[d] Ft	Winter Col. 4 Median of Annual Extremes	99%	97½%	Col. 5 Coincident Wind Velocity[e]	Summer Col. 6 Design Dry-Bulb 1%	2½%	5%	Col. 7 Outdoor Daily Range[f]	Col. 8 Design Wet-Bulb 1%	2½%	5%
INDIANA (continued)														
La Porte	41	3	810	−10	− 4	0	M	93	91	88	22	77	76	74
Marion	40	3	791	− 8	− 2	2	M	93	91	88	23	78	76	75
Muncie	40	1	955	− 8	− 2	2	M	93	91	88	22	78	77	75
Peru, Bunker Hill AFB	40	4	804	− 9	− 3	1	M	91	89	86	22	77	76	74
Richmond AP	39	5	1138	− 7	− 1	3	M	93	91	88	22	78	77	75
Shelbyville	39	3	765	− 4	2	6	M	94	92	89	22	78	77	76
South Bend AP	41	4	773	− 6	− 2	3	M	92	89	87	22	77	76	74
Terre Haute AP	39	3	601	− 3	3	7	M	95	93	91	22	79	78	77
Valparaiso	41	2	801	−12	− 6	− 2	M	92	90	87	22	78	76	75
Vincennes	38	4	420	− 1	5	9	M	96	94	91	22	79	78	77
IOWA														
Ames	42	0	1004	−17	−11	− 7	M	94	92	89	23	79	78	76
Burlington AP	40	5	694	−10	− 4	0	M	95	92	89	22	80	78	77
Cedar Rapids AP	41	5	863	−14	− 8	− 4	M	92	90	87	23	78	76	75
Clinton	41	5	595	−13	− 7	− 3	M	92	90	87	23	78	77	76
Council Bluffs	41	2	1210	−14	− 7	− 3	M	97	94	91	22	79	78	76
Des Moines AP	41	3	948r	−13	− 7	− 3	M	95	92	89	23	79	77	76
Dubuque	42	2	1065	−17	−11	− 7	M	92	90	87	22	78	76	75
Fort Dodge	42	3	1111	−18	−12	− 8	M	94	92	89	23	78	77	75
Iowa City	41	4	645	−14	− 8	− 4	M	94	91	88	22	79	77	76
Keokuk	40	2	526	− 9	− 3	1	M	95	93	90	22	79	78	77
Marshalltown	42	0	898	−16	−10	− 6	M	93	91	88	23	79	77	76
Mason City AP	43	1	1194	−20	−13	− 9	M	91	88	85	24	77	75	74
Newton	41	4	946	−15	− 9	− 5	M	95	93	90	23	79	77	76
Ottumwa AP	41	1	842	−12	− 6	− 2	M	95	93	90	22	79	78	76
Sioux City AP	42	2	1095	−17	−10	− 6	M	96	93	90	24	79	77	76
Waterloo	42	3	868	−18	−12	− 8	M	91	89	86	23	78	76	75
KANSAS														
Atchison	39	3	945	− 9	− 2	2	M	97	95	92	23	79	78	77
Chanute AP	37	4	977	− 3	3	7	H	99	97	95	23	79	78	77
Dodge City AP	37	5	2594	− 5	3	7	M	99	97	95	25	74	73	72
El Dorado	37	5	1282	− 3	4	8	H	101	99	96	24	78	77	76
Emporia	38	2	1209	− 4	3	7	H	99	97	94	25	78	77	76
Garden City AP	38	0	2882	−10	− 1	3	M	100	98	96	28	74	73	72
Goodland AP	39	2	3645	−10	− 2	4	M	99	96	93	31	71	70	69
Great Bend	38	2	1940	− 5	2	6	M	101	99	96	28	77	76	75
Hutchinson AP	38	0	1524	− 5	2	6	H	101	99	96	28	77	76	75
Liberal	37	0	2838	− 4	4	8	M	102	100	99	28	74	73	71
Manhattan, Fort Riley	39	0	1076	− 7	− 1	4	H	101	98	95	24	79	78	77
Parsons	37	2	908	− 2	5	9	H	99	97	94	23	79	78	77
Russell AP	38	5	1864	− 7	0	4	M	102	100	97	29	78	76	75
Salina	38	5	1271	− 4	3	7	H	101	99	96	26	78	76	75
Topeka AP	39	0	877	− 4	3	6	M	99	96	94	24	79	78	77
Wichita AP	37	4	1321	− 1	5	9	H	102	99	96	23	77	76	75
KENTUCKY														
Ashland	38	3	551	1	6	10	L	94	92	89	22	77	76	75
Bowling Green AP	37	0	535	1	7	11	L	97	95	93	21	79	78	77
Corbin AP	37	0	1175	0	5	9	L	93	91	89	23	79	77	76
Covington AP	39	0	869	− 3	3	8	L	93	90	88	22	77	76	75
Hopkinsville, Campbell AFB	36	4	540	4	10	14	L	97	95	92	21	79	78	77
Lexington AP	38	0	979	0	6	10	M	94	92	90	22	78	77	76
Louisville AP	38	1	474	1	8	12	L	96	93	91	23	79	78	77
Madisonville	37	2	439	1	7	11	L	96	94	92	22	79	78	77
Owensboro	37	5	420	0	6	10	L	96	94	92	23	79	78	77
Paducah AP	37	0	398	4	10	14	L	97	95	94	20	80	79	78
LOUISIANA														
Alexandria AP	31	2	92	20	25	29	L	97	95	94	20	80	80	79
Baton Rouge AP	30	3	64	22	25	30	L	96	94	92	19	81	80	79
Bogalusa	30	5	103	20	24	28	L	96	94	93	19	80	79	78
Houma	29	3	13	25	29	33	L	94	92	91	15	81	80	79
Lafayette AP	30	1	38	23	28	32	L	95	93	92	18	81	81	80
Lake Charles AP	30	1	14	25	29	33	M	95	93	91	17	80	79	79
Minden	32	4	250	17	22	26	L	98	96	95	20	81	80	79

Reprinted by permission from ASHRAE Handbook of Fundamentals (1967).

TABLE 3–33   Climatic Conditions for United States and Canada

Col. 1 State and Station[b]	Col. 2 Latitude[c] ° '	Col. 3 Elev,[d] Ft	Winter Col. 4 Median of Annual Extremes	99%	97½%	Col. 5 Coincident Wind Velocity[e]	Summer Col. 6 Design Dry-Bulb 1%	2½%	5%	Col. 7 Outdoor Daily Range[f]	Col. 8 Design Wet-Bulb 1%	2½%	5%
LOUISIANA (continued)													
Monroe AP	32 3	78	18	23	27	L	98	96	95	20	81	81	80
Natchitoches	31 5	120	17	22	26	L	99	97	96	20	81	80	79
New Orleans AP	30 0	3	29	32	35	M	93	91	90	16	81	81	79
Shreveport AP	32 3	252	18	22	26	M	99	96	94	20	81	80	79
MAINE													
Augusta AP	44 2	350	−13	− 7	− 3	M	88	86	83	22	74	73	71
Bangor, Dow AFB	44 5	162	−14	− 8	− 4	M	88	85	81	22	75	73	71
Caribou AP	46 5	624	−24	−18	−14	L	85	81	78	21	72	70	68
Lewiston	44 0	182	−14	− 8	− 4	M	88	86	83	22	74	73	71
Millinocket AP	45 4	405	−22	−16	−12	L	87	85	82	22	74	72	70
Portland AP	43 4	61	−14	− 5	0	M	88	85	81	22	75	73	71
Waterville	44 3	89	−15	− 9	− 5	M	88	86	82	22	74	73	71
MARYLAND													
Baltimore AP	39 1	146	8	12	15	M	94	91	89	21	79	78	77
Baltimore CO	39 2	14	12	16	20	M	94	92	89	17	79	78	77
Cumberland	39 4	945	0	5	9	L	94	92	89	22	76	75	74
Frederick AP	39 2	294	2	7	11	M	94	92	89	22	78	77	76
Hagerstown	39 4	660	1	6	10	L	94	92	89	22	77	76	75
Salisbury	38 2	52	10	14	18	M	92	90	87	18	79	78	77
MASSACHUSETTS													
Boston AP	42 2	15	− 1	6	10	H	91	88	85	16	76	74	73
Clinton	42 2	398	− 8	− 2	2	M	87	85	82	17	75	74	72
Fall River	41 4	190	− 1	5	9	H	88	86	83	18	75	74	73
Framingham	42 2	170	− 7	− 1	3	M	91	89	86	17	76	74	73
Gloucester	42 3	10	− 4	2	6	H	86	84	81	15	74	73	72
Greenfield	42 3	205	−12	− 6	− 2	M	89	87	84	23	75	74	73
Lawrence	42 4	57	− 9	− 3	1	M	90	88	85	22	76	74	72
Lowell	42 3	90	− 7	− 1	3	M	91	89	86	21	76	74	72
New Bedford	41 4	70	3	9	13	H	86	84	81	19	75	73	72
Pittsfield AP	42 3	1170	−11	− 5	− 1	M	86	84	81	23	74	72	71
Springfield, Westover AFB	42 1	247	− 8	− 3	2	M	91	88	85	19	76	74	73
Taunton	41 5	20	− 9	− 4	0	H	88	86	83	18	76	75	74
Worcester AP	42 2	986	− 8	− 3	1	M	89	87	84	18	75	73	71
MICHIGAN													
Adrian	41 5	754	− 6	0	4	M	93	91	88	23	76	75	74
Alpena AP	45 0	689	−11	− 5	− 1	M	87	85	82	27	74	73	71
Battle Creek AP	42 2	939	− 6	1	5	M	92	89	86	23	76	74	73
Benton Harbor AP	42 1	649	− 7	− 1	3	M	90	88	85	20	76	74	73
Detroit Met. CAP	42 2	633	0	4	8	M	92	88	85	20	76	75	74
Escanaba	45 4	594	−13	− 7	− 3	M	82	80	77	17	73	71	69
Flint AP	43 0	766	− 7	− 1	3	M	89	87	84	25	76	75	74
Grand Rapids AP	42 5	681	− 3	2	6	M	91	89	86	24	76	74	73
Holland	42 5	612	− 4	2	6	M	90	88	85	22	76	74	73
Jackson AP	42 2	1003	− 6	0	4	M	92	89	86	23	76	75	74
Kalamazoo	42 1	930	− 5	1	5	M	92	89	86	23	76	75	74
Lansing AP	42 5	852	− 4	2	6	M	89	87	84	24	76	75	73
Marquette CO	46 3	677	−14	− 8	− 4	L	88	86	83	18	73	71	69
Mt. Pleasant	43 4	796	− 9	− 3	1	M	89	87	84	24	75	74	73
Muskegon AP	43 1	627	− 2	4	8	M	87	85	82	21	75	74	73
Pontiac	42 4	974	− 6	0	4	M	90	88	85	21	76	75	73
Port Huron	43 0	586	− 6	− 1	3	M	90	88	85	21	76	74	73
Saginaw AP	43 3	662	− 7	− 1	3	M	88	86	83	23	76	75	73
Sault Ste. Marie AP	46 3	721	−18	−12	− 8	L	83	81	78	23	73	71	69
Traverse City AP	44 4	618	− 6	0	4	M	89	86	83	22	75	73	72
Ypsilanti	42 1	777	− 3	− 1	5	M	92	89	86	22	76	74	73
MINNESOTA													
Albert Lea	43 4	1235	−20	−14	−10	M	91	89	86	24	77	76	74
Alexandria AP	45 5	1421	−26	−19	−15	L	90	88	85	24	76	74	72
Bemidji AP	47 3	1392	−38	−32	−28	L	87	84	81	24	73	72	71
Brainerd	46 2	1214	−31	−24	−20	L	88	85	82	24	74	73	72
Duluth AP	46 5	1426	−25	−19	−15	M	85	82	79	22	73	71	69
Faribault	44 2	1190	−23	−16	−12	L	90	88	85	24	77	75	74
Fergus Falls	46 1	1210	−28	−21	−17	L	92	89	86	24	75	74	72
International Falls AP	48 3	1179	−35	−29	−24	L	86	82	79	26	72	69	68

Reprinted by permission from ASHRAE Handbook of Fundamentals (1967).

68

TABLE 3-33 Climatic Conditions for United States and Canada

Col. 1 State and Station[b]	Col. 2 Latitude[c] ° '		Col. 3 Elev.[d] Ft	Winter Col. 4 Median of Annual Extremes	99%	97½%	Col. 5 Coincident Wind Velocity[e]	Summer Col. 6 Design Dry-Bulb 1%	2½%	5%	Col. 7 Outdoor Daily Range[f]	Col. 8 Design Wet-Bulb 1%	2½%	5%
MINNESOTA (continued)														
Mankato	44	1	785	−23	−16	−12	L	91	89	86	24	77	75	74
Minneapolis/St. Paul AP	44	5	822	−19	−14	−10	L	92	89	86	22	77	75	74
Rochester AP	44	0	1297	−23	−17	−13	M	90	88	85	24	77	75	74
St. Cloud AP	45	4	1034	−26	−20	−16	L	90	88	85	24	77	75	73
Virginia	47	3	1435	−32	−25	−21	L	86	83	80	23	73	71	69
Willmar	45	1	1133	−25	−18	−14	L	91	88	85	24	77	75	73
Winona	44	1	652	−19	−12	− 8	M	91	89	86	24	77	76	74
MISSISSIPPI														
Biloxi, Keesler AFB	30	2	25	26	30	32	M	93	92	90	16	82	81	80
Clarksdale	34	1	178	14	20	24	L	98	96	95	21	81	80	79
Columbus AFB	33	4	224	13	18	22	L	97	95	93	22	79	79	78
Greenville AFB	33	3	139	16	21	24	L	98	96	94	21	81	80	79
Greenwood	33	3	128	14	19	23	L	98	96	94	21	81	80	79
Hattiesburg	31	2	200	18	22	26	L	97	95	94	21	80	79	78
Jackson AP	32	2	330	17	21	24	L	98	96	94	21	79	78	78
Laurel	31	4	264	18	22	26	L	97	95	94	21	80	79	78
McComb AP	31	2	458	18	22	26	L	96	94	93	18	80	79	79
Meridian AP	32	2	294	15	20	24	L	97	95	94	22	80	79	78
Natchez	31	4	168	18	22	26	L	96	94	93	21	80	80	79
Tupelo	34	2	289	13	18	22	L	98	96	95	22	80	79	78
Vicksburg CO	32	2	234	18	23	26	L	97	95	94	21	80	80	79
MISSOURI														
Cape Girardeau	37	1	330	2	8	12	M	98	96	94	21	80	79	78
Columbia AP	39	0	778	− 4	2	6	M	97	95	92	22	79	78	77
Farmington AP	37	5	928	− 2	4	8	M	97	95	93	22	79	78	77
Hannibal	39	4	489	− 7	− 1	4	M	96	94	91	22	79	78	77
Jefferson City	38	4	640	− 4	2	6	M	97	95	93	23	79	78	77
Joplin AP	37	1	982	1	7	11	M	97	95	93	24	79	78	77
Kansas City AP	39	1	742	− 2	4	8	M	100	97	94	20	79	77	76
Kirksville AP	40	1	966	−13	− 7	− 3	M	96	94	91	24	79	78	77
Mexico	39	1	775	− 7	− 1	3	M	96	94	91	22	79	78	77
Moberly	39	3	850	− 8	− 2	2	M	96	94	91	23	79	78	77
Poplar Bluff	36	5	322	3	9	13	M	98	96	94	22	80	79	78
Rolla	38	0	1202	− 3	3	7	M	97	95	93	22	79	78	77
St. Joseph AP	39	5	809	− 8	− 1	3	M	97	95	92	23	79	78	77
St. Louis AP	38	5	535	− 2	4	8	M	98	95	92	21	79	78	77
St. Louis CO	38	4	465	1	7	11	M	96	94	92	18	79	78	77
Sedalia, Whiteman AFB	38	4	838	− 2	4	9	M	97	94	92	22	79	77	76
Sikeston	36	5	318	4	10	14	L	98	96	94	21	80	79	78
Springfield AP	37	1	1265	0	5	10	M	97	94	91	23	78	77	76
MONTANA														
Billings AP	45	5	3567	−19	−10	− 6	L	94	91	88	31	68	66	65
Bozeman	45	5	4856	−25	−15	−11	L	88	85	82	32	61	60	59
Butte AP	46	0	5526r	−34	−24	−16	VL	86	83	80	35	60	59	57
Cut Bank AP	48	4	3838r	−32	−23	−17	L	89	86	82	35	63	61	61
Glasgow AP	48	1	2277	−33	−25	−20	L	96	93	89	29	69	67	65
Glendive	47	1	2076	−28	−20	−16	L	96	93	90	29	71	69	68
Great Falls AP	47	3	3664r	−29	−20	−16	L	91	88	85	28	64	63	61
Havre	48	3	2488	−32	−22	−15	M	91	87	84	33	66	64	63
Helena AP	46	4	3893	−27	−17	−13	L	90	87	84	32	65	63	61
Kalispell AP	48	2	2965	−17	− 7	− 3	VL	88	84	81	34	65	63	62
Lewiston AP	47	0	4132	−27	−18	−14	L	89	86	83	30	65	63	62
Livingston AP	45	4	4653	−26	−17	−13	L	91	88	85	32	63	62	61
Miles City AP	46	3	2629	−27	−19	−15	L	97	94	91	30	71	69	68
Missoula AP	46	5	3200	−16	− 7	− 3	VL	92	89	86	36	65	63	61
NEBRASKA														
Beatrice	40	2	1235	−10	− 3	1	M	99	97	94	24	78	77	76
Chadron AP	42	5	3300	−21	−13	− 9	M	97	95	92	30	72	70	69
Columbus	41	3	1442	−14	− 7	− 3	M	98	96	93	25	78	76	75
Fremont	41	3	1203	−14	− 7	− 3	M	99	97	94	22	78	77	76
Grand Island AP	41	0	1841	−14	− 6	− 2	M	98	95	92	28	76	75	74
Hastings	40	4	1932	−11	− 3	1	M	98	96	94	27	77	75	74
Kearney	40	4	2146	−14	− 6	− 2	M	97	95	92	28	76	75	74
Lincoln CO	40	5	1150	−10	− 4	0	M	100	96	93	24	78	77	76

Reprinted by permission from ASHRAE Handbook of Fundamentals (1967).

TABLE 3–33 Climatic Conditions for United States and Canada

Col. 1 State and Station[b]	Col. 2 Latitude[c] °	′	Col. 3 Elev,[d] Ft	Col. 4 Median of Annual Extremes	99%	97½%	Col. 5 Coincident Wind Velocity[e]	Col. 6 Design Dry-Bulb 1%	2½%	5%	Col. 7 Outdoor Daily Range[f]	Col. 8 Design Wet-Bulb 1%	2½%	5%
NEBRASKA (continued)														
McCook	40	1	2565	−12	− 4	0	M	99	97	94	28	74	72	71
Norfolk	42	0	1532	−18	−11	− 7	M	97	95	92	30	78	76	75
North Platte AP	41	1	2779	−13	− 6	− 2	M	97	94	90	28	74	73	72
Omaha AP	41	2	978	−12	− 5	− 1	M	97	94	91	22	79	78	76
Scottsbluff AP	41	5	3950	−16	− 8	− 4	M	96	94	91	31	70	69	67
Sidney AP	41	1	4292	−15	− 7	− 2	M	95	92	89	31	70	69	67
NEVADA†														
Carson City	39	1	4675	− 4	3	7	VL	93	91	88	42	62	61	60
Elko AP	40	5	5075	−21	−13	− 7	VL	94	92	90	42	64	62	61
Ely AP	39	1	6257	−15	− 6	− 2	VL	90	88	86	39	60	59	58
Las Vegas AP	36	1	2162	18	23	26	VL	108	106	104	30	72	71	70
Lovelock AP	40	0	3900	0	7	11	VL	98	96	93	42	65	64	62
Reno AP	39	3	4404	− 2	2	7	VL	95	92	90	45	64	62	61
Reno CO	39	3	4490	8	12	17	VL	94	92	89	45	64	62	61
Tonopah AP	38	0	5426	2	9	13	VL	95	92	90	40	64	63	62
Winnemucca AP	40	5	4299	− 8	1	5	VL	97	95	93	42	64	62	61
NEW HAMPSHIRE														
Berlin	44	3	1110	−25	−19	−15	L	87	85	82	22	73	71	70
Claremont	43	2	420	−19	−13	− 9	L	89	87	84	24	74	73	72
Concord AP	43	1	339	−17	−11	− 7	M	91	88	85	26	75	73	72
Keene	43	0	490	−17	−12	− 8	M	90	88	85	24	75	73	72
Laconia	43	3	505	−22	−16	−12	M	89	87	84	25	74	73	72
Manchester, Grenier AFB	43	0	253	−11	− 5	1	M	92	89	86	24	76	74	73
Portsmouth, Pease AFB	43	1	127	− 8	− 2	3	M	88	86	83	22	75	73	72
NEW JERSEY														
Atlantic City CO	39	3	11	10	14	18	H	91	88	85	18	78	77	76
Long Branch	40	2	20	4	9	13	H	93	91	88	18	77	76	75
Newark AP	40	4	11	6	11	15	M	94	91	88	20	77	76	75
New Brunswick	40	3	86	3	8	12	M	91	89	86	19	77	76	75
Paterson	40	5	100	3	8	12	M	93	91	88	21	77	76	75
Phillipsburg	40	4	180	1	6	10	L	93	91	88	21	77	76	75
Trenton CO	40	1	144	7	12	16	M	92	90	87	19	78	77	76
Vineland	39	3	95	7	12	16	M	93	90	87	19	78	77	76
NEW MEXICO														
Alamagordo, Holloman AFB	32	5	4070	12	18	22	L	100	98	96	30	70	69	68
Albuquerque AP	35	0	5310	6	14	17	L	96	94	92	27	66	65	64
Artesia	32	5	3375	9	16	19	L	101	99	97	30	71	70	69
Carlsbad AP	32	2	3234	11	17	21	L	101	99	97	28	72	71	70
Clovis AP	34	3	4279	2	14	17	L	99	97	95	28	70	69	68
Farmington AP	36	5	5495	− 3	6	9	VL	95	93	91	30	66	65	64
Gallup	35	3	6465	−13	− 5	− 1	VL	92	90	87	32	64	63	62
Grants	35	1	6520	−15	− 7	− 3	VL	91	89	86	32	64	63	62
Hobbs AP	32	4	3664	9	15	19	L	101	99	96	29	72	71	70
Las Cruces	32	2	3900	13	19	23	L	102	100	97	30	70	69	68
Los Alamos	35	5	7410	− 4	5	9	L	88	86	83	32	64	63	62
Raton AP	36	5	6379	−11	− 2	2	L	92	90	88	34	66	65	64
Roswell, Walker AFB	33	2	3643	5	16	19	L	101	99	97	33	71	70	69
Santa Fe CO	35	4	7045	− 2	7	11	L	90	88	85	28	65	63	62
Silver City AP	32	4	5373	8	14	18	VL	95	93	91	30	68	67	66
Socorro AP	34	0	4617	6	13	17	L	99	97	94	30	67	66	65
Tucumcari AP	35	1	4053	1	9	13	L	99	97	95	28	71	70	69
NEW YORK														
Albany AP	42	5	277	−14	− 5	0	L	91	88	85	23	76	74	73
Albany CO	42	5	19	− 5	1	5	L	91	89	86	20	76	74	73
Auburn	43	0	715	−10	− 2	2	M	89	87	84	22	75	73	72
Batavia	43	0	900	− 7	− 1	3	M	89	87	84	22	75	74	72
Binghamton CO	42	1	858	− 8	− 2	2	L	91	89	86	20	74	72	71
Buffalo AP	43	0	705r	− 3	3	6	M	88	86	83	21	75	73	72
Cortland	42	4	1129	−11	− 5	− 1	L	90	88	85	23	75	73	72
Dunkirk	42	3	590	− 2	4	8	M	88	86	83	18	75	74	72
Elmira AP	42	1	860	− 5	1	5	L	92	90	87	24	75	73	72
Geneva	42	5	590	− 8	− 2	2	M	91	89	86	22	75	73	72
Glens Falls	43	2	321	−17	−11	− 7	L	88	86	83	23	74	72	71
Gloversville	43	1	770	−12	− 6	− 2	L	89	87	84	23	75	73	71
Hornell	42	2	1325	−15	− 9	− 5	L	87	85	82	24	74	72	71

Reprinted by permission from ASHRAE Handbook of Fundamentals (1967).

TABLE 3–33 Climatic Conditions for United States and Canada

Col. 1 State and Station[b]	Col. 2 Latitude[e] ° '		Col. 3 Elev,[d] Ft	Col. 4 Winter — Median of Annual Extremes	99%	97½%	Col. 5 Coincident Wind Velocity[e]	Col. 6 Design Dry-Bulb 1%	2½%	5%	Col. 7 Outdoor Daily Range[f]	Col. 8 Design Wet-Bulb 1%	2½%	5%
NEW YORK (continued)														
Ithaca	42	3	950	−10	−4	0	L	91	88	85	24	75	73	72
Jamestown	42	1	1390	−5	1	5	M	88	86	83	20	75	73	72
Kingston	42	0	279	−8	−2	2	L	92	90	87	22	76	74	73
Lockport	43	1	520	−4	2	6	M	87	85	82	21	75	74	72
Massena AP	45	0	202r	−22	−16	−12	M	86	84	81	20	75	74	72
Newburgh-Stewart AFB	41	3	460	−4	2	6	M	92	89	86	21	78	76	74
NYC-Central Park	40	5	132	6	11	15	H	94	91	88	17	77	76	75
NYC-Kennedy AP	40	4	16	12	17	21	H	91	87	84	16	77	76	75
NYC-LaGuardia AP	40	5	19	7	12	16	H	93	90	87	16	77	76	75
Niagara Falls AP	43	1	596	−2	4	7	M	88	86	83	20	75	74	73
Olean	42	1	1420	−13	−8	−3	L	87	85	82	23	74	72	71
Oneonta	42	3	1150	−13	−7	−3	L	89	87	84	24	74	72	71
Oswego CO	43	3	300	−4	2	6	M	86	84	81	20	75	74	72
Plattsburg AFB	44	4	165	−16	−10	−6	L	86	84	81	22	74	73	71
Poughkeepsie	41	4	103	−6	−1	3	L	93	90	87	21	77	75	74
Rochester AP	43	1	543	−5	2	5	M	91	88	85	22	75	74	72
Rome-Griffiss AFB	43	1	515	−13	−7	−3	L	90	87	84	22	76	74	73
Schenectady	42	5	217	−11	−5	−1	L	90	88	85	22	75	73	72
Suffolk County AFB	40	5	57	4	9	13	H	87	84	81	16	76	75	74
Syracuse AP	43	1	424	−10	−2	2	M	90	87	85	20	76	74	73
Utica	43	1	714	−12	−6	−2	L	89	87	84	22	75	73	72
Watertown	44	0	497	−20	−14	−10	M	86	84	81	20	75	74	72
NORTH CAROLINA														
Asheville AP	35	3	2170r	8	13	17	L	91	88	86	21	75	74	73
Charlotte AP	35	1	735	13	18	22	L	96	94	92	20	78	77	76
Durham	36	0	406	11	15	19	L	94	92	89	20	78	77	76
Elizabeth City AP	36	2	10	14	18	22	M	93	91	89	18	80	79	78
Fayetteville, Pope AFB	35	1	95	13	17	20	L	97	94	92	20	80	79	78
Goldsboro, Seymour-Johnson AFB	35	2	88	14	18	21	M	95	92	90	18	80	79	78
Greensboro AP	36	1	897	9	14	17	L	94	91	89	21	77	76	75
Greenville	35	4	25	14	18	22	M	95	93	90	19	81	80	79
Henderson	36	2	510	8	12	16	L	94	92	89	20	79	78	77
Hickory	35	4	1165	9	14	18	L	93	91	88	21	77	76	75
Jacksonville	34	5	24	17	21	25	M	94	92	89	18	81	80	79
Lumberton	34	4	132	14	18	22	L	95	93	90	20	81	80	79
New Bern AP	35	1	17	14	18	22	L	94	92	89	18	81	80	79
Raleigh/Durham AP	35	5	433	13	16	20	L	95	92	90	20	79	78	77
Rocky Mount	36	0	81	12	16	20	L	95	93	90	19	80	79	78
Wilmington AP	34	2	30	19	23	27	L	93	91	89	18	82	81	80
Winston-Salem AP	36	1	967	9	14	17	L	94	91	89	20	77	76	75
NORTH DAKOTA														
Bismarck AP	46	5	1647	−31	−24	−19	VL	95	91	88	27	74	72	70
Devil's Lake	48	1	1471	−30	−23	−19	M	93	89	86	25	73	71	69
Dickinson AP	46	5	2595	−31	−23	−19	L	96	93	90	25	72	70	68
Fargo AP	46	5	900	−28	−22	−17	L	92	88	85	25	76	74	72
Grand Forks AP	48	0	832	−30	−26	−23	L	91	87	84	25	74	72	70
Jamestown AP	47	0	1492	−29	−22	−18	L	95	91	88	26	75	73	71
Minot AP	48	2	1713	−31	−24	−20	M	91	88	84	25	72	70	68
Williston	48	1	1877	−28	−21	−17	M	94	90	87	25	71	69	67
OHIO														
Akron/Canton AP	41	0	1210	−5	1	6	M	89	87	84	21	75	73	72
Ashtabula	42	0	690	−3	3	7	M	89	87	84	18	76	75	74
Athens	39	2	700	−3	3	7	M	93	91	88	22	77	76	75
Bowling Green	41	3	675	−7	−1	3	M	93	91	88	23	77	75	74
Cambridge	40	0	800	−6	0	4	M	91	89	86	23	77	76	75
Chillicothe	39	2	638	−1	5	9	M	93	91	88	22	77	76	75
Cincinnati CO	39	1	761	2	8	12	L	94	92	90	21	78	77	76
Cleveland AP	41	2	777r	−2	2	7	M	91	89	86	22	76	75	74
Columbus AP	40	0	812	−1	2	7	M	92	88	86	24	77	76	75
Dayton AP	39	5	997	−2	0	6	M	92	90	87	20	77	75	74
Defiance	41	2	700	−7	−1	1	M	93	91	88	24	77	76	74
Findlay AP	41	0	797	−6	0	4	M	92	90	88	24	77	76	75

Reprinted by permission from ASHRAE Handbook of Fundamentals (1967).

TABLE 3–33   Climatic Conditions for United States and Canada

Col. 1 State and Station[b]	Col. 2 Latitude[c] ° '	Col. 3 Elev,[d] Ft	Winter Col. 4 Median of Annual Extremes	99%	97½%	Col. 5 Coincident Wind Velocity[e]	Summer Col. 6 Design Dry-Bulb 1%	2½%	5%	Col. 7 Outdoor Daily Range[f]	Col. 8 Design Wet-Bulb 1%	2½%	5%
OHIO (continued)													
Fremont	41 2	600	− 7	− 1	3	M	92	90	87	24	76	75	74
Hamilton	39 2	650	− 2	4	8	M	94	92	90	22	78	77	76
Lancaster	39 4	920	− 5	1	5	M	93	91	88	23	77	76	75
Lima	40 4	860	− 6	0	4	M	93	91	88	24	77	76	75
Mansfield AP	40 5	1297	− 7	1	3	M	91	89	86	22	76	75	74
Marion	40 4	920	− 5	1	6	M	93	91	88	23	77	76	75
Middletown	39 3	635	− 3	3	7	M	93	91	88	22	77	76	75
Newark	40 1	825	− 7	− 1	3	M	92	90	87	23	77	76	75
Norwalk	41 1	720	− 7	− 1	3	M	92	90	87	22	76	75	74
Portsmouth	38 5	530	0	5	9	L	94	92	89	22	77	76	75
Sandusky CO	41 3	606	− 2	4	8	M	92	90	87	21	76	75	74
Springfield	40 0	1020	− 3	3	7	M	93	90	88	21	77	76	75
Steubenville	40 2	992	− 2	4	9	M	91	89	86	22	76	75	74
Toledo AP	41 4	676r	− 5	1	5	M	92	90	87	25	77	75	74
Warren	41 2	900	− 6	0	4	M	90	88	85	23	75	74	73
Wooster	40 5	1030	− 7	− 1	3	M	90	88	85	22	76	75	74
Youngstown AP	41 2	1178	− 5	1	6	M	89	86	84	23	75	74	73
Zanesville AP	40 0	881	− 7	− 1	3	M	92	89	87	23	77	76	75
OKLAHOMA													
Ada	34 5	1015	6	12	16	H	102	100	98	23	79	78	77
Altus AFB	34 4	1390	7	14	18	H	103	101	99	25	77	76	75
Ardmore	34 2	880	9	15	19	H	103	101	99	23	79	78	77
Bartlesville	36 5	715	− 1	5	9	H	101	99	97	23	79	78	77
Chickasha	35 0	1085	5	12	16	H	103	101	99	24	77	76	75
Enid-Vance AFB	36 2	1287	3	10	14	H	103	100	98	24	78	77	76
Lawton AP	34 3	1108	6	13	16	H	103	101	98	24	78	77	76
McAlester	34 5	760	7	13	17	H	102	100	98	23	79	78	77
Muskogee AP	35 4	610	6	12	16	M	102	99	96	23	79	78	77
Norman	35 1	1109	5	11	15	H	101	99	97	24	78	77	76
Oklahoma City AP	35 2	1280	4	11	15	H	100	97	95	23	78	77	76
Ponca City	36 4	996	1	8	12	H	102	100	97	24	78	77	76
Seminole	35 2	865	6	12	16	H	102	100	98	23	78	77	76
Stillwater	36 1	884	2	9	13	H	101	99	97	24	78	77	76
Tulsa AP	36 1	650	4	12	16	H	102	99	96	22	79	78	77
Woodward	36 3	1900	− 3	4	8	H	103	101	98	26	76	74	73
OREGON													
Albany	44 4	224	17	23	27	VL	91	88	84	31	69	67	65
Astoria AP	46 1	8	22	27	30	M	79	76	72	16	61	60	59
Baker AP	44 5	3368	−10	− 3	1	VL	94	92	89	30	66	65	63
Bend	44 0	3599	− 7	0	4	VL	89	87	84	33	64	62	61
Corvallis	44 3	221	17	23	27	VL	91	88	84	31	69	67	65
Eugene AP	44 1	364	16	22	26	VL	91	88	84	31	69	67	65
Grants Pass	42 3	925	16	22	26	VL	94	92	89	33	68	66	65
Klamath Falls AP	42 1	4091	− 5	1	5	VL	89	87	84	36	63	62	61
Medford AP	42 2	1298	15	21	23	VL	98	94	91	35	70	68	66
Pendleton AP	45 4	1492	− 2	3	10	VL	97	94	91	29	66	65	63
Portland AP	45 4	21	17	21	24	L	89	85	81	23	69	67	66
Portland CO	45 3	57	21	26	29	L	91	88	84	21	69	68	67
Roseburg AP	43 1	505	19	25	29	VL	93	91	88	30	69	67	65
Salem AP	45 0	195	15	21	25	VL	92	88	84	31	69	67	66
The Dalles	45 4	102	7	13	17	VL	93	91	88	28	70	68	67
PENNSYLVANIA													
Allentown AP	40 4	376	− 2	3	5	M	92	90	87	22	77	75	74
Altoona CO	40 2	1468	− 4	1	5	L	89	87	84	23	74	73	72
Butler	40 4	1100	− 8	− 2	2	L	91	89	86	22	75	74	73
Chambersburg	40 0	640	0	5	9	L	94	92	89	23	76	75	74
Erie AP	42 1	732	1	7	11	M	88	85	82	18	76	74	73
Harrisburg AP	40 1	335	4	9	13	L	92	89	86	21	76	75	74
Johnstown	40 2	1214	− 4	1	5	L	91	87	85	23	74	73	72
Lancaster	40 1	255	− 3	2	6	L	92	90	87	22	77	76	75
Meadville	41 4	1065	− 6	0	4	M	88	86	83	21	75	73	72
New Castle	41 0	825	− 7	− 1	4	M	91	89	86	23	75	74	73

Reprinted by permission from ASHRAE Handbook of Fundamentals (1967).

TABLE 3-33 Climatic Conditions for United States and Canada

Col. 1 State and Station[b]	Col. 2 Latitude[c] °	'	Col. 3 Elev,[d] Ft	Winter Col. 4 Median of Annual Extremes	99%	97½%	Col. 5 Coincident Wind Velocity[e]	Summer Col. 6 Design Dry-Bulb 1%	2½%	5%	Col. 7 Outdoor Daily Range[f]	Col. 8 Design Wet-Bulb 1%	2½%	5%
PENNSYLVANIA (continued)														
Philadelphia AP	39	5	7	7	11	15	M	93	90	87	21	78	77	76
Pittsburgh AP	40	3	1137	− 1	5	9	M	90	87	85	22	75	74	73
Pittsburgh CO	40	3	749 r	1	7	11	M	90	88	85	19	75	74	73
Reading CO	40	2	226	1	6	9	M	92	90	87	19	77	76	75
Scranton/Wilkes-Barre	41	2	940	− 3	2	6	L	89	87	84	19	75	74	73
State College	40	5	1175	− 3	2	6	L	89	87	84	23	74	73	72
Sunbury	40	5	480	− 2	3	7	L	91	89	86	22	76	75	74
Uniontown	39	5	1040	− 1	4	8	L	90	88	85	22	75	74	73
Warren	41	5	1280	− 8	− 3	1	L	89	87	84	24	75	73	72
West Chester	40	0	440	4	9	13	M	92	90	87	20	77	76	75
Williamsport AP	41	1	527	− 5	1	5	L	91	89	86	23	76	75	74
York	40	0	390	− 1	4	8	L	93	91	88	22	77	76	75
RHODE ISLAND														
Newport	41	3	20	1	5	11	H	86	84	81	16	75	74	73
Providence AP	41	4	55	0	6	10	M	89	86	83	19	76	75	74
SOUTH CAROLINA														
Anderson	34	3	764	13	18	22	L	96	94	91	21	77	76	75
Charleston AFB	32	5	41	19	23	27	L	94	92	90	18	81	80	79
Charleston CO	32	5	9	23	26	30	L	95	93	90	13	81	80	79
Columbia AP	34	0	217	16	20	23	L	98	96	94	22	79	79	78
Florence AP	34	1	146	16	21	25	L	96	94	92	21	80	79	78
Georgetown	33	2	14	19	23	26	L	93	91	88	18	81	80	79
Greenville AP	34	5	957	14	19	23	L	95	93	91	21	77	76	75
Greenwood	34	1	671	15	19	23	L	97	95	92	21	78	77	76
Orangeburg	33	3	244	17	21	25	L	97	95	92	20	80	79	78
Rock Hill	35	0	470	13	17	21	L	97	95	92	20	78	77	76
Spartanburg AP	35	0	816	13	18	22	L	95	93	90	20	77	76	75
Sumter-Shaw AFB	34	0	291	18	23	26	L	96	94	92	21	80	79	78
SOUTH DAKOTA														
Aberdeen AP	45	3	1296	−29	−22	−18	L	95	92	89	27	77	75	74
Brookings	44	2	1642	−26	−19	−15	M	93	90	87	25	77	75	74
Huron AP	44	3	1282	−24	−16	−12	L	97	93	90	28	77	75	74
Mitchell	43	5	1346	−22	−15	−11	M	96	94	91	28	77	76	74
Pierre AP	44	2	1718 r	−21	−13	− 9	M	98	96	93	29	76	74	73
Rapid City AP	44	0	3165	−17	− 9	− 6	M	96	94	91	28	72	71	69
Sioux Falls AP	43	4	1420	−21	−14	−10	M	95	92	89	24	77	75	74
Watertown AP	45	0	1746	−27	−20	−16	L	93	90	87	26	76	74	73
Yankton	43	0	1280	−18	−11	− 7	M	96	94	91	25	78	76	75
TENNESSEE														
Athens	33	3	940	10	14	18	L	96	94	91	22	77	76	75
Bristol-Tri City AP	36	3	1519	6	11	16	L	92	90	88	22	76	75	74
Chattanooga AP	35	0	670	11	15	19	L	97	94	92	22	78	78	77
Clarksville	36	4	470	6	12	16	L	98	96	94	21	79	78	77
Columbia	35	4	690	8	13	17	L	97	95	93	21	79	78	77
Dyersburg	36	0	334	7	13	17	L	98	96	94	21	80	79	78
Greenville	35	5	1320	5	10	14	L	93	91	88	22	76	75	74
Jackson AP	35	4	413	8	14	17	L	97	95	94	21	80	79	78
Knoxville AP	35	5	980	9	13	17	L	95	92	90	21	77	76	75
Memphis AP	35	0	263	11	17	21	L	98	96	94	21	80	79	78
Murfreesboro	35	5	608	7	13	17	L	97	94	92	22	79	78	77
Nashville AP	36	1	577	6	12	16	L	97	95	92	21	79	78	77
Tullahoma	35	2	1075	7	13	17	L	96	94	92	22	79	78	77
TEXAS														
Abilene AP	32	3	1759	12	17	21	M	101	99	97	22	76	75	74
Alice AP	27	4	180	26	30	34	M	101	99	97	20	81	80	79
Amarillo AP	35	1	3607	2	8	12	M	98	96	93	26	72	71	70
Austin AP	30	2	597	19	25	29	M	101	98	96	22	79	78	77
Bay City	29	0	52	25	29	33	M	95	93	91	16	81	80	79
Beaumont	30	0	18	25	29	33	M	96	94	93	19	81	80	79
Beeville	28	2	225	24	28	32	M	99	97	96	18	81	80	79
Big Spring AP	32	2	2537	12	18	22	M	100	98	96	26	75	73	72
Brownsville AP	25	5	16	32	36	40	M	94	92	91	18	80	80	79
Brownwood	31	5	1435	15	20	25	M	102	100	98	22	76	75	74
Bryan AP	30	4	275	22	27	31	M	100	98	96	20	79	78	78

Reprinted by permission from ASHRAE Handbook of Fundamentals (1967).

TABLE 3-33 Climatic Conditions for United States and Canada

Col. 1 State and Station[b]	Col. 2 Latitude[e] ° '		Col. 3 Elev,[d] Ft	Winter Col. 4 Median of Annual Extremes	99%	97½%	Col. 5 Coincident Wind Velocity[e]	Summer Col. 6 Design Dry-Bulb 1%	2½%	5%	Col. 7 Outdoor Daily Range[f]	Col. 8 Design Wet-Bulb 1%	2½%	5%
TEXAS (continued)														
Corpus Christi AP	27	5	43	28	32	36	M	95	93	91	19	81	80	80
Corsicana	32	0	425	16	21	25	M	102	100	98	21	79	78	77
Dallas AP	32	5	481	14	19	24	H	101	99	97	20	79	78	78
Del Rio, Laughlin AFB	29	2	1072	24	28	31	M	101	99	98	24	79	77	76
Denton	33	1	655	12	18	22	H	102	100	98	22	79	78	77
Eagle Pass	28	5	743	23	27	31	L	106	104	102	24	80	79	78
El Paso AP	31	5	3918	16	21	25	L	100	98	96	27	70	69	68
Fort Worth AP	32	5	544r	14	20	24	H	102	100	98	22	79	78	77
Galveston AP	29	2	5	28	32	36	M	91	89	88	10	82	81	81
Greenville	33	0	575	13	19	24	H	101	99	97	21	79	78	78
Harlingen	26	1	37	30	34	38	M	96	95	94	19	80	80	79
Houston AP	29	4	50	23	28	32	M	96	94	92	18	80	80	79
Houston CO	29	5	158r	24	29	33	M	96	94	92	18	80	80	79
Huntsville	30	4	494	22	27	31	M	99	97	96	20	80	79	78
Killeen-Gray AFB	31	0	1021	17	22	26	M	100	99	97	22	78	77	76
Lamesa	32	5	2965	7	14	18	M	100	98	96	26	74	73	72
Laredo AFB	27	3	503	29	32	36	L	103	101	100	23	79	78	78
Longview	32	2	345	16	21	25	M	100	98	96	20	81	80	79
Lubbock AP	33	4	3243	4	11	15	M	99	97	94	26	73	72	71
Lufkin AP	31	1	286	19	24	28	M	98	96	95	20	81	80	79
McAllen	26	1	122	30	34	38	M	102	100	98	21	80	79	78
Midland AP	32	0	2815r	13	19	23	M	100	98	96	26	74	73	72
Mineral Wells AP	32	5	934	12	18	22	H	102	100	98	22	78	77	76
Palestine CO	31	5	580	16	21	25	M	99	97	96	20	80	79	78
Pampa	35	3	3230	0	7	11	M	100	98	95	26	73	72	71
Pecos	31	2	2580	10	15	19	L	102	100	97	27	72	71	70
Plainview	34	1	3400	3	10	14	M	100	98	95	26	73	72	71
Port Arthur AP	30	0	16	25	29	33	M	94	92	91	19	81	80	80
San Angelo, Goodfellow AFB	31	2	1878	15	20	25	M	101	99	97	24	76	75	74
San Antonio AP	29	3	792	22	25	30	L	99	97	96	19	77	77	76
Sherman-Perrin AFB	33	4	763	12	18	23	H	101	99	97	22	79	78	77
Snyder	32	4	2325	9	15	19	M	102	100	97	26	75	74	73
Temple	31	1	675	18	23	27	M	101	99	97	22	79	78	77
Tyler AP	32	2	527	15	20	24	M	99	97	96	21	80	79	78
Vernon	34	1	1225	7	14	18	H	103	101	99	24	77	76	75
Victoria AP	28	5	104	24	28	32	M	98	96	95	18	80	79	79
Waco AP	31	4	500	16	21	26	M	101	99	98	22	79	78	78
Wichita Falls AP	34	0	994	9	15	19	H	103	100	98	24	77	76	75
UTAH														
Cedar City AP	37	4	5613	−10	− 1	6	VL	94	91	89	32	65	64	62
Logan	41	4	4775	− 7	3	7	VL	93	91	89	33	66	65	63
Moab	38	5	3965	2	12	16	VL	100	98	95	30	66	65	64
Ogden CO	41	1	4400	− 3	7	11	VL	94	92	89	33	66	65	64
Price	39	4	5580	− 7	3	7	L	93	91	88	33	65	64	63
Provo	40	1	4470	− 6	2	6	L	96	93	91	32	67	66	65
Richfield	38	5	5300	−10	− 1	3	L	94	92	89	34	66	65	64
St. George CO	37	1	2899	13	22	26	VL	104	102	99	33	71	70	69
Salt Lake City AP	40	5	4220	− 2	5	9	L	97	94	92	32	67	66	65
Vernal AP	40	3	5280	−20	−10	− 6	VL	90	88	84	32	64	63	62
VERMONT														
Barre	44	1	1120	−23	−17	−13	L	86	84	81	23	73	72	70
Burlington AP	44	3	331	−18	−12	− 7	M	88	85	83	23	74	73	71
Rutland	43	3	620	−18	−12	− 8	L	87	85	82	23	74	73	71
VIRGINIA														
Charlottsville	38	1	870	7	11	15	L	93	90	88	23	79	77	76
Danville AP	36	3	590	9	13	17	L	95	92	90	21	78	77	76
Fredericksburg	38	2	50	6	10	14	M	94	92	89	21	79	78	76
Harrisonburg	38	3	1340	0	5	9	L	92	90	87	23	78	77	76
Lynchburg AP	37	2	947	10	15	19	L	94	92	89	21	77	76	75
Norfolk AP	36	5	26	18	20	23	M	94	91	89	18	79	78	78
Petersburg	37	1	194	10	15	18	L	96	94	91	20	80	79	78

Reprinted by permission from ASHRAE Handbook of Fundamentals (1967).

74

TABLE 3–33 Climatic Conditions for United States and Canada

Col. 1 State and Station[b]	Col. 2 Latitude[e] ° ′		Col. 2 Elev.[d] Ft	Winter			Col. 5 Coincident Wind Velocity[e]	Summer						
				Col. 4 Median of Annual Extremes	99%	97½%		Col. 6 Design Dry-Bulb 1%	2½%	5%	Col. 6 Outdoor Daily Range[f]	Col. 8 Design Wet-Bulb 1%	2½%	5%
VIRGINIA (continued)														
Richmond AP	37	3	162	10	14	18	L	96	93	91	21	79	78	77
Roanoke AP	37	2	1174r	9	15	18	L	94	91	89	23	76	75	74
Staunton	38	2	1480	3	8	12	L	92	90	87	23	78	77	75
Winchester	39	1	750	1	6	10	L	94	92	89	21	78	76	75
WASHINGTON														
Aberdeen	47	0	12	19	24	27	M	83	80	77	16	62	61	60
Bellingham AP	48	5	150	8	14	18	L	76	74	71	19	67	65	63
Bremerton	47	3	162	17	24	29	L	85	81	77	20	68	66	65
Ellensburg AP	47	0	1729	− 5	2	6	VL	91	89	86	34	67	65	63
Everett-Paine AFB	47	5	598	13	19	24	L	82	78	74	20	67	65	63
Kennewick	46	0	392	4	11	15	VL	98	96	93	30	69	68	66
Longview	46	1	12	14	20	24	L	88	86	83	30	68	66	65
Moses Lake, Larson AFB	47	1	1183	−14	− 7	− 1	VL	96	93	90	32	68	66	65
Olympia AP	47	0	190	15	21	25	L	85	83	80	32	67	65	63
Port Angeles	48	1	99	20	26	29	M	75	73	70	18	60	58	57
Seattle-Boeing Fld	47	3	14	17	23	27	L	82	80	77	24	67	65	64
Seattle CO	47	4	14	22	28	32	L	81	79	76	19	67	65	64
Seattle-Tacoma AP	47	3	386	14	20	24	L	85	81	77	22	66	64	63
Spokane AP	47	4	2357	− 5	− 2	4	VL	93	90	87	28	66	64	63
Tacoma-McChord AFB	47	1	350	14	20	24	L	85	81	78	22	68	66	64
Walla Walla AP	46	1	1185	5	12	16	VL	98	96	93	27	69	68	66
Wenatchee	47	2	634	− 2	5	9	VL	95	92	89	32	68	66	64
Yakima AP	46	3	1061	− 1	6	10	VL	94	92	89	36	69	67	65
WEST VIRGINIA														
Beckley	37	5	2330	− 4	0	6	L	91	88	86	22	74	73	72
Bluefield AP	37	2	2850	1	6	10	L	88	86	83	22	74	73	72
Charleston AP	38	2	939	1	9	14	L	92	90	88	20	76	75	74
Clarksburg	39	2	977	− 2	3	7	L	92	90	87	21	76	75	74
Elkins AP	38	5	1970	− 4	1	5	L	87	84	82	22	74	73	72
Huntington CO	38	2	565r	4	10	14	L	95	93	91	22	77	76	75
Martinsburg AP	39	2	537	1	6	10	L	96	94	91	21	78	77	76
Morgantown AP	39	4	1245	− 2	3	7	L	90	88	85	22	76	74	73
Parkersburg CO	39	2	615r	2	8	12	L	93	91	88	21	77	76	75
Wheeling	40	1	659	0	5	9	L	91	89	86	21	76	75	74
WISCONSIN														
Appleton	44	2	742	−16	−10	− 6	M	89	87	84	23	75	74	72
Ashland	46	3	650	−27	−21	−17	L	85	83	80	23	73	71	69
Beloit	42	3	780	−13	− 7	− 3	M	92	90	87	24	77	76	75
Eau Claire AP	44	5	888	−21	−15	−11	L	90	88	85	23	76	74	72
Fond du Lac	43	5	760	−17	−11	− 7	M	89	87	84	23	76	74	73
Green Bay AP	44	3	683	−16	−12	− 7	M	88	85	82	23	75	73	72
La Crosse AP	43	5	652	−18	−12	− 8	M	90	88	85	22	78	76	75
Madison AP	43	1	858	−13	− 9	− 5	M	92	88	85	22	77	75	73
Manitowoc	44	1	660	−11	− 5	− 1	M	88	86	83	21	75	74	72
Marinette	45	0	605	−14	− 8	− 4	M	88	86	83	20	74	72	70
Milwaukee AP	43	0	672	−11	− 6	− 2	M	90	87	84	21	77	75	73
Racine	42	4	640	−10	− 4	0	M	90	88	85	21	77	75	73
Sheboygan	43	4	648	−10	− 4	0	M	89	87	84	20	76	74	72
Stevens Point	44	3	1079	−22	−16	−12	M	89	87	84	23	75	73	71
Waukesha	43	0	860	−12	− 6	− 2	M	91	89	86	22	77	75	74
Wausau AP	44	6	1196	−24	−18	−14	M	89	86	83	23	74	72	70
WYOMING														
Casper AP	42	5	5319	−20	−11	− 5	L	92	90	87	31	63	62	60
Cheyenne AP	41	1	6126	−15	− 6	− 2	M	89	86	83	30	63	62	61
Cody AP	44	3	5090	−23	−13	− 9	L	90	87	84	32	61	60	59
Evanston	41	2	6860	−22	−12	− 8	VL	84	82	79	32	58	57	56
Lander AP	42	5	5563	−26	−16	−12	VL	92	90	87	32	63	62	60
Laramie AP	41	2	7266	−17	− 6	− 2	M	82	80	77	28	61	59	58
Newcastle	43	5	4480	−18	− 9	− 5	M	92	89	86	30	68	67	66
Rawlins	41	5	6736	−24	−15	−11	L	86	84	81	40	62	61	60
Rock Springs AP	41	4	6741	−16	− 6	− 1	VL	86	84	82	32	58	57	56
Sheridan AP	44	5	3942	−21	−12	− 7	L	95	92	89	32	67	65	64
Torrington	42	0	4098	−20	−11	− 7	M	94	92	89	30	68	67	66

Reprinted by permission from ASHRAE Handbook of Fundamentals (1967).

TABLE 3–33   Climatic Conditions for United States and Canada

CANADA

Col. 1 Province and Station[b]	Col. 2 Latitude[c] ° '		Col. 3 Elev,[d] Ft	Winter Col. 4 Average Annual Minimum	99%	97½%	Col. 5 Coincident Wind Velocity[e]	Summer Col. 6 Design Dry-Bulb 1%	2½%	5%	Col. 7 Outdoor Daily Range[f]	Col. 8 Design Wet-Bulb 1%	2½%	5%
ALBERTA														
Calgary AP	51	1	3540	−30	−29	−25	M	87	85	82	26	66	64	63
Edmonton AP	53	3	2219	−30	−29	−26	VL	86	83	80	23	69	67	65
Grande Prairie AP	55	1	2190	−44	−43	−37	VL	84	81	78	23	66	64	63
Jasper CO	52	5	3480	−38	−32	−28	VL	87	84	81	28	66	64	63
Lethbridge AP	49	4	3018	−31	−31	−24	M	91	88	85	28	68	66	64
McMurray AP	56	4	1216	−44	−42	−39	VL	87	84	81	28	69	67	65
Medicine Hat AP	50	0	2365	−33	−30	−26	M	96	93	90	28	72	69	67
Red Deer AP	52	1	2965	−38	−33	−28	VL	88	86	83	25	67	65	64
BRITISH COLUMBIA														
Dawson Creek	55	5	2200	−47	−40	−35	L	84	81	78	25	66	64	63
Fort Nelson AP	58	5	1230	−43	−44	−41	VL	87	84	81	23	66	64	63
Kamloops CO	50	4	1150	−15	−16	−10	VL	97	94	91	31	71	69	68
Nanaimo CO	49	1	100	16	17	20	VL	81	78	75	20	66	64	62
New Westminster CO	49	1	50	12	15	19	VL	86	84	82	20	68	66	65
Penticton AP	49	3	1121	0	− 1	3	L	94	91	88	31	71	69	68
Prince George AP	53	5	2218	−38	−37	−31	VL	85	82	79	26	68	65	63
Prince Rupert CO	54	2	170	9	11	15	L	73	71	69	13	62	60	59
Trail	49	1	1400	− 3	− 2	3	VL	94	91	88	30	70	68	67
Vancouver AP	49	1	16	13	15	19	L	80	78	76	17	68	66	65
Victoria CO	48	3	228	20	20	23	M	80	76	72	16	64	62	60
MANITOBA														
Brandon CO	49	5	1200	−36	−29	−26	M	90	87	84	26	75	73	71
Churchill AP	58	5	115	−43	−40	−38	H	79	75	72	18	68	66	63
Dauphin AP	51	1	999	−35	−29	−26	M	89	86	83	24	74	72	70
Flin Flon CO	54	5	1098	−38	−40	−36	L	85	81	78	19	71	69	67
Portage la Prairie AP	49	5	867	−28	−25	−22	M	90	87	84	22	75	74	72
The Pas AP	54	0	894	−41	−35	−32	M	85	81	78	20	73	71	69
Winnipeg AP	49	5	786	−31	−28	−25	M	90	87	84	23	75	74	72
NEW BRUNSWICK														
Campbellton CO	48	0	25	−20	−18	−14	L	87	84	81	20	74	71	69
Chatham AP	47	0	112	−17	−15	−10	M	90	87	84	22	74	71	69
Edmundston CO	47	2	500	−29	−20	−16	M	84	81	78	21	75	72	70
Fredericton AP	45	5	74	−19	−16	−10	L	89	86	83	23	73	70	68
Moncton AP	46	1	248	−16	−12	− 7	H	88	85	82	21	74	71	69
Saint John AP	45	2	352	−15	−12	− 7	M	81	79	77	18	71	68	66
NEWFOUNDLAND														
Corner Brook CO	49	0	40	− 9	−10	− 5	H	84	81	79	18	69	68	66
Gander AP	49	0	482	− 5	− 5	− 1	H	85	82	79	20	69	68	66
Goose Bay AP	53	2	144	−28	−27	−25	M	86	81	77	18	69	67	65
St. John's AP	47	4	463	1	2	6	H	79	77	75	17	69	68	66
Stephenville	48	3	44	− 4	− 6	− 1	H	79	76	74	13	69	68	66
NORTHWEST TERRITORIES														
Fort Smith AP	60	0	665	−51	−49	−46	VL	85	83	80	25	67	65	64
Frobisher Bay AP	63	5	68	−45	−45	−42	H	63	59	56	14			
Inuvik	68	2	75	−54	−50	−48	VL	80	77	75	23	63	61	60
Resolute AP	74	4	209	−52	−49	−47	M	54	51	49	10			
Yellowknife AP	62	3	682	−51	−49	−47	VL	78	76	74	17	65	63	62
NOVA SCOTIA														
Amherst	45	5	63	−15	−10	− 5	H	85	82	79	21	72	70	68
Halifax AP	44	4	136	− 4	0	4	H	83	80	77	16	69	68	67
Kentville CO	45	0	50	− 8	− 4	0	M	86	83	80	23	72	70	69
New Glasgow	45	4	317	−16	−10	− 5	H	84	81	79	21	72	70	68
Sydney AP	46	1	197	− 3	0	5	H	84	82	80	20	72	70	68
Truro CO	45	2	77	−17	−12	− 7	M	84	81	79	22	72	70	69
Yarmouth AP	43	5	136	2	5	9	H	76	73	71	15	69	68	67
ONTARIO														
Belleville CO	44	1	250	−15	−11	− 7	M	89	86	84	21	77	75	73
Chatham CO	42	2	600	− 1	3	6	M	92	90	88	20	77	75	74
Cornwall	45	0	210	−22	−14	− 9	M	89	86	84	23	77	75	74
Fort William AP	48	2	644	−31	−27	−23	L	86	83	80	23	72	70	68
Hamilton	43	2	303	− 2	0	3	M	91	88	86	21	77	75	73
Kapuskasing AP	49	3	752	−37	−31	−28	M	87	84	81	23	73	71	69
Kenora AP	49	5	1345	−33	−31	−28	M	86	83	80	20	75	73	71
Kingston CO	44	2	300	−16	−10	− 7	M	85	82	80	20	77	75	73

Reprinted by permission from ASHRAE Handbook of Fundamentals (1967).

TABLE 3–33 Climatic Conditions for United States and Canada

Col. 1 Province and Station[b]	Col. 2 Latitude[c] ° ′	Col. 3 Elev,[d] Ft	Col. 4 Winter — Average Annual Minimum	99%	97½%	Col. 5 Coincident Wind Velocity[e]	Col. 6 Design Dry-Bulb 1%	2½%	5%	Col. 7 Outdoor Daily Range[f]	Col. 8 Design Wet-Bulb 1%	2½%	5%
ONTARIO (continued)													
Kitchener	43 3	1125	−11	− 3	1	M	88	85	83	24	76	75	74
London AP	43 0	912	− 9	− 1	3	M	90	88	86	22	76	75	74
North Bay AP	46 2	1210	−27	−21	−17	M	87	84	82	18	71	70	69
Oshawa	43 5	370	−11	− 5	− 2	M	90	87	85	21	77	75	73
Ottawa AP	45 2	339	−21	−17	−13	M	90	87	84	21	75	74	73
Owen Sound	44 3	597	− 9	− 5	− 1	M	87	84	82	21	74	72	71
Peterborough CO	44 2	648	−20	−13	− 9	M	90	87	85	22	76	74	73
St. Catharines CO	43 1	325	1	2	5	M	91	88	86	20	77	75	73
Sarnia	43 0	625	− 6	2	6	M	92	90	88	19	76	74	73
Sault Ste. Marie CO	46 3	675	−21	−20	−15	M	88	85	83	22	72	70	68
Sudbury	46 3	850	−25	−20	−15	VL	89	86	84	25	72	70	69
Timmins CO	48 3	1100	−37	−33	−28	M	90	87	84	24	73	71	69
Toronto AP	43 4	578	−10	− 3	1	M	90	87	85	22	77	75	73
Windsor AP	42 2	637	− 1	4	7	M	92	90	88	20	77	75	74
PRINCE EDWARD ISLAND													
Charlottetown AP	46 2	186	−11	− 6	− 3	H	84	81	79	16	72	70	68
Summerside AP	46 3	78	−10	− 8	− 3	H	84	81	79	16	72	70	68
QUEBEC													
Bagotville	48 2	536	−35	−26	−22	VL	88	84	81	20	72	71	69
Chicoutimi CO	48 3	150	−31	−24	−20	VL	87	83	80	20	72	71	69
Drummondville CO	45 5	270	−26	−18	−13	M	88	85	82	22	76	74	72
Granby	45 2	550	−23	−17	−12	L	87	84	82	21	76	74	72
Hull	45 3	200	−21	−17	−13	M	90	87	84	21	75	74	73
Mégantic AP	45 4	1362	−27	−20	−16	M	84	81	78	19	75	73	71
Montréal AP	45 3	98	−20	−16	−10	M	88	86	84	18	76	74	73
Québec AP	46 5	245	−25	−19	−13	M	86	82	79	21	75	73	71
Rimouski	48 3	117	−18	−16	−12	H	78	74	71	18	71	69	68
St. Jean	45 2	129	−21	−15	−10	M	87	85	83	20	76	74	73
St. Jérôme	45 5	310	−30	−18	−13	L	87	84	82	23	76	74	73
Sept Îles AP	50 1	190	−29	−27	−22	L	80	78	75	17	66	64	63
Shawinigan	46 3	306	−27	−20	−15	L	88	85	83	21	76	74	72
Sherbrooke CO	45 2	595	−25	−18	−13	L	87	84	81	20	75	73	71
Thetford Mines	46 0	1020	−25	−19	−14	M	86	83	80	22	75	73	71
Trois Rivières CO	46 2	200	−30	−18	−13	M	88	85	82	23	76	74	72
Val d'Or AP	48 0	1108	−37	−31	−27	L	88	85	82	22	72	71	69
Valleyfield	45 2	150	−20	−14	− 9	M	87	85	83	21	76	74	73
SASKATCHEWAN													
Estevan AP	49 0	1884	−32	−30	−25	M	93	89	86	25	75	73	71
Moose Jaw AP	50 2	1857	−33	−32	−27	M	93	89	86	27	73	71	69
North Battleford AP	52 5	1796	−33	−33	−29	L	90	86	83	25	71	69	67
Prince Albert AP	53 1	1414	−45	−41	−35	VL	88	84	81	25	72	70	68
Regina AP	50 3	1884	−38	−34	−29	M	92	88	85	27	73	71	69
Saskatoon AP	52 1	1645	−37	−34	−30	M	90	86	83	25	71	69	67
Swift Current AP	50 2	2677	−31	−29	−25	M	93	89	86	24	72	70	68
Yorkton AP	51 2	1653	−38	−33	−28	M	89	85	82	23	74	72	70
YUKON TERRITORY													
Whitehorse AP	60 4	2289	−45	−45	−42	VL	78	75	72	22	62	60	59

* Data for U. S. stations extracted from *Evaluated Weather Data for Cooling Equipment Design, Addendum No. 1, Winter and Summer Data*, with the permission of the publisher, Fluor Products Company, Inc., Box 1267, Santa Rosa, California.

ᵃ Data compiled from official weather stations, where hourly weather observations are made by trained observers, and from other sources. Table 1 prepared by ASHRAE Technical Committee 2.2, Weather Data and Design Conditions. Percentage of *winter* design data show the percent of 3-month period, December through February. Canadian data are based on January only. Percentage of *summer* design data show the percent of 4-month period, June through September. Canadian data are based on July only. Also see References 1 to 7.

ᵇ When airport temperature observations were used to develop design data, "AP" follows station name, and "AFB" follows Air Force Bases. Data for stations followed by "CO" came from office locations within an urban area and generally reflect an influence of the surrounding area. Stations without designation can be considered semirural and may be directly compared with most airport data.

ᶜ Latitude is given to the nearest 10 minutes, for use in calculating solar loads. For example, the latitude for Anniston, Alabama is given as 33 4, or 33°40′.

ᵈ Elevations are ground elevations for each station as of 1964. Temperature readings are generally made at an elevation of 5 ft above ground, except for locations marked r, indicating roof exposure of thermometer.

ᵉ Coincident wind velocities derived from approximately coldest 600 hours out of 20,000 hours of December through February data per station. Also see References 5 and 6. The four classifications are:
VL = Very Light, 70 percent or more of cold extreme hours ≤7 mph. M = Moderate, 50 to 74 percent cold extreme hours >7 mph.
L = Light, 50 to 69 percent cold extreme hours ≤7 mph. H = High, 75 percent or more cold extreme hours >7 mph., and 50 percent are >12 mph.
ᶠ The difference between the average maximum and average minimum temperatures during the warmest month.

† More detailed data on Arizona, California, and Nevada may be found in *Recommended Design Temperatures, Northern, California*, published by the Golden Gate Chapter; and *Recommended Design Temperatures, Southern California, Arizona, Nevada*, published by the Southern California Chapter.

Reprinted by permission from ASHRAE Handbook of Fundamentals (1967).

TABLE 3-34 Wall Construction Group Description

Group No.	Description of Construction	Weight(lb/ft²)	U-value (Btu/(h·ft²·F))	Code Numbers of Layers (see Table 8)
4-in. Face Brick+(*Brick*)				
C	Air Space+4-in. Face Brick	83	0.358	A0, A2, B1, A2, E0
D	4-in. Common Brick	90	0.415	A0, A2, C4, E1, E0
C	1-in. Insulation or Air space+4-in. Common Brick	90	0.174-0.301	A0, A2, C4, B1/B2, E1, E0
B	2-in. Insulation+4-in. Common Brick	88	0.111	A0, A2, B2, C4, E1, E0
B	8-in. Common Brick	130	0.302	A0, A2, C9, E1, E0
A	Insulation or Air space+8-in. Common Brick	130	0.154-0.243	A0, A2, C9, B1/B2, E1, E0
4-in. Face Brick+(*H.W. Concrete*)				
C	Air Space+2-in. Concrete	94	0.350	A0, A2, B1, C5, E1, E0
B	2-in. Insulation+4-in. concrete	97	0.116	A0, A2, B3, C5, E1, E0
A	Air Space or Insulation+8-in. or more Concrete	143-190	0.110-0.112	A0, A2, B1, C10/11, E1, E0
4-in. Face Brick+(*L.W. or H.W. Concrete Block*)				
E	4-in. Block	62	0.319	A0, A2, C2, E1, E0
D	Air Space or Insulation+4-in. Block	62	0.153-0.246	A0, A2, C2, B1/B2, E1, E0
D	8-in. Block	70	0.274	A0, A2, C7, A6, E0
C	Air Space or 1-in. Insulation+6-in. or 8-in. Block	73-89	0.221-0.275	A0, A2, B1, C7/C8, E1, E0
B	2-in. Insulation+8-in. Block	89	0.096-0.107	A0, A2, B3, C7/C8, E1, E0
4-in Face Brick+(*Clay Tile*)				
D	4-in. Tile	71	0.381	A0, A2, C1, E1, E0
D	Air Space+4-in. Tile	71	0.281	A0, A2, C1, B1, E1, E0
C	Insulation+4-in. Tile	71	0.169	A0, A2, C1, B2, E1, E0
C	8-in. Tile	96	0.275	A0, A2, C6, E1, E0
B	Air Space or 1-in. Insulation+8-in. Tile	96	0.142-0.221	A0, A2, C6, B1/B2, E1, E0
A	2-in. Insulation+8-in. Tile	97	0.097	A0, A2, B3, C6, E1, E0
H.W. Concrete Wall+(*Finish*)				
E	4-in. Concrete	63	0.585	A0, A1, C5, E1, E0
D	4-in. Concrete+1-in. or 2-in. Insulation	63	0.119-0.200	A0, A1, C5, B2/B3, E1, E0
C	2-in. Insulation+4-in. Concrete	63	0.119	A0, A1, B6, C5, E1, E0
C	8-in. Concrete	109	0.490	A0, A1, C10, E1, E0
B	8-in. Concrete+1-in. or 2-in. Insulation	110	0.115-0.187	A0, A1, C10, B5/B6, E1, E0
A	2-in. Insulation+8-in. Concrete	110	0.115	A0, A1, B3, C10, E1, E0
B	12-in. Concrete	156	0.421	A0, A1, C11, E1, E0
A	12-in. Concrete+Insulation	156	0.113	A0, C11, B6, A6, E0
L.W. and H.W. Concrete Block+(*Finish*)				
F	4-in. Block+Air Space/Insulation	29	0.161-0.263	A0, A1, C2, B1/B2, E1, E0
E	2-in. Insulation+4-in. Block	29-37	0.105-0.114	A0, A1, B1, C2/C3, E1, E0
E	8-in. Block	47-51	0.294-0.402	A0, A1, C7/C8, E1, E0
D	8-in. Block+Air Space/Insulation	41-57	0.149-0.173	A0, A1, C7/C8, B2, E1, E0
Clay *Tile*+(*Finish*)				
F	4-in. Tile	39	0.419	A0, A1, C1, E1, E0
F	4-in. Tile+Air space	39	0.303	A0, A1, C1, B1, E1, E0
E	4-in. Tile+1-in. Insulation	39	0.175	A0, A1, C1, B2, E1, E0
D	2-in. Insulation+4-in. Tile	40	0.110	A0, A1, B1, C1, E1, E0
D	8-in. Tile	63	0.296	A0, A1, C6, E1, E0
C	8-in. Tile+Air Space/1-in. Insulation	63	0.151-0.231	A0, A1, C6, B1/B2, E1, E0
B	2-in. Insulation+8-in. Tile	63	0.099	A0, A1, B1, C6, E1, E0
Metal Curtain Wall				
G	With/without Air Space+1-in./2-in./3-in. Insulation	5-6	0.091-0.230	A0, A3, B5/B6/B12, A3, E0
Frame Wall				
G	1-in. to 3-in. Insulation	16	0.081-0.178	A0, A1, B1, B2/B3/B4, E1, E0

TABLE 3-35 Cooling Load Temp Differences for Calculating Cooling Load from Sunlit Walls

Wall Facing	1	2	3	4	5	6	7	8	9	10	11	12	13	14	15	16	17	18	19	20	21	22	23	24	Hr of Maximum CLTD	Minimum CLTD	Maximum CLTD	Difference CLTD
North Latitude																												
Group A Walls																												
N	14	14	14	13	13	13	12	12	11	11	10	10	10	10	10	10	11	11	12	12	13	13	14	14	2	10	14	4
NE	19	19	19	18	17	17	16	15	15	15	15	15	16	16	17	18	18	19	19	20	20	20	20	22	15	15	20	5
E	24	24	23	23	22	21	20	19	19	18	19	19	20	21	22	23	24	24	25	25	25	25	25	22	18	18	25	7
SE	24	23	23	22	21	20	20	19	18	18	18	18	18	19	20	21	22	23	23	24	24	24	24	22	18	18	24	6
S	20	20	19	19	18	18	17	16	16	15	14	14	14	14	14	15	16	17	18	19	19	20	20	23	14	14	20	6
SW	25	25	25	24	24	23	22	21	20	19	19	18	17	17	17	17	18	19	20	22	23	24	25	24	17	17	25	8
W	27	27	26	26	25	24	24	23	22	21	20	19	19	18	18	18	18	19	20	22	23	25	26	26	1	18	27	9
NW	21	21	21	20	20	19	19	18	17	16	16	15	15	14	14	14	15	15	16	17	18	19	20	21	1	14	21	7
Group B Walls																												
N	15	14	14	13	12	11	11	10	9	9	9	8	9	9	9	10	11	12	13	14	14	15	15	15	24	8	15	7
NE	19	18	17	16	15	14	13	12	12	13	14	15	16	17	18	19	19	20	20	21	21	21	20	20	21	12	21	9
E	23	22	21	20	18	17	16	15	15	15	17	19	21	22	24	25	26	26	27	27	26	26	25	24	20	15	27	12
SE	23	22	21	20	18	17	16	15	14	14	15	16	18	20	21	23	24	25	26	26	26	26	25	24	21	14	26	12
S	21	20	19	18	17	15	14	13	12	11	11	11	11	12	14	15	17	19	20	21	22	22	22	21	23	11	22	11
SW	27	26	25	24	22	21	19	18	16	15	14	14	13	13	14	15	17	20	22	25	27	28	28	28	24	13	28	15
W	29	28	27	26	24	23	21	19	18	17	16	15	14	14	14	15	17	19	22	25	27	29	29	30	24	14	30	16
NW	23	22	21	20	19	18	17	15	14	13	12	12	12	11	12	12	13	15	17	19	21	22	23	23	24	11	23	9
Group C Walls																												
N	15	14	13	12	11	10	9	8	8	7	7	8	8	9	10	12	13	14	15	16	17	17	17	16	22	7	17	10
NE	19	17	16	14	13	11	10	10	11	13	15	17	19	20	21	22	22	23	23	23	23	22	21	20	20	10	23	13
E	22	21	19	17	15	14	12	12	14	16	19	22	25	27	29	29	30	30	30	29	28	27	26	24	18	12	30	18
SE	22	21	19	17	15	14	12	12	12	13	16	19	22	24	26	28	29	29	29	29	28	27	26	24	19	12	29	17
S	21	19	18	16	15	13	12	10	9	9	9	10	11	14	17	20	22	24	25	26	25	25	24	22	20	9	26	17
SW	29	27	25	22	20	18	16	15	13	12	11	11	11	13	15	18	22	26	29	32	33	33	32	31	22	11	33	22
W	31	29	27	25	22	20	18	16	14	13	12	12	12	13	14	16	20	24	29	32	35	35	35	33	22	12	35	23
NW	25	23	21	20	18	16	14	13	11	10	10	10	10	11	12	13	15	18	22	25	27	27	27	26	22	10	27	17
Group D Walls																												
N	15	13	12	10	9	7	6	6	6	6	6	7	8	10	12	13	15	17	18	19	19	19	18	16	21	6	19	13
NE	17	15	13	11	10	8	7	8	10	14	17	20	22	23	23	24	24	25	25	24	23	22	20	18	19	7	25	18
E	19	17	15	13	11	9	8	9	12	17	22	27	30	32	33	33	32	32	31	30	28	26	24	22	16	8	33	25
SE	20	17	15	13	11	10	8	8	10	13	17	22	26	29	31	32	32	32	31	30	28	26	24	22	17	8	32	24
S	19	17	15	13	11	9	8	7	6	6	7	9	12	16	20	24	27	29	29	29	27	26	24	22	19	6	29	23
SW	28	25	22	19	16	14	12	10	9	8	8	8	10	12	16	21	27	32	36	38	38	37	34	31	21	8	38	30
W	31	27	24	21	18	15	13	11	10	9	9	9	10	11	14	18	24	30	36	40	41	40	38	34	21	9	41	32
NW	25	22	19	17	14	12	10	9	8	7	7	7	8	9	10	12	14	18	22	27	31	32	32	30	22	7	32	25
Group E Walls																												
N	12	10	8	7	5	4	3	4	5	6	7	9	11	13	15	17	19	20	21	23	20	18	16	14	20	3	22	19
NE	13	11	9	7	6	4	5	9	15	20	24	25	25	26	26	26	26	26	25	24	22	19	17	15	16	4	26	22
E	14	12	10	8	6	5	6	11	18	26	33	36	38	37	36	34	33	32	30	28	25	22	20	17	13	5	38	33
SE	15	12	10	8	7	5	5	8	12	19	25	31	35	37	37	36	34	33	31	28	26	23	20	17	15	5	37	32
S	15	12	10	8	7	5	4	3	4	5	9	13	19	24	29	32	34	33	31	29	26	23	20	17	17	3	34	31
SW	22	18	15	12	10	8	6	5	5	5	6	7	9	12	18	24	32	38	43	45	44	40	35	30	19	5	45	40
W	25	21	17	14	11	9	7	6	6	6	7	9	11	14	20	27	36	43	49	49	45	40	34	29	20	6	49	43
NW	20	17	14	11	9	7	6	5	5	5	6	8	10	13	16	20	26	32	37	38	36	32	28	24	20	5	38	33
Group F Walls																												
N	8	6	5	3	2	1	2	4	6	7	9	11	14	17	19	21	22	23	24	23	20	16	13	11	19	1	23	23
NE	9	7	5	3	2	1	5	14	23	28	30	29	28	27	27	27	26	24	22	19	16	13	11	11	11	1	30	29
E	10	7	6	4	3	2	6	17	28	38	44	45	43	39	36	34	32	30	27	24	21	17	15	12	12	2	45	43
SE	10	7	6	4	3	2	4	10	19	28	36	41	43	42	39	36	34	31	28	25	21	18	15	12	13	2	43	41
S	10	8	6	4	3	2	1	1	3	7	13	20	27	34	38	39	38	35	31	26	22	18	15	12	16	1	39	38
SW	15	11	9	6	5	3	2	2	4	5	8	11	17	26	35	44	50	53	52	45	37	28	23	18	18	2	53	48
W	17	13	10	7	5	4	3	3	4	6	8	11	14	20	28	39	49	57	60	54	43	34	27	21	19	3	60	57
NW	14	10	8	6	4	3	2	2	3	5	8	10	13	15	21	27	35	42	46	43	35	28	22	18	19	2	46	44
Group G Walls																												
N	3	2	1	0	-1	2	7	8	9	12	15	18	21	23	24	24	25	26	22	15	11	9	7	5	18	-1	26	27
NE	3	2	1	0	-1	9	27	36	39	35	30	26	26	27	27	26	25	22	18	14	11	9	7	5	9	-1	39	40
E	4	2	1	0	-1	11	31	47	54	55	50	40	33	31	30	29	27	24	19	15	12	10	8	6	10	-1	55	56
SE	4	2	1	0	-1	5	18	32	42	49	51	48	42	36	32	30	27	24	19	15	12	10	8	6	11	-1	51	52
S	4	2	1	0	-1	0	1	5	12	22	31	39	45	46	43	37	31	25	20	15	12	10	8	5	14	-1	46	47
SW	5	4	3	1	0	0	2	5	8	12	16	26	38	50	59	63	61	52	37	24	17	13	10	8	16	0	63	63
W	6	5	3	2	1	1	2	5	8	11	15	19	27	41	56	67	72	67	48	29	20	15	11	8	17	1	72	71
NW	5	3	2	1	0	0	2	5	8	11	15	18	21	27	37	47	55	55	41	25	17	13	10	7	18	0	55	55

1. Application: These values may be used for all normal air-conditioning estimates; usually without correction (except as noted below) when the load is calculated for the hottest weather.

2. Corrections: The values in the table were calculated for an inside temperature of 78 F and an outdoor daily range of 21 deg F. The table remains approximately correct for other outdoor maximums (93-102 F) and other outdoor daily ranges (16-34 deg F), provided the outdoor daily average temperature remains approximately 85 F. If the room temperature is different from 78 F and/or the outdoor daily average temperature is different from 85 F, the following rules apply: (a) For room air temperature less than 78 F, add the difference between 78 F and room air temperature; if greater than 78 F, subtract the difference. (b) For outdoor daily average temperature less than 85 F, subtract the difference between 85 F and the daily average temperature; if greater than 85 F, add the difference. The table values will be approximately correct for the east or west wall in any latitude (o° to 50° North or South) during the hottest weather.

3. Color of exterior surface of wall: For light colors, multiply the Wall-Cooling Load Temperature Difference in the tables by 0.65. Use temperature differences for light walls only when the permanence of the light wall is established by experience. For cream colors, use the light wall values. For medium colors, interpolate half-way between the dark and light values. Medium colors are medium blue and green, bright red, light brown, unpainted wood, and natural color concrete. Dark blue, red, brown, and green are considered dark colors.

4. Correction for other months and latitudes: The CLTD is calculated by adding this correction factor to the CLTD of the above table: $[(I_{DT-new}/I_{DT-Table\ 22})-1]\,(t_{ea}-t_{oa})$ where I_{DT-new} is the sum of two appropriate half-day totals of solar heat gain in Tables 17 to 25 of Chapter 26.

5. For each 7 increase in R-value due to insulation added to the wall structures in Table 6, use the CLTD for the wall group with the next higher letter in the alphabet. When the insulation is added to the exterior of the construction rather than the interior, use the CLTD for the wall group two letters higher. If this is not possible, due to having already selected a wall in Group A, use a Δt in the load calculation equal to the difference between the 24 hr average sol-air temperature and the room air temperature.

TABLE 3-36 Cooling Load Temp Differences for Calculating Cooling Load from Flat Roofs

Roof No	Description of Construction	Weight lb/ft²	U-value Btu/(h·ft²·°F)	1	2	3	4	5	6	7	8	9	10	11	12	13	14	15	16	17	18	19	20	21	22	23	24 CLTD	Hour of Maximum CLTD	Minimum CLTD	Maximum CLTD	Difference CLTD
Without Suspended Ceiling																														
1	Steel sheet with 1-in. (or 2-in.) insulation	7 (8)	0.213 (0.124)	1	-2	-3	-3	-5	-3	6	19	34	49	61	71	78	79	77	70	59	45	30	18	12	8	5	3	14	-5	79	84
2	1-in. wood with 1-in. insulation	8	0.170	6	3	0	-1	-3	-3	-2	4	14	27	39	52	62	70	74	74	70	62	51	38	28	20	14	9	16	-3	74	77
3	4-in. l.w. concrete	18	0.213	9	5	2	0	-2	-3	-3	1	9	20	32	44	55	64	70	73	71	66	57	45	34	25	18	13	16	-3	73	76
4	2-in. h.w. concrete with 1-in. (or 2-in.) insulation	29	0.206 (0.122)	12	8	5	3	0	-1	-1	3	11	20	30	41	51	59	65	66	66	62	54	45	36	29	22	17	16	-1	67	68
5	1-in. wood with 2-in. insulation	19	0.109	3	0	-3	-4	-5	-7	-6	-3	5	16	27	39	49	57	63	64	62	57	48	37	26	18	11	7	16	-7	64	71
6	6-in. l.w. concrete	24	0.158	22	17	13	9	6	3	1	1	3	7	15	23	33	43	51	58	62	64	62	57	50	42	35	28	18	1	54	63
7	2.5-in. wood with 1-insulation	13	0.130	29	24	20	16	13	10	7	6	6	9	13	20	27	34	42	48	53	55	56	54	49	44	39	34	19	6	56	50
8	8-in. l.w. concrete	31	0.126	35	30	26	22	18	14	11	9	7	7	9	13	19	25	33	39	46	50	53	54	53	49	45	40	20	7	54	47
9	4-in. h.w. concrete with 1-in. (or 2-in.) insulation	52 (52)	0.200 (0.120)	25	22	18	15	12	9	8	8	10	14	20	26	33	40	46	50	53	53	52	48	43	38	34	30	18	8	53	45
10	2.5-in. wood with 2-in. insulation	13	0.093	30	26	23	19	16	13	10	9	8	9	13	17	23	29	36	41	46	49	51	50	47	43	39	35	19	8	51	43
11	Roof terrace system	75	0.106	34	31	28	25	22	19	16	14	13	13	15	18	22	26	31	36	40	44	45	46	45	43	40	37	20	13	46	33
12	6-in. h.w. concrete with 1-in. (or 2-in.) insulation	(75) 75	0.192 (0.117)	31	28	25	22	20	17	15	14	14	16	18	22	26	31	36	40	43	45	45	44	42	40	37	34	19	14	45	31
13	4-in. wood with 1-in. (or 2-in) insulation	17 (18)	0.106 (0.078)	38	36	33	30	28	25	22	20	18	17	16	17	18	21	24	28	32	36	39	41	43	43	42	40	22	16	43	27
With Suspended Ceiling																														
1	Steel Sheet with 1-in. (or 2-in.) insulation	9 (10)	0.134 (0.092)	2	0	-2	-3	-4	-4	-1	9	23	37	50	62	71	77	78	74	67	56	42	28	18	12	8	5	15	-4	78	82
2	1-in. wood with 1-in. insulation	10	0.115	20	15	11	8	5	3	2	3	7	13	21	30	40	48	55	60	62	61	58	51	44	37	30	25	17	2	62	60
3	4-in. l.w. concrete	20	0.134	19	14	10	7	4	2	0	0	4	10	19	29	39	48	56	62	65	64	61	54	46	38	30	24	17	0	65	65
4	2-in. h.w. concrete with 1-in. insulation	30	0.131	28	25	23	20	17	15	13	13	14	16	20	25	30	35	39	43	46	47	46	44	41	38	35	32	18	13	47	34
5	1-in. wood with 2-in. insulation	10	0.083	25	20	16	13	10	7	5	5	7	12	18	25	33	41	48	53	57	57	56	52	46	40	34	29	18	5	57	52
6	6-in. l.w. concrete	26	0.109	32	28	23	19	16	13	10	8	7	8	11	16	22	29	36	42	48	52	54	54	51	47	42	37	20	7	54	47
7	2.5-in. wood with 1-in. insulation	15	0.096	34	31	29	26	23	21	18	16	15	15	16	18	21	25	30	34	38	41	43	44	44	42	40	37	21	15	44	29
8	8-in. l.w. concrete	33	0.093	39	36	33	29	26	23	20	18	15	14	14	15	17	20	25	29	34	38	42	45	46	45	44	42	21	14	46	32
9	4-in. h.w. concrete with 1-in. (or 2-in.) insulation	53 (54)	0.128 (0.090)	30	29	27	26	24	22	21	20	20	21	22	24	27	29	32	34	36	38	38	38	37	36	34	33	19	20	38	18
10	2.5-in. wood with 2-in. insulation	15	0.072	35	33	30	28	26	24	22	20	18	18	18	20	22	25	28	32	35	38	40	41	41	40	39	37	21	18	41	23
11	Roof terrace system	77	0.082	30	29	28	27	26	25	24	23	22	22	22	23	23	25	26	28	29	31	32	33	33	33	33	32	22	22	33	11
12	6-in. h.w. concrete with 1-in. (or 2-in) insulation	77 (77)	0.125 (0.088)	29	28	27	26	25	24	23	22	21	21	22	23	25	26	28	30	32	33	34	34	34	33	32	31	20	21	34	13
13	4-in. wood with 1-in (or 2-in.) insulation	19 (20)	0.082 (0.064)	35	34	33	32	31	29	27	26	24	23	22	21	22	22	24	25	27	30	32	34	35	36	37	36	23	21	37	16

1 Application: These values may be used for all normal air-conditioning estimates; usually without correction (except as noted below) in latitude 0° to 50° North or South when the load is calculated for the hottest weather.

2 Corrections: The values in the table were calculated for an inside temperature of 78 F and an outdoor maximum temperature of 95 F, with an outdoor daily range of 21 deg F. The table remains approximately correct for other outdoor maximums (93-102 F) and other outdoor daily ranges (16-34 deg F), provided the outdoor daily average temperature remains approximately 85 F. If the room air temperature is different from 78 F and/or the outdoor daily average temperature is different from 85 F, the following rules apply: (a) For room air temperature less than 78 F, add the difference between 78 F and room air temperature; if greater than 78 F, subtract the difference. (b) For outdoor daily average temperature less than 85 F, subtract the difference between 85 F and the daily average temperature; if greater than 85 F, add the difference.

3 Attics or other spaces between the roof and ceiling: If the ceiling is insulated and a fan is used for positive ventilation in the space between the ceiling and roof, the total temperature difference for calculating the room load may be decreased by 25%. If the attic space contains a return duct or other air plenum, care should be taken in determining the portion of the heat gain that reaches the ceiling.

4 Light Colors: Multiply the CLTD's in the table by 0.5. Credit should not be taken for light-colored roofs except where the permanence of light color is established by experience, as in rural areas or where there is little smoke.

5 For solar transmission in other months: The table values of temperature differences calculated for July 21 will be approximately correct for a roof in the following months:

North Latitude	
Latitude	**Months**
0°	All Months
10°	All Months
20°	All Months except Nov., Dec., Jan.
30°	Mar., Apr., May, June, July, Aug., Sept.
40°	April, May, June, July, Aug.
50°	May, June, July

South Latitude	
Latitude	**Months**
0°	All Months
10°	All Months
20°	All Months except May, June, July
30°	Sept., Oct., Nov., Dec., Jan., Feb., Mar.
40°	Oct., Nov., Dec., Jan., Feb.
50°	Nov., Dec., Jan.

6 For each 7 increase in R-value due to insulation added to the roof structures (Table 4), use a CLTD for a roof whose weight is approximately the same but whose CLTD has a maximum value 2 hr later. If this is not possible, due to having already selected the roof with the longest time lag, use a Δt in the load calculation equal to the difference between the 24 hr average sol-air temperature and the room air temperature.

TABLE 3-37 Maximum Solar Heat Gain Factors But/(h ft²)

	N	NE/NW	E/W	SE/SW	S	HOR		N	NE/NW	E/W	SE/SW	S	HOR
	0°N Latitude							**8°N Latitude**					
Jan.	34	88	234	235	118	296		32	71	224	242	162	275
Feb.	36	132	245	210	67	306		34	114	239	219	110	294
Mar.	38	170	242	170	38	303		37	156	241	184	55	300
Apr.	71	193	221	118	37	284		44	184	225	134	39	289
May	113	203	201	80	37	265		74	198	209	97	38	277
June	129	206	190	65	37	255		90	200	200	82	39	269
July	115	201	195	77	38	260		77	195	204	93	39	272
Aug.	75	187	212	112	38	276		47	179	216	128	41	282
Sep.	40	163	231	163	40	293		38	149	230	176	56	290
Oct.	37	129	236	202	66	299		35	112	231	211	108	288
Nov.	35	88	230	230	117	293		33	71	220	238	160	273
Dec.	34	71	226	241	138	288		31	54	215	247	180	264
	16° N Latitude							**24° N Latitude**					
Jan.	30	55	210	251	199	248		27	41	190	253	227	214
Feb.	33	96	231	233	154	275		30	80	220	243	192	249
Mar.	35	140	239	197	93	291		34	124	234	214	137	275
Apr.	39	172	227	150	45	289		37	159	228	169	75	283
May	52	189	215	115	41	282		43	178	218	132	46	282
June	66	194	207	99	41	277		56	184	212	117	43	279
July	55	187	210	111	42	277		45	176	213	129	46	278
Aug.	41	168	219	143	46	282		38	156	220	162	72	277
Sep.	36	134	227	191	93	282		35	119	222	206	134	266
Oct.	33	95	223	225	150	270		31	79	211	235	187	244
Nov.	30	55	206	247	196	246		27	42	187	249	224	213
Dec.	29	41	198	254	213	234		25	29	179	252	237	199
	32° N Latitude							**40° N Latitude**					
Jan.	24	29	175	249	246	176		20	20	154	241	254	133
Feb.	27	65	205	248	221	217		24	50	186	246	241	180
Mar.	32	107	227	227	176	252		29	93	218	236	206	223
Apr.	36	146	227	187	115	271		34	140	224	203	154	252
May	38	170	220	155	74	277		37	165	220	175	113	265
June	47	176	214	139	60	276		48	172	215	161	95	268
July	40	167	215	150	72	273		38	163	216	170	109	262
Aug.	37	141	219	181	111	265		35	135	216	196	149	247
Sep.	33	103	215	218	171	244		30	87	205	226	200	215
Oct.	28	63	195	239	215	213		25	49	180	238	234	177
Nov.	24	29	173	245	243	175		18	20	151	237	250	132
Dec.	22	22	162	246	252	158		18	18	135	232	253	112
	48° N Latitude							**56° N Latitude**					
Jan.	15	15	118	216	245	85		10	10	74	169	205	40
Feb.	20	36	168	242	250	138		16	21	139	223	244	91
Mar.	26	80	204	239	228	188		22	65	185	238	241	149
Apr.	31	132	219	215	186	226		28	123	211	223	210	195
May	35	158	218	192	150	247		36	149	215	206	181	222
June	47	165	215	180	134	252		53	161	213	195	167	231
July	37	156	214	187	146	244		37	147	211	201	177	221
Aug.	33	128	211	208	180	223		30	119	203	215	203	193
Sep.	27	72	191	228	220	182		23	58	171	227	231	144
Oct.	21	35	161	233	242	136		16	20	132	213	234	91
Nov.	15	15	115	212	240	85		10	10	72	165	200	40
Dec.	13	13	91	195	233	64		7	7	46	135	170	23
	64° N Latitude												
Jan.	3	3	15	67	96	8							
Feb.	11	11	89	177	210	45							
Mar.	18	47	159	226	239	105							
Apr.	25	113	201	225	224	160							
May	48	150	211	215	204	192							
June	62	162	213	208	193	203							
July	49	148	207	211	200	192							
Aug.	27	109	193	217	217	159							
Sep.	19	43	148	213	227	101							
Oct.	11	11	83	167	199	46							
Nov.	4	4	15	66	93	8							
Dec.	0	0	1	10	14	1							

TABLE 3-38 Cooling Load Factors for Glass without Interior Shading
(includes reflective and heat absorbing glass)

N. Latitude Fenestration Facing	Room Construction	Solar Time, hr																							
		1	2	3	4	5	6	7	8	9	10	11	12	13	14	15	16	17	18	19	20	21	22	23	24
N	L	0.17	0.14	0.11	0.09	0.08	0.33	0.42	0.48	0.56	0.63	0.71	0.76	0.80	0.82	0.82	0.79	0.80	0.84	0.61	0.48	0.38	0.31	0.25	0.20
	M	0.23	0.20	0.18	0.16	0.14	0.34	0.41	0.46	0.52	0.59	0.65	0.70	0.73	0.75	0.76	0.74	0.75	0.79	0.61	0.50	0.42	0.36	0.31	0.27
	H	0.25	0.23	0.21	0.20	0.19	0.38	0.45	0.50	0.55	0.60	0.65	0.69	0.72	0.73	0.72	0.70	0.70	0.74	0.57	0.46	0.39	0.34	0.31	0.28
NE	L	0.04	0.04	0.03	0.02	0.02	0.23	0.41	0.51	0.51	0.45	0.39	0.36	0.33	0.31	0.28	0.26	0.23	0.19	0.15	0.12	0.10	0.08	0.06	0.05
	M	0.07	0.06	0.06	0.05	0.04	0.21	0.36	0.44	0.45	0.40	0.36	0.33	0.31	0.30	0.28	0.26	0.23	0.21	0.17	0.15	0.13	0.11	0.09	0.08
	H	0.09	0.08	0.08	0.07	0.07	0.23	0.37	0.44	0.44	0.39	0.34	0.31	0.29	0.27	0.26	0.24	0.22	0.20	0.16	0.14	0.13	0.12	0.11	0.10
E	L	0.04	0.04	0.03	0.02	0.02	0.19	0.37	0.51	0.57	0.57	0.51	0.42	0.36	0.32	0.29	0.25	0.22	0.19	0.14	0.12	0.09	0.08	0.06	0.05
	M	0.07	0.06	0.06	0.05	0.04	0.18	0.33	0.44	0.50	0.51	0.45	0.39	0.35	0.32	0.29	0.26	0.23	0.21	0.17	0.15	0.13	0.11	0.10	0.08
	H	0.09	0.09	0.08	0.08	0.07	0.21	0.34	0.45	0.50	0.49	0.43	0.36	0.32	0.29	0.26	0.24	0.22	0.19	0.17	0.15	0.13	0.12	0.11	0.10
SE	L	0.05	0.04	0.04	0.03	0.02	0.13	0.28	0.43	0.55	0.62	0.63	0.57	0.48	0.42	0.37	0.33	0.28	0.24	0.19	0.15	0.12	0.10	0.08	0.07
	M	0.09	0.08	0.07	0.06	0.05	0.14	0.26	0.38	0.48	0.54	0.55	0.51	0.45	0.40	0.36	0.33	0.29	0.25	0.21	0.18	0.16	0.14	0.12	0.10
	H	0.11	0.10	0.10	0.09	0.08	0.17	0.28	0.40	0.49	0.53	0.53	0.48	0.41	0.36	0.33	0.30	0.27	0.24	0.20	0.18	0.16	0.14	0.13	0.12
S	L	0.08	0.07	0.05	0.04	0.04	0.06	0.09	0.14	0.22	0.34	0.48	0.59	0.65	0.65	0.59	0.50	0.43	0.36	0.28	0.22	0.18	0.15	0.12	0.10
	M	0.12	0.11	0.09	0.08	0.07	0.08	0.11	0.14	0.21	0.31	0.42	0.52	0.57	0.58	0.53	0.47	0.41	0.36	0.29	0.25	0.21	0.18	0.16	0.14
	H	0.13	0.12	0.12	0.11	0.10	0.12	0.14	0.17	0.24	0.33	0.43	0.51	0.56	0.55	0.50	0.43	0.38	0.32	0.26	0.22	0.20	0.18	0.16	0.15
SW	L	0.12	0.10	0.08	0.06	0.05	0.06	0.08	0.10	0.12	0.14	0.16	0.24	0.36	0.49	0.60	0.66	0.66	0.58	0.43	0.33	0.27	0.22	0.18	0.14
	M	0.15	0.13	0.12	0.10	0.09	0.09	0.10	0.12	0.13	0.15	0.17	0.23	0.33	0.44	0.53	0.58	0.59	0.53	0.41	0.33	0.28	0.24	0.21	0.18
	H	0.15	0.14	0.13	0.12	0.11	0.12	0.13	0.14	0.16	0.17	0.19	0.25	0.34	0.44	0.52	0.56	0.56	0.49	0.37	0.30	0.25	0.21	0.19	0.17
W	L	0.12	0.10	0.08	0.07	0.05	0.06	0.07	0.08	0.10	0.11	0.13	0.14	0.20	0.32	0.45	0.57	0.64	0.61	0.44	0.34	0.27	0.22	0.18	0.14
	M	0.15	0.13	0.11	0.10	0.09	0.09	0.09	0.10	0.11	0.12	0.13	0.14	0.19	0.29	0.40	0.50	0.56	0.55	0.41	0.33	0.27	0.23	0.20	0.17
	H	0.14	0.13	0.12	0.11	0.10	0.11	0.12	0.13	0.13	0.14	0.15	0.16	0.21	0.30	0.40	0.49	0.54	0.52	0.38	0.30	0.24	0.21	0.18	0.16
NW	L	0.11	0.09	0.08	0.06	0.05	0.06	0.08	0.10	0.12	0.14	0.16	0.17	0.19	0.23	0.33	0.47	0.59	0.60	0.43	0.33	0.26	0.21	0.17	0.14
	M	0.14	0.12	0.11	0.09	0.08	0.09	0.10	0.11	0.13	0.14	0.16	0.17	0.18	0.21	0.30	0.42	0.51	0.53	0.39	0.32	0.26	0.22	0.19	0.16
	H	0.14	0.12	0.11	0.11	0.10	0.11	0.12	0.13	0.15	0.16	0.18	0.19	0.19	0.22	0.30	0.41	0.50	0.51	0.36	0.29	0.23	0.20	0.17	0.15
HOR	L	0.11	0.09	0.07	0.06	0.05	0.07	0.14	0.24	0.36	0.48	0.58	0.66	0.72	0.74	0.73	0.67	0.59	0.47	0.37	0.30	0.24	0.19	0.16	0.13
	M	0.16	0.14	0.12	0.11	0.09	0.11	0.16	0.24	0.33	0.43	0.52	0.59	0.64	0.66	0.66	0.62	0.55	0.47	0.38	0.32	0.28	0.24	0.21	0.18
	H	0.17	0.16	0.15	0.14	0.13	0.15	0.20	0.27	0.36	0.45	0.52	0.59	0.62	0.64	0.62	0.58	0.51	0.42	0.35	0.29	0.26	0.23	0.21	0.19

L=Light construction: frame exterior wall, 2-in. concrete floor slab, approximately 30 lb of material/square feet of floor area.
M=Medium construction: 4-in. concrete exterior wall, 4-in. concrete floor slab, approximately 70 lb of building material/square feet of floor area.
H=Heavy construction: 6-in. concrete exterior wall, 6-in. concrete floor slab, approximately 130 lb of building materials/square feet of floor area.

TABLE 3-39 Cooling Load Factors for Glass w/Interior Shading
(includes reflective and heat absorbing glass)

N. Latitude Fenestration Facing	Room Construction	Solar Time, hr																							
		1	2	3	4	5	6	7	8	9	10	11	12	13	14	15	16	17	18	19	20	21	22	23	24
N	L	0.07	0.05	0.04	0.04	0.05	0.70	0.65	0.65	0.74	0.81	0.87	0.91	0.91	0.88	0.84	0.77	0.80	0.92	0.27	0.19	0.15	0.12	0.10	0.08
	M	0.08	0.07	0.06	0.06	0.07	0.73	0.66	0.65	0.73	0.80	0.86	0.89	0.89	0.86	0.82	0.75	0.78	0.91	0.24	0.18	0.15	0.13	0.11	0.09
	H	0.09	0.09	0.08	0.07	0.09	0.75	0.67	0.66	0.74	0.80	0.86	0.89	0.88	0.85	0.80	0.73	0.76	0.88	0.23	0.17	0.14	0.13	0.11	0.10
NE	L	0.02	0.01	0.01	0.01	0.02	0.55	0.76	0.75	0.60	0.39	0.31	0.28	0.27	0.25	0.23	0.20	0.16	0.12	0.06	0.05	0.04	0.03	0.02	0.02
	M	0.03	0.02	0.02	0.02	0.02	0.56	0.76	0.74	0.58	0.37	0.29	0.27	0.26	0.24	0.22	0.20	0.16	0.12	0.06	0.05	0.04	0.04	0.03	0.03
	H	0.03	0.03	0.03	0.03	0.04	0.57	0.77	0.74	0.58	0.36	0.28	0.26	0.25	0.23	0.21	0.19	0.16	0.11	0.06	0.05	0.05	0.04	0.04	0.04
E	L	0.02	0.01	0.01	0.01	0.01	0.45	0.71	0.80	0.77	0.64	0.43	0.29	0.25	0.23	0.20	0.17	0.14	0.10	0.06	0.05	0.04	0.03	0.02	0.02
	M	0.03	0.02	0.02	0.02	0.02	0.47	0.72	0.80	0.76	0.62	0.41	0.27	0.24	0.22	0.20	0.17	0.14	0.11	0.06	0.05	0.04	0.04	0.03	0.03
	H	0.04	0.03	0.03	0.03	0.03	0.48	0.72	0.80	0.75	0.61	0.40	0.25	0.22	0.21	0.19	0.16	0.14	0.10	0.06	0.05	0.05	0.04	0.04	0.04
SE	L	0.02	0.02	0.01	0.01	0.01	0.29	0.56	0.74	0.82	0.81	0.70	0.52	0.35	0.30	0.26	0.22	0.18	0.13	0.08	0.06	0.05	0.04	0.03	0.03
	M	0.03	0.03	0.02	0.02	0.02	0.30	0.56	0.74	0.81	0.79	0.68	0.49	0.33	0.28	0.25	0.22	0.18	0.13	0.08	0.07	0.06	0.05	0.04	0.04
	H	0.04	0.04	0.04	0.03	0.04	0.31	0.57	0.74	0.81	0.79	0.67	0.48	0.31	0.27	0.23	0.20	0.17	0.13	0.07	0.07	0.06	0.05	0.05	0.05
S	L	0.03	0.03	0.02	0.02	0.02	0.08	0.15	0.22	0.37	0.58	0.75	0.84	0.82	0.71	0.53	0.37	0.29	0.20	0.11	0.09	0.07	0.06	0.05	0.04
	M	0.04	0.04	0.03	0.03	0.03	0.09	0.16	0.22	0.38	0.58	0.75	0.83	0.80	0.68	0.50	0.35	0.27	0.19	0.11	0.09	0.08	0.07	0.06	0.05
	H	0.05	0.05	0.04	0.04	0.04	0.11	0.17	0.24	0.39	0.59	0.75	0.82	0.79	0.67	0.49	0.33	0.26	0.18	0.10	0.08	0.07	0.06	0.06	0.05
SW	L	0.05	0.04	0.03	0.02	0.02	0.06	0.10	0.13	0.16	0.18	0.22	0.38	0.59	0.76	0.84	0.83	0.72	0.48	0.18	0.13	0.11	0.08	0.07	0.06
	M	0.06	0.05	0.04	0.04	0.03	0.07	0.11	0.14	0.16	0.19	0.22	0.38	0.59	0.75	0.83	0.81	0.69	0.45	0.15	0.12	0.10	0.08	0.07	0.06
	H	0.06	0.05	0.05	0.04	0.04	0.08	0.12	0.15	0.18	0.20	0.23	0.39	0.59	0.75	0.82	0.80	0.68	0.43	0.14	0.11	0.09	0.08	0.07	0.06
W	L	0.05	0.04	0.03	0.02	0.02	0.05	0.08	0.11	0.13	0.14	0.15	0.17	0.30	0.53	0.72	0.83	0.83	0.63	0.19	0.14	0.11	0.08	0.07	0.06
	M	0.05	0.05	0.04	0.04	0.03	0.06	0.09	0.11	0.13	0.15	0.16	0.17	0.31	0.53	0.72	0.82	0.81	0.61	0.16	0.12	0.10	0.08	0.07	0.06
	H	0.05	0.05	0.04	0.04	0.04	0.07	0.10	0.12	0.14	0.16	0.17	0.18	0.31	0.54	0.71	0.81	0.80	0.59	0.15	0.11	0.09	0.07	0.06	0.06
NW	L	0.04	0.04	0.03	0.02	0.02	0.06	0.10	0.13	0.16	0.19	0.20	0.21	0.22	0.30	0.52	0.73	0.83	0.71	0.19	0.13	0.10	0.08	0.07	0.05
	M	0.05	0.04	0.04	0.03	0.03	0.07	0.11	0.14	0.17	0.19	0.20	0.21	0.22	0.30	0.52	0.73	0.82	0.69	0.16	0.12	0.09	0.08	0.07	0.06
	H	0.05	0.04	0.04	0.04	0.04	0.08	0.12	0.15	0.18	0.20	0.21	0.22	0.23	0.30	0.52	0.73	0.81	0.67	0.15	0.11	0.08	0.07	0.06	0.05
HOR	L	0.04	0.03	0.03	0.02	0.02	0.10	0.26	0.43	0.59	0.72	0.81	0.87	·0.87	0.83	0.74	0.60	0.44	0.27	0.15	0.12	0.09	0.08	0.06	0.05
	M	0.06	0.05	0.04	0.04	0.03	0.12	0.27	0.44	0.59	0.72	0.81	0.85	0.85	0.81	0.71	0.58	0.42	0.25	0.14	0.12	0.10	0.08	0.07	0.06
	H	0.06	0.06	0.05	0.05	0.05	0.13	0.29	0.45	0.60	0.72	0.81	0.85	0.84	0.79	0.70	0.56	0.40	0.23	0.13	0.11	0.09	0.08	0.08	0.07

L=Light construction: frame exterior wall, 2-in. concrete floor slab, approximately 30 lb of material/sq ft of floor area.
M=Medium construction: 4-in. concrete exterior wall, 4-in. concrete floor slab, approximately 70 lb of building material/sq ft of floor area.
H=Heavy construction: 6-in. concrete exterior wall, 6-in. concrete floor slab, approximately 130 lb of building materials/sq ft of floor area.

TABLE 3-40 Design Values of a Coefficient Features of Room Furnishings, Light Fixtures and Ventilation Arrangements

a	Furnishings	Air Supply and Return	Type of Light Fixture
0.45	Heavyweight, simple furnishings, no carpet	Low rate; supply and return below ceiling ($V \leqslant 0.5$)*	Recessed, not vented
0.55	Ordinary furniture, no carpet	Medium to high ventilation rate; supply and return below ceiling or through ceiling grill and space ($V \geqslant 0.5$)*	Recessed, not vented
0.65	Ordinary furniture, with or without carpet	Medium to high ventilation rate or fan coil or induction type air-conditioning terminal unit; supply through ceiling or wall diffuser; return around light fixtures and through ceiling space. ($V \geqslant 0.5$)*	Vented
0.75 or greater	Any type of furniture	Ducted returns through light fixtures	Vented or free-hanging in air stream with ducted returns

*V is room air supply rate in cfm/ft^2 of floor area.

TABLE 3-41 The b Classification Values Calculated for Different Envelope Constructions and Room Air Condition Rates

Room Envelope Construction* (mass of floor area, lb/ft^2)	Room Air Circulation and Type of Supply and Return**			
	Low	Medium	High	Very High
2-in. Wood Floor (10)	B	A	A	A
3-in. Concrete Floor (40)	B	B	B	A
6-in. Concrete Floor (75)	C	C	C	B
8-in. concrete Floor (120)	D	D	C	C
12-in. concrete Floor (160)	D	D	D	D

*Floor covered with carpet and rubber pad; for a floor covered only with floor tile take next classification to the right in the same row.

**Low: Low ventilation rate—minimum required to cope with cooling load due to lights and occupants in interior zone. Supply through floor, wall or ceiling diffuser. Ceiling space not vented and h=0.4 Btu/(h·ft^2·deg F) (where h=inside surface convection coefficient used in calculation of b classification).

Medium: Medium ventilation rate, supply through floor, wall or ceiling diffuser. Ceiling space not vented and h=0.6 Btu/(h·ft^2·deg F).

High: Room air circulation induced by primary air of induction unit or by fan coil unit. Return through ceiling space and h=0.8 Btu/(h·ft^2·deg F).

Very High: High room air circulation used to minimize temperature gradients in a room. Return through ceiling space and h=1.2 Btu/(h·ft^2·deg F).

TABLE 3-42A Cooling Load Factors When Lights Are on for 8 Hours

"a" Coefficients	"b" Classification	Number of hours after lights are turned on																							
		0	1	2	3	4	5	6	7	8	9	10	11	12	13	14	15	16	17	18	19	20	21	22	23
0.45	A	0.02	0.46	0.57	0.65	0.72	0.77	0.82	0.85	0.88	0.46	0.37	0.30	0.24	0.19	0.15	0.12	0.10	0.08	0.06	0.05	0.04	0.03	0.03	0.02
	B	0.07	0.51	0.56	0.61	0.65	0.68	0.71	0.74	0.77	0.34	0.31	0.28	0.25	0.22	0.20	0.18	0.16	0.15	0.13	0.12	0.11	0.10	0.09	0.08
	C	0.11	0.55	0.58	0.60	0.63	0.65	0.67	0.69	0.71	0.28	0.26	0.25	0.23	0.22	0.20	0.19	0.18	0.17	0.16	0.15	0.14	0.13	0.12	0.12
	D	0.14	0.58	0.60	0.61	0.62	0.63	0.64	0.65	0.66	0.22	0.22	0.21	0.20	0.20	0.19	0.19	0.18	0.18	0.17	0.16	0.16	0.16	0.15	0.15
0.55	A	0.01	0.56	0.65	0.72	0.77	0.82	0.85	0.88	0.90	0.37	0.30	0.24	0.19	0.16	0.13	0.10	0.08	0.07	0.05	0.04	0.03	0.03	0.02	0.02
	B	0.06	0.60	0.64	0.68	0.71	0.74	0.76	0.79	0.81	0.28	0.25	0.23	0.20	0.18	0.16	0.15	0.13	0.12	0.11	0.10	0.09	0.08	0.07	0.06
	C	0.09	0.63	0.66	0.68	0.70	0.71	0.73	0.75	0.76	0.23	0.21	0.20	0.19	0.18	0.17	0.16	0.15	0.14	0.13	0.12	0.11	0.11	0.10	0.10
	D	0.11	0.66	0.67	0.68	0.69	0.70	0.71	0.72	0.72	0.18	0.18	0.17	0.17	0.16	0.16	0.15	0.15	0.14	0.14	0.13	0.13	0.13	0.12	0.12
0.65	A	0.01	0.66	0.73	0.78	0.82	0.86	0.88	0.91	0.93	0.29	0.23	0.19	0.15	0.12	0.10	0.08	0.06	0.05	0.04	0.03	0.03	0.02	0.02	0.01
	B	0.04	0.69	0.72	0.75	0.77	0.80	0.82	0.84	0.85	0.22	0.19	0.18	0.16	0.14	0.13	0.12	0.10	0.09	0.08	0.08	0.07	0.06	0.06	0.05
	C	0.07	0.72	0.73	0.75	0.76	0.78	0.79	0.80	0.82	0.18	0.17	0.16	0.15	0.14	0.13	0.12	0.11	0.11	0.10	0.10	0.09	0.08	0.08	0.07
	D	0.09	0.73	0.74	0.75	0.76	0.77	0.77	0.78	0.79	0.14	0.14	0.13	0.13	0.13	0.12	0.12	0.11	0.11	0.11	0.10	0.10	0.10	0.10	0.09
0.75	A	0.01	0.76	0.80	0.84	0.87	0.90	0.92	0.93	0.95	0.21	0.17	0.13	0.11	0.09	0.07	0.06	0.05	0.04	0.03	0.02	0.02	0.02	0.01	0.01
	B	0.03	0.78	0.80	0.82	0.84	0.85	0.87	0.88	0.89	0.15	0.14	0.13	0.11	0.10	0.09	0.08	0.07	0.07	0.06	0.05	0.05	0.04	0.04	0.04
	C	0.05	0.80	0.81	0.82	0.83	0.84	0.85	0.86	0.87	0.13	0.12	0.11	0.10	0.10	0.09	0.09	0.08	0.08	0.07	0.07	0.06	0.06	0.06	0.05
	D	0.06	0.81	0.82	0.82	0.83	0.83	0.84	0.84	0.85	0.10	0.10	0.10	0.09	0.09	0.09	0.08	0.08	0.08	0.08	0.07	0.07	0.07	0.07	0.07

TABLE 3-42B Cooling Load Factors when Lights Are on for 10 Hours

"a" Coefficients	"b" Classification	Number of hours after lights are turned on																							
		0	1	2	3	4	5	6	7	8	9	10	11	12	13	14	15	16	17	18	19	20	21	22	23
0.45	A	0.03	0.47	0.58	0.66	0.73	0.78	0.82	0.86	0.88	0.91	0.93	0.49	0.39	0.32	0.26	0.21	0.17	0.13	0.11	0.09	0.07	0.06	0.05	0.04
	B	0.10	0.54	0.59	0.63	0.66	0.70	0.73	0.76	0.78	0.80	0.82	0.39	0.35	0.32	0.28	0.26	0.23	0.21	0.19	0.17	0.15	0.14	0.12	0.11
	C	0.15	0.59	0.61	0.64	0.66	0.68	0.70	0.72	0.73	0.75	0.76	0.33	0.31	0.29	0.27	0.26	0.24	0.23	0.21	0.20	0.19	0.18	0.17	0.16
	D	0.18	0.62	0.63	0.64	0.66	0.67	0.68	0.69	0.69	0.70	0.71	0.27	0.26	0.26	0.25	0.24	0.23	0.23	0.22	0.21	0.21	0.20	0.19	0.19
0.55	A	0.02	0.57	0.65	0.72	0.78	0.82	0.85	0.88	0.91	0.92	0.94	0.40	0.32	0.26	0.21	0.17	0.14	0.11	0.09	0.07	0.06	0.05	0.04	0.03
	B	0.08	0.62	0.66	0.69	0.73	0.75	0.78	0.80	0.82	0.84	0.85	0.32	0.29	0.26	0.23	0.21	0.19	0.17	0.15	0.14	0.12	0.11	0.10	0.09
	C	0.12	0.66	0.68	0.70	0.72	0.74	0.75	0.77	0.78	0.79	0.81	0.27	0.25	0.24	0.22	0.21	0.20	0.19	0.17	0.16	0.15	0.14	0.14	0.13
	D	0.15	0.69	0.70	0.71	0.72	0.73	0.73	0.74	0.75	0.76	0.76	0.22	0.22	0.21	0.20	0.20	0.19	0.18	0.18	0.17	0.17	0.16	0.16	0.15
0.65	A	0.02	0.66	0.73	0.78	0.83	0.86	0.89	0.91	0.93	0.94	0.95	0.31	0.25	0.20	0.16	0.13	0.11	0.08	0.07	0.05	0.04	0.04	0.03	0.02
	B	0.06	0.71	0.74	0.76	0.79	0.81	0.83	0.84	0.86	0.87	0.89	0.25	0.22	0.20	0.18	0.16	0.15	0.13	0.12	0.11	0.10	0.09	0.08	0.07
	C	0.09	0.74	0.75	0.77	0.78	0.80	0.81	0.82	0.83	0.84	0.85	0.21	0.20	0.18	0.17	0.16	0.15	0.14	0.14	0.13	0.12	0.11	0.11	0.10
	D	0.11	0.76	0.77	0.77	0.78	0.79	0.79	0.80	0.81	0.81	0.82	0.17	0.17	0.16	0.16	0.15	0.15	0.14	0.14	0.14	0.13	0.13	0.12	0.12
0.75	A	0.01	0.76	0.81	0.84	0.88	0.90	0.92	0.93	0.95	0.96	0.97	0.22	0.18	0.14	0.12	0.09	0.08	0.06	0.05	0.04	0.03	0.03	0.02	0.02
	B	0.04	0.79	0.81	0.83	0.85	0.86	0.88	0.89	0.90	0.91	0.92	0.18	0.16	0.14	0.13	0.12	0.10	0.09	0.08	0.08	0.07	0.06	0.06	0.05
	C	0.07	0.81	0.82	0.83	0.84	0.85	0.86	0.87	0.88	0.89	0.89	0.15	0.14	0.13	0.12	0.12	0.11	0.10	0.10	0.09	0.09	0.08	0.08	0.07
	D	0.08	0.83	0.83	0.84	0.84	0.85	0.85	0.86	0.86	0.87	0.87	0.12	0.12	0.12	0.11	0.11	0.11	0.10	0.10	0.10	0.09	0.09	0.09	0.09

TABLE 3-42C Cooling Load Factors when Lights Are on for 12 Hours

		1	2	3	4	5	6	7	8	9	10	11	12	13	14	15	16	17	18	19	20	21	22	23	24
0.45	A	0.05	0.49	0.59	0.67	0.73	0.78	0.83	0.86	0.89	0.91	0.93	0.94	0.95	0.51	0.41	0.33	0.27	0.22	0.17	0.14	0.11	0.09	0.07	0.06
	B	0.13	0.57	0.61	0.65	0.69	0.72	0.75	0.77	0.79	0.82	0.83	0.85	0.87	0.43	0.39	0.35	0.31	0.28	0.25	0.23	0.21	0.18	0.17	0.15
	C	0.19	0.63	0.65	0.67	0.69	0.71	0.73	0.74	0.76	0.77	0.79	0.80	0.81	0.37	0.35	0.33	0.31	0.29	0.27	0.26	0.24	0.23	0.21	0.20
	D	0.22	0.66	0.67	0.68	0.69	0.70	0.71	0.72	0.73	0.74	0.74	0.75	0.76	0.32	0.31	0.30	0.29	0.28	0.27	0.26	0.26	0.25	0.24	0.23
0.55	A	0.04	0.58	0.66	0.73	0.78	0.82	0.86	0.89	0.91	0.93	0.94	0.95	0.96	0.42	0.34	0.27	0.22	0.18	0.14	0.11	0.09	0.07	0.06	0.05
	B	0.11	0.65	0.68	0.72	0.74	0.77	0.79	0.81	0.83	0.85	0.86	0.88	0.89	0.35	0.32	0.28	0.26	0.23	0.21	0.19	0.17	0.15	0.14	0.12
	C	0.15	0.69	0.71	0.73	0.75	0.76	0.78	0.79	0.80	0.81	0.83	0.84	0.85	0.30	0.29	0.27	0.25	0.24	0.22	0.21	0.20	0.19	0.17	0.16
	D	0.18	0.72	0.73	0.74	0.75	0.76	0.76	0.77	0.78	0.78	0.79	0.80	0.80	0.26	0.25	0.24	0.24	0.23	0.22	0.22	0.21	0.20	0.20	0.19
0.65	A	0.03	0.67	0.74	0.79	0.83	0.86	0.89	0.91	0.93	0.94	0.95	0.96	0.96	0.33	0.26	0.21	0.17	0.14	0.11	0.09	0.07	0.06	0.05	0.04
	B	0.09	0.73	0.75	0.78	0.80	0.82	0.84	0.85	0.87	0.88	0.89	0.90	0.91	0.27	0.25	0.22	0.20	0.18	0.16	0.15	0.13	0.12	0.11	0.10
	C	0.12	0.76	0.78	0.79	0.80	0.81	0.83	0.84	0.85	0.86	0.86	0.87	0.88	0.24	0.22	0.21	0.20	0.19	0.17	0.16	0.15	0.14	0.14	0.13
	D	0.14	0.79	0.79	0.80	0.80	0.81	0.82	0.82	0.83	0.83	0.84	0.84	0.85	0.20	0.20	0.19	0.18	0.18	0.17	0.17	0.16	0.16	0.15	0.15
0.75	A	0.02	0.77	0.81	0.85	0.88	0.90	0.92	0.94	0.95	0.96	0.97	0.97	0.98	0.23	0.19	0.15	0.12	0.10	0.08	0.06	0.05	0.04	0.03	0.03
	B	0.06	0.81	0.82	0.84	0.86	0.87	0.88	0.90	0.91	0.92	0.92	0.93	0.94	0.19	0.18	0.16	0.14	0.13	0.12	0.10	0.09	0.08	0.08	0.07
	C	0.09	0.83	0.84	0.85	0.86	0.87	0.88	0.88	0.89	0.90	0.90	0.91	0.91	0.17	0.16	0.15	0.14	0.13	0.12	0.12	0.11	0.10	0.10	0.09
	D	0.10	0.85	0.85	0.86	0.86	0.86	0.87	0.87	0.88	0.88	0.88	0.89	0.89	0.14	0.14	0.13	0.13	0.12	0.12	0.12	0.11	0.11	0.11	

TABLE 3-42D Cooling Load Factors when Lights Are on for 14 Hours

		1	2	3	4	5	6	7	8	9	10	11	12	13	14	15	16	17	18	19	20	21	22	23	24
0.45	A	0.07	0.51	0.61	0.68	0.74	0.79	0.83	0.87	0.89	0.91	0.93	0.94	0.95	0.96	0.97	0.53	0.42	0.34	0.27	0.22	0.18	0.14	0.12	0.09
	B	0.18	0.61	0.65	0.68	0.72	0.74	0.77	0.79	0.81	0.83	0.85	0.86	0.88	0.89	0.90	0.46	0.41	0.37	0.34	0.30	0.27	0.24	0.22	0.20
	C	0.24	0.67	0.69	0.71	0.73	0.74	0.76	0.77	0.79	0.80	0.81	0.82	0.83	0.84	0.85	0.41	0.39	0.36	0.34	0.32	0.30	0.28	0.27	0.25
	D	0.26	0.71	0.72	0.72	0.73	0.74	0.75	0.76	0.77	0.78	0.78	0.79	0.80	0.80	0.80	0.36	0.35	0.34	0.33	0.32	0.31	0.30	0.29	0.28
0.55	A	0.06	0.69	0.68	0.74	0.79	0.83	0.86	0.89	0.91	0.93	0.94	0.95	0.96	0.97	0.98	0.43	0.35	0.28	0.22	0.18	0.15	0.12	0.09	0.08
	B	0.15	0.68	0.71	0.74	0.77	0.79	0.81	0.83	0.85	0.86	0.88	0.89	0.90	0.91	0.92	0.38	0.34	0.31	0.27	0.25	0.22	0.20	0.18	0.16
	C	0.19	0.73	0.75	0.76	0.78	0.79	0.80	0.81	0.83	0.84	0.85	0.86	0.86	0.87	0.88	0.34	0.32	0.30	0.28	0.26	0.25	0.23	0.22	0.21
	D	0.22	0.76	0.77	0.77	0.78	0.79	0.79	0.80	0.81	0.81	0.82	0.82	0.83	0.83	0.84	0.29	0.28	0.28	0.27	0.26	0.25	0.24	0.24	0.23
0.65	A	0.05	0.69	0.75	0.80	0.84	0.87	0.89	0.92	0.93	0.95	0.96	0.96	0.97	0.98	0.98	0.34	0.27	0.22	0.17	0.14	0.11	0.09	0.07	0.06
	B	0.11	0.75	0.78	0.80	0.82	0.64	0.85	0.87	0.88	0.89	0.90	0.91	0.92	0.93	0.94	0.29	0.26	0.24	0.21	0.19	0.17	0.16	0.14	0.13
	C	0.15	0.79	0.80	0.82	0.83	0.84	0.85	0.86	0.86	0.87	0.88	0.89	0.89	0.90	0.91	0.26	0.25	0.23	0.22	0.20	0.19	0.18	0.17	0.16
	D	0.17	0.81	0.82	0.82	0.83	0.83	0.84	0.84	0.85	0.85	0.86	0.86	0.87	0.87	0.87	0.23	0.22	0.21	0.21	0.20	0.20	0.19	0.18	0.18
0.75	A	0.03	0.78	0.82	0.86	0.88	0.91	0.92	0.94	0.95	0.96	0.97	0.97	0.98	0.98	0.99	0.24	0.19	0.16	0.12	0.10	0.08	0.07	0.05	0.04
	B	0.08	0.82	0.84	0.86	0.87	0.88	0.90	0.91	0.92	0.92	0.93	0.94	0.94	0.95	0.96	0.21	0.19	0.17	0.15	0.14	0.12	0.11	0.10	0.09
	C	0.11	0.85	0.86	0.87	0.88	0.88	0.89	0.90	0.90	0.91	0.91	0.92	0.92	0.93	0.93	0.19	0.18	0.17	0.16	0.15	0.14	0.13	0.12	0.11
	D	0.12	0.87	0.87	0.87	0.88	0.88	0.89	0.89	0.89	0.90	0.90	0.90	0.90	0.91	0.91	0.16	0.16	0.15	0.15	0.14	0.14	0.14	0.13	0.13

TABLE 3-42E Cooling Load Factors when Lights Are on for 16 Hours

		1	2	3	4	5	6	7	8	9	10	11	12	13	14	15	16	17	18	19	20	21	22	23	24
0.45	A	0.12	0.54	0.63	0.70	0.76	0.81	0.85	0.88	0.90	0.92	0.94	0.95	0.96	0.97	0.97	0.98	0.98	0.54	0.43	0.35	0.28	0.23	0.18	0.15
	B	0.23	0.66	0.69	0.72	0.75	0.78	0.80	0.82	0.84	0.85	0.87	0.88	0.89	0.90	0.91	0.92	0.93	0.49	0.44	0.39	0.35	0.32	0.29	0.26
	C	0.29	0.72	0.74	0.75	0.77	0.78	0.80	0.81	0.82	0.83	0.84	0.85	0.86	0.87	0.88	0.88	0.89	0.45	0.42	0.39	0.37	0.35	0.33	0.31
	D	0.31	0.75	0.76	0.77	0.77	0.78	0.79	0.79	0.80	0.81	0.81	0.81	0.82	0.82	0.83	0.83	0.84	0.84	0.40	0.39	0.37	0.36	0.35	0.34	0.33
0.55	A	0.10	0.63	0.70	0.76	0.81	0.84	0.87	0.90	0.92	0.93	0.95	0.96	0.97	0.97	0.98	0.98	0.99	0.44	0.35	0.28	0.23	0.18	0.15	0.12
	B	0.19	0.72	0.75	0.77	0.80	0.82	0.84	0.85	0.87	0.88	0.89	0.90	0.91	0.92	0.93	0.94	0.94	0.40	0.36	0.32	0.29	0.26	0.24	0.21
	C	0.24	0.77	0.79	0.80	0.81	0.82	0.83	0.84	0.85	0.86	0.87	0.88	0.88	0.89	0.90	0.90	0.91	0.37	0.34	0.32	0.30	0.29	0.27	0.25
	D	0.26	0.80	0.80	0.81	0.82	0.82	0.83	0.83	0.84	0.84	0.85	0.85	0.86	0.86	0.86	0.87	0.87	0.33	0.32	0.31	0.30	0.29	0.28	0.27
0.65	A	0.07	0.71	0.77	0.81	0.85	0.88	0.90	0.92	0.94	0.95	0.96	0.97	0.97	0.98	0.98	0.99	0.99	0.34	0.27	0.22	0.18	0.14	0.12	0.09
	B	0.15	0.78	0.81	0.82	0.84	0.86	0.87	0.88	0.90	0.91	0.92	0.92	0.93	0.94	0.94	0.95	0.96	0.31	0.28	0.25	0.23	0.20	0.18	0.16
	C	0.18	0.82	0.83	0.84	0.85	0.86	0.87	0.88	0.89	0.89	0.90	0.90	0.91	0.92	0.92	0.93	0.93	0.28	0.27	0.25	0.24	0.22	0.21	0.20
	D	0.20	0.84	0.85	0.85	0.86	0.86	0.87	0.87	0.87	0.88	0.88	0.88	0.89	0.89	0.89	0.90	0.90	0.25	0.25	0.24	0.23	0.22	0.22	0.21
0.75	A	0.05	0.79	0.83	0.87	0.89	0.91	0.93	0.94	0.95	0.96	0.97	0.98	0.98	0.98	0.99	0.99	0.99	0.24	0.20	0.16	0.13	0.10	0.08	0.07
	B	0.11	0.85	0.86	0.87	0.89	0.90	0.91	0.92	0.93	0.93	0.94	0.95	0.95	0.96	0.96	0.96	0.97	0.22	0.20	0.18	0.16	0.15	0.13	0.12
	C	0.13	0.87	0.88	0.89	0.89	0.90	0.91	0.91	0.92	0.92	0.93	0.93	0.94	0.94	0.94	0.95	0.95	0.20	0.19	0.18	0.17	0.16	0.15	0.14
	D	0.14	0.89	0.89	0.89	0.90	0.90	0.90	0.91	0.91	0.91	0.91	0.92	0.92	0.92	0.92	0.93	0.93	0.18	0.18	0.17	0.17	0.16	0.16	0.15

TABLE 3-43 Sensible Heat Cooling Load Factors for People

Total Hours in Space	Hours after Each Entry Into Space																							
	1	2	3	4	5	6	7	8	9	10	11	12	13	14	15	16	17	18	19	20	21	22	23	24
2	0.49	0.58	0.17	0.13	0.10	0.08	0.07	0.06	0.05	0.04	0.04	0.03	0.03	0.02	0.02	0.02	0.02	0.01	0.01	0.01	0.01	0.01	0.01	0.01
4	0.49	0.59	0.66	0.71	0.27	0.21	0.16	0.14	0.11	0.10	0.08	0.07	0.06	0.06	0.05	0.04	0.04	0.03	0.03	0.02	0.02	0.02	0.02	0.01
6	0.50	0.60	0.67	0.72	0.76	0.79	0.34	0.26	0.21	0.18	0.15	0.13	0.11	0.10	0.08	0.07	0.06	0.06	0.05	0.04	0.04	0.03	0.03	0.03
8	0.51	0.61	0.67	0.72	0.76	0.80	0.82	0.84	0.38	0.30	0.25	0.21	0.18	0.15	0.13	0.12	0.10	0.09	0.08	0.07	0.06	0.05	0.05	0.04
10	0.53	0.62	0.69	0.74	0.77	0.80	0.83	0.85	0.87	0.89	0.42	0.34	0.28	0.23	0.20	0.17	0.15	0.13	0.11	0.10	0.09	0.08	0.07	0.06
12	0.55	0.64	0.70	0.75	0.79	0.81	0.84	0.86	0.88	0.89	0.91	0.92	0.45	0.36	0.30	0.25	0.21	0.19	0.16	0.14	0.12	0.11	0.09	0.08
14	0.58	0.66	0.72	0.77	0.80	0.83	0.85	0.87	0.89	0.90	0.91	0.92	0.93	0.94	0.47	0.38	0.31	0.26	0.23	0.20	0.17	0.15	0.13	0.11
16	0.62	0.70	0.75	0.79	0.82	0.85	0.87	0.88	0.90	0.91	0.92	0.93	0.94	0.94	0.95	0.96	0.49	0.39	0.33	0.28	0.24	0.20	0.18	0.16
18	0.66	0.74	0.79	0.82	0.85	0.87	0.89	0.90	0.92	0.93	0.94	0.94	0.95	0.96	0.96	0.97	0.97	0.97	0.50	0.40	0.33	0.28	0.24	0.21

Time	CLF (Table 17)	Sensible Load No. x q_s x CLF (Btuh)	Latent Load No. x q_l (Btuh)
1200	0.67	680	680
1400	0.76	775	775
1600	0.82	835	835

TABLE 3-44 Sensible Heat Cooling Load Factors for Appliances — Hooded

Total Operational Hours	Hours after appliances are on																							
	1	2	3	4	5	6	7	8	9	10	11	12	13	14	15	16	17	18	19	20	21	22	23	24
2	0.27	0.40	0.25	0.18	0.14	0.11	0.09	0.08	0.07	0.06	0.05	0.04	0.04	0.03	0.03	0.03	0.02	0.02	0.02	0.02	0.01	0.01	0.01	0.01
4	0.28	0.41	0.51	0.59	0.39	0.30	0.24	0.19	0.16	0.14	0.12	0.11	0.09	0.08	0.07	0.06	0.05	0.05	0.04	0.04	0.03	0.03	0.02	0.02
6	0.29	0.42	0.52	0.59	0.65	0.70	0.48	0.37	0.30	0.25	0.21	0.18	0.16	0.14	0.12	0.11	0.09	0.08	0.07	0.06	0.05	0.05	0.04	0.04
8	0.31	0.44	0.54	0.61	0.66	0.71	0.75	0.78	0.55	0.43	0.35	0.30	0.25	0.22	0.19	0.16	0.14	0.13	0.11	0.10	0.08	0.07	0.06	0.06
10	0.33	0.46	0.55	0.62	0.68	0.72	0.76	0.79	0.81	0.84	0.60	0.48	0.39	0.33	0.28	0.24	0.21	0.18	0.16	0.14	0.12	0.11	0.09	0.08
12	0.36	0.49	0.58	0.64	0.69	0.74	0.77	0.80	0.82	0.85	0.87	0.88	0.64	0.51	0.42	0.36	0.31	0.26	0.23	0.20	0.18	0.15	0.13	0.12
14	0.40	0.52	0.61	0.67	0.72	0.76	0.79	0.82	0.84	0.86	0.88	0.89	0.91	0.92	0.67	0.54	0.45	0.38	0.32	0.28	0.24	0.21	0.19	0.16
16	0.45	0.57	0.65	0.70	0.75	0.78	0.81	0.84	0.86	0.87	0.89	0.90	0.92	0.93	0.94	0.94	0.69	0.56	0.46	0.39	0.34	0.29	0.25	0.22
18	0.52	0.63	0.70	0.75	0.79	0.82	0.84	0.86	0.88	0.89	0.91	0.92	0.93	0.94	0.95	0.95	0.96	0.96	0.71	0.58	0.48	0.41	0.35	0.30

TABLE 3-45 Sensible Heat Cooling Load Factors for Appliances — Unhooded

Total Operational Hours	Hours after appliances are on																							
	1	2	3	4	5	6	7	8	9	10	11	12	13	14	15	16	17	18	19	20	21	22	23	24
2	0.56	0.64	0.15	0.11	0.08	0.07	0.06	0.05	0.04	0.04	0.03	0.03	0.02	0.02	0.02	0.02	0.01	0.01	0.01	0.01	0.01	0.01	0.01	0.01
4	0.57	0.65	0.71	0.75	0.23	0.18	0.14	0.12	0.10	0.08	0.07	0.06	0.05	0.05	0.04	0.04	0.03	0.03	0.02	0.02	0.02	0.02	0.01	0.01
6	0.57	0.65	0.71	0.76	0.79	0.82	0.29	0.22	0.18	0.15	0.13	0.11	0.10	0.08	0.07	0.06	0.06	0.05	0.04	0.04	0.03	0.03	0.03	0.02
8	0.58	0.66	0.72	0.76	0.80	0.82	0.85	0.87	0.33	0.26	0.21	0.18	0.15	0.13	0.11	0.10	0.09	0.08	0.07	0.06	0.05	0.04	0.04	0.03
10	0.60	0.68	0.73	0.77	0.81	0.83	0.85	0.87	0.89	0.90	0.36	0.29	0.24	0.20	0.17	0.15	0.13	0.11	0.10	0.08	0.07	0.07	0.06	0.05
12	0.62	0.69	0.75	0.79	0.82	0.84	0.86	0.88	0.89	0.91	0.92	0.93	0.38	0.31	0.25	0.21	0.18	0.16	0.14	0.12	0.11	0.09	0.08	0.07
14	0.64	0.71	0.76	0.80	0.83	0.85	0.87	0.89	0.90	0.92	0.93	0.94	0.95	0.95	0.40	0.32	0.27	0.23	0.19	0.17	0.15	0.13	0.11	0.10
16	0.67	0.74	0.79	0.82	0.85	0.87	0.89	0.90	0.91	0.92	0.93	0.94	0.95	0.96	0.96	0.97	0.42	0.34	0.28	0.24	0.20	0.18	0.15	0.13
18	0.71	0.78	0.82	0.85	0.87	0.89	0.90	0.92	0.93	0.94	0.94	0.95	0.96	0.96	0.97	0.97	0.97	0.98	0.43	0.35	0.29	0.24	0.21	0.18

TABLE 3-46 Typical Electric Motor Efficiencies

Motor Name Plate (Hp)	Motor Type	Nominal rpm	Full Load Motor Efficiency (%)
0.05	Shaded Pole	1500	35
0.08	Shaded Pole	1500	35
0.125	Shaded Pole	1500	35
0.16	Shaded Pole	1500	35
0.25	Split Phase	1750	54
0.33	Split Phase	1750	56
0.5	Split Phase	1750	60
0.75	3-Phase	1750	72
1	3-Phase	1750	75
1.5	3-Phase	1750	77
2	3-Phase	1750	79
3	3-Phase	1750	81
5	3-Phase	1750	82
7.5	3-Phase	1750	84
10	3-Phase	1750	85
15	3-Phase	1750	86
20	3-Phase	1750	87
25	3-Phase	1750	88
30	3-Phase	1750	89
40	3-Phase	1750	89
50	3-Phase	1750	89
60	3-Phase	1750	89
75	3-Phase	1750	90
100	3-Phase	1750	90
125	3-Phase	1750	90
150	3-Phase	1750	91
200	3-Phase	1750	91
250	3-Phase	1750	91

CHAPTER 4

Anatomy of Heating

GAUL IS NOT the only thing that is divided into three principal parts. Heating systems also consist of three different parts: the heat generator; the heat transmitter; and the heat emitter. Each of these parts is also used to classify the types of heating systems. For instance, if we wish to classify them by the generator, we can speak of steam systems, hot water systems, warm air systems, and electric systems. Under the transmission system we can have one pipe, two pipe, three pipe, pressure, vacuum, one duct, two duct,

etc. Then, under the emission devices we can have radiator systems, convector systems, panel systems, etc. Let us now examine each of these parts in detail.

GENERATING SYSTEMS

Heat can be generated in boilers as hot water or steam, in furnaces as warm air, and as electricity in a power plant. Or it can be extracted from a heat source by a heat pump.

Boilers for heating are either cast iron or steel. Cast iron boilers are built of cast iron sections which are assembled on the job. Water circulates inside of the casting, and fire on the outside. The great advantage of cast iron boilers is that even large boilers can be carried through a doorway and assembled on the job. Typical cast iron boilers are shown in Figs. 4–1 and 4–2. In this way, boilers can be replaced without tearing out walls or roofs. On new construction it is not

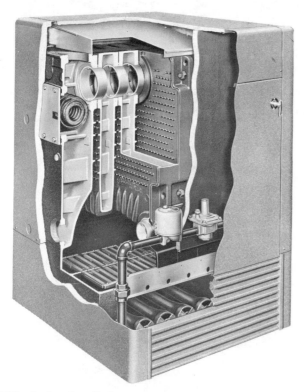

FIG. 4–1 Small Cast Iron Boiler (Courtesy: Weil-McLain Co.)

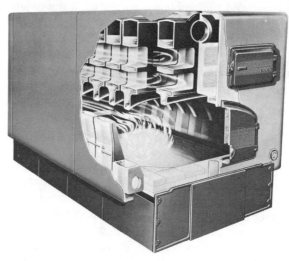

FIG. 4–2 Large Cast Iron Boiler (Courtesy: Weil-McLain Co.)

necessary to get the boiler into the boiler room before walls, floors, or roofs are erected. Within limits, cast iron boilers can be expanded by merely adding sections. Because of the thickness of the metal, corrosion causes less trouble than in steel boilers.

Steel boilers are built in a variety of shapes, but they have in common the fact that they are built of tubes attached to headers called tube sheets. Most common on heating jobs is the fire tube boiler (Figs. 4–3, 4–4), in which fire is on the inside of a tube and water on the outside. Generally restricted to process or very large central heating installations, is the water tube boiler. Here, water is inside of the tube, while the fire is on the outside. Several manufacturers build water tube boilers in the smaller size range to compete with the fire tube (Fig. 4–5).

The steel boiler lends itself to power firing and packaging as a heat generator. It has the advantage of being assembled at the factory with a savings in field labor. After assembly, it is fire-tested and the consumer is assured of a heat generator that needs only power and pipe connections for immediate efficient firing. Because of the thinness of the metal, the tubes are more subject to corrosion than cast iron sections. On the other hand, if a tube springs a leak in the winter, it is not necessary to shut down for a long period for repairs. In a matter of hours the leaking tube is plugged and not used. When convenient, it is replaced.

This is probably a good place to discuss draft. In order to have combustion, we must have oxygen, and after combustion, we must get rid of the products of combustion. A chimney serves this purpose. Its height and the difference in temperature between the flue gases and the ambient air determine the amount of air that will be induced into the boiler for combustion. This is called draft.

The amount of air required and the friction in the combustion chamber, flue passages, breeching, and chimney determine the chimney size. As you can see, the variation in outside temperature will cause a variation in the draft. Wind can increase the draft, but, unfortunately, it can also decrease the draft or cause downdrafts.

Underwriters' rules should be followed in designing chimneys. To reduce downdrafts, extend chimneys 2 ft above the high point of roof. The chimney should be higher than any building within 20 ft of the chimney.

I, for one, prefer forced or induced draft, which makes a boiler independent of nature's whims. In forced draft a fan pushes combustion air into the boiler with sufficient pressure to force the products of combustion out the stack. In induced draft firing, the

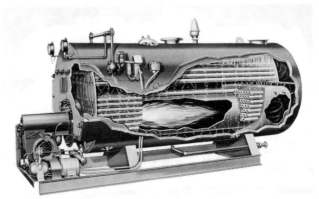

FIG. 4–3 Scotch Marine Package Steel Boiler (Courtesy: Kewanee Co.)

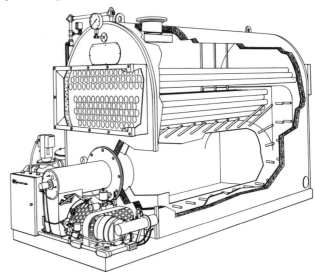

FIG. 4–4 High Firebox Package Steel Boiler (Courtesy: Kewanee Co.)

FIG. 4–5 Water Tube Boiler (Courtesy: Bryan Steam Corp.)

fan is at the discharge end of the boiler and pulls combusion air into the combustion chamber, pulls the products of combustion out of the boiler, and then pushes them out the stack.

Furnaces are similar to the fire tube boiler. Fire is circulated through tubes and air forced around the tubes and, in this way, heated. Commercial and industrial units are called direct-fired space heaters (Fig. 4–6) and are generally built of stainless steel for durability. Two companies still build a cast iron furnace. While the cast iron furnace is rugged and lasts forever, it has the disadvantage of large size by comparison to steel, and a higher initial cost.

The last heat source, electricity, is purchased from a utility, and either is used in resistance heaters, or is used to drive a heat pump. The heat pump will be discussed in a later chapter.

HEAT EMISSION UNITS

We must skip a bit. We will consider the emission units before we consider the transmitting system. This part of our system is the part most associated with comfort because it is in the space to be controlled.

Steam or hot water can be circulated through cast iron radiators, convectors, finned pipe radiation,

and baseboard radiators. The radiator comes in several shapes, one of which is shown in Fig. 4–7. The bulk of the heat is given off by radiation and the rest by convection. To transmit heat by radiation requires much surface and, consequently, the radiator is large and bulky.

Some early pioneers noticed that if an enclosure were built about a radiator, its output increased. Output increased because now the heat was being transmitted by convection. The cold air entered the bottom of the enclosure and as it made contact with the hot surface of the radiator, it expanded, became lighter, and rose. As it rose, it was further heated until it was discharged into the room. Now the heating element could be reduced in size and the convector (Fig. 4–8) was born. The convector is furnished in a variety of sizes, styles, and shapes. It can be floor mounted, wall hung, front outlet, top outlet, fully or semi-recessed, and even have a plaster front. A damper is generally furnished, which controls the flow of air over the heating element.

The next development was the finned radiator (Figs. 4–9, 4–10). This is simply a steel or copper pipe with steel or aluminum fins fastened to the pipe. The fins provide an extended surface for heat transfer by convection. The enclosures come in a variety of shapes, from simple expanded metal to solid front covers with baked enamel finishes. Finned radiators, like convectors, transmit the bulk of their heat by convection.

While both convectors and finned radiators are

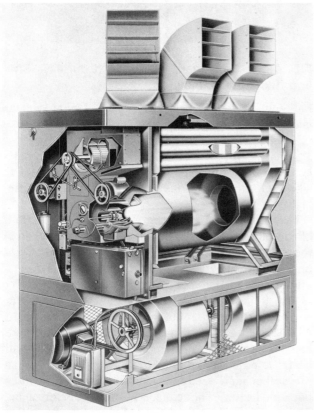

FIG. 4–6 Direct-Fired Space Heater (Courtesy: Lennox Industries Inc.)

FIG. 4–7 Cast Iron Radiator (Courtesy: American Standard, Plbg. & Htg. Div.)

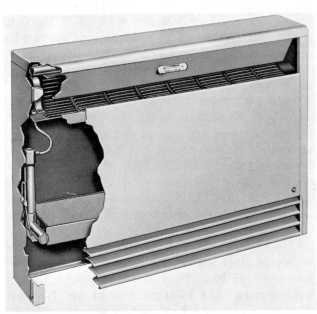

FIG. 4–8 Convector (Courtesy: Modine Mfg. Co.)

FIG. 4–9 Finned Radiator (Courtesy: Sterling Radiator Co. Inc.)

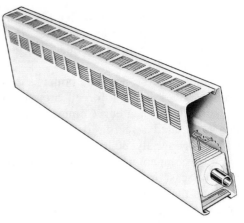

FIG. 4–10 Finned Radiator (Courtesy: Sterling Radiator Co. Inc.)

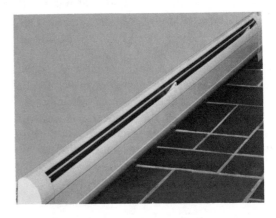

FIG. 4–11 Baseboard Radiator (Courtesy: Sterling Radiator Co. Inc.)

convection units, the convector is used as a concentrated heater and the finned unit as an extended heater. The average 10-ft × 10-ft room can be heated by a 38-in. H × 40-in. L × 6-in. D convector. It can also be heated by an 8-ft-long $1\frac{1}{4}$-in. finned pipe with a 12-in.-high enclosure. The advantage of the finned radiator is that it blankets all the exterior wall with heat, instead of concentrating it in one spot.

A variation of the finned radiator is the baseboard radiator. It, too, consists of a finned pipe with an enclosure. You might say that the baseboard is a scaled down version. The finned radiator uses a $1\frac{1}{4}$-in. pipe with $4\frac{1}{4}$ or $3\frac{1}{4}$ square fins, and the baseboard uses a $\frac{3}{4}$-in. pipe with smaller fins (Fig. 4–11). The enclosure size varies from eight to twelve inches high.

Everytime I think of propeller unit heaters, I think of the old chestnut, "What came first, the chicken or the egg?" What came first—the car radiator or the unit heater? There is a resemblance, and I like to think that some pioneer saw a car radiator and decided to use it for heating. The propeller type can be installed to blow horizontally or vertically (Figs. 4–12, 4–13). The vertical units have a variety of diffusers to compensate for various mounting heights. The unit heater is primarily an industrial heating unit.

Another motorized emission unit is the centrifugal fan unit heater. This can be a small floor or ceiling mounted unit (Figs. 4–14, 4–15), popular for use in heating entrances. A unit 36-in. H × 40-in. L × 12-in. D can produce 50,000 Btuh, in contrast to 12,000 Btuh

FIG. 4–12 Vertical Blow Unit Heater (Courtesy: Modine Mfg. Co.)

FIG. 4–14 Cabinet Unit Heater (Courtesy: Modine Mfg. Co.)

FIG. 4–13 Horizontal Blow Unit Heater (Courtesy: Modine Mfg. Co.)

FIG. 4–15 Ceiling Cabinet Heater (Courtesy: Modine Mfg. Co.)

for a convector of the same physical dimensions. The units are also made in larger sizes for industrial heating (Figs. 4–16, 4–17). Notice that the floor unit in Fig. 4–17 has discharge nozzles for directing the flow of air.

When the large unit heaters shown in Figs. 4–16 and 4–17 have an outside air connection, they are called heating and ventilating units. The outside air provides ventilation. All outside air may be used, or a set of outside-air and return-air dampers may be installed to proportion the amount of outside air. These dampers may be installed in the ductwork, or a mixing box with dampers may be purchased from the manufacturer.

The fans for these heating and ventilating units can be an industrial type, or quiet commercial or institutional type. The units can be floor or ceiling mounted. Centrifugal fan heating and ventilating units can be used with ductwork to distribute heating, or heating and ventilating, over a large area. When a cooling coil and drain pan are added, the units are called air conditioning units. A typical horizontal unit is shown in Fig. 4–18. Notice the similarity to the heating and ventilating unit shown in Fig. 4–16. A refrigeration system can be incorporated into the unit and then we have a self-contained air conditioning unit. These units can be air cooled or water cooled.

The next emission units to be considered are probably the oldest we will discuss. Radiant heating or, more properly, panel heating. The ancient Romans heated their buildings by circulating the products of combustion from charcoal fires through passages in the walls and floors. These gases warmed the floors and walls which, in turn, warmed the room and the occupant by convection and radiation. In effect, the floor and walls became low temperature radiators.

In the early 1900's this principle was rediscovered by Barker, Adlam, and other English engineers. They started by circulating warm air through passages in a fashion similar to the Roman method. They also embedded pipe coils in and under floors, walls, and ceilings. Low temperature hot water (130° maximum) is used in the floor coils, while higher temperature (180° maximum) is used in wall and ceiling coils. The lower temperature is used in the floor to keep the floor surface temperature at a maximum of 90°. Our 85° skin temperature means that 90° will feel warm. It has to be the maximum because higher temperatures cause feet to swell and asphalt tile to melt.

FIG. 4–17 Vertical Heating and Ventilating Unit (Courtesy: American Standard, Industrial Div.)

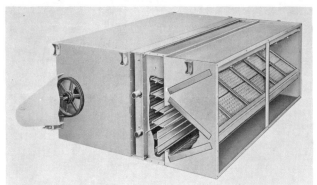

FIG. 4–16 Horizontal Heating and Ventilating Unit (Courtesy: American Standard, Industrial Div.)

FIG. 4–18 Horizontal Air Conditioning Unit (Courtesy: American Standard, Industrial Div.)

Higher temperatures are used for walls and ceilings. These systems are very comfortable for two reasons. One, the heat is introduced at a low temperature and over a large area. Secondly, our bodies lose most of their heat by radiation and so, greater comfort results if the heat lost is replaced in the same fashion as it is lost. At the peak of its popularity, some extreme benefits were claimed. Some claimed that rooms could be kept at lower temperatures because the bulk of the heat was introduced by radiation. They jumped to this conclusion because the English areas heated by panel heating were kept at lower temperatures than American rooms. But this was due to custom and differences in clothing habits between Americans and Englishmen.

The system has one large drawback. The heating panels, because they are part of the structure, have a large mass. Once a floor is heated, for instance, it is a gigantic flywheel. If the outdoor temperature warms suddenly, or if a large internal heat load, such as people, enters the room, severe overheating results. The system just cannot adjust quickly. To overcome this, panel heaters should be used with air systems which can adjust quickly to changing loads. The extra burden of installing an air system causes the combination system to be expensive.

A development for overcoming the problem of the flywheel effect is the suspended metal pan radiant ceiling. Perforated ceiling panels are fastened to pipes that carry hot water or chilled water. Fig. 4–19 shows a diagram of the panel parts, while Fig. 4–20 shows a partially completed installation. Fig. 4–21 is a new panel for use with lay-in ceilings. In this way, we have a unit that both heats and cools. If cooling is desired, then an air system must be used along with the panels to provide additional cooling and dehumidification

in order to prevent condensation on the panel surface. The air system also provides ventilation.

The unit ventilator was a natural outgrowth of the unit heater. Classrooms, in particular, needed a unit which could both heat and ventilate. As can be seen in Fig. 4–22, the unit ventilator adds to the cabinet unit heater an outside-air connection, and return-air and outside-air dampers. With this unit, heat is provided along with a minimum amount of outside air.

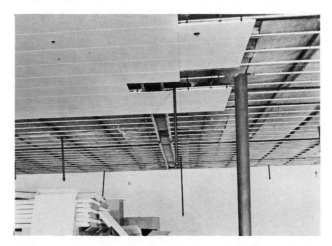

FIG. 4–20 Partially Completed Radiant Panel (Courtesy: Burgess Manning Co.)

FIG. 4–21 Radiant Panel For Use With Lay-In Ceiling (Courtesy: Burgess Manning Co.)

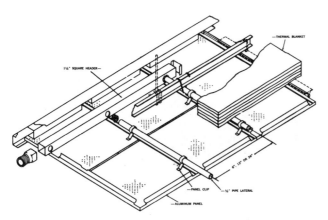

FIG. 4–19 Ceiling Radiant Panel Diagram (Courtesy: Burgess Manning Co.)

If the room needs cooling, more outside air is brought in by opening the outside-air damper. The concentration of all of the heat (for a room with 30 ft of outside wall) in a 5-ft-long unit, left much of the wall without heat.

Drafts resulted from the cold unheated wall and the body radiated heat to this wall. Various things were done to overcome this. Some manufacturers furnish finned radiation to be installed along with the unit ventilator (Fig. 4–23). Other manufacturers discharge part of the heated air through wings that extend from the unit to the partitions (Fig. 4–24). Instead of supplying air to these wings, some manufacturers use the wings to pull cold air off the window and keep it from hitting the floor (Fig. 4–25). Unit ventilators can be provided with either steam, hot water, or electric coils. The use of a chilled water and the addition of a drain pan makes it an air conditioning unit. They are also suspended and used with and without ductwork.

Cousins to the unit ventilator are the fancoil units (Fig. 4–26), which can be small enough to use in a residence, or large enough for a large office. They generally have only a manual damper instead of the motorized outside-air damper of the unit ventilator. They can be used for heating only, or heating and air conditioning, by circulating hot water or chilled water. A compressor and condenser can be added to the fan-coil unit to make a self-contained air conditioner. Figure 4–27 shows an exploded view of a typical air-cooled unit. Item No. 1 is the room cabinet, No. 2 is the air-cooled condensing unit, No. 3 is the heater section, consisting of fans and a hot water or steam coil, No. 4 is the wall sleeve, or box, and No. 5, the outside-air louver. Figure 4–28 shows another similar unit with an electric heating coil. In the smaller sizes, 15,000 Btuh cooling or less, the units bring in only a minimum amount of outside air. In the larger units, 24,000 Btuh and 48,000 Btuh, the units are available

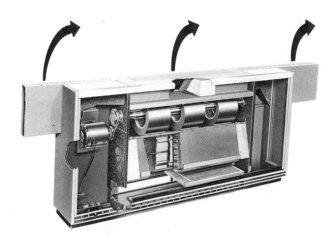

FIG. 4–24   Unit Ventilator With Discharge Wings (Courtesy: The Trane Co., La Crosse, Wis.)

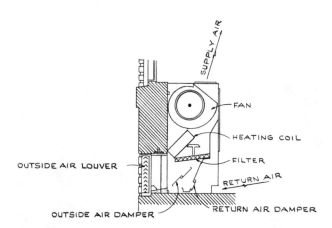

FIG. 4–22   Section Through Unit Ventilator

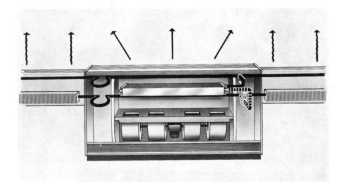

FIG. 4–23   Unit Ventilator With Radiation Wings (Courtesy: ITT Nesbitt, Inc.)

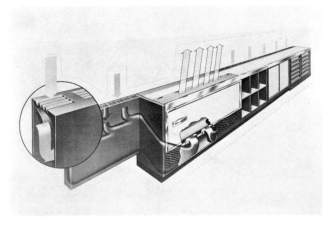

FIG. 4–25   Unit Ventilator With Air Intake Wings (Courtesy: American Air Filter Co., Inc.)

with 100% outside air. These larger units are really unit ventilators with refrigeration units. Both the large and small units can be purchased without the refrigeration unit. The refrigeration can then be installed at a future date.

THE TRANSMISSION SYSTEM

Steam and hot water are transmitted in pipes, warm air in ductwork, and electricity in wires. Steam, water, and air systems can be subclassified by the piping or ductwork arrangement, or pressure conditions.

STEAM–HEATING SYSTEMS

Steam-heating systems may be classified according to any one of, or a combination of, the following fea-

FIG. 4–26 Fancoil Unit (Courtesy: American Standard, Industrial Div.)

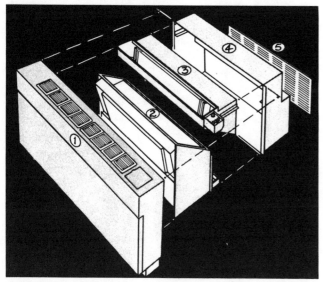

FIG. 4–27 Remington Self-Contained Air Conditioning Unit (Courtesy: The Singer Co., Climate Control Div.)

tures: (A) by the piping arrangement; (B) by the pressure or vacuum conditions obtained in operation; (C) by the method of returning condensate to the boiler.

  A. *By the Piping Arrangement:*

    1. A steam-heating system is known as a *one-pipe system* when a single main serves the dual purpose of supplying steam to the heating unit and conveying the condensate from it. Ordinarily there is but one connection between the main and the heating unit which must serve as both the supply and the return, although separate supply and return connection may be used.

    2. A steam-heating system is known as a *two-pipe system* when steam and condensate flow to and from the heating unit in separate mains and branches.

    3. The systems may also be described as *up-flow* and *down-flow,* depending on the direction of steam flow in the risers, and as a *dry-return system* or a *wet-return system,* depending on whether the condensate mains are above the water line in the boiler or condensate receiver.

  B. *By Pressure or Vacuum Conditions:*

    1. Steam-heating systems may also be classified as *high-pressure, low-pressure, vapor,* and *vacuum systems,* depending on the pressure conditions under which the system is designed to operate.

    2. A system is known as a *high-pressure system* when the operating pressures employed are above 15 psi; as a *low-pressure system* when pressures range from 0 to 15 psi; as a *vapor system* when the system operates under both vacuum and low-pressure conditions without the use of a vacuum pump; and as a *vacuum*

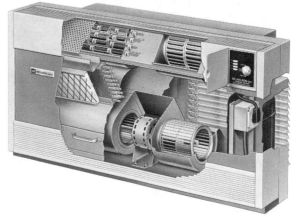

FIG. 4–28 Self-Contained Air Conditioning Unit (Courtesy: American Air Filter Co., Inc., Herman Nelson Div.)

system when the system operates under low-pressure and vacuum conditions with the use of a vacuum pump.

3. When automatic controls are employed to vary the pressure conditions in the system in accordance with outside weather conditions, the system may be known as a *sub-atmospheric, differential,* or *synchronized system.* These latter classifications are proprietary designations.

4. When orifices are employed on the inlets to the heating units, the system may be known as an *orifice system.*

C. *By Method of Returning Condensate:*

1. When condensation can be returned to the boiler by gravity, the system is known as a *gravity-return system.* All heating units in a gravity-return system must be located sufficiently above the water line of the boiler so that the condensation can flow freely to the boiler. The elevation of heating units above the water line must be sufficient to overcome pressure drop due to flow, as well as pressure differences between supply and return piping due to operating conditions.

2. When conditions are such that condensation cannot be returned to the boiler by the action of gravity, either traps or pumps must be employed to return the condensation to the boiler, and the system is known as a *mechanical-return system.* There are three general types of mechanical-return devices in common use with mechanical-return systems today. These are the alternating-return trap, the condensate-return pump, and the vacuum-return pump.

3. In systems where pressure conditions in the system vary between that of a gravity-return and a forced-return system, a boiler return trap or alternating receiver is employed, and the system may be known as an *alternating-return system.*

4. When condensate is pumped to the boiler under atmospheric pressure or above, the system is known as a *condensate-return pump system.*

5. When condensate is pumped to the boiler under vacuum conditions, the system is known as a *vacuum pump-return system.* In either the condensate or vacuum pump-return systems, it is highly desirable to arrange for gravity flow to a receiver and to the pump. The pump then forces the condensate into the boiler against boiler pressure.

ONE-PIPE STEAM-HEATING SYSTEMS: (Fig. 4–29)

A. *Description*

In the one-pipe steam system, steam and condensate run in the same pipe. Steam fills the radiator, driving out the air through the air valve. When the steam condenses, it runs back down the steam riser to the main. In the riser, steam and condensate are flowing in opposite directions. Care must be taken to correctly size and install the riser, or water hammer will result. Generally, the steam and condensate flow in the same direction in the main—away from the boiler. At the end of the main an air vent is installed and a return pipe returns the condensate to the boiler. It can be either a dry-return or a wet-return.

B. *Advantages*

1. Low first cost.
2. Simple in principle.

C. *Disadvantages*

1. Very poor control.
2. Unless properly designed and installed, system can be very noisy and heat in a spotty fashion.
3. Heating can only be provided above the boiler.
4. Because the system is essentially a gravity system, piping is difficult to conceal and can cause architectural problems.

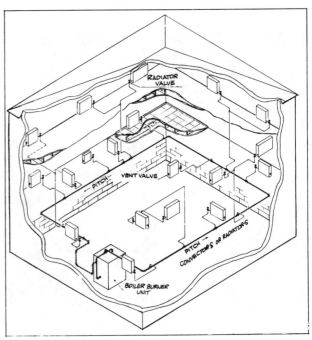

FIG. 4–29 One-Pipe Steam System

5. For best results, only large cast iron radiators should be used.

6. Limited to small buildings.

7. Because of the large amounts of make-up water, corrosion is accelerated.

8. Humidity cannot be added.

9. Cannot be used for air conditioning.

D. *Applications*

The one-pipe system is rarely used today. In the past it was used in residences, small commercial, and small institutional buildings.

TWO-PIPE LOW-PRESSURE STEAM SYSTEMS
(Fig. 4–30)

A. *Description*

These systems operate at pressures of from 0 to 15 psi. As the name implies, two pipes are used—one to carry steam and the other condensate.

Separate connections are made to the heat emission units. A steam trap is employed on the return to be sure that the steam cannot enter the return system. Traps can be thermostatic type, in which a thermostatic element closes the outlet port until the steam is condensed; float type, in which a float operates the outlet port; and the float and thermostatic type, which has both a float and a thermostatic element. Pressure-type air valves should be

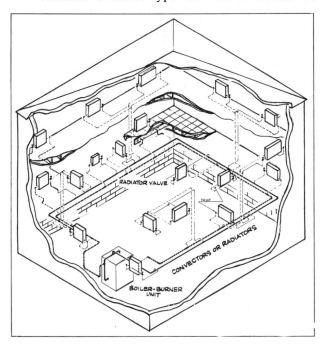

FIG. 4–30 Two-Pipe Steam System

provided on supply and return mains. Quick-vent type should be used on supply, and thermostatic type on returns.

B. *Advantages*

1. Better control than one-pipe steam.

2. Can be used in any size building.

3. Can use all kinds of heating emission units.

C. *Disadvantages*

1. Although the control is better than in a one-pipe steam system, it still is not as good as a hot-water system.

2. Unless properly designed and installed, system can be very noisy and heat in a spotty fashion.

3. Heating can only be provided above the boiler unless the condensate pump is used to lift the condensate back to the boiler.

4. Because the system is essentially a gravity system, piping is difficult to conceal and can cause architectural problems.

5. Because of the large amounts of make-up water, corrosion is accelerated.

6. Humidity cannot be added.

7. Cannot be used for air conditioning.

D. *Applications*

Two-pipe steam systems are not used as much as in former years. They can be used in all types of buildings that do not require that the same piping system be used for heating and cooling. In some cases, a two-pipe steam system is used to transport heating from a central plant to remote buildings where a heat exchanger is introduced to heat hot water with steam. However, when this is done, generally higher pressures are employed.

TWO-PIPE VAPOR AND VACUUM SYSTEMS

A. *Description*

Again, separate mains are used for steam and condensate. Vapor and vacuum systems differ from the pressure system because they operate under vacuum conditions. Vapor systems do not require a vacuum pump. A vapor system creates its own vacuum.

Steam fills the radiator, driving out the air through air valves which do not permit the re-entry of air. When the steam condenses, a vacuum is created. The vacuum lowers the boiling point of water and the condensate flashes back to steam. Vacuums as high as 30 in. of mercury are created. The vacuum will be retained for as long as five hours in a tight system. Thermostatic, or float and thermostatic traps are used on each heat emission element. In a vacuum system, a vacuum pump is used.

B. *Advantages*

1. More comfortable heating than either one-pipe or two-pipe steam.

2. Steam is distributed equally to all parts of the building, reducing the possibility of short circuiting which can cause a cold room.

3. Can use all kinds of heating emission units.

C. *Disadvantages*

1. Control, although better than one-pipe or two-pipe systems, is not as good as hot-water systems.

2. More expensive to install than one-pipe and two-pipe systems.

3. Heating can only be provided above the boiler unless the condensate pump is used to lift the condensate back to the boiler.

4. Because the system is essentially a gravity system, piping is difficult to conceal and can cause architectural problems.

5. Because of the large amounts of make-up water, corrosion is accelerated.

6. Humidity cannot be added.

7. Cannot be used for air conditioning.

D. *Applications*

Vapor systems, like all steam systems, are not used too much today. They can be used in residential and small commercial and institutional buildings. Vacuum systems can be used in any type of building.

VARIABLE VACUUM SYSTEMS

This is a specialized vacuum system used in larger buildings of all types. During very cold weather, steam is circulated under pressure conditions. As the outside temperature rises, a vacuum is introduced. The higher the outside temperature, the higher the vacuum. In this way, during mild weather, a small amount of steam is generated and the high vacuum stretches it to fill the entire system. This permits equal heating throughout the building at low steam temperatures of 180°.

HARTFORD LOOP (Fig. 4–31)

A boiler and its return piping is nothing more than a large manometer. With no steam, both ends are at the same level. If the burner is turned on and steam generated, a pressure develops. This pressure not

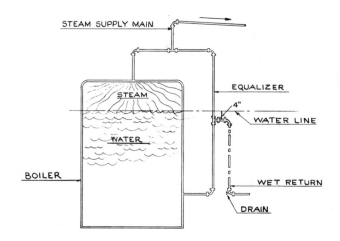

FIG. 4–31 Hartford Loop

only pushes up to force steam out of the boiler, but, because steam is a gas, it exerts pressure equally in all directions. So a pressure is exerted against the sides of the boiler and the boiler must be strong enough to contain the pressure. Pressure is also exerted down on the water. If a pressure of 1 psig is developed, it will push down and force the water up 28 in. in the return main. Until some water is condensed in the radiator and flows to the return main and raises the level in the return main, no water can enter the boiler. Of course, if the boiler shuts down, the pressure drops, water enters the boiler, and the levels are again equal. As you can see, if a leak or break develops in the return main, the boiler will lose all its water. By installing the Hartford Loop, the pressures at the supply and return side of the boiler are balanced, and the boiler is protected against a break or a leak. The water line is not depressed by the boiler pressure, but a height of water greater than the boiler pressure must build up in order for water to return to the boiler.

WATER SYSTEMS

Hot-water heating systems can be subdivided as follows:

A. *Low-pressure and temperature (under 250°)*

1. Gravity.
   a. One-pipe.
   b. Two-pipe.
2. Forced.
   a. One-pipe.
   b. Two-pipe.
   c. Three-pipe.
   d. Series loop.

B. *High-pressure and temperature (above 250°)*

GRAVITY

The early gravity systems were, when properly designed and installed, engineering gems. They are no longer used. In fact, the ASHRAE discontinued publishing design data in the 1957 issue of the yearly *Guide and Data Book*.

ONE-PIPE FORCED HOT WATER (Fig. 4–32)

The one-pipe hot-water system utilizes a single main with supply and return connections to each heat emission device. Special tees are installed which cause some of the water to be diverted to the radiator or convector. One-pipe systems are used primarily in residences and small commercial and institutional buildings.

A. *Advantages*

1. Excellent control, which provides maximum comfort.

2. Because only small amounts of make-up water are added to the system, corrosion is held to a minimum, and if properly designed and installed, corrosion should be non-existent.

3. System is inexpensive, although more expensive than a warm-air system.

4. Because system is a pumped system, it is easier to conceal piping and it is, therefore, more desirable architecturally.

B. *Disadvantages*

1. System cannot be used for air conditioning.

2. Humidity cannot be added.

3. Can only be used with gravity convection heating units.

TWO-PIPE FORCED HOT WATER (Fig. 4–33)

In this system two pipes are used—one for supply, and the other for return water. A connection is made from each main to the heat emission unit. The two-pipe system can be either direct-return or reverse-return. In the direct-return system, the first heat emitter has a shorter path than the second, and so on. This means that unless a balancing device is introduced to add pressure drop or friction, the first units will overheat, while the further ones will be starved for water and underheat. While balancing devices can be used, if there are a large number of units to balance, the work can be quite long, involved, and tedious. A better solution is to reverse the return, or supply, for that matter, and in this way provide equal paths of approximately equal length for each heat emitter in the system. A system like this is virtually self-balancing.

We have now arrived at the first system to be considered that can be used for both heating and cooling. By the use of fancoil units and the circulation of chilled water in summer, we can have an "almost"

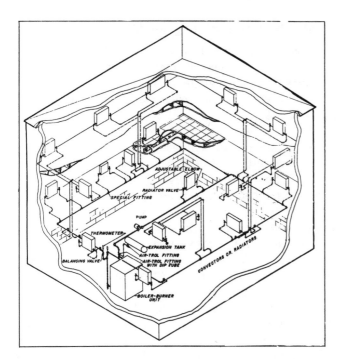

FIG. 4–32 One-Pipe Forced Hot Water System

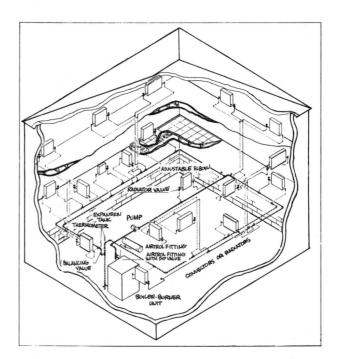

FIG. 4–33 Two-Pipe Forced Hot Water System

air conditioning system. I say "almost" because it does everything but humidify in the winter. The system can be used in all building types.

A. *Advantages*

1. Excellent control which provides maximum comfort.

2. Because only small amounts of make-up water are added to the system, corrosion is held to a minimum and, if properly designed and installed, corrosion should be non-existent.

3. System is inexpensive, although more expensive than warm-air system.

4. Because the system is a pumped system, it is easier to conceal piping and it is, therefore, more desirable architecturally.

B. *Disadvantages*

1. Humidity cannot be added.

THREE-PIPE SYSTEMS

Modern architecture brought on the three-pipe system, which uses one pipe with hot or warm water, one with cold or cool water, and the third with mixed water.

Why three pipes? Modern buildings have large amounts of glass. While this glass is a source of much heat loss, it can, conversely, be the source of much heat gain if the sun is shining, even on a cold winter day. Coupling this with the large amount of lighting used in modern buildings, we can have a situation where one side of the building needs heating and the other side cooling. With a two-pipe system you can only satisfy one side of the building; with three pipes, both sides can be satisfied. This system has all the advantages of the two-pipe system, plus the flexibility just mentioned. The system can be used in all building types.

SERIES-LOOP SYSTEM (Fig. 4–34)

This system was developed for use in residential work after the baseboard radiator was introduced. In this system, the heat emission units, the baseboard radiators, are used as the main and are interconnected by short pieces of pipe. It derives its name from the fact that the mains are in series with the heat emitter, while in the other system they are in parallel. While primarily used in residential work, it is also used in small commercial and institutional work.

AIR SYSTEMS

Early warm-air systems were gravity type. These systems were limited in use to residences and, when

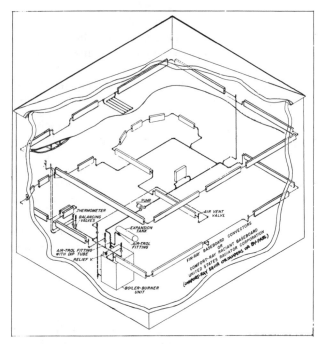

FIG. 4–34   Series Loop Hot Water System

fired with coal, were a most comfortable system. They did not adapt well to gas and oil firing. This, along with the large ducts and the fact that the furnace had to be in the middle of the basement, using up much space, caused this system to be abandoned in favor of the forced warm-air system.

FORCED WARM AIR (Fig. 4–35)

The early forced-air systems added only a fan to the gravity unit, leaving the furnance location and ductwork the same as the gravity system. The supply registers were on the inside wall and the return on the outside.

While this was necessary for gravity systems to insure circulation, forced systems did not need to be tied down to this location. Many men tried to change the location of the supply registers to the outside wall, but failed until an enterprising pioneer glamourized the move by calling systems with registers on the outside wall "perimeter systems."

The use of forced systems enabled the heating people to install the furnace in a corner of the basement or in a furnace room. Also, flat ductwork could be installed with an increase in headroom. Now, for the first time, basements could be heated and used for recreation rooms. Figure 4–36 shows a typical furnace unit with a cooling coil. These systems are used primarily in residences, but are also used in small commercial buildings.

A large furnace suitable for use with or without duct-work is shown back in Fig. 4–6. The units shown back in Figs. 4–16, 4–17, and 4–18 can be used with ductwork to heat, heat and ventilate, or air condition. These single-zone systems are controlled by a single thermostat. While this may be fine in a small building,

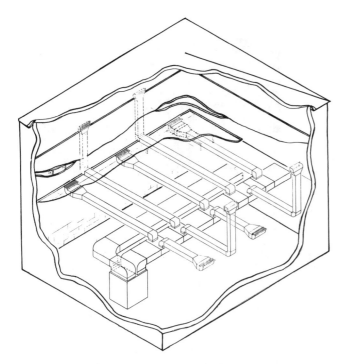

FIG. 4–35 Warm Air System

A. *Advantages*
  1. Least expensive system.
  2. Least expensive method of providing year-round air conditioning.
  3. Humidity can be added in the winter.
  4. If electronic air cleaners are used in conjunction with the air system, the dust in the house is reduced considerably, which reduces housekeeping. The electronic air cleaner is also a blessing to people who suffer from allergies because it removes dust and pollen.
  5. Although the ductwork is difficult to conceal, the system is extremely flexible and enables the architect to run windows right down to the floor.
B. *Disadvantages*
  1. Unless properly designed, system can be noisy.
  2. Unless properly designed to provide continuous air circulation, the system will not be as comfortable as the hot-water system.
  3. Large ductwork to conceal presents architectural problems.

Big brothers to the residential warm-air system are a variety of air systems. The source of heat can be a gas-fired unit heater (Fig. 4–37) used in the same manner as hot-water and steam units. The gas-fired duct furnace (Fig. 4–38) is installed in ductwork where the air is moved by either a separate centrifugal fan or a fan that is part of a package air conditioner.

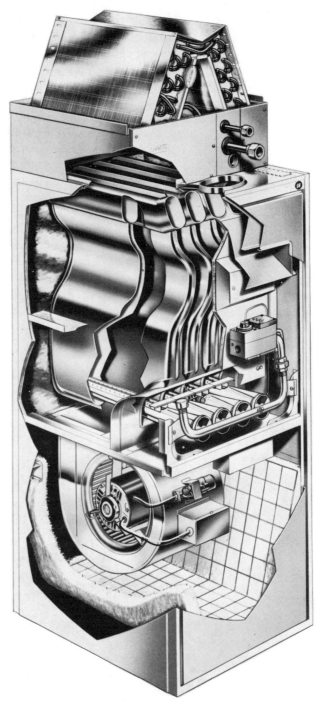

FIG. 4–36 Warm Air Furnace (Courtesy: Lennox Industries Inc.)

it is not satisfactory in most buildings because of the requirements of multiple zones or individual room control.

Rooms can be grouped together into zones and we have the multizone system. To provide individual room control, we can use reheat systems, induction systems, and dual duct systems.

The multizone system gathers the rooms into zones. Zones are generally arranged by exposure. All of the rooms facing north are grouped into one zone, and so on. Further zoning might be by floors. A private office, or a conference room with its varying loads, might be set up as single zones. The zoning is accomplished at the unit. A blow-through unit, where the fan blows through the coils, is used. Coils are arranged in parallel, generally one above the other. The coils discharge into hot and cold plenums. Attached to the plenums are zone dampers, arranged so that as the hot plenum damper opens, its respective cold damper closes. A room thermostat positions these dampers. There can be as many as eight or even twelve zones. The heating does not have to be provided by hot-water or steam coils. Sometimes, gas-fired duct heaters are used instead of heating coils, as shown in Fig. 4–39. This figure gives a good cutaway of a multizone. Figure 4–40 shows a typical multizone with a heating coil and cool-

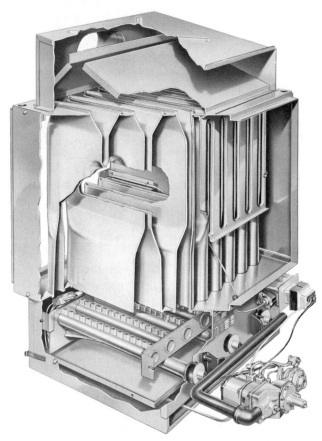

FIG. 4–38 Gas-Fired Duct Furnace (Courtesy: Modine Mfg. Co.)

FIG. 4–37 Gas-Fired Unit Heater (Courtesy: Modine Mfg. Co.)

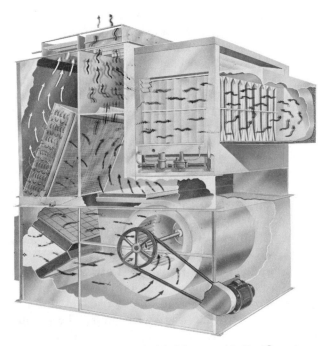

FIG. 4–39 Gas-Fired Multizone Unit (Courtesy: Airfan Engineering Co.)

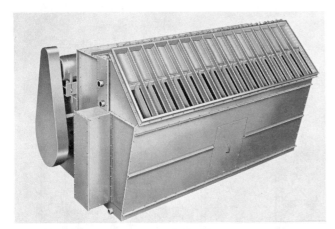

FIG. 4–40 Air Conditioning Multizone Unit (Courtesy: American Standard, Industrial Div.)

FIG. 4–41 Heating And Ventilating Multizone Unit (Courtesy: American Standard, Industrial Div.)

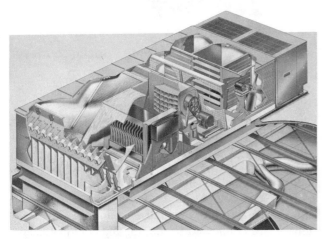

FIG. 4–42 Rooftop Multizone Unit (Courtesy: Lennox Industries Inc.)

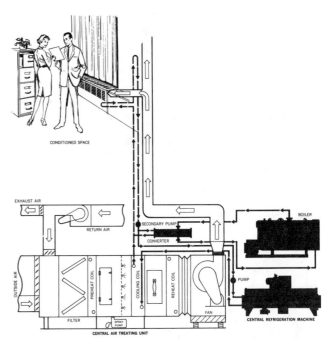

FIG. 4–43 Typical Induction System (Courtesy: American Standard, Industrial Div.)

ing coil, while Fig. 4–41 is a heating and ventilating only multizone. In this unit, the cold deck is only a bypass. A rooftop unit is shown in Fig. 4–42.

In the reheat system, air at a temperature cold enough to satisfy the cooling requirements of the worst room is transported to each room. In the room, the air is taken into a reheat cabinet where it is heated to a temperature that satisfies that room. These cabinets can be located above a ceiling or on the floor under the windows. Reheating can be provided by hot-water coils, steam coils, or electric coils.

An induction unit saves space by reducing the amount of air to be circulated. Air is cooled to a low dewpoint in a central unit. This cold dry air is carried in risers to the terminal units. The terminal unit contains a secondary coil which must be supplied with hot or chilled water, depending on the season. Figure 4–43 shows a typical induction system. Notice that the water serving the secondary coils first goes through the primary coil before it is taken to the secondary coil. I should point out that this is during the cooling season; in the heating season, the water bypasses the coils. The reason to take the water through the primary coil first, is to assure that the water entering the secondary coil is warmer than the air entering the unit, reducing the possibility of condensation on the secondary coil. Dehumidification is accomplished by the dry primary air, and the secondary coil accom-

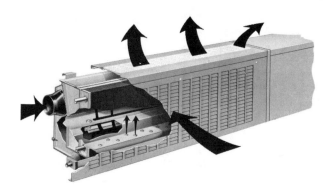

FIG. 4–44 Induction Unit (Courtesy: American Standard, Industrial Div.)

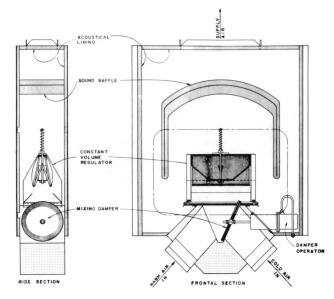

FIG. 4–46 Dual Duct Mixing Box (Courtesy: Buensod-Stacey Corp.)

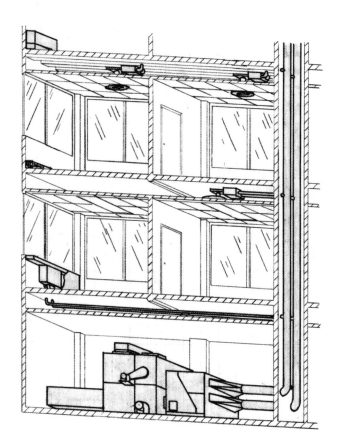

FIG. 4–45 Dual Duct System (Courtesy: Buensod-Stacey Corp.)

FIG. 4–47 Open End Mixing Box (Courtesy: Barber-Colman Co.)

plishes only additional cooling. In Fig. 4–44 we have a section through a typical induction unit. The high-pressure primary air enters the unit and is discharged through nozzles behind the secondary coil. These high-pressure jets induce room air through the grille, and then the secondary coil, where it mixes with the primary air and is discharged through the top grille into the room. In general, the induction rate is one-to-four, so for every one cfm of primary air, four cfm is induced from the room. You can see that the amount of air circulated is considerably reduced.

In a dual duct system (Fig. 4–45), a hot duct extending from the hot plenum and a cold duct extending from the cold plenum circulate air through the building. Mixing boxes that contain hot and cold dampers are connected to the hot and cold mains and this mixed air is extended to the rooms. This air is introduced from above through diffusers or through

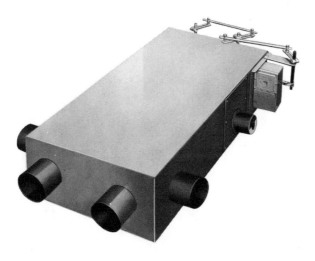

FIG. 4–48 Octopus Mixing Box (Courtesy: Barber-Colman Co.)

FIG. 4–50 Floor-Type Mixing Box (Courtesy: Buensod-Stacey Corp.)

FIG. 4–49 Mixing Box With Diffuser (Courtesy: Barber-Colman Co.)

registers located on walls. Many times the mixing box itself is placed on the floor under a window and the air enters the room directly from the mixing box. Figure 4–46 shows a section through a typical dual duct mixing box. A room thermostat operates the damper motor to position the damper to provide the required mixture of hot and cold air. The constant-volume regulator keeps the total air quantity constant by adjusting to changing system pressures. The sound baffle and acoustical lining attenuate the noise created within the dampers and constant-volume regulator. Figure 4–47 shows an open end box. A duct is connected to this open end and serves a number of diffusers or registers. An octopus box is shown in Fig. 4–48. Individual, round, flexible ducts are run from the box to ceiling diffusers or other terminals. Figure 4–49 shows a dual duct box, complete with a diffuser, and Fig. 4–50 shows a box built into a cabinet. This box is generally built into window sills.

CHAPTER 5

Ventilation

"UNTIL QUITE recently ventilation has been generally regarded as a luxury rather than as an absolute necessity. The discomfort of a poorly ventilated room has been realized with sufficient vividness, but the difficulty of substituting for the debilitating atmosphere one that is pure and invigorating has in many cases been so far beyond the power of ordinary methods to accomplish that a crowded apartment and a vitiated atmosphere have been looked on as inseparable. But such an atmosphere is more than uncomfortable and disagreeable; it is positively and undeniably injurious and continued exposure to it is certain to lead to serious consequences.

"The evil effects of lack of ventilation are made only too evident by such facts as that death rates have been reduced by the introduction of efficient ventilating systems in children's hospitals from 50 to 5%. . . ."

The above quotation is certainly interesting isn't it? Even more interesting is that it appeared in a book called *Ventilation and Heating,* published by B. F. Sturtevant Company in 1896. Now, some sixty-eight years later, people still forget ventilation.

Surprisingly enough, the amount of outside air required for ventilation is not dictated by our breathing requirements, but by (you should forgive the expression) body odor. It takes less than one cfm per person to replace the oxygen consumed in breathing, and less than four to keep the carbon dioxide concentration within limits. However, if the occupancy rate is one person per 150 cu ft, we require 20 cfm per person to remove objectionable body odors. This is for a sedentary adult; for moderate activity, 36 cfm per person is

required. As the space per person increases, the ventilation required decreases, as follows:

Air Space cu ft per person	cfm per person sedentary	cfm per person mod. act.
200	16	24
300	12	18
400	9	13
500	7	12
750	5	7

The average classroom would require 12 cfm per person × 30 people = 360 cfm of outside air. If you depended on infiltration, a 5-mph wind would supply the average classroom with 10 cfm. The average wind velocity for the Detroit area, for instance, is about 12 mph, so the infiltration will average 40 cfm. This is a long way from 300, isn't it? So a schoolroom that depends on infiltration alone for ventilation can get, shall we say, "ripe."

But while odor is bad, an even more serious problem can arise when ventilation is missing. Moisture! That's right, moisture. Moisture levels in the room can get quite high. In fact, it is possible to reach nearly 100% relative humidity. This is very uncomfortable and not very healthy. It is even bad for building materials. Where does all this moisture come from? The occupants of the room. Twenty-five youngsters and a teacher give off two pounds of water per hour. The twelve pounds given off in six hours can, as I said earlier, give us near 100% relative humidity.

What is the effect on building materials? Well, we can get visible condensation on windows. This can run down to the window sill, and if it is wood—you know what happens. But hidden condensation is an even greater evil. All material is porous to varying degrees. With a poor vapor barrier on the room side, and a high moisture content, moisture will condense inside the building material. This will cause dry rot, oxidation, or other forms of failure. Remember those old cellars that were full of mildew and were green in color and odor? I have seen schoolrooms where hidden condensation caused the schoolroom to look and smell like an old cellar. Not only can hidden condensation cause structural damage, it can also destroy the insulating effect of building materials.

Toilet rooms that do not have windows must, by code, be exhausted. The Detroit Plumbing Code, for instance, requires one air change at least every seven minutes. It has been our experience that even toilet rooms with windows need exhausting.

A very important reason for exhaust ventilation is to remove the heat from a process. The process can be an industrial one where you also remove harmful contaminants. Other examples of heat producing processes are cooking, laundering, or the mysterious rituals in beauty parlors. In the case of kitchen hoods, 100 cfm per sq ft of hood should be exhausted. Island hoods should exhaust 150 cfm per sq ft of hood area. Also, consult the local code.

Too often forgotten is the fact that to exhaust air you must have air. If you do not supply tempered make-up air to the space, exhausted air will be pulled into the building through every crack or crevice, creating a most uncomfortable situation. Not only are objectionable drafts created, but, unless the heating system has the capacity to handle 100 times the planned infiltration, cold rooms will result. In addition, fans that are starved for air cannot work efficiently and the air quantity exhausted can be reduced 50% or more. I remember a building, a bar to be specific—well it was more high class than a bar, it was a cocktail lounge. Anyhow, this lounge was planned around a mammoth fireplace. It was called the Fireplace Lounge. Everytime they built a fire in the fireplace, the building would fill with smoke. So we had the Fireplace Lounge with a useless fireplace. What was the problem? NO MAKE-UP AIR! The kitchen exhaust hood pulled air down the chimney and, of course, filled the building with smoke.

Another reason for ventilation comes to mind—to provide air for combustion. A good rule of thumb is one sq in. of louver per 1000 Btu/hr input for boilers or heating equipment with atmospheric burners. All combustion devices in a boiler room must be considered in calculating the louver size—boiler, water heater, and incinerator. If forced or induced draft is used, the louver size can be reduced by one half. In this way we reduce the louver size to something a little more compatible with the architecture. Another reason is that we can mix outside air with room air to a point above freezing, so we don't freeze any piping in the path of the air.

Remember to ventilate in order to:

Replace	Oxygen
Dilute	Carbon Dioxide
Dilute	Odors
Dilute	Moisture
Remove	Heat
Dilute	Noxious gases or dusts
and Provide	Make-up air.

TABLE 5-1
Outdoor Air Requirements

Application	Smoking	Cfm per Person Recommended	Cfm per Person Minimum	Cfm per Sq Ft of Floor Minimum
Apartment				
Average	Some	20	10	....
DeLuxe	Some	20	10	....
Banking space	Occasional	10	7½	....
Barber shops	Considerable	15	10	....
Beauty parlors	Occasional	10	7½	....
Brokers' board rooms	Very heavy	50	20	....
Cocktail bars		40	25	....
Corridors (supply or exhaust)		...	...	0.25
Department stores	None	7½	5	0.05
Directors' rooms	Extreme	50	30	....
Drug stores	Considerable	10	7½	....
Hotel rooms	Heavy	30	25	0.33
Kitchens				
Restaurant		...	...	4.0
Residence		...	...	2.0
Laboratories	Some	20	15	....
Meeting rooms	Very heavy	50	30	1.25
Offices				
General	Some	15	10	0.25
Private	None	25	15	0.25
Private	Considerable	30	25	0.25
Restaurants				
Cafeteria	Considerable	12	10	....
Dining room	Considerable	15	12	....
Schoolrooms	None	...	...	....
Shop, retail	None	10	7½	....
Theater	None	7½	5	....
Theater	Some	15	10	....
Toilets (exhaust)		...	...	2.0

Extracted by permission from ASHRAE Handbook of Fundamentals, 1967.

CHAPTER 6

Refrigeration

VISUALIZE A hot room. Now, let's pretend that the heat in this room is something viscous or liquid that we can scoop up in a bucket and pour out a window (Fig. 6–1). But, we will soon get tired of carting these buckets, so why don't we replace it with a bucket machine (Fig. 6–2) to make our work easier? Let's assign some terms at this point. The room is the "source" of heat, and the "outdoors" is the sink. So any place where we pick up heat is a source, and where we dispose of it is a sink.

Our bucket machine would work if the heat in the room were a liquid, but of course it isn't. So what can

we do? We can hang a coil in the space and use a fan to blow air across the coil. Now, let's put the same thing outdoors. Interconnect the coils with piping and use a pump to circulate water through the two coils (Fig. 6–3). Now, what heat we pick up in the room (our source) heats the water, which we cool with outdoor air. The outdoor air is our sink. This will work fine as long as our sink is colder than our source, but what do we do if it isn't?

Well, we could do it with a compressed gas. How? Have you ever noticed that when you spray with an "aerosol" can of insecticide the gas feels cool? The

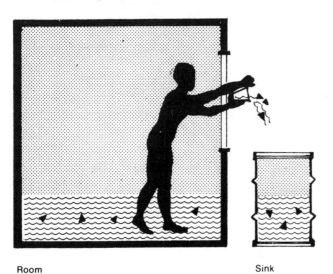

Room   Sink

FIG. 6–1

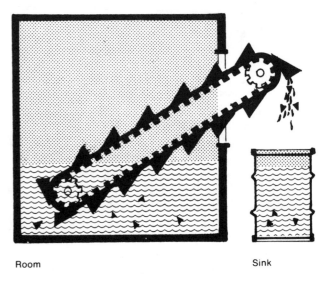

Room   Sink

FIG. 6–2

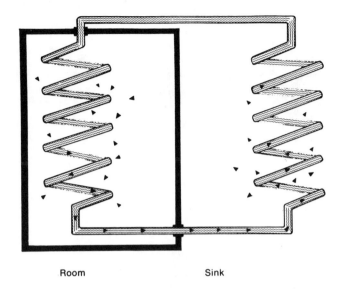

Room   Sink

FIG. 6–3

can contains a liquid under a high pressure and at room temperature. When this liquid is suddenly released from the can, the pressure is reduced, which causes it to boil. But in order to boil, it must absorb heat from somewhere, and it generally gets it from itself, and so it is cooled.

This use of the expansion of a high pressure liquid is an interesting one. Sometimes it is hard to understand, isn't it? Well, look at it like this. The raising of pressure also raises the boiling point, doesn't it? This is why a pressure cooker is used in cooking (Fig. 6–4). By cooking under a pressure of 15 psig, we raise the boiling point of water from 212°F to almost 250°. Conversely, if we reduce pressure, we lower the boiling point, don't we? In fact, we can make water boil at 40° by reducing the pressure to 29.673 inch Hg vacuum. But always remember that we must add heat to a liquid to change it to a gas. When water is at 212° and atmospheric pressure, it is at its boiling point. BUT IT WON'T BOIL UNTIL YOU ADD 1000 BTU PER POUND. After it changes to steam it is still at 212°. The heat was needed to give the molecules the necessary energy to break away. Conversely, you must remove 1000 Btu per pound to change steam back to water.

Let's go a step further. You wish to do some cooling, so you buy a tank of liquid R-12 refrigerant. This refrigerant is at a pressure of 126.55 psig and at room temperature, say 80°. If you look at some R-12 refrigerant tables, you will find that the boiling point is 105° at this pressure. But, look at the same table,—if you don't have a complete table, look at Table 6–1, which is an abbreviated table for R–12 (Dichlorodifluorone-

thane)—and notice that at 40° the pressure is 36.971 psig.

TABLE 6–1
Properties of R–12

Temp. F	Pressure psig	Enthalpy Liquid	Btu/lb Vapor
40°	36.971	17.273	81.436
50°	46.698	19.507	82.433
60°	57.737	21.766	83.409
100°	117.16	31.100	87.029
105°	126.55	32.310	87.442
110°	136.41	33.531	87.844

We have this tank of liquid at 80° and 126.55 psig and we want to do some cooling. The room is at 80°, so we need something colder than 80°. If our tank of liquid had a temperature of 60, it would do some cooling, obviously. Back to the chart (Table 6–1). If we reduce the pressure to 57.737, the boiling point will drop to 60°. Assume this tank is so perfectly insulated that no heat can enter the tank from the room. Now reduce the pressure to 57.737 psig. The liquid will want to boil, won't it? It must boil, by all the laws of science. But to boil, it needs heat. It needs 83.409 − 21.766 = 61.643 Btu/lb to boil. If it can't get this from the room, it gets it from itself. It's at 80° and wants to be at 60°, so if 7% of this liquid boils, then the remaining 93% of the liquid will cool to 60° or so. Let's call the heat needed to boil, Q_1, and the heat given up by the liquid as it cools, Q_2. In our closed, perfectly insulated container, $Q_1 = Q_2$. Now, $Q_1 = X$ (61.643) where X is the amount of liquid that boils. Also, $Q_2 = (1 − X)$ (Specific heat) $(t_2 − t_1)$ where the specific heat is 0.22 Btu/lb/deg, $(1 − X)$ is the amount of liquid that does not boil, and $t_2 − t_1$ is the change in temperature. If $Q_2 = Q_1$, then

$$Q_1 = Q_2$$
$$X (61.643) = (1 − X) (0.22) (80° − 60°)$$
$$X = 0.067$$
$$X = 6.7\%$$

Here is a method we can use. All we need is a tank of high-pressure liquid refrigerant, and a coil plus fan

FIG. 6–4 Boiling Point Varies With Pressure

to absorb heat, and we can cool the room (Fig. 6–5). But we do have to reduce the pressure, don't we? We said that as we reduced the pressure in the tank, boiling would take place, but didn't say how to reduce the pressure. An orifice can be used. What is an orifice? A small hole. If a liquid or gas is flowing in a pipe there is a drop in pressure because of friction. But, if the cross sectional area is suddenly reduced, it takes more work to get through this smaller area, and a large pressure drop results (Fig. 6–6). So we can reduce pressure by putting an orifice between our tank and coil. We could also use a capillary tube, which is a long orifice. Or, we could use a globe valve (Fig. 6–7). Notice that a globe valve gives a variable orifice. The area can be changed merely by raising or lowering the stem. And it can be used as a shut-off valve. When a globe valve is used to cause a pressure drop in a refrigeration system, it is called an expansion valve (Fig. 6–8).

So, let's put the expansion valve between the tank and the coil. A thermometer is installed in the pipe between the expansion valve and the coil. Now, by checking the thermometer, we can manually set the expansion valve to maintain the desired temperature in the coil. But, just as our room is cooling off, we run out of gas and must go out to buy another tank. This can get expensive and, if the gas remains in the room, it can be dangerous.

But, if we could keep this gas out of the room and reuse it, this might be the answer. Let's go back to our two coil idea. This time, let's circulate a refrigerant with a low boiling point instead of water. To circulate this refrigerant we need a pump. We call this pump a compressor. Add the expansion valve and we have all of the elements needed in a refrigeration system. Figure 6–9 shows the relative positions of the pieces of equipment that make up a refrigeration system. The coil at the source is the evaporator, the one at the sink is the condenser; between condenser and evaporator we have the expansion valve, and between the evaporator and condenser is the compressor.

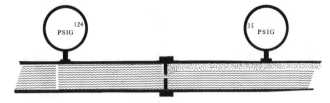

Mixture 19 percent gas 81 percent liquid

FIG. 6–6 Orifice

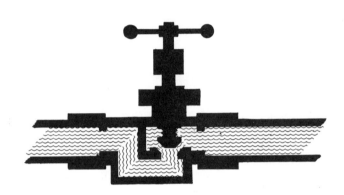

FIG. 6–7 Globe Valve

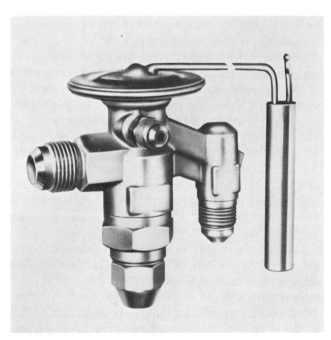

FIG. 6–8 Thermostatic Expansion Valve (Courtesy: Alco Controls Corp.)

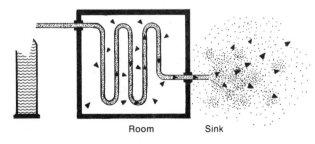

Room   Sink

FIG. 6–5

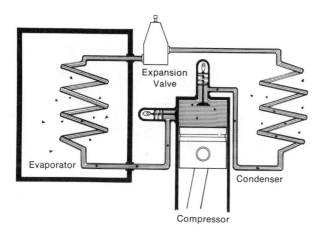

FIG. 6-9

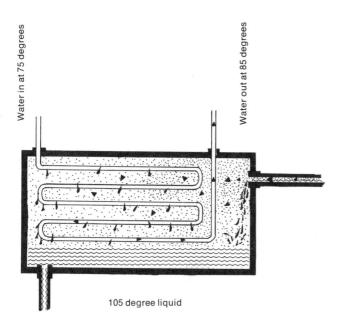

Water in at 75 degrees

Water out at 85 degrees

105 degree liquid

FIG. 6-10 Condenser

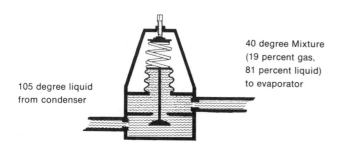

105 degree liquid from condenser

40 degree Mixture (19 percent gas, 81 percent liquid) to evaporator

FIG. 6-11 Expansion Valve

The compressor elevates the pressure of the refrigerant to a point where it can be condensed to a liquid by the available sink temperature. A high-pressure gas enters the condenser and a high-pressure liquid leaves the condenser (Fig. 6-10). This liquid is hot, say 105°. At the expansion valve, the pressure is reduced. As the pressure is reduced, the boiling point is reduced and some of the liquid flashes or boils. About 19% of the liquid boils, cooling the remaining liquid to 40°. We entered the expansion valve with a hot high-pressure liquid, and we leave with a mixture of about 19% of gas, and the rest, a low-pressure, cold liquid (Fig. 6-11). At the evaporator, heat is absorbed or flows from the hot source to the cold refrigerant, causing it all to boil. We leave the evaporator with a cold low-pressure gas and return to the compressor (Fig. 6-12). We now have a system that picks up heat where it is not wanted and disposes of it. Check your home refrigerator. The evaporator is inside the box and it is cold. Put your hand behind the box and, if the unit is running, you will feel heat. This is the heat absorbed inside the box. Because a refrigeration system absorbs heat at one point, and pumps it to another point to be rejected, it can be called a heat pump.

The condenser can take many forms. It can be a shell and tube heat exchanger (Fig. 6-13), where water flows through the tubes and the refrigerant flows through the shell. The water absorbs heat, liquifying the gaseous refrigerant. The water is either wasted to the sewer, or taken to a cooling tower (Fig. 6-14) or spray pond (Fig. 6-15) to be cooled so it can be re-used.

Another condensor is the air-cooled type (Fig. 6-16). Here, the refrigerant flows through a finned coil over which outside air is blown. This air removes heat and liquifies the refrigerant, which then flows to a liquid receiver for storage. Problems can occur at high and low air temperatures. At high air temperatures, the condenser capacity decreases and so decreases the system capacity. It is important to be generous in selection of this equipment. On the other hand, all this surface causes problems at low air temperatures. It can condense gas at so low a pressure that the machine cannot operate properly and safety controls will shut it off. Fortunately, there are accessories available which can prevent this low-pressure condition from occurring.

Another type of condenser is the evaporative condenser (Fig. 6-17). This is an air-cooled condenser whose efficiency is increased by spraying water on the condenser coil. The evaporation of water increases the cooling effect.

The compressor can be a reciprocating machine,

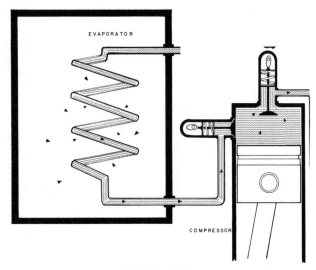

FIG. 6–13 Water Cooled Condenser (Courtesy: ITT Bell & Gossett)

FIG. 6–12

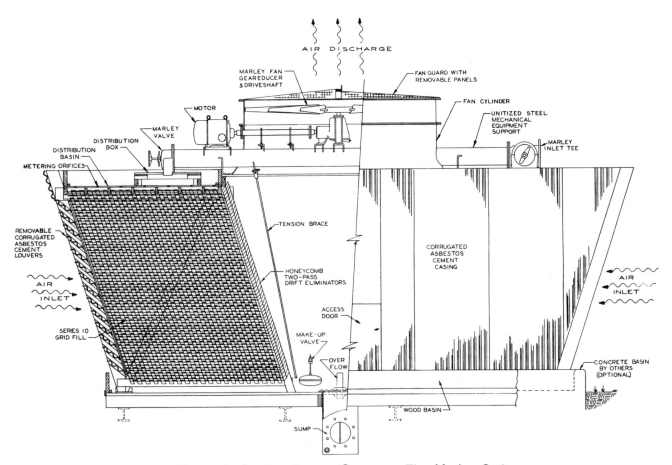

FIG. 6–14 Cooling Tower (Courtesy: The Marley Co.)

FIG. 6–15 Spray Pond (Begrow & Brown, Architects; J. B. Olivieri & Assocs., Inc.)

FIG. 6–17 Evaporative Condenser (Courtesy: Acme Industries, Inc.)

FIG. 6–16 Air Cooled Condenser (Courtesy: Airtemp Div., Chrysler Corp.)

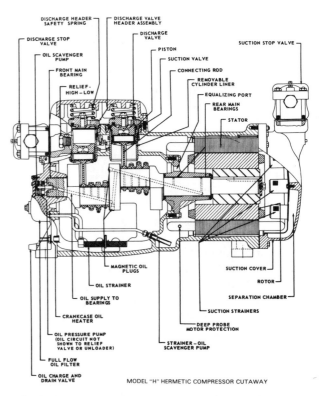

FIG. 6–18 Section Through Hermetic Compressor (Courtesy: Airtemp Div., Chrysler Corp.)

which is similar to an automotive engine. A cutaway view of an hermetic reciprocating unit is shown in Fig. 6–18. A typical reciprocating condensing unit is shown in Fig. 6–19, while a reciprocating water chiller is shown in Fig. 6–20. Reciprocating equipment is generally used up to 100-ton sizes. Between 100 and 200 tons, both reciprocating and centrifugal units are used, while over 200 tons, centrifugal units are generally selected. A cutaway view of an hermetic centrifugal unit is shown in Fig. 6–21. In a centrifugal unit compression takes place by the centrifugal action of a turbine. These units can be driven by electric motors or steam turbines. A typical unit, including cabinet, is shown in Fig. 6–22.

Most modern systems use fluorinated hydrocarbons. Reciprocating equipment uses R–12 and R–22, while R–11 is used in centrifugal machines. You will notice that I have avoided the use of the word "Freon." The word "Freon" is a trademark of the du Pont Company and should not be used as the generic name for this family of refrigerants. These refrigerants are generally safe and non-toxic, but in the presence of flame or excessive heat, will break down into a poisonous gas and/or hydrochloric acid.

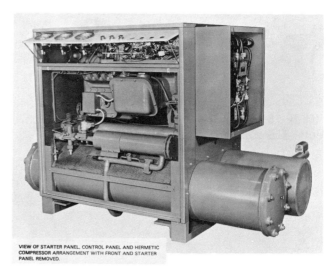

FIG. 6–20 Hermetic Reciprocating Water Chiller (Courtesy: Airtemp Div., Chrysler Corp.)

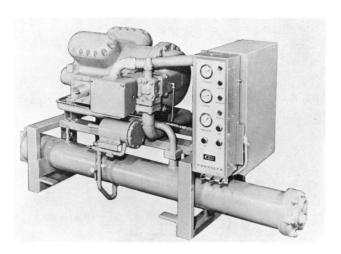

FIG. 6–19 Hermetic Condensing Unit (Courtesy: Airtemp Div., Chrysler Corp.)

The source of power for the reciprocating compressor can be an electric motor or natural gas engine. Figure 6–23 shows a refrigeration unit with a gas engine. Centrifugal machines can be driven by electric motors, steam turbines, or gas turbines. The selection of the source of power depends on an economic analysis of owning and operating costs, availability of fuels, service facilities, and familiarity of operating personnel with equipment.

Early refrigerants were ammonia, sulphur dioxide, carbon dioxide, or methyl chloride. With the exception of ammonia, these refrigerants are rarely, if ever, used today because of their toxicity and other disadvantages. Ammonia, even though it is toxic, is still being used in product and process refrigeration. True, its use is declining, but many operators prefer it and continue to install ammonia systems.

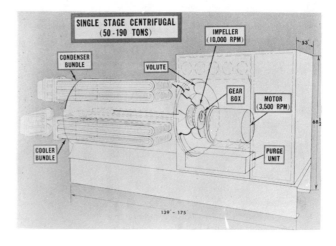

FIG. 6–21 Section Through Centrifugal Machine (Courtesy: Airtemp Div., Chrysler Corp.)

An evaporator can be a finned coil (Fig. 6–24) installed in an air-handling unit which cools and/or dehumidifies air. Another form is a shell and tube heat exchanger (similar to the type used as a condenser) used to chill water or brine. (In refrigeration work, the term brine is used to signify any anti-freeze solution and does not necessarily mean a salt water solution.)

Let's skip back to the compressor for a moment. Remember, we stated that the compressor raised the

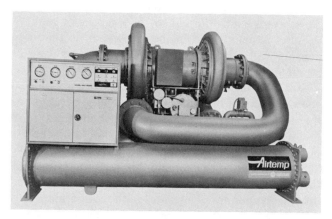

FIG. 6–22 Centrifugal Machine (Courtesy: Airtemp Div., Chrysler Corp.)

FIG. 6–23 Gas Engine Drive Unit (Courtesy: Ready Power Co.)

pressure of the refrigerant and its boiling point or, if it is easier to visualize, its condensing point so that we could condense the gas into a liquid. I, in my ignorance, once asked a refrigeration serviceman to show me the control which set this high pressure. He pointed to the condenser. I persisted and asked where the "pressure trol" was. He again pointed to the condenser, but this time also explained that the temperature of the condensing media—water, air, or water and air—plus the amount of heat transfer surface set the temperature at which the gas condensed. This in turn determined to what pressure the compressor would raise the gas. Think of it this way. With all variables such as surface fixed, the only variable is the temperature of the condensing media.

Visualize the compressor as a single cylinder with a movable piston and inlet and outlet ports. The outlet is connected to the receiver by a long pipe. Let's start with an empty system. Open the inlet valve and fill the cylinder with gas. Now, open the outlet valve and push the gas into the condenser with the piston. Oh, yes, we are using R–22. Read the gauge—not very high is it? If we now sprayed 106° water on the condenser and receiver, we would find that the gas would not change to a liquid. Close the outlet valve and pull the piston back. Open the inlet valve and let more gas into the cylinder. Close the inlet and open the outlet and push the gas into the condenser and receiver. Notice the pressure increase? Repeat this until the gauge reads 216 psig. Ignoring the resistance to heat transfer of the pipe and receiver material, the gas would condense to a liquid when sprayed with 106° water. If our water temperature had been 96° we would have condensed at 186.5 psig. The above is what takes place in a compressor and condenser combination. Of

FIG. 6–24 Cooling Coil

course, we have many cylinders and our valves are not hand valves but the type we have in our automotive engines. Basically, the discharge valves open as the piston compresses the gas and so increases the pressure. If the gas in the condenser does not condense, this pressure is exerted on the top of the discharge valve. So the compressor must compress the gas to a pressure higher than this to lift the valves. As long as the gas does not condense, it exerts an increasing back pressure and the compressor must compress to a higher pressure to lift the valve. When a point is reached at which some of the gas condenses, the compressor will have found the equilibrium pressure at which it will have to work. So the high pressure is determined by the temperature of the condensing media.

Similarly, the low pressure is determined by the temperature of the media being cooled. Visualize the expansion valve as a hand needle valve or variable orifice. The refrigerant, in order to cool, must be colder than what must be cooled, at least 5° cooler than, say, the leaving air temperature. As we explained earlier, reducing the pressure by passing the liquid through the orifice causes some of the liquid to boil and so cools the rest of the liquid. The amount the pressure is reduced determines how much will boil and so how cool the remaining liquid will be. To maintain 40° refrigerant temperature with R–22, we must reduce the pressure to 69 psig; for 20°, 43.3 psig. So you can see again that the temperature of what needs to be cooled determines the low side pressure. The operation of the expansion valve is automatic, but its position is determined by the temperature of what has to be cooled.

ABSORPTION REFRIGERATION

Here the refrigerant is water. If we had a tank partially filled with water, some of the water would evaporate, wouldn't it? What happens when water evaporates? It changes from liquid water to the vapor phase or steam. (See page 108, Chapter 7.) But to change from liquid to vapor takes heat, doesn't it? If the tank is perfectly insulated and heat is not supplied by external means, then the heat necessary for evaporation must come from the liquid water itself. This will cool the water, won't it?

But, in a short time, equilibrium is reached and the cooling stops. If we could continuously evaporate water, we could continuously cool, couldn't we? To continuously evaporate, we must continuously remove the vapor. This can be done by connecting this water tank to a salt tank. Now you have seen salt shakers

clog up on a damp day because they have absorbed water. In a similar fashion, the salt in the salt or absorber tank absorbs the water as it evaporates. The water tank can also be put under a vacuum to speed evaporation.

Eventually the salt solution would get quite dilute and would have to be replaced. Now wouldn't this be expensive and a nuisance? So a generator or concentrator section is used, in which the dilute salt solution is concentrated by boiling off the water. The concentrated salt solution is returned to the absorber tank, and the water boiled off is condensed in a condenser tank and returned to the evaporator tank. A vacuum is maintained in the condenser so that condensing will take place at as low a temperature as possible.

Steam or high temperature water is used as a heat source. Lithium bromide is the salt in general use as the absorber, and water is the refrigerant. Because of the high vacuum necessary in the evaporator, the use of this water, directly, can cause service problems. To keep from having to maintain this vacuum throughout the complete system, a secondary coil is introduced into the evaporator. The water in the evaporator cools this secondary water, which is circulated to the room cooling equipment. Figure 6–25 shows a flow diagram of an absorption system. In Fig. 6–26 we see a diagrammatic section and flow diagram of an absorption cycle. The evaporator pump circulates the refrigerant (water) through the sprays. The refrigerant is vaporized and cools the air conditioning water circulated through a finned coil in the evaporator. The vapor flows to the absorber. Lithium bromide solution in the

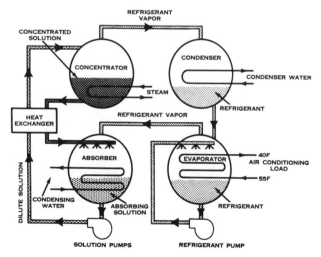

FIG. 6–25   Flow Diagram Absorption Unit (Courtesy: The Trane Co., LaCrosse, Wis.)

absorber maintains a pressure lower than the evaporator, so that the vapor will flow from the evaporator to the absorber. Condenser water cools this vapor and condenses it. As the condensed vapor mixes with the absorbing solution, it dilutes it, so the solution must be pumped up to the concentrator, passing through the heat exchanger on the way. The heat exchanger is an economizer which cools the concentrated solu-

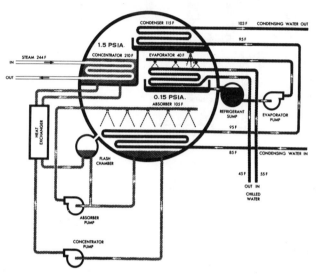

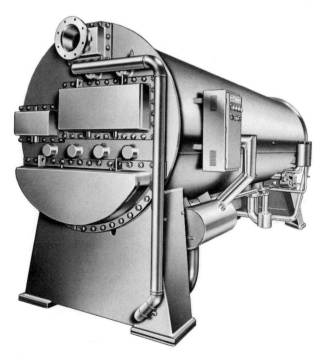

FIG. 6—26 Diagramming Section Through Absorption Unit (Courtesy: The Trane Co., LaCrosse, Wis.)

FIG. 6—27 Absorption Unit (Courtesy: The Trane Co., LaCrosse, Wis.)

tion with the cooler, dilute solution. In the concentrator, the refrigerant (water) is boiled off by a steam coil, concentrating the solution. The concentrated solution flows to the absorber, passing through the heat exchanger on the way. It enters the absorber through sprays. Some of the solution vaporizes because of the lower pressure in the absorber, cooling the remaining solution which mixes with the solution in the absorber. The vapor boiled off in the concentrator flows to the condenser, where it is condensed. The condensed liquid is at a pressure of 1.5 psig and a temperature of about 115°. When it enters the evaporator, which is at 0.15 psig, enough liquid flashes to cool the rest of the liquid to 40°. This completes the cycle.

HEAT PUMP

As we stated earlier, the refrigeration cycle is a transportation system that removes heat from a source and disposes of it in a sink. The process is generally used to provide cooling but can be used to provide heating, or heating and cooling. All we need is a source of heat—water, ground, or air; any source, just so long as we have something to cool. If we have something to cool, we will then have heat to dispose of, and this heat can be used for heating instead of being poured down the drain.

Why not use the electricity directly in resistance heating and save all the machinery? For two reasons: first, the heat pump provides cooling and heating; secondly, it is more economical to operate. When you heat with resistance heating, 1 kw gives 3400 Btu. But use that 1 kw to drive a compressor and you get, at 40° refrigerant temperature, about 13,000 Btu. This is almost four times as much heat.

Then what's the gimmick? Why isn't this used more

FIG. 6—28 Package Water Source Heat Pump (Courtesy: American Standard, Industrial Div.)

HEAT PUMP PRINCIPLE

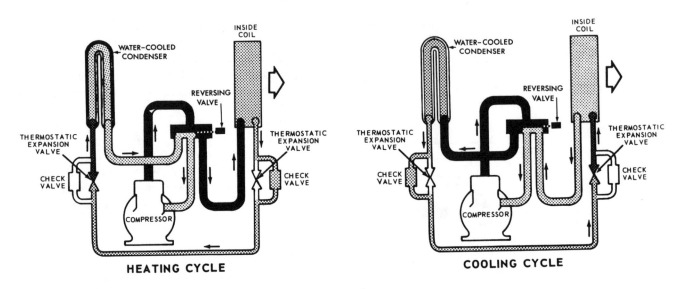

FIG. 6-29   Heat Pump Principle (Courtesy: American Standard, Industrial Div.)

often? One reason is that unless you are interested in cooling, the initial investment is more than that of conventional means of heating. Another reason is the problem of matching summer and winter loads. In many areas, if the heat pump is selected on the basis of the summer load, it will not have the capacity for the winter load. Selecting the machine on the basis of winter load provides too large a unit for summer, which increases the initial investment. The biggest problem, however, is the source.

If we have a source warm enough so we can use 40° refrigerant, our heat output is about four times our energy input. This, by the way, is called coefficient of performance, or COP. Unfortunately, our COP drops as our source temperature does and so our refrigerant temperature drops. You see, a compressor can pump a fixed volume of gas. As our refrigerant temperature drops, its specific volume increases, requiring the compressor to move more gas. It cannot, so obviously, capacity is reduced. But what causes our source temperature to drop, producing this problem? If the source is well water, the temperature is pretty stable and these problems do not occur. But when outdoor air is used as a source, then we have a continually changing source temperature.

As you can see, the capacity is least when the requirements for heating are highest. An average commercial building in a —10° design area would have a winter design load of 60 Btuh/sq ft and a summer load

of 50 Btuh/sq ft. Selecting a unit on the basis of the summer load, we could provide, at 40° suction, about 55 Btuh/sq ft heating. These conditions would occur at about 55° outdoor temperature when our heat requirements would be only 11 Btuh/sq ft. But at 10° outdoors, our heating capacity would be 27 Btuh/sq ft and our requirement would be 45 Btuh/sq ft. The loads would balance at about 25° outdoor temperature. But what do we do from 25° to —10°? We make up the difference with resistance heating which, as we have shown, is more expensive. Does this mean outside-air heat-source heat pumps cannot be used in severe winter heating areas? No, it does not. If the number of hours that the temperature is below 25° is small in comparison to the number of hours above 25°, then the high cost of resistance heating may not be significant. An Owning and Operating cost analysis must be made, as shown in Chapter 11.

Heat pumps can be custom designed for large projects, but are available as complete packages in the smaller sizes. Fig. 6-28 shows a typical packaged water source heat pump. Fig. 6-29 shows a flow diagram for the same unit. Notice that the change from cooling to heating is accomplished by changing the functions of the evaporator and condenser. In winter, the evaporator functions as a condenser, providing heat for the space, and the condenser functions as the evaporator, cooling outside air to provide heat. This is accomplished by the reversing valve.

CHAPTER 7

Psychrometrics

IN THE STUDY of air conditioning we sometimes spend too much time on conditioning and not enough on air. To design a proper air conditioning system requires that we know and understand air and how it behaves. The study of air and how it behaves is called psychrometrics.

First of all, what is air? Air is a mixture. What is a mixture, and how does it differ from a compound? A mixture is like a bowl of jelly beans. There are black ones, red ones, yellow ones, and so forth. We have one bowl, but in that bowl are individuals, each with their own identities and properties. In a compound, the individual parts surrender their identities and properties and merge into a new entity. If you took the bowl of jelly beans and melted them down, you would have a new mass in which you would not be able to identify red, green, or black jelly beans. The jelly beans would no longer act as individuals.

Air is like our bowl of jelly beans. Instead of jelly beans of different colors, we have different molecules. The gases, which mixed together make air, are 78% nitrogen, 20% oxygen, 1% argon, 0.04% carbon dioxide, minute amounts of hydrogen, neon, helium, krypton, xenon, and varying amounts of water vapor. This variation in the amount of water vapor affects our comfort and so is important to us.

What is water vapor? Well, water vapor is steam. But how can steam exist at room temperature when the boiling point of water is 212°? Would it be easier to accept water vapor as steam if we turned it around and said that steam is water vapor? So now when you put the pan of water on the stove and turn on the heat, remember that it is water vapor that comes off when the water boils. By the way, what you see is not water vapor or steam. It is vapor that didn't quite make it

into space and condensed back into liquid water. We are aware of the change from liquid to vapor when water boils because of all the activity, the violent bubbling, sizzling, gurgling, etc., plus the condensation of some of the vapor. This also happens at room temperature, but so quietly that we are not aware of it.

Matter exists in three phases: solid, liquid, and gaseous. A solid is crystalline, rigid, retains its shape, and resists deformation. A solid has a definite boundary which separates it from the other phases. A liquid is dense, though less dense than a solid, but is not rigid. It will assume the shape of the container in which it is placed. Liquids do, however, have boundaries. Gases are not rigid, have no boundaries, and expand to fill the container in which they are placed. If a bottle of gas is opened it will expand to fill the room in which the bottle is opened.

Matter can exist in one, two, or three phases at one time. Fig. 7–1 shows a container of water; ice cubes are floating in the water and water vapor exists above the liquid. I just changed terms, didn't I? I said vapor instead of gas. Well, a vapor is a gas. Scientifically, gases exist above the critical temperature and vapors below the critical temperature. In common usage, matter that exists in the gaseous phase at 70° and atmospheric pressure is called a gas. If it is solid or liquid at these temperatures, it is called vapor when in its gaseous phase.

But, how can we have steam at 70°? As you know, matter is made up of molecules. These little devils cannot sit still. They must keep moving. If we place a pan of water on a table at room temperature we see no activity. But actually, the molecules are dashing about, bumping one another and the sides of the container. Those at the surface sometimes make a wrong turn and find themselves free and soaring into space. Here they begin to bump into the molecules of the gases that make up air. Some get bumped up further into space, some sideways, and some are bumped down into the pan to rejoin their liquid friends. We call this evaporation.

At room temperature relatively few escape to mix with the air. Temperature is a measure of molecular activity. The higher the temperature, the faster the molecules move. The more a liquid is heated, the more frenzied the molecules become. They move faster and faster, and more and more, and make their way to the surface to escape and join the vapor state. So at a given temperature the molecules have a given state of activity, and this determines how many will join the vapor phase.

If a bowl of water is placed under a container we will reach the condition where for each molecule that escapes, one will return, and we are in equilibrium. This is called a saturated condition. We have at this point a saturated vapor. Do not think of air as a sponge that absorbs water! The air does not control the water. The water is master of its own fate. Its temperature determines the amount of vapor that mixes with the air. This vapor adds to the air pressure. Remember

that pressure is a measure of the number of collisions and the force of the collisions that molecules make with a container. So you can see, if you introduce water vapor molecules into the air mixture they are going to get in their "two cents" worth. This leads right to Dalton's law of partial pressures. Dalton said that the pressure of a mixture of gases is equal to the sum of the pressures of the individual components, each taken at the temperature and volume of the mixture. Just what does he mean?

Air is a mixture of gases and each of these gases contributes to the pressure. The composition of air is consistently uniform throughout the world, so it is proper to think in terms of air pressure or, more properly, dry air pressure. Let's look at it this way. Air is a mixture of air and varying amounts of water vapor. So air pressure is the pressure of dry air plus the pressure of water vapor. If you took a cubic foot of air and cooled it to −176°, all of the water vapor would freeze. If we then removed the air and allowed the water to thaw and go back to its original temperature, we could then measure the pressure exerted by the water vapor only. If we also measured the pressure of the dry air, we would find that the original pressure equaled the sum of the pressure of dry air and water vapor. All of us, at one time or another, have wondered why the vapor in the air does not freeze at temperatures below 32°. Some of us have even found it hard to conceive of water as vapor in the air at temperatures below 32°. The water vapor does not freeze or go into the solid state for the same reason it does not go into the liquid state between 32° and 212°. As long as the vapor is below its saturation pressure it will stay in its vapor state.

Take a sample of warm moist air and cool it. When the saturation temperature is reached some of the vapor will condense. Not all of it! Only enough to maintain the saturation condition. This happens even at temperatures below 32°, except here the vapor goes directly from vapor to solid (ice). A good example is frost on a window in winter. The reverse also takes place. Water molecules escape from ice and add to the water vapor in the air. Other things go from the solid state directly to the vapor state, without passing through the liquid state. An excellent, vivid example is a moth ball. This is called sublimation.

As the vapor condenses, the amount of vapor is reduced, which reduces the vapor pressure. At each temperature we have a saturated vapor pressure, which is a measure of the maximum amount of water vapor that exists in the air vapor mixture. As long as we are below this vapor pressure, the vapor will not condense or freeze. The following table gives the

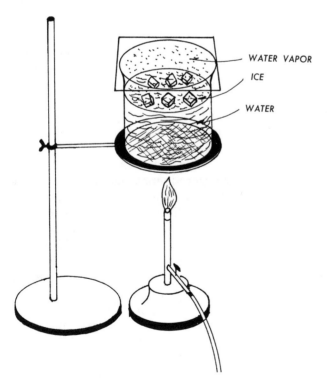

WATER VAPOR

ICE

WATER

FIG. 7–1 Three States Of Water Existing At One Time

vapor pressure and amount of vapor for several temperatures to illustrate the point.

Sat. Temp.	Sat. Press.	Wgt. of Vapor
80°	0.50701 psia	0.02233 lb/lb dry air
50°	0.17799 psia	0.00766 lb/lb dry air
32°	0.08859 psia	0.00379 lb/lb dry air
20°	0.05045 psia	0.00215 lb/lb dry air

Up to this point we have been talking about saturated vapor conditions. Actually, most of the time the vapor exists in a superheated condition. Don't let this word superheat fool you. It has nothing to do with the amount of heat. It is a term that indicates that the vapor has been heated to a temperature higher than the saturation temperature. For example, at 67° saturated, an air-vapor mixture would have 0.01419 lb vapor/lb dry air and a saturated vapor pressure of 0.3276 psia. At this condition, just a small amount of cooling and some of the vapor would have to condense. But if we heated this mixture of water vapor and air to 80°, the amount of water vapor would still be 0.01419 lb/lb dry air, and so the vapor pressure would be unchanged. This air-vapor mixture could be cooled from 80° to 67° without any condensation.

Dr. Carrier developed the Carrier Equation which is used to determine the vapor pressure of a superheated vapor at the dew point temperature.

$$P_v = P_{vwb} - \frac{(P - P_{vwb})(t_{db} - t_{wb})}{2830 - 1.3\, t_{wb}}$$

P_v = vapor pressure of the superheated vapor
P_{vwb} = vapor pressure at the wet-bulb temperature
P = barometric pressure
t_{db} = dry-bulb temperature
t_{wb} = wet-bulb temperature

It looks like its about time to redefine dry-bulb temperature, wet-bulb temperature, and dew point.

DRY BULB

Dry-bulb temperature is the temperature measured by a thermometer and is a measure of sensible heat.

WET BULB

Wet-bulb temperature is harder to define. It is easier to tell you about it. If you wrapped a wet sack about the bulb of a thermometer and blew air across the bulb, some of the water would evaporate. Remember that, in order for water to evaporate, heat must be removed from it. This heat removal depresses the temperature on the thermometer. The temperature read on this thermometer is the wet-bulb temperature. The drier

the air blown across the bulb, the more water evaporates and the greater the depression in temperature. If moisture continues to evaporate, why does the temperature reach an equilibrium? Well, if you lower the temperature of the bulb, it will absorb heat from the warmer surrounding air. When the amount of heat lost by evaporation is equaled by the heat absorbed from the surrounding air, the temperature will stop dropping. In taking wet-bulb temperatures, be sure the air velocity across the bulb is between 500 and 2000 fpm and shield the bulb from any high temperature sources to prevent radiation heat gain.

DEW POINT

If you cooled a mixture of air and water slowly you would reach temperature at which beads of moisture would begin to form on the cooling device. This temperature is the dew point temperature. This is also called saturation. The vapor pressure at this point is the saturated vapor pressure.

Example #1

What is the vapor pressure of air at 80° DB and 67° WB and 14.7 psia barometric pressure?
From Table 7-2, the saturated vapor pressure at the wet-bulb temperature (P_{vwb}) is 0.32750 psia.

$$P_v = P_{vwb} - \frac{(P - P_{vwb})(t_{db} - t_{wb})}{2830 - 1.3\, t_{wb}}$$

$$P_v = 0.32750 - \frac{(14.7 - 0.32750)(80 - 67)}{2830 - 1.3\,(67)}$$

$P_v = 0.32750 - 0.0725$
$P_v = 0.255$ psia

ABSOLUTE TEMPERATURE AND PRESSURE

What is meant by absolute temperature and pressure? A pressure gage on a boiler indicates only the pressure generated inside the vessel. If there is no fire the gage will read 0. But is the pressure in the boiler 0? No. The pressure inside the boiler is the same as the pressure outside the boiler, which is atmospheric. Normal atmospheric pressure is 14.7 psi. So the true pressure inside the boiler is not 0 but 14.7 psi. We call this absolute pressure. Gage pressure is given as psi gage, or psig for short. Absolute pressure is given as psi absolute, or psia. If we turn on the burner in our boiler and generate steam until the gage reads 5 psi, what will our gage and absolute pressures be? The gage pressure will be what we read on the gage, 5 psig, and the absolute will be 5 + 14.7 = 19.7 psia.

As we stated earlier, temperature is a measure of molecular activity. Reason tells us that if temperature measures molecular activity, then the colder it gets, the slower the molecules will move. At some point molecular activity will cease. This point we call absolute zero. It is 460° below zero. We must add 460° to the fahrenheit temperature to get the absolute or Rankine temperature. (By the way, science now feels that even at absolute zero there is some molecular activity.)

SENSIBLE AND LATENT HEAT

While we are redefining, let's get some others out of the way. Sensible heat is the heat which, when added to a substance, changes only its temperature (dry-bulb). Latent heat is the heat which must be added or subtracted from a substance to change its state. When we change a liquid to a vapor, we speak of the latent heat of vaporization. The heat removed to change water to ice is the latent heat of fusion. When you heat a pound of water from 211° to 212°, you add about 1 Btu and this is sensible heat because you have only raised the temperature. But when you change water at 212° to steam at 212°, you add about 1000 Btu to change the state. This heat is called the latent heat. Heating steam from 212° to 213° requires, once again, sensible heat.

PERFECT GAS LAW

The perfect gas law states that

$Pv = RT$
P = absolute pressure lb/sq ft
v = specific volume cu ft/lb
R = the gas constant which varies for each gas
T = absolute temperature in degrees Rankine.

Manipulating vapor pressure, saturated vapor pressure, and barometric pressure in formula gives us useful data. These pressures and the perfect gas law enable us to determine the humidity ratio and the relative humidity.

If you had air at 80° saturated, what would its vapor pressure be? The specific volume of water vapor at this temperature is 633 cu ft/lb and the gas constant for water is 85.76.

$$p = \frac{RT}{v} = \frac{85.76\,(80° + 460°)}{633}$$

$$p = 72.5 \text{ psfa}; \; 0.5 \text{ psia}$$

Sometimes it is more convenient to work with volume, rather than specific volume, so multiply both sides of the equation by the mass and:

$PMv = MRT$
$V = Mv$
$PV = MRT$
$M = \dfrac{PV}{RT}$
M = mass in pounds
V = volume in cubic feet

HUMIDITY RATIO (SPECIFIC HUMIDITY)

Humidity ratio is the ratio of the mass of water vapor to the mass of air.

$$w = \frac{\text{lbs water}}{\text{lb dry air}} = \frac{M_v}{M_a} \text{ where } M_v = \text{mass of water vapor and}$$
$$M_a = \text{mass of dry air.}$$

If this mixture occupies one cubic foot, then we have

$$w = \frac{\dfrac{M_v}{\text{cu ft}}}{\dfrac{M_a}{\text{cu ft}}}$$

The reciprocal of mass per cubic foot is specific volume, so

$$M_v = \frac{1}{v_v} \qquad\qquad M_a = \frac{1}{v_a}$$

$$w = \frac{M_v}{M_a} = \frac{\dfrac{1}{v_v}}{\dfrac{1}{v_a}}$$

$$Pv = RT$$

$$v_a = \frac{R_a T}{P_a} \qquad\qquad v_v = \frac{R_v T}{P_v}$$

$$w = \frac{\dfrac{R_a T}{P_a}}{\dfrac{R_v T}{P_v}} = \frac{R_a P_v}{R_v P_a} = \frac{53.34\, P_v}{85.76\, P_a}$$

$$w = 0.622\,\frac{P_v}{P_a} = 0.622\,\frac{P_v}{P - P_v}$$

Example #2

What is the humidity ratio of air at 80° DB and 67° WB? From Example 1, the vapor pressure $P_v = 0.26$ psia. Barometric pressure = 14.7 psia.

$$w = 0.622\, P_v/P - P_v$$

$$w = 0.622\,\frac{0.26}{14.7 - 0.26}$$

$$w = 0.0112 \text{ lb vapor/lb dry air}$$

What is the humidity ratio of air at 67° saturated? From Table 7-2, the vapor pressure $P_v = 0.32750$ psia.

$$w = 0.622 \; P_v/P - P_v$$

$$w = 0.622 \; \frac{0.32750}{14.7 - 0.32750}$$

$$w = 0.0142 \text{ lb vapor/lb dry air}$$

RELATIVE HUMIDITY

Relative humidity is the ratio of the vapor pressure in an air-water mixture to the vapor pressure at saturation.

$$\emptyset = \frac{P_v}{P_{vs}}$$

$\emptyset$ = relative humidity
P_v = partial pressure of water vapor
P_{vs} = saturated partial pressure of water vapor

Example #3

What is the relative humidity of 80° DB and 67° WB air? From example #1, the vapor pressure at 80° DB and 67° WB is 0.255 psia and from Table 7-2, the vapor pressure at 80° sat. is 0.51 psia.

$$\emptyset = \frac{0.255}{0.51} = .50 \text{ or } 50\%$$

You may occasionally read that relative humidity is equal to the ratio of the mass of vapor in an air-vapor mixture to the mass of vapor at saturation. Where mass ratios are used, the result is called degree of saturation. The difference between the two, in normal air conditioning work, is less than $\frac{1}{2}\%$ and so is generally insignificant. The difference occurs because the perfect gas law is correct only at low pressures and large specific volumes. The greatest error occurs at and near saturation. The relative humidity and degree of saturation are related as follows:

U = degree of saturation

$$U = \frac{W \text{ (weight of vapor)}}{W_s \text{ (weight of vapor at saturation)}}$$

but weight of vapor is the humidity ratio.

$$W = 0.622 \; \frac{P_v}{P_a} = 0.622 \; \frac{P_v}{P - P_v}$$

$$W_s = 0.622 \; \frac{P_{vs}}{P_a} = 0.622 \; \frac{P_{vs}}{P - P_{vs}}$$

$$U = \frac{0.622 \; P_v/P - P_v}{0.622 \; P_{vs}/P - P_{vs}}$$

$$U = \frac{P_v}{P_{vs}} \frac{(P - P_{vs})}{(P - P_v)}$$

$$\emptyset = \frac{P_v}{P_{vs}}$$

$$U = \emptyset \frac{(P - P_{vs})}{(P - P_v)}$$

Because of the small error involved, we will use relative humidity and degree of saturation interchangeably.

ENTHALPY

The next property too often makes cowards of us all. Enthalpy. There, I said it and I'm glad.

This may be heresy, but don't worry about what enthalpy is; just use it. Most students are puzzled because enthalpy, unlike temperature, pressure, or volume, cannot be measured or seen. It is a mathematical property developed to make thermodynamic calculations simpler. Enthalpy is equal to the internal energy of a substance, plus the pressure times the volume.

$$h = u + pv$$

Internal energy is the energy stored in the molecular and atomic structure of a substance.

For the subject of psychrometrics we consider enthalpy to be the total heat, sensible and latent, of air and water vapor present in the air. We must have a starting point, so even though air at 0° has enthalpy, we call the enthalpy zero at this point. Because we don't care what the enthalpy actually is, only what the difference in enthalpy is, calling the enthalpy 0 at 0° is perfectly proper. The enthalpy of water is 0 at 32°.

Enthalpy enables us to determine the amount of heat transferred in air conditioning processes. By knowing the enthalpy of the air-vapor mixture entering an air conditioning unit, and the enthalpy leaving, we can determine the amount of heat that enters or leaves the system.

The enthalpy of a mixture of gases is equal to the sum of the individual enthalpies.

$$h = h_a + wh_v$$

where

h = enthalpy of air and water vapor mixture
h_a = enthalpy of dry air
h_v = enthalpy of water vapor
w = humidity ratio lb water/lb dry air

The enthalpy of dry air for a constant pressure process is equal to the specific heat at constant pressure, multiplied by the air temperature.

$$h_a = C_{pa}t_{db}$$
C_{pa} = specific heat of air at constant pressure
t_{db} = dry-bulb temperature °F

The enthalpy of the vapor for a constant pressure process is equal to the enthalpy of the vapor at its dew point, plus the heat required to superheat the vapor.

$$h_v = h_{vdp} + C_{pv}(t_{db} - t_{dp})$$

C_{pv} = specific heat of water vapor at constant pressure
t_{dp} = dew point temperature
t_{db} = dry-bulb temperature

So, if $h = h_a + wh_v$, then:

$$h = C_{pa}t_{db} + w(h_{vdp} + C_{pv}(t_{db} - t_{dp}))$$

using average values for C_{pa} and C_{pv}

$$h = .24\, t_{db} + w(h_{vdp} + .45(t_{db} - t_{dp}))$$

In the absence of steam tables, a very close approximation can be made by:

$$h = .24\, t_{db} + w(1061 + .45\, t_{db})$$

where 1061 is the enthalpy of saturated vapor at 0°F.

Example #4

What is the enthalpy of air at 80° DB and 67° WB?
The enthalpy of the air alone is

$$h_a = C_{pa}(t_{db})$$
$$h_a = 0.24(80°)$$
$$h_a = 19.20\ \text{Btu/lb}$$

The enthalpy of the vapor is

$$h_v = 1061 + 0.45\, t_{db}$$
$$h_v = 1061 + 0.45(80)$$
$$h_v = 1061 + 36.0$$
$$h_v = 1097\ \text{Btu/lb}$$

So the enthalpy of the air is

$$h = h_a + wh_v$$
$w = 0.0112$ lb vapor/lb dry air (Example 2)
$$h = 19.20 + 0.0112(1096.12)$$
$$h = 19.20 + 12.28$$
$$h = 31.48$$

What is the enthalpy of air at 67° saturated?

$$h_a = 0.24(67)$$
$$h_a = 16.08\ \text{Btu/lb}$$
$$h_v = 1061 + 0.45(67)$$
$$h_v = 1061 + 30.15$$
$$h_v = 1091.15\ \text{Btu/lb}$$
$$h = h_a + wh_v$$
$w = 0.0142$ lb vapor/lb dry air (Example 2)
$$h = 16.08 + 0.0142(1091.15)$$
$$h = 32.02$$

SPECIFIC VOLUME

The property specific volume is the volume occupied by one pound of a substance. The reciprocal of specific volume is a familiar friend, density, pounds per cubic foot.

The specific volume of dry air would be cubic feet of dry air per pound of dry air. For water vapor it would be cubic feet of water vapor per pound of water vapor. And for a mixture of air and water it would be cubic feet of a mixture of air and water per pound of air and water mixture.

But our tables and the psychrometric chart give all properties in terms of one pound of dry air, so to avoid complications, the true specific volume is not used in psychrometry, but an apparent specific volume is introduced. This gives us cubic feet of an air and water mixture *per pound of dry air*.

Our tables and charts introduce an apparent contradiction of Dalton's Law by reducing the volume of air as water vapor is introduced. One would have to assume that the water vapor is driving off the air or that the air is sponge-like.

The tables and the chart are based on a constant barometric pressure of 14.7 psia. To keep this pressure constant, we must reduce the air volume in order to reduce the air pressure as water vapor and its pressure are introduced. While this is a false assumption, it is one that must be made to keep our calculations within reason.

Let's examine what really happens in an air and water mixture. If we started with dry air at 70° we would have (see Table 7-1) a pressure of 14.7 psia and a specific volume of 13.348 cu ft/lb. The reciprocal of specific volume is density, which gives us 0.075 lb/cu ft.

If we now introduce sufficient water vapor to give us 50% RH, we will have 0.00058 lb of water vapor per cubic foot.

$$Pv = RT$$
$$P = 0.50\, P_{sat} = 0.50(0.36304) = 0.18152\ \text{psia}$$
$$R = 85$$
$$T = 460 + 70 = 530°$$

$$v = \frac{85(530)}{0.18105(144)} = 1725\ \text{cu ft/lb}$$

$$\varrho = \frac{1}{v} = \frac{1}{1725} = 0.00058\ \text{lb/cu ft.}$$

Our cubic foot now contains 0.075 lb of air and 0.00058 lb of water vapor, or 0.07558 lb of mixture per cubic foot of mixture. Our pressure, of course, would have to increase because of the increased number of molecules. The new pressure would be equal to the pressure of the dry air, plus the pressure of the water vapor.

$$p = P_a + P_v$$
$$p = 14.7 + 0.182$$
$$p = 14.882\ \text{psia}$$

TABLE 7-1 Thermodynamic Properties Of MOIST AIR (Standard Atmospheric Pressure, 29.921 in. Hg)

Fahr. Temp. t(F)	Humidity Ratio $W_s \times 10^3$	Volume cu ft/lb dry air			Enthalpy Btu/lb dry air			Entropy Btu per (°F) (lb dry air)			Condensed Water			Fahr. Temp. t(F)
		v_a	v_{as}	v_s	h_a	h_{as}	h_s	s_a	s_{as}	s_s	Enthalpy Btu/Lb h_w	Entropy Btu per (°F)(Lb) s_w	Vap. Press In. Hg $p_s \times 10^2$	
7	1.130	11.756	0.021	11.777	1.681	1.202	2.883	0.00364	0.00271	0.00635	−155.61	−0.3172	5.4022	7
8	1.189	11.781	0.022	11.803	1.922	1.266	3.188	0.00415	0.00285	0.00700	−155.13	−0.3162	5.6832	8
9	1.251	11.806	0.024	11.830	2.162	1.332	3.494	0.00467	0.00299	0.00766	−154.65	−0.3152	5.9776	9
10	1.315	11.831	0.025	11.856	2.402	1.401	3.803	0.00518	0.00314	0.00832	−154.17	−0.3141	6.2858	10
11	1.383	11.857	0.026	11.883	2.642	1.474	4.116	0.00569	0.00330	0.00899	−153.69	−0.3131	6.6085	11
12	1.454	11.882	0.028	11.910	2.882	1.550	4.432	0.00620	0.00346	0.00966	−153.21	−0.3121	6.9462	12
13	1.528	11.907	0.029	11.936	3.123	1.620	4.753	0.00671	0.00363	0.01034	−152.73	−0.3111	7.2997	13
14	1.606	11.933	0.030	11.963	3.363	1.713	5.076	0.00721	0.00380	0.01101	−152.24	−0.3100	7.6696	14
15	1.687	11.958	0.032	11.990	3.603	1.800	5.403	0.00772	0.00399	0.01171	−151.76	−0.3090	8.0565	15
16	1.772	11.983	0.034	12.017	3.843	1.892	5.735	0.00822	0.00418	0.01240	−151.27	−0.3080	8.4612	16
17	1.861	12.009	0.035	12.044	4.083	1.988	6.071	0.00873	0.00438	0.01311	−150.78	−0.3070	8.8843	17
18	1.953	12.034	0.038	12.072	4.324	2.088	6.412	0.00923	0.00459	0.01382	−150.29	−0.3059	9.3267	18
19	2.051	12.059	0.040	12.099	4.564	2.192	6.756	0.00973	0.00481	0.01454	−149.80	−0.3049	9.7889	19
20	2.152	12.084	0.042	12.126	4.804	2.302	7.106	0.01023	0.00504	0.01527	−149.31	−0.3039	10.272	20
21	2.258	12.110	0.044	12.154	5.044	2.416	7.460	0.01073	0.00528	0.01601	−148.82	−0.3029	10.777	21
22	2.369	12.135	0.046	12.181	5.284	2.536	7.820	0.01123	0.00553	0.01676	−148.33	−0.3018	11.305	22
23	2.485	12.160	0.049	12.209	5.525	2.661	8.186	0.01173	0.00579	0.01752	−147.84	−0.3008	11.856	23
24	2.606	12.186	0.051	12.237	5.765	2.792	8.557	0.01223	0.00607	0.01830	−147.34	−0.2998	12.431	24
25	2.733	12.211	0.054	12.265	6.005	2.929	8.934	0.01273	0.00635	0.01908	−146.85	−0.2988	13.032	25
26	2.865	12.236	0.057	12.293	6.245	3.072	9.317	0.01322	0.00665	0.01987	−146.35	−0.2977	13.659	26
27	3.003	12.262	0.059	12.321	6.485	3.221	9.706	0.01372	0.00696	0.02068	−145.85	−0.2967	14.313	27
28	3.147	12.287	0.062	12.349	6.726	3.377	10.103	0.01421	0.00728	0.02149	−145.36	−0.2957	14.966	28
29	3.297	12.312	0.065	12.377	6.966	3.540	10.506	0.01470	0.00761	0.02231	−144.86	−0.2947	15.709	29
30	3.454	12.338	0.068	12.406	7.206	3.709	10.915	0.01519	0.00796	0.02315	−144.36	−0.2936	16.452	30
31	3.617	12.363	0.071	12.434	7.446	3.887	11.333	0.01568	0.00832	0.02400	−143.86	−0.2926	17.227	31
32	3.788	12.388	0.075	12.463	7.686	4.072	11.758	0.01617	0.00870	0.02487	−143.36	−0.2916	18.035	32
32*	3.788	12.388	0.075	12.463	7.686	4.072	11.758	0.01617	0.00870	0.02487	0.04	0.0000	18.037	32*
33	3.944	12.413	0.079	12.492	7.927	4.242	12.169	0.01666	0.00904	0.02570	1.05	0.0020	18.778	33
34	4.107	12.438	0.082	12.520	8.167	4.418	12.585	0.01715	0.00940	0.02655	2.06	0.0041	19.546	34
35	4.275	12.464	0.085	12.549	8.407	4.601	13.008	0.01764	0.00977	0.02741	3.06	0.0061	20.342	35
36	4.450	12.489	0.089	12.578	8.647	4.791	13.438	0.01812	0.01016	0.02828	4.07	0.0081	21.166	36
37	4.631	12.514	0.093	12.607	8.887	4.987	13.874	0.01861	0.01056	0.02917	5.07	0.0102	22.020	37

Fahr. Temp. t(F)	Humidity Ratio $W_s \times 10^3$	Volume cu ft/lb dry air			Enthalpy Btu/lb dry air			Entropy Btu per (°F) (lb dry air)			Condensed Water			Fahr. Temp. t(F)
		v_a	v_{as}	v_s	h_a	h_{as}	h_s	s_a	s_{as}	s_s	Enthalpy Btu/Lb h_w	Entropy Btu per (°F)(Lb) s_w	Vap. Press In. Hg p_s	
38	4.818	12.540	0.097	12.637	9.128	5.191	14.319	0.01909	0.01097	0.03006	6.08	0.0122	0.22904	38
39	5.012	12.565	0.101	12.666	9.368	5.403	14.771	0.01957	0.01139	0.03096	7.08	0.0142	0.23819	39
40	5.213	12.590	0.105	12.695	9.608	5.622	15.230	0.02005	0.01183	0.03188	8.09	0.0162	0.24767	40
41	5.421	12.616	0.109	12.725	9.848	5.849	15.697	0.02053	0.01228	0.03281	9.09	0.0182	0.25748	41
42	5.638	12.641	0.114	12.755	10.088	6.084	16.172	0.02101	0.01275	0.03376	10.09	0.0202	0.26763	42
43	5.860	12.666	0.119	12.785	10.329	6.328	16.657	0.02149	0.01323	0.03472	11.10	0.0222	0.27813	43
44	6.091	12.691	0.124	12.815	10.569	6.580	17.149	0.02197	0.01373	0.03570	12.10	0.0242	0.28899	44
45	6.331	12.717	0.129	12.846	10.809	6.841	17.650	0.02245	0.01425	0.03670	13.10	0.0262	0.30023	45
46	6.578	12.742	0.134	12.876	11.049	7.112	18.161	0.02293	0.01478	0.03771	14.10	0.0282	0.31185	46
47	6.835	12.767	0.140	12.907	11.289	7.391	18.680	0.02340	0.01534	0.03874	15.11	0.0302	0.32386	47
48	7.100	12.792	0.146	12.938	11.530	7.681	19.211	0.02387	0.01591	0.03978	16.11	0.0321	0.33629	48
49	7.374	12.818	0.151	12.969	11.770	7.981	19.751	0.02434	0.01650	0.04084	17.11	0.0341	0.34913	49
50	7.658	12.843	0.158	13.001	12.010	8.291	20.301	0.02481	0.01711	0.04192	18.11	0.0361	0.36240	50
51	7.952	12.868	0.164	13.032	12.250	8.612	20.862	0.02528	0.01774	0.04302	19.11	0.0381	0.37611	51
52	8.256	12.894	0.170	13.064	12.491	8.945	21.436	0.02575	0.01839	0.04414	20.11	0.0400	0.39028	52
53	8.569	12.919	0.178	13.097	12.731	9.289	22.020	0.02622	0.01906	0.04528	21.12	0.0420	0.40492	53
54	8.894	12.944	0.185	13.129	12.971	9.644	22.615	0.02669	0.01976	0.04645	22.12	0.0439	0.42004	54
55	9.229	12.970	0.192	13.162	13.211	10.01	23.22	0.02716	0.02047	0.04763	23.12	0.0459	0.43565	55
56	9.575	12.995	0.200	13.195	13.452	10.39	23.84	0.02762	0.02121	0.04883	24.12	0.0478	0.45176	56
57	9.934	13.020	0.208	13.228	13.692	10.79	24.48	0.02809	0.02197	0.05006	25.12	0.0497	0.46840	57
58	10.30	13.045	0.216	13.261	13.932	11.19	25.12	0.02855	0.02276	0.05131	26.12	0.0517	0.48558	58
59	10.69	13.071	0.224	13.295	14.172	11.61	25.78	0.02902	0.02357	0.05259	27.12	0.0536	0.50330	59
60	11.08	13.096	0.233	13.329	14.413	12.05	26.46	0.02948	0.02441	0.05389	28.12	0.0555	0.52159	60
61	11.49	13.121	0.242	13.363	14.653	12.50	27.15	0.02994	0.02527	0.05521	29.12	0.0574	0.54047	61
62	11.91	13.147	0.251	13.398	14.893	12.96	27.85	0.03040	0.02616	0.05656	30.12	0.0594	0.55994	62
63	12.35	13.172	0.261	13.433	15.134	13.44	28.57	0.03086	0.02708	0.05794	31.12	0.0613	0.58002	63
64	12.80	13.197	0.271	13.468	15.374	13.94	29.31	0.03132	0.02803	0.05935	32.12	0.0632	0.60073	64
65	13.26	13.222	0.282	13.504	15.614	14.45	30.06	0.03177	0.02901	0.06078	33.11	0.0651	0.62209	65
66	13.74	13.247	0.292	13.539	15.855	14.98	30.83	0.03223	0.03002	0.06225	34.11	0.0670	0.64411	66
67	14.24	13.273	0.303	13.576	16.095	15.53	31.62	0.03269	0.03106	0.06375	35.11	0.0689	0.66681	67
68	14.75	13.298	0.315	13.613	16.335	16.09	32.42	0.03314	0.03213	0.06527	36.11	0.0708	0.69019	68
69	15.28	13.323	0.327	13.650	16.576	16.67	33.25	0.03360	0.03323	0.06683	37.11	0.0727	0.71430	69

[a] Compiled by John A. Goff and S. Gratch.
* Extrapolated to represent metastable equilibrium with undercooled liquid.

Extracted by permission from ASHRAE Guide and Data Book, 1965

Fahr. Temp. t(F)	Humidity Ratio $W_s \times 10^2$	Volume cu ft/lb dry air			Enthalpy Btu/lb dry air			Entropy Btu per (°F) (lb dry air)			Condensed Water			Fahr. Temp. t(F)
		v_a	v_{as}	v_s	h_a	h_{as}	h_s	s_a	s_{as}	s_s	Enthalpy Btu/Lb h_w	Entropy Btu per (°F)(Lb) s_w	Vap. Press In. Hg p_s	
70	1.582	13.348	0.339	13.687	16.816	17.27	34.09	0.03405	0.03437	0.06842	38.11	0.0746	0.73915	70
71	1.639	13.373	0.351	13.724	17.056	17.89	34.95	0.03450	0.03554	0.07004	39.11	0.0765	0.76475	71
72	1.697	13.398	0.364	13.762	17.297	18.53	35.83	0.03495	0.03675	0.07170	40.11	0.0784	0.79112	72
73	1.757	13.424	0.377	13.801	17.537	19.20	36.74	0.03540	0.03800	0.07340	41.11	0.0803	0.81828	73
74	1.819	13.449	0.392	13.841	17.778	19.88	37.66	0.03585	0.03928	0.07513	42.10	0.0821	0.84624	74
75	1.882	13.474	0.407	13.881	18.018	20.59	38.61	0.03630	0.04060	0.07690	43.10	0.0840	0.87504	75
76	1.948	13.499	0.422	13.921	18.259	21.31	39.57	0.03675	0.04197	0.07872	44.10	0.0859	0.90470	76
77	2.016	13.525	0.437	13.962	18.499	22.07	40.57	0.03720	0.04337	0.08057	45.10	0.0877	0.93523	77
78	2.086	13.550	0.453	14.003	18.740	22.84	41.58	0.03765	0.04482	0.08247	46.10	0.0896	0.96665	78
79	2.158	13.575	0.470	14.045	18.980	23.64	42.62	0.03810	0.04631	0.08441	47.10	0.0914	0.99899	79
80	2.233	13.601	0.486	14.087	19.221	24.47	43.69	0.03854	0.04784	0.08638	48.10	0.0933	1.0323	80
81	2.310	13.626	0.504	14.130	19.461	25.32	44.78	0.03899	0.04942	0.08841	49.09	0.0952	1.0665	81
82	2.389	13.651	0.523	14.174	19.702	26.20	45.90	0.03943	0.05105	0.09048	50.09	0.0970	1.1017	82
83	2.471	13.676	0.542	14.218	19.942	27.10	47.04	0.03987	0.05273	0.09260	51.09	0.0989	1.1379	83
84	2.555	13.702	0.560	14.262	20.183	28.04	48.22	0.04031	0.05446	0.09477	52.09	0.1007	1.1752	84
85	2.642	13.727	0.581	14.308	20.423	29.01	49.43	0.04075	0.05624	0.09699	53.09	0.1025	1.2135	85
86	2.731	13.752	0.602	14.354	20.663	30.00	50.66	0.04119	0.05807	0.09926	54.08	0.1043	1.2529	86
87	2.824	13.777	0.624	14.401	20.904	31.03	51.93	0.04163	0.05995	0.10158	55.08	0.1062	1.2934	87
88	2.919	13.803	0.645	14.448	21.144	32.09	53.23	0.04207	0.06189	0.10396	56.08	0.1080	1.3351	88
89	3.017	13.828	0.668	14.496	21.385	33.18	54.56	0.04251	0.06389	0.10640	57.08	0.1098	1.3779	89
90	3.118	13.853	0.692	14.545	21.625	34.31	55.93	0.04295	0.06596	0.10890	58.08	0.1116	1.4219	90
91	3.223	13.879	0.716	14.595	21.865	35.47	57.33	0.04339	0.06807	0.11146	59.07	0.1135	1.4671	91
92	3.330	13.904	0.741	14.645	22.106	36.67	58.78	0.04382	0.07025	0.11407	60.07	0.1153	1.5135	92
93	3.441	13.929	0.768	14.697	22.346	37.90	60.25	0.04426	0.07249	0.11675	61.07	0.1171	1.5612	93
94	3.556	13.954	0.795	14.749	22.587	39.18	61.77	0.04469	0.07480	0.11949	62.07	0.1188	1.6102	94
95	3.673	13.980	0.822	14.802	22.827	40.49	63.32	0.04513	0.07718	0.12231	63.07	0.1206	1.6606	95
96	3.795	14.005	0.851	14.856	23.068	41.85	64.92	0.04556	0.07963	0.12519	64.06	0.1224	1.7123	96
97	3.920	14.030	0.881	14.911	23.308	43.24	66.55	0.04600	0.08215	0.12815	65.06	0.1242	1.7654	97
98	4.049	14.056	0.911	14.967	23.548	44.68	68.23	0.04643	0.08474	0.13117	66.06	0.1260	1.8199	98
99	4.182	14.081	0.942	15.023	23.789	46.17	69.96	0.04686	0.08741	0.13427	67.06	0.1278	1.8759	99
100	4.319	14.106	0.975	15.081	24.029	47.70	71.73	0.04729	0.09016	0.13745	68.06	0.1296	1.9333	100
101	4.460	14.131	1.009	15.140	24.270	49.28	73.55	0.04772	0.09299	0.14071	69.05	0.1314	1.9923	101
102	4.606	14.157	1.043	15.200	24.510	50.91	75.42	0.04815	0.09591	0.14406	70.05	0.1332	2.0528	102
103	4.756	14.182	1.079	15.261	24.751	52.59	77.34	0.04858	0.09891	0.14749	71.05	0.1350	2.1149	103
104	4.911	14.207	1.117	15.324	24.991	54.32	79.31	0.04900	0.1020	0.1510	72.05	0.1367	2.1786	104

Fahr. Temp. t(F)	Humidity Ratio $W_s \times 10$	Volume cu ft/lb dry air			Enthalpy Btu/lb dry air			Entropy Btu per (°F) (lb dry air)			Condensed Water			Fahr. Temp. t(F)
		v_a	v_{as}	v_s	h_a	h_{as}	h_s	s_a	s_{as}	s_s	Enthalpy Btu/Lb h_w	Entropy Btu per (°F)(Lb) s_w	Vap. Press In. Hg p_s	
105	0.5070	14.232	1.155	15.387	25.232	56.11	81.34	0.04943	0.1052	0.1546	73.04	0.1385	2.2439	105
106	0.5234	14.258	1.194	15.452	25.472	57.95	83.42	0.04985	0.1085	0.1584	74.04	0.1403	2.3109	106
107	0.5404	14.283	1.235	15.518	25.713	59.85	85.56	0.05028	0.1118	0.1621	75.04	0.1421	2.3797	107
108	0.5578	14.308	1.278	15.586	25.953	61.80	87.76	0.05070	0.1153	0.1660	76.04	0.1438	2.4502	108
109	0.5758	14.333	1.321	15.654	26.194	63.82	90.03	0.05113	0.1189	0.1700	77.04	0.1456	2.5225	109
110	0.5944	14.359	1.365	15.724	26.434	65.91	92.34	0.05155	0.1226	0.1742	78.03	0.1472	2.5966	110
111	0.6135	14.384	1.412	15.796	26.675	68.05	94.72	0.05197	0.1264	0.1784	79.03	0.1491	2.6726	111
112	0.6333	14.409	1.460	15.869	26.915	70.27	97.18	0.05239	0.1302	0.1826	80.03	0.1508	2.7505	112
113	0.6536	14.435	1.509	15.944	27.156	72.55	99.71	0.05281	0.1342	0.1870	81.03	0.1525	2.8304	113
114	0.6746	14.460	1.560	16.020	27.397	74.91	102.31	0.05323	0.1384	0.1916	82.03	0.1543	2.9123	114
115	0.6962	14.485	1.613	16.098	27.637	77.34	104.98	0.05365	0.1426	0.1963	83.02	0.1560	2.9962	115
116	0.7185	14.510	1.668	16.178	27.878	79.85	107.73	0.05407	0.1470	0.2011	84.02	0.1577	3.0821	116
117	0.7415	14.536	1.723	16.259	28.119	82.43	110.55	0.05449	0.1515	0.2060	85.02	0.1595	3.1701	117
118	0.7652	14.561	1.782	16.343	28.359	85.10	113.46	0.05490	0.1562	0.2111	86.02	0.1612	3.2603	118
119	0.7897	14.586	1.842	16.428	28.600	87.86	116.46	0.05532	0.1610	0.2163	87.02	0.1629	3.3527	119
120	0.8149	14.611	1.905	16.516	28.841	90.70	119.54	0.05573	0.1659	0.2216	88.01	0.1646	3.4474	120
121	0.8410	14.637	1.968	16.605	29.082	93.64	122.72	0.05615	0.1710	0.2272	89.01	0.1664	3.5443	121
122	0.8678	14.662	2.034	16.696	29.322	96.66	125.98	0.05656	0.1763	0.2329	90.01	0.1681	3.6436	122
123	0.8955	14.687	2.103	16.790	29.563	99.79	129.35	0.05698	0.1817	0.2387	91.01	0.1698	3.7452	123
124	0.9242	14.712	2.174	16.886	29.804	103.0	132.8	0.05739	0.1872	0.2446	92.01	0.1715	3.8493	124
125	0.9537	14.738	2.247	16.985	30.044	106.4	136.4	0.05780	0.1930	0.2508	93.01	0.1732	3.9558	125
126	0.9841	14.763	2.323	17.086	30.285	109.8	140.1	0.05821	0.1989	0.2571	94.01	0.1749	4.0649	126
127	1.016	14.788	2.401	17.189	30.526	113.4	143.9	0.05862	0.2050	0.2636	95.00	0.1766	4.1765	127
128	1.048	14.813	2.482	17.295	30.766	117.0	147.8	0.05903	0.2113	0.2703	96.00	0.1783	4.2907	128
129	1.082	14.839	2.565	17.404	31.007	120.8	151.8	0.05944	0.2178	0.2772	97.00	0.1800	4.4076	129
130	1.116	14.864	2.652	17.516	31.248	124.7	155.9	0.05985	0.2245	0.2844	98.00	0.1817	4.5272	130
131	1.152	14.889	2.742	17.631	31.489	128.8	160.3	0.06026	0.2314	0.2917	99.00	0.1834	4.6495	131
132	1.189	14.915	2.834	17.749	31.729	133.0	164.7	0.06067	0.2386	0.2993	100.00	0.1851	4.7747	132
133	1.227	14.940	2.930	17.870	31.970	137.3	169.3	0.06108	0.2459	0.3070	101.00	0.1868	4.9028	133
134	1.267	14.965	3.029	17.994	32.211	141.8	174.0	0.06148	0.2536	0.3151	102.00	0.1885	5.0337	134

[a] Compiled by John A. Goff and S. Gratch.

Extracted by permission from ASHRAE Guide and Data Book, 1965

TABLE 7-2   Thermodynamic Properties Of WATER At Saturation

Fahr. Temp. t(F)	Absolute Pressure p		Specific Volume, cu ft per lb			Enthalpy, Btu per lb			Entropy, Btu per (Lb) (°F)			Fahr. Temp. t(F)
	Lb/Sq In.	In. Hg	Sat. Solid v_i	Evap. v_{ig}	Sat. Vapor v_g	Sat. Solid h_i	Evap. h_{ig}	Sat. Vapor h_g	Sat. Solid s_i	Evap. s_{ig}	Sat. Vapor s_g	
7	0.02653	0.05402	0.01743	10480	10480	−155.66	1219.84	1064.18	−0.3172	2.6138	2.2966	7
8	0.02791	0.05683	0.01743	9979	9979	−155.18	1219.80	1064.62	−0.3162	2.6081	2.2919	8
9	0.02936	0.05977	0.01744	9507	9507	−154.70	1219.76	1065.06	−0.3152	2.6025	2.2873	9
10	0.03087	0.06286	0.01744	9060	9060	−154.22	1219.72	1065.50	−0.3142	2.5969	2.2827	10
11	0.03246	0.06608	0.01744	8636	8636	−153.74	1219.68	1065.94	−0.3131	2.5912	2.2781	11
12	0.03412	0.06946	0.01744	8234	8234	−153.26	1219.64	1066.38	−0.3121	2.5857	2.2736	12
13	0.03585	0.07300	0.01744	7851	7851	−152.77	1219.59	1066.82	−0.3111	2.5801	2.2690	13
14	0.03767	0.07669	0.07144	7489	7489	−152.29	1219.55	1067.26	−0.3101	2.5746	2.2645	14
15	0.03957	0.08056	0.01744	7144	7144	−151.80	1219.50	1067.70	−0.3090	2.5690	2.2600	15
16	0.04156	0.08461	0.01745	6817	6817	−151.32	1219.46	1068.14	−0.3080	2.5635	2.2555	16
17	0.04363	0.08884	0.01745	6505	6505	−150.83	1219.41	1068.58	−0.3070	2.5581	2.2511	17
18	0.04581	0.09326	0.01745	6210	6210	−150.34	1219.36	1069.02	−0.3060	2.5526	2.2466	18
19	0.04808	0.09789	0.01745	5929	5929	−149.85	1219.31	1069.46	−0.3049	2.5471	2.2422	19
20	0.05045	0.1027	0.01745	5662	5662	−149.36	1219.26	1069.90	−0.3039	2.5417	2.2378	20
21	0.05293	0.1078	0.01745	5408	5408	−148.87	1219.21	1070.34	−0.3029	2.5364	2.2335	21
22	0.05552	0.1130	0.01746	5166	5166	−148.38	1219.16	1070.78	−0.3019	2.5310	2.2291	22
23	0.05823	0.1186	0.01746	4936	4936	−147.88	1219.10	1071.22	−0.3008	2.5256	2.2248	23
24	0.06105	0.1243	0.01746	4717	4717	−147.39	1219.05	1071.66	−0.2998	2.5203	2.2205	24
25	0.06400	0.1303	0.01746	4509	4509	−146.89	1218.98	1072.09	−0.2988	2.5150	2.2162	25
26	0.06708	0.1366	0.01746	4311	4311	−146.40	1218.93	1072.53	−0.2978	2.5097	2.2119	26
27	0.07030	0.1431	0.01746	4122	4122	−145.90	1218.87	1072.97	−0.2968	2.5045	2.2077	27
28	0.07365	0.1500	0.01746	3943	3943	−145.40	1218.81	1073.41	−0.2957	2.4991	2.2034	28
29	0.07715	0.1571	0.01747	3771	3771	−144.90	1218.75	1073.85	−0.2947	2.4939	2.1992	29
30	0.08080	0.1645	0.01747	3608	3608	−144.40	1218.69	1074.29	−0.2937	2.4887	2.1950	30
31	0.08461	0.1723	0.01747	3453	3453	−143.90	1218.63	1074.73	−0.2927	2.4835	2.1908	31
32	0.08858	0.1803	0.01747	3305	3305	−143.40	1218.56	1075.16	−0.2916	2.4783	2.1867	32
*32	0.088586	0.18036	0.01602	3304.6	3304.6	0.00	1075.16	1075.16	0.00000	2.1867	2.1867	32*

Fahr. Temp. t(F)	Absolute Pressure p_s		Specific Volume, cu ft per lb			Enthalpy, Btu per lb			Entropy, Btu per (Lb) (°F)			Fahr. Temp. t(F)
	Lb/Sq In.	In. Hg	Sat. Liquid v_f	Evap. v_{fg}	Sat. Vapor v_g	Sat. Liquid h_f	Evap. h_{fg}	Sat. Vapor h_g	Sat. Liquid s_f	Evap. s_{fg}	Sat. Vapor s_g	
33	0.092227	0.18778	0.01602	3180.5	3180.5	1.01	1074.59	1075.60	0.00205	2.1811	2.1831	33
34	0.095999	0.19546	0.01602	3061.7	3061.7	2.01	1074.03	1076.04	0.00409	2.1755	2.1796	34
35	0.099908	0.20342	0.01602	2947.8	2947.8	3.02	1073.46	1076.48	0.00612	2.1700	2.1761	35
36	0.10396	0.21166	0.01602	2838.7	2838.7	4.02	1072.90	1076.92	0.00815	2.1644	2.1726	36
37	0.10815	0.22020	0.01602	2734.1	2734.1	5.03	1072.33	1077.36	0.01018	2.1589	2.1691	37
38	0.11249	0.22904	0.01602	2633.8	2633.8	6.03	1071.77	1077.80	0.01220	2.1535	2.1657	38
39	0.11699	0.23819	0.01602	2537.6	2537.6	7.04	1071.20	1078.24	0.01422	2.1480	2.1622	39
40	0.12164	0.24767	0.01602	2445.4	2445.4	8.04	1070.64	1078.68	0.01623	2.1426	2.1588	40
41	0.12646	0.25748	0.01602	2356.9	2356.9	9.05	1070.06	1079.11	0.01824	2.1372	2.1554	41
42	0.13145	0.26763	0.01602	2272.0	2272.0	10.05	1069.50	1079.55	0.02024	2.1318	2.1520	42
43	0.13660	0.27813	0.01602	2190.5	2190.5	11.05	1068.94	1079.99	0.02224	2.1265	2.1487	43
44	0.14194	0.28899	0.01602	2112.3	2112.3	12.06	1068.37	1080.43	0.02423	2.1211	2.1453	44
45	0.14746	0.30023	0.01602	2037.3	2037.3	13.06	1067.81	1080.87	0.02622	2.1158	2.1420	45
46	0.15317	0.31185	0.01602	1965.2	1965.2	14.06	1067.24	1081.30	0.02820	2.1105	2.1387	46
47	0.15907	0.32387	0.01602	1896.0	1896.0	15.06	1066.68	1081.74	0.03018	2.1052	2.1354	47
48	0.16517	0.33629	0.01602	1829.5	1829.5	16.07	1066.11	1082.18	0.03216	2.0999	2.1321	48
49	0.17148	0.34913	0.01602	1765.7	1765.7	17.07	1065.55	1082.62	0.03413	2.0947	2.1288	49
50	0.17799	0.36240	0.01602	1704.3	1704.3	18.07	1064.99	1083.06	0.03610	2.0895	2.1256	50
51	0.18473	0.37611	0.01602	1645.4	1645.4	19.07	1064.42	1083.49	0.03806	2.0842	2.1223	51
52	0.19169	0.39028	0.01602	1588.7	1588.7	20.07	1063.86	1083.93	0.04002	2.0791	2.1191	52
53	0.19888	0.40402	0.01603	1534.3	1534.3	21.07	1063.30	1084.37	0.04197	2.0739	2.1159	53
54	0.20630	0.42003	0.01603	1481.9	1481.9	22.08	1062.72	1084.80	0.04392	2.0688	2.1127	54
55	0.21397	0.43564	0.01603	1431.5	1431.5	23.08	1062.16	1085.24	0.04587	2.0637	2.1096	55
56	0.22188	0.45176	0.01603	1383.1	1383.1	24.08	1061.60	1085.68	0.04781	2.0586	2.1064	56
57	0.23006	0.46840	0.01603	1336.5	1336.5	25.08	1061.04	1086.12	0.04975	2.0535	2.1033	57
58	0.23849	0.49658	0.01603	1291.7	1291.7	26.08	1060.47	1086.55	0.05168	2.0485	2.1002	58
59	0.24720	0.50330	0.01603	1248.6	1248.6	27.08	1059.91	1086.99	0.05361	2.0434	2.0970	59
60	0.25618	0.52160	0.01603	1207.1	1207.1	28.08	1059.34	1087.42	0.05553	2.0385	2.0940	60
61	0.26545	0.54047	0.01604	1167.2	1167.2	29.08	1058.78	1087.86	0.05746	2.0334	2.0909	61
62	0.27502	0.55994	0.01604	1128.7	1128.7	30.08	1058.22	1088.30	0.05937	2.0284	2.0878	62
63	0.28488	0.58002	0.01604	1091.7	1091.7	31.08	1057.65	1088.73	0.06129	2.0235	2.0848	63
64	0.29505	0.60073	0.01604	1056.1	1056.1	32.08	1057.09	1089.17	0.06320	2.0186	2.0818	64
65	0.30554	0.62209	0.01604	1021.7	1021.7	33.08	1056.52	1089.60	0.06510	2.0136	2.0787	65
66	0.31636	0.64411	0.01604	988.63	988.65	34.07	1055.97	1090.04	0.06700	2.0087	2.0757	66
67	0.32750	0.66681	0.01605	956.76	956.78	35.07	1055.40	1090.47	0.06890	2.0039	2.0728	67
68	0.33900	0.69021	0.01605	926.06	926.08	36.07	1054.84	1090.91	0.07080	1.9990	2.0698	68
69	0.35084	0.71432	0.01605	896.47	896.49	37.07	1054.27	1091.34	0.07269	1.9941	2.0668	69
70	0.36304	0.73916	0.01605	867.95	867.97	38.07	1053.71	1091.78	0.07458	1.9893	2.0639	70
71	0.37561	0.76476	0.01605	840.45	840.47	39.07	1053.14	1092.21	0.07646	1.9845	2.0610	71

[a] Compiled by John A. Goff and S. Gratch.
* Extrapolated to represent metastable equilibrium with undercooled liquid.

Extracted by permission from ASHRAE Guide and Data Book, 1965

Fahr. Temp. t(F)	Absolute Pressure p		Specific Volume, cu ft per lb			Enthalpy, Btu per lb			Entropy, Btu per (lb) (°F)			Fahr. Temp. t(F)
	Lb/Sq In.	In. Hg	Sat. Liquid v_f	Evap. v_{fg}	Sat. Vapor v_g	Sat. Liquid h_f	Evap. h_{fg}	Sat. Vapor h_g	Sat. Liquid s_f	Evap. s_{fg}	Sat. Vapor s_g	
72	0.38856	0.79113	0.01606	813.95	813.97	40.07	1052.58	1092.65	0.07834	1.9797	2.0580	72
73	0.40190	0.81829	0.01606	788.38	788.40	41.07	1052.01	1093.08	0.08022	1.9749	2.0551	73
74	0.41564	0.84626	0.01606	763.73	763.75	42.06	1051.46	1093.52	0.08209	1.9701	2.0522	74
75	0.42979	0.87506	0.01606	739.95	739.97	43.06	1050.89	1093.95	0.08396	1.9654	2.0494	75
76	0.44435	0.90472	0.01606	717.01	717.03	44.06	1050.32	1094.38	0.08582	1.9607	2.0465	76
77	0.45935	0.93524	0.01607	694.88	694.90	45.06	1049.76	1094.82	0.08769	1.9560	2.0437	77
78	0.47478	0.96666	0.01607	673.52	673.54	46.06	1049.19	1095.25	0.08954	1.9513	2.0408	78
79	0.49066	0.99900	0.01607	652.91	652.93	47.06	1048.62	1095.68	0.09140	1.9466	2.0380	79
80	0.50701	1.0323	0.01607	633.01	633.03	48.05	1048.07	1096.12	0.09325	1.9419	2.0352	80
81	0.52382	1.0665	0.01608	613.80	613.82	49.05	1047.50	1096.55	0.09510	1.9373	2.0324	81
82	0.54112	1.1017	0.01608	595.25	595.27	50.05	1046.93	1096.98	0.09694	1.9328	2.0297	82
83	0.55892	1.1380	0.01608	577.34	577.36	51.05	1046.37	1097.42	0.09878	1.9281	2.0269	83
84	0.57722	1.1752	0.01608	560.04	560.06	52.05	1045.80	1097.85	0.10062	1.9236	2.0242	84
85	0.59604	1.2136	0.01609	543.33	543.35	53.05	1045.23	1098.28	0.10246	1.9189	2.0214	85
86	0.61540	1.2530	0.01609	527.19	527.21	54.04	1044.67	1098.71	0.10429	1.9144	2.0187	86
87	0.63530	1.2935	0.01609	511.60	511.62	55.04	1044.10	1099.14	0.10611	1.9099	2.0160	87
88	0.65575	1.3351	0.01610	496.52	496.54	56.04	1043.54	1099.58	0.10794	1.9054	2.0133	88
89	0.67678	1.3779	0.01610	481.96	481.98	57.04	1042.97	1100.01	0.10976	1.9008	2.0106	89
90	0.69838	1.4219	0.01610	467.88	467.90	58.04	1042.40	1100.44	0.11158	1.8963	2.0079	90
91	0.72059	1.4671	0.01610	454.26	454.28	59.03	1041.84	1100.87	0.11339	1.8919	2.0053	91
92	0.74340	1.5136	0.01611	441.10	441.12	60.03	1041.27	1101.30	0.11520	1.8874	2.0026	92
93	0.76684	1.5613	0.01611	428.38	428.40	61.03	1040.70	1101.73	0.11701	1.8830	2.0000	93
94	0.79091	1.6103	0.01611	416.07	416.09	62.03	1040.13	1102.16	0.11881	1.8786	1.9974	94
95	0.81564	1.6607	0.01612	404.17	404.19	63.03	1039.56	1102.59	0.12061	1.8741	1.9947	95
96	0.84103	1.7124	0.01612	392.65	392.67	64.02	1039.00	1103.02	0.12241	1.8698	1.9922	96
97	0.86711	1.7655	0.01612	381.51	381.53	65.02	1038.43	1103.45	0.12420	1.8654	1.9896	97
98	0.89388	1.8200	0.01613	370.73	370.75	66.02	1037.86	1103.88	0.12600	1.8610	1.9870	98
99	0.92137	1.8759	0.01613	360.30	360.32	67.02	1037.29	1104.31	0.12778	1.8566	1.9844	99
100	0.94959	1.9334	0.01613	350.20	350.22	68.02	1036.72	1104.74	0.12957	1.8523	1.9819	100
101	0.97854	1.9923	0.01614	340.42	340.44	69.01	1036.16	1105.17	0.13135	1.8480	1.9793	101
102	1.0083	2.0529	0.01614	330.96	330.98	70.01	1035.58	1105.59	0.13313	1.8437	1.9768	102
103	1.0388	2.1149	0.01614	321.80	321.82	71.01	1035.01	1106.02	0.13490	1.8394	1.9743	103
104	1.0700	2.1786	0.01614	312.93	312.95	72.01	1034.44	1106.45	0.13667	1.8351	1.9718	104
105	1.1021	2.2440	0.01615	304.34	304.36	73.01	1033.87	1106.88	0.13844	1.8309	1.9693	105
106	1.1351	2.3110	0.01615	296.02	296.04	74.01	1033.29	1107.30	0.14021	1.8266	1.9668	106
107	1.1688	2.3798	0.01616	287.96	287.98	75.00	1032.73	1107.73	0.14197	1.8224	1.9644	107

Fahr. Temp. t(F)	Absolute Pressure P_s		Specific Volume, cu ft per lb			Enthalpy, Btu per lb			Entropy, Btu per (lb) (°F)			Fahr. Temp. t(F)
	Lb/Sq In.	In. Hg	Sat. Liquid v_f	Evap. v_{fg}	Sat. Vapor v_g	Sat. Liquid h_f	Evap. h_{fg}	Sat. Vapor h_g	Sat. Liquid s_f	Evap. s_{fg}	Sat. Vapor s_g	
108	1.2035	2.4503	0.01616	280.14	280.16	76.00	1032.16	1108.16	0.14373	1.8182	1.9619	108
109	1.2390	2.5226	0.01616	272.58	272.60	77.00	1031.58	1108.58	0.14549	1.8140	1.9595	109
110	1.2754	2.5968	0.01617	265.24	265.26	78.00	1031.01	1109.01	0.14724	1.8098	1.9570	110
111	1.3128	2.6728	0.01617	258.14	258.16	79.00	1030.44	1109.44	0.14899	1.8056	1.9546	111
112	1.3510	2.7507	0.01617	251.25	251.27	80.00	1029.86	1109.86	0.15074	1.8015	1.9522	112
113	1.3902	2.8306	0.01618	244.57	244.59	80.99	1029.30	1110.29	0.15248	1.7973	1.9498	113
114	1.4305	2.9125	0.01618	238.10	238.12	81.99	1028.72	1110.71	0.15423	1.7932	1.9474	114
115	1.4717	2.9963	0.01618	231.82	231.84	82.99	1028.15	1111.14	0.15596	1.7890	1.9450	115
116	1.5139	3.0823	0.01619	225.73	225.75	83.99	1027.57	1111.56	0.15770	1.7849	1.9426	116
117	1.5571	3.1703	0.01619	219.83	219.85	84.99	1026.99	1111.98	0.15943	1.7809	1.9403	117
118	1.6014	3.2606	0.01620	214.10	214.12	85.99	1026.42	1112.41	0.16116	1.7767	1.9379	118
119	1.6468	3.3530	0.01620	208.54	208.56	86.98	1025.85	1112.83	0.16289	1.7727	1.9356	119
120	1.6933	3.4477	0.01620	203.16	203.18	87.98	1025.28	1113.26	0.16461	1.7687	1.9333	120
121	1.7409	3.5446	0.01621	197.93	197.95	88.98	1024.70	1113.68	0.16634	1.7647	1.9310	121
122	1.7897	3.6439	0.01621	192.85	192.87	89.98	1024.12	1114.10	0.16805	1.7606	1.9286	122
123	1.8396	3.7455	0.01622	187.93	187.95	90.98	1023.54	1114.52	0.16977	1.7566	1.9264	123
124	1.8907	3.8496	0.01622	183.15	183.17	91.98	1022.96	1114.94	0.17148	1.7526	1.9241	124
125	1.9430	3.9561	0.01622	178.51	178.53	92.98	1022.39	1115.37	0.17319	1.7486	1.9218	125
126	1.9966	4.0651	0.01623	174.00	174.02	93.98	1021.81	1115.79	0.17490	1.7446	1.9195	126
127	2.0514	4.1768	0.01623	169.63	169.65	94.97	1021.24	1116.21	0.17660	1.7407	1.9173	127
128	2.1075	4.2910	0.01624	165.38	165.40	95.97	1020.66	1116.63	0.17830	1.7367	1.9150	128
129	2.1649	4.4078	0.01624	161.26	161.28	96.97	1020.08	1117.05	0.18000	1.7328	1.9128	129
130	2.2237	4.5274	0.01625	157.25	157.27	97.97	1019.50	1117.47	0.18170	1.7289	1.9106	130
131	2.2838	4.6498	0.01625	153.36	153.38	98.97	1018.92	1117.89	0.18339	1.7250	1.9084	131
132	2.3452	4.7750	0.01626	149.58	149.60	99.97	1018.34	1118.31	0.18508	1.7211	1.9062	132
133	2.4081	4.9030	0.01626	145.91	145.93	100.97	1017.76	1118.73	0.18676	1.7172	1.9040	133
134	2.4725	5.0340	0.01626	142.34	142.36	101.97	1017.18	1119.15	0.18845	1.7134	1.9018	134
135	2.5382	5.1679	0.01627	138.87	138.89	102.97	1016.59	1119.56	0.19013	1.7095	1.8996	135
136	2.6055	5.3049	0.01627	135.50	135.52	103.97	1016.01	1119.98	0.19181	1.7056	1.8974	136
137	2.6743	5.4450	0.01628	132.22	132.24	104.97	1015.43	1120.40	0.19348	1.7018	1.8953	137
138	2.7446	5.5881	0.01628	129.04	129.06	105.97	1014.85	1120.82	0.19516	1.6979	1.8931	138
139	2.8165	5.7345	0.01629	125.94	125.96	106.97	1014.26	1121.23	0.19683	1.6942	1.8910	139
140	2.8900	5.8842	0.01629	122.94	122.96	107.96	1013.69	1121.65	0.19850	1.6903	1.8888	140
141	2.9651	6.0371	0.01630	120.01	120.03	108.96	1013.11	1122.07	0.20016	1.6865	1.8867	141
142	3.0419	6.1934	0.01630	117.16	117.18	109.96	1012.52	1122.48	0.20182	1.6828	1.8846	142
143	3.1204	6.3532	0.01631	114.40	114.42	110.96	1011.94	1122.90	0.20348	1.6790	1.8825	143

[a] Compiled by John A. Goff and S. Gratch.

Extracted by permission from ASHRAE Guide and Data Book, 1965

As stated earlier, the chart is based on a total pressure of 14.7 psia, so we would have to reduce the pressure of the dry air: $14.7 = p_a + 0.182$, $p_a = 14.7 - 0.182$, $p_a = 14.518$ psia. The reduced pressure, of course, means an increase in volume.

$$P_1v_1 = P_2v_2$$
$$v_2 = v_1P_1/P_2$$
$$v_2 = (13.348)\frac{14.7}{14.518}$$
$$v_2 = 13.51 \text{ cu ft/lb}$$

This means that one pound of dry air occupies 13.51 cu ft. By Dalton's law, the water vapor in the air occupies the same volume. So we have 13.51 cu ft of mixture per lb of dry air. This is the apparent specific volume. The true specific volume is equal to 13.51 cu ft, divided by the weight of the mixture. At 70° sat., the humidity ratio is 0.01582 lb vapor/lb dry air. At 50% RH, the humidity ratio is 0.00791 lb/lb dry air. So the weight of the mixture is 1 lb of dry air +0.00791 lb vapor/lb dry air = 1.00791 lb. Dividing 13.51 by 1.00791 gives 13.4 cu ft/lb, the true specific volume.

TABULATED THERMODYNAMIC PROPERTIES

Table 7-1 lists the thermodynamic properties of moist air. Properties are listed for the various air temperatures. Given are the specific volume, enthalpy, and entropy. The subscripts "a," "as," and "s" are used. Dry air is indicated by "a," moist air at saturation by "s," and "as" indicates the difference between saturated moist air and dry air. For instance, at 40° the volume of dry air V_a is 12.590 cu ft/lb, the volume of moist air at saturation V_s is 12.695 cu ft/lb, and V_{as} equals $V_a - V_s$ or 0.105 cu ft/lb. Also given are the enthalpy, entropy, and vapor pressure for condensed water.

Table 7-2 gives the vapor pressure, specific volume, enthalpy, and entropy of water at saturation. The subscripts "i," "f," and "g" are used to indicate solid water (ice), liquid water, and vapor (steam), respectively. The subscripts "ig," "fg" indicate the difference between the solid or liquid and vapor phase.

GENERAL ENERGY EQUATION

The general energy equation is an expression of the first law of thermodynamics. The first law states that if any system goes through a process in which work or heat is added or removed from the system, none of the work or heat is destroyed in the system or created in the system. Heat can be converted to work and work to heat. The classic example given in most thermodynamics texts is an apparatus with fluid streams flow-

ing in and out of it. Heat enters the apparatus and shaft work leaves it. What it amounts to is that the sum of all the forms of energy entering the system must be equal to all the energy leaving.

$$Mu_1 + MPV_1 + \frac{MV_1^2}{2g} + MZ_1 + q =$$
$$Mu_2 + MPV_2 + \frac{MV_2^2}{2g} + MZ_2 + W$$

where

u = internal energy
PV = flow work
$\frac{V^2}{2g}$ = kinetic energy
Z = potential energy
q = heat
w = work
M = mass

In the field of air conditioning, work, kinetic energy, and potential energy are insignificant and can be disregarded. This leaves

$$M(u_1 + PV_1) + q = M(u_2 + PV_2)$$

Earlier we stated that enthalpy, h, was equal to $(u + PV)$, so substituting:

$$Mh_1 + q = Mh_2$$
$$q = M(h_2 - h_1) \qquad \text{(Formula 7-1)}$$

With this equation we can solve all air conditioning problems.

THERMODYNAMIC WET BULB

Thermodynamic wet bulb, also called the adiabatic saturation temperature, is also easier described than defined. Take a long, perfectly insulated duct or tunnel. If it is perfectly insulated, no heat can be lost or gained by air passing through the duct. In thermodynamics, a condition such as this, where no heat is added or subtracted from a system, is called adiabatic. In this duct place a long pan of water at the same temperature as the air leaving the duct. Provision is made to add water as required. Now blow air through this duct.

If no heat is added or subtracted from the air, then the air leaving the duct must have the same amount of heat as the air entering the duct. But the enthalpy of the air-vapor mixture entering the duct is less than the enthalpy of the air-vapor mixture leaving the duct. The case of the missing enthalpy. Where did the additional enthalpy come from? From the water added to the mixture as the air traveled down the duct. And where did the heat come from to evaporate the water entering the air? The duct is perfectly insulated, so it can come from only one place—the air itself. And sure enough, if you measured the dry-bulb temperature of the air entering the duct, you would find it higher than the dry-bulb of the air leaving. So we can write a heat balance.

Enthalpy Entering + Enthalpy Added = Enthalpy Leaving

$$h_{a1} + W_1h_{v1} + (W_1 - W_2) h_f = h_{a2} + W_2h_{v2}$$

h_a = enthalpy of the air
h_v = enthalpy of the vapor
W = mass of vapor
1 = denotes entering condition
2 = denotes leaving condition
h_f = enthalpy of the liquid

On first examination there appears to be an error in the above formula. Do you see it yet? Notice that I used h_f, the enthalpy of the liquid, and not h_{fg}, the enthalpy of vaporization. But, you say, the water had to be vaporized to enter our air. And it takes heat to vaporize the water. Yes, but the heat necessary to vaporize came from cooling the air, so all we added to the air was the heat or enthalpy that the liquid brought with it. Let's do an example:

Example #5

Air at 80° DB and 67° WB enters a duct and it leaves at 67° DB and 67° WB. The water in the pan is at 67°.
From Examples 1, 2, 3, 4

h_{a1} = 19.2 Btu/lb enthalpy of dry air entering duct
h_{v1} = 1096.12 Btu/lb enthalpy of vapor entering duct
h_{a2} = 16.08 Btu/lb enthalpy of dry air leaving duct
h_{v2} = 1091.15 Btu/lb enthalpy of vapor leaving duct
W_1 = 0.01120 lb vapor/lb dry air amount of vapor in entering air
W_2 = 0.01420 lb vapor/lb dry air amount of vapor in leaving air

From Table 7–2

h_f = 35.07 Btu/lb enthalpy of water
$h_{a1} + W_1h_{v1} + (W_2 - W_1)h_f = h_{a2} + W_2h_{v2}$
$19.2 + 0.01120 (1096.12) + (0.01420 - 0.01120) 35.07 =$
$16.08 + 0.01420 (1091.15)$
$19.2 + 12.28 + 0.10 = 16.08 + 15.49$
$31.58 = 31.57$

Let's go one step further. Let's prove that the heat required to vaporize the water added is equal to the loss in sensible heat of the air-vapor mixture.

$(h_{a1} - h_{a2}) + W_1 (h_{v1} - h_{v2}) = (W_2 - W_1) (h_{fg})$
$(19.2 - 16.08) + 0.01120 (1096.2 - 1091.15) =$
$(0.01420 - 0.0112) (1055.4)$
$3.12 + 0.050 = 3.17$
$3.17 = 3.17$

Notice in Example 5 that the only difference in enthalpy between the air entering the duct at 80° DB and 67° WB, and the air leaving the duct at 67° saturated, is the enthalpy of the liquid. Dr. Carrier noted that if the enthalpy of the liquid is subtracted, all water-vapor mixtures at the same wet-bulb temperature would have the same heat content. He called this new property the sigma heat. Early psychrometric charts used the sigma function instead of enthalpy. For normal air conditioning work this produced an insignificant error. This error was not insignificant in some work, so present charts use enthalpy and give curves for enthalpy deviation.

HEATING

Example #6

How much heat is required to heat 1000 cfm from 50° and 50% RH to 100°? The humidity ratio is constant because we are only heating and not humidifying or de-humidifying.

h_{a1} = 12.010 Btu/lb of dry air (enthalpy of dry air at 50°)
h_{a2} = 24.029 Btu/lb of dry air (enthalpy of dry air at 100°)
w = 0.50 (0.007658) lb vapor/lb dry air (Table 7–1).
h_{v1} = 1083.06 Btu/lb of vapor
h_{v2} = 1104.74 Btu/lb of vapor
v = 12.843 cu ft/lb dry air
$P_v = \emptyset P_{vs} = 0.50 (0.17799) = 0.08899$ psia

$$v' = 12.843 \frac{(14.7)}{14.7 - 0.08899}$$

$v' = 12.86$ (apparent specific volume)

$$\frac{\text{lb of air}}{\text{hr}} = \frac{\dfrac{ft^3}{min} \times \dfrac{60 \text{ min}}{hr}}{\dfrac{ft^3}{lb}}$$

$$\frac{\text{lb of air}}{\text{hr}} = \frac{1000 \times 60}{12.86} = 4,670$$

From the general energy equation

$q = m(h_2 - h_1)$
$h_2 = h_{a2} + wh_{v2}$
$h_2 = 24.029 + 0.003829 (1104.74)$
$h_2 = 24.029 + 4.22$
$h_2 = 28.25$ Btu/lb
$h_1 = 12.010 + 0.003829 (1083.06)$
$h_1 = 12.010 + 4.14$
$h_1 = 16.15$ Btu/lb
$h_2 - h_1 = 28.25 - 16.15$
$h_2 - h_1 = 12.10$ Btu/lb
$q = m (h_2 - h_1)$
$q = 4,670 (12.10)$
$q = 56,500$ Btu/hr

As you can see, the amount of heat required to heat the vapor in the air from 50° to 100° was only 4.22 —

4.14 = 0.08 Btu/lb of dry air, or 388 Btu/hr. In normal work, ignoring this would result in an insignificant error. As long as it is remembered that an error is introduced in the ignoring of the superheating of the water vapor, the following formula may be used:

$$q \frac{(Btu)}{(hr)} =$$

$$\frac{cfm \times 60 \text{ min/hr} \times \text{specific heat (Btu/lb °F)} \times \text{temp. rise}}{\text{specific volume}}$$

$$q \frac{(Btu)}{(hr)} = \frac{cfm (60) (0.24) (t_2 - t_1)}{13.35}$$

$$q \frac{(Btu)}{(hr)} = 1.08 \text{ cfm } (t_2 - t_1) \qquad \text{(Formula 7-2)}$$

In the heating of high temperature air and water mixtures, the ignoring of the superheating of the vapor can introduce a significant error. Let's determine the amount of heat required to raise the temperature of 1000 cfm of 150° saturated air to 200°.

h_{a1} = 36.063 Btu/lb of dry air
h_{a2} = 48.119 Btu/lb of dry air
h_{v1} = 1125.79 Btu/lb of vapor
h_{v2} = 1145.78 Btu/lb of vapor
w = 0.2125 lb vapor/lb dry air

$$v' = 15.3 \frac{14.7}{14.7 - 3.7} = 15.3 \frac{14.7}{11}$$

v' = 20.580 cu ft/lb (apparent specific volume of air)
q = m ($h_2 - h_1$)

$$m = \frac{1000 \text{ cfm} \times 60 \text{ min}}{20.580 \text{ cu ft/lb}}$$

m = 2900 lb of dry air
h_1 = $h_{a1} + wh_{v1}$
h_1 = 36.063 + 0.2125 (1125.79)
h_1 = 36.063 + 240
h_1 = 276.063
h_2 = $h_{a2} + wh_{v2}$
h_2 = 48.119 + 0.2125 (1145.78)
h_2 = 48.119 + 244.2
h_2 = 292.319
q = m ($h_2 - h_1$)
q = 2900 (292.319 − 276.063)
q = 2900 (16.256)
q = 47,000 Btuh

Note that the enthalpy of the vapor is 244.2 − 240.0 = 4.2 Btu/lb of dry air. This is 26% of the total enthalpy difference of 16.256, so if it is ignored, a serious error can result.

HUMIDIFICATION

Example #7

How much heat is required to raise the relative humidity of 1000 cfm at 80° from 20% to 70%, using

water at 80°? Again, we use the general energy equation, remembering to include all energy terms.

Enthalpy of entering air + enthalpy of liquid water added + heat added = enthalpy of leaving air.

m ($h_{a1} + w_1 h_{v1}$) + m ($w_2 - w_1$) h_f + q = m ($h_{a2} + w_2 h_{v2}$)
q = m (($h_{a2} + w_2 h_{v2}$) − ($h_{a1} + w_1 h_{v1}$) − ($w_2 - w_1$) h_f)
h_2 = $h_{a2} + w_2 h_{v2}$
h_1 = $h_{a1} + w_1 h_{v1}$
q = m ($h_2 - h_1 - (w_2 - w_1) h_f$)
v = 13.60 cu ft/lb dry air
P_v = $\emptyset P_{vs}$ = 0.20 (0.50701) = 0.1014

$$v' = 13.60 \frac{(14.7)}{(14.7 - 0.1014)} = 13.65$$

h_{a1} = h_{a2} = 19.221 Btu/lb dry air
h_{v1} = h_{v2} = 1096.12 Btu/lb of vapor
w_1 = .20 × 0.02233 lb/lb dry air = 0.004466
w_2 = .70 × 0.02233 lb/lb dry air = 0.015631
h_f = 48.05 Btu/lb of water

$$m = \frac{1000 \text{ cfm} (60 \text{ min/hr})}{13.65 \text{ cu ft/lb}} = 4400 \text{ lb of dry air}$$

h_1 = 19.221 + 0.004466 (1096.12)
h_1 = 24
h_2 = 19.221 + 0.015631 (1096.12)
h_2 = 36.421
q = 4400 (36.421 − 24 − (0.015631 − 0.004466) 48.05)
q = 4400 (36.421 − 24 − 0.53)
q = 4400 (11.891)
q = 52,100 Btuh

HEATING AND HUMIDIFICATION

Example #8

How much heat is required to raise 1000 cfm of air from 50° and 50% RH to 80° and 70% RH? Assume water added to the air is at 80°.

h_{a1} = 12.010 Btu/lb of dry air
h_{v1} = 1083.06 Btu/lb of vapor
w_1 = 0.50 (0.007658) lb vapor/lb dry air
h_{a2} = 19.221 Btu/lb of dry air
h_{v2} = 1096.12 Btu/lb of vapor
w_2 = 0.70 (0.02233) lb vapor/lb dry air
v' = 12.86 cu ft/lb dry air (from Example #6)
h_f = 48.05 Btu/lb of water

Again, using the general energy equation and remembering to add all energy, we have:
Enthalpy of the entering air + the enthalpy of the liquid water added + heat added = enthalpy of the leaving air.

m ($h_{a1} + w_1 h_{v1}$) + m ($w_2 - w_1$) h_f + q = m ($h_{a2} + w_2 h_{v2}$)
q = m ($h_{a2} + w_2 h_{v2} - h_{a1} - w_1 h_{v1}$) − m ($w_2 - w_1$) h_f
q = m ($h_2 - h_1 - (w_2 - w_1) h_f$)

$$m = \frac{1000 \text{ cfm} \times 60 \text{ min/hr}}{12.86 \text{ cu ft/lb}} = 4,680 \text{ lb of dry air}$$

$h_1 = 12.010 + 0.003829\ (1083.06)$
$h_1 = 16.16$
$h_2 = 19.221 + 0.015631\ (1096.12)$
$h_2 = 36.35$
$q = 4680\ (36.35 - 16.16 - (0.015631 - 0.003829)\ 48.05)$
$q = 4680\ (36.35 - 16.16 - 0.54)$
$q = 91,982\ \text{Btuh}$

COOLING

Example #9

How much heat must be extracted from 1000 cfm at 95° DB and 75° WB to cool it to 70°?

First, we must determine the humidity ratio of the air at the entering air condition.

$$w = 0.622\ \frac{P_v}{P - P_v}$$

Ah! But we don't know what P_v, the vapor pressure, is; however, we do know the Carrier equation and can determine it from this:

$$P_v = P_{vwb} - \frac{(P - P_{vwb})\ (t_{db} - t_{wb})}{2830 - 1.3 t_{wb}}$$

$P_{vwb} = 0.42979$ (from Table 7–2)
$P = 14.7$ psia
$t_{db} = 95°$
$t_{wb} = 75°$
$$P_v = 0.42979 - \frac{(14.7 - 0.42979)\ (95 - 75)}{2830 - 1.3\ (75)}$$

$P_v = 0.325$
$$w = 0.622\ \frac{0.325}{14.7 - 0.325}$$
$w = 0.0139$ lb/lb dry air
$h_{a1} = 22.827$ Btu/lb dry air
$h_{a2} = 16.816$ Btu/lb dry air
$h_{v1} = 1102.59$ Btu/lb water vapor
$h_{v2} = 1091.78$ Btu/lb water vapor
$v = 13.980$ cu ft/lb
$q = m\ (h_2 - h_1)$

$$v' = 13.980\ \frac{(14.7)}{(14.7 - 0.325)} = 14.30\ \text{ft}^3/\text{lb}$$

$$m = \frac{1000\ (60)}{14.30} = 4,200\ \text{lb/hr}$$

$h_1 = 22.827 + 0.0139\ (1102.59) = 38.2$ Btu/lb
$h_2 = 16.816 + 0.0139\ (1091.78) = 32.0$ Btu/lb
$q = 4200\ (32.0 - 38.2)$
$q = -26,000$ Btuh

The minus sign indicates that heat goes out of the system.

COOLING AND DEHUMIDIFICATION

Example #10

Determine how much heat must be extracted from 1000 cfm of air at 95° and 40% RH to cool it to 50°

saturated. This problem can also be solved through the use of the general energy equation. Remember that in cooling the air from 95° and 40% RH to 50° saturated, the air is cooled below its dew point and, therefore, some of the vapor will condense and be removed from the air stream. In setting up the general equation the enthalpy of the liquid removed must be included.

$$m\ (h_{a1} + w_1 h_{v1}) = -q + m\ (h_{a2} + w_2 h_{v2}) + m\ (w_1 - w_2)\ h_f$$
$v' = 14.30$ cu ft/lb

$$m = \frac{1000\ \text{cfm}\ (60\ \text{min})}{14.30\ \text{cu ft/lb}} = 4,200\ \text{lb/hr}$$

$h_{a1} = 22.827$ Btu/lb dry air
$h_{v1} = 1102.59$ Btu/lb vapor
$w_1 = .40\ (0.03673) = 0.014701$ lb vapor
$h_{a2} = 12.010$ Btu/lb dry air
$h_{v2} = 1083.06$ Btu/lb vapor
$w_2 = 0.007658$ lb vapor
$h_f = 18.07$ Btu/lb water
h_f is the enthalpy of the liquid taken at the temperature of the leaving air.
$4200[22.827 + 0.014224\ (1102.59)] =$
$-q + 4200[12.010 + 0.007658\ (1083.06) + (0.014701 - 0.007658)18.07]$
$q = 4200\ (20.31 + 0.118 - 38.4)$
$q = -76,000$ Btuh

EVAPORATIVE COOLING

Example #11

1000 cfm of air at 95° DB and 75° WB is drawn through a 95% efficient spray washer. What will be the final air temperature? How much water must be evaporated?

In evaporative cooling, air is drawn through a series of water sprays and moisture is added to the air stream. The heat required to evaporate the moisture comes from the air itself. This process is similar to the process of adiabatic saturation discussed earlier. To be adiabatic, no heat may enter or leave the system. Obviously, it would be difficult to obtain water at exactly the wet-bulb temperature of the entering air. But if a large amount of water is re-circulated and a small amount evaporated, then the water would soon reach a temperature close to that of the entering air wet-bulb temperature.

The leaving air temperature will be at 75° WB. Because the washer is not 100% efficient the leaving dry-bulb will not be the same as the wet-bulb, as occurred in our adiabatic saturater. The new DB temperature will be equal to the entering DB − % eff($t_{db} - t_{wb}$). In this case 95° − 0.95(95° − 75°) = 76° DB

$m = 4200$ lb (from Example #9)

To determine w_2, first find P_{v2} by the Carrier equation.

$$P_v = P_{vs} - \frac{(P - P_{vs})(t_{db} - t_{wb})}{2800 - 1.3t_{wb}}$$

$P_{vs} = 0.42979$
$P = 14.7$ psia
$t_{db} = 76°$
$t_{wb} = 75°$

$$P_v = 0.42979 - \frac{(14.7 - 0.42979)(76 - 75)}{2800 - 1.3(75)}$$

$$P_v = 0.425$$

$$W_2 = 0.622 \frac{(0.425)}{(14.7 - 0.425)} = 0.0185 \text{ lb/lb dry air}$$

$W_2 - W_1 = 0.0185 - 0.0139$
$W_2 - W_1 = 0.0046$ lb of water/lb dry air
$0.0046 \times 4200 = 19.8$ lb of water

PSYCHROMETRIC CHART

While accurate, the solving of problems by the general energy equation is time consuming. The development of the psychrometric chart not only made problem solving easier, but at the same time, gave a pictorial representation of the process. A psychrometric chart is shown on Fig. 7–2. Dry-bulb temperatures are plotted as almost vertical lines, humidity ratios are plotted as horizontal lines. Wet-bulb and enthalpy lines run diagonally. Note that they are almost parallel. Relative humidity lines are curves fanning out from the left-hand corner.

Now let's practice a bit. Find 95° DB and 75° WB. Got it? Fine. Now read 40% relative humidity, 38.4 Btu/lb enthalpy, 0.0142 lb/lb humidity ratio, and 14.3 cu ft/lb specific volume. Easy, isn't it. Now let's do Examples 6 through 11 using the psychrometric chart. Figs. 7–3 through 7–8 show Examples 6 through 11 plotted on the psychrometric chart.

The upper left hand corner contains an auxiliary chart that is used to find the enthalpy of the liquid and enthalpy deviation. The vertical lines are wet-bulb or water temperatures. The horizontal lines are humidity ratio lines. The solid curved lines are the enthalpy of the liquid, while the dotted curved lines are lines of enthalpy deviation. Instead of reading enthalpy directly, you may find it easier to follow the wet-bulb line to the saturation line and read the enthalpy. To this, add or subtract the deviation. For example, at 95° DB and 75° WB, follow the wet-bulb and read 38.6 Btu/lb. Now, on the deviation chart at the intersection of 75° WB and 0.014 humidity ratio, read −0.21 Btu/lb. The enthalpy is then 38.6 − 0.21 = 38.39 Btu/lb.

To determine the enthalpy of the liquid, locate the intersection of the amount of water added or removed and the temperature of the water, and follow the solid curved line to the top and read the enthalpy of the liquid in Btu per lb of dry air. For instance, in Example #8, page 120,

$$W_2 - W_1 = 0.015631 - 0.003829 = 0.01180 .$$

Locate the intersection of 0.01180 and 80° and read 0.54 Btu/lb of dry air.

ADIABATIC MIXING OF TWO STREAMS OF MOIST AIR

Air conditioning processes generally mix room air with outside air. This can be shown graphically by locating the room- and outside-air state points and joining them with a straight line. The state point of the mixture lies on this line. The distance from the mixed-air point to either the outside-air or room-air is proportional to the mass, respectively, of the outside air or room air.

Example #12 (Fig. 7–9)

Locate the state point of a mixture of 2000 cfm of 95° DB and 75° WB air, and 8000 cfm of 80° DB and 67° WB air.

$$m_1 = \frac{2000}{14.3} = 140 \text{ lb dry air/min}$$

$$m_2 = \frac{8000}{13.85} = 575 \text{ lb dry air/min}$$

The mixed air state point lies $140/(140 + 575) = 0.196$ of the distance from point 2 to 1, or $575/(140 + 575) = 0.804$ of the distance from point 1 to 2. This can be measured with a ruler and the point located 82.9° DB and 68.6° WB.

APPARATUS DEW POINT

Up to this point we have specified both the entering and leaving condition in our problems. Generally, we only know what our load and entering or room conditions are, and must determine what the leaving conditions and air quantity are. In heating or cooling this is simple, because you can select a desired leaving-air temperature and plug it into Formula 7–2 to get cfm. The leaving-air temperature for heating is generally about 120°. On the other hand, if you need a certain air quantity, then you plug this air quantity into Formula 7–2 to get the leaving-air temperature.

When simultaneous cooling and dehumidification

are desired, it is necessary to find a saturated leaving-air temperature that will satisfy both the cooling and dehumidification. This is called the apparatus dew point.

$$cfm = \frac{Btuh \ (sensible)}{1.08 \ (t_1 - t_2)} \qquad (Formula \ 7-2)$$

$$cfm = \frac{Btuh \ (latent)}{0.075 \ (60) \ (1076) \ (W_2 - W_1)}$$

$$cfm = \frac{Btuh \ (latent)}{4850 \ (W_1 - W_2)} \qquad (Formula \ 7-3)$$

$0.075 =$ average density of air
$60 \quad\; =$ min per hr
$1076 \; =$ average enthalpy of evaporation for water Btu/lb
$W_1 \quad =$ humidity ratio of entering air
$W_2 \quad =$ humidity ratio of leaving air
$t_1 \quad\; =$ dry-bulb temperature of entering air
$t_2 \quad\; =$ dry-bulb temperature of leaving air

By varying the leaving-air temperature, there are an infinite number of air quantities (cfm) that will remove the sensible heat load. By varying the leaving humidity ratio, there are also an infinite number of air quantities that will remove the latent heat. But there is only one temperature that will remove both the sensible heat and latent. This is called the apparatus dew point.

How do we find the apparatus dew point? Well, we could do it by trial and error, using Formula 7–2 and 7–3 as follows:

Example #13 (Fig. 7–10)

A room has a heat gain of 100,000 Btuh sensible and 20,000 Btuh latent. What temperature air and what air quantity are required to maintain 75° DB, and 40% RH?

Apparatus Dew Point	Temp Diff	cfm	Humidity Ratio at ADP	$W_2 - W_1$	cfm
48	27°	3430	0.00710	0.00030	13800
46	29°	3200	0.00658	0.00082	5020
45	30°	3080	0.00633	0.00107	4250
44	31°	3000	0.00609	0.00130	3160
43	32°	2880	0.00586	0.00154	2670

As you can see, somewhere between 43° and 44°, we would have a condition that would satisfy both the sensible and latent room conditions. Also, it can be seen that this is a very tedious and time consuming method.

The psychrometric chart and the room sensible heat/room total heat ratio do away with trial and error and substitute a graphical solution. On the psychro-

metric chart you will see a protractor which plots around its perimeter:

$$\frac{Room \ Sensible \ Heat}{Room \ Total \ Heat} = \frac{\Delta H_s}{\Delta H_t}$$

and

$$\frac{Enthalpy}{Humidity \ Ratio} = \frac{\Delta h}{\Delta w}$$

In our problem, the sensible heat/total heat ratio is 100,000 Btuh/120,000 Btuh = 0.833. Locate on the psychrometric chart the room conditions of 75° DB, 40% RH. Now, using two triangles, draw a line through the room conditions that parallels a line through the center of the protractor and 0.833 on the left side of the protractor. This line will intersect the saturation line at 43.5°. Any point along this line will provide the exact simultaneous cooling and dehumidification required.

For example: At 48° DB and 43.5 ADP, the humidity ratio is 0.0062. Substituting in Formula 7-2 and 7-3, we find the required air quantity to be 3430 cfm.

$$cfm = \frac{100,000}{1.08 \ (75 - 48)} = 3430$$

$$cfm = \frac{20,000}{4850 \ (0.0074 - 0.0062)} = 3430$$

In our discussion of the apparatus dew point we stated that any point on the line connecting the room conditions and the apparatus dew point would provide the proper simultaneous cooling and dehumidification. But where on that line will we be? The leaving conditions of the air would be the same as the refrigerant or chilled-water temperature if the coil were infinitely deep. Consider a coil with 500 fpm face velocity and 45° refrigerant temperature. On Fig. 7–11, the leaving conditions for 3, 4, 5, and 6-row coils with 83° DB, 68.6° WB entering conditions, and 500 fpm face velocity, have been plotted and a curve drawn through the points terminating at the refrigerant temperature 45°. Most manufacturers present leaving-air conditions for direct expansion coils in tabular form. So all you have to do is simply select a coil that can cool air to a point on or below the $\Delta H_s / \Delta H_t$ line. Notice on Fig. 7–11 that a 4-row coil will meet our conditions.

Chilled-water coils are not so easy to select. In selecting a chilled-water coil an empirical formula is used to determine the required rows of tubes. To solve the formula, you must know the leaving conditions.

Some authorities recommend that an arbitrary dry-bulb temperature be selected and the leaving wet-bulb temperature be determined by the intersection of the dry-bulb line and the $\Delta H_s/\Delta H_t$ line.

Others use the bypass factor method. This was first presented by Dr. Carrier to the ASME in 1936. The simplest explanation that I can give you is as follows: first, visualize the cooling and dehumidifying coil—rows of tubes with fins spaced from 48 to the foot to 96 or more. It is hard to visualize that some of the air may not contact all of this surface, but some does not.

The term bypass factor is applied to the percentage of the air that does not come in contact with the surface. The deeper or the more rows of tubes in the direction of air flow, the smaller the bypass factor. Average values that can be safely used are 0.2 for a 4-row coil, 0.15 for 5-rows, 0.10 for 6-rows, and 0.05 for 8-rows. So, if you select a 6-row coil with a 0.10 bypass factor, then if the ADP is 55°, the leaving dry-bulb is $80° - (1 - 0.10)(80° - 55°) = 57.5°$. Find the intersection of 57.5° DB with the $\Delta H_s/\Delta H_t$ and read 56.2° WB. So the required leaving conditions are 57.5° DB and 56.2° WB. With a 4-row coil $(1 - 0.2)(80 - 55) = 20$; $80 - 20 = 60°$ DB; plotting, we get 57.5° WB.

What happens if we arbitrarily raise or lower the ADP.? If we raise it we will not be able to properly dehumidify. For example, raise the ADP to 58° for a space load of 100,000 Btuh sensible and 33,000 Btuh latent. The leaving conditions with a 4-row coil will be 62° DB, 58.5° WB and 0.0104 humidity ratio.

$$cfm = \frac{100,000}{1.08 (80 - 62)} = 5150$$

Btuh (latent) = 4850 (5150) (0.011 − 0.0104)
Btuh (latent) = 15,000

So if we can only pick up 15,000 Btuh of the 33,000 Btuh latent load, we cannot maintain 80° and 50% RH.

On the other hand, if we lower the ADP we will be able to remove more moisture and so maintain a lower relative humidity. But to get this bonus, we must pay for it. At 50°ADP with a 6-row coil:

$$cfm = \frac{RSH}{1.08(CF)(t_1 - ADP)} \quad \text{(Formula 7-4)}$$

where   RSH = room sensible heat
        CF   = contact factor
        t_1 = entering air DB temp.

$$cfm = \frac{100,000}{1.08 (0.9) (80 - 50)} = 3420$$

$$Btuh \text{ (total)} = \frac{3420(60) (31.2 - 20.4)}{13.2}$$

Btuh (total) = 168,000

By dropping the ADP from 55° to 50° we increase the total load by $168,000 - 133,000 = 35,000$ Btuh.

There are times when the $\Delta H_s/\Delta H_t$ does not intersect the saturation line and we have what is called a reheat condition. What does this mean? It means that you cannot find a temperature to which you can cool the air and, while removing all the room sensible heat, remove sufficient latent heat. Conversely, if the air is cooled enough, to a point where all the room latent load is removed, too much sensible heat is removed and the room is too cold.

Example #14 (Fig. 7-12)

A room has a heat gain of 100,000 Btuh sensible and 80,000 Btuh latent. What is the ADP if the room conditions are 80° and 50% RH? Notice that $\Delta H_s/\Delta H_t = 0.56$. Draw the $\Delta H_s/\Delta H_t$ through 80° and 50% RH and notice that it does not intercept the saturation line. So we have no ADP. The dotted line represents the coil cooling path starting at point 3, the coil entering conditions, and ending at the refrigerant temperature of 40°. Notice that the coil line does not intercept the $\Delta H_s/\Delta H_t$ line. Point 5 represents the leaving conditions for an 8-row coil at 500 fpm face velocity and 40° refrigerant temperature. These conditions are 46° DB, 45.8° WB, 0.0062 lb/lb humidity ratio. At these conditions, we'd need to circulate:

$$cfm = \frac{100,000}{1.08 (80 - 46)} = 2720$$

We would only pick up:

Btuh (latent) = 2720 (4850) (0.0102 − 0.0062)
Btuh (latent) = 53,000

But our latent load is 80,000 Buth. This means that our room will not be maintained at 80° and 50% RH, but at 80° and about 60%. But then if we supply enough air at 0.0062 humidity ratio we will overcool:

$$cfm = \frac{80,000 \text{ Btuh (latent)}}{4850 (0.0102 - 0.0062)} = 4100$$

Btuh (sensible) = 4100 (1.08) (80 − 46)
Btuh (sensible) = 150,000

Because this is 50% more than we need, our room will be at a temperature considerably lower than 80° DB, something around 65°.

The best way to handle this is to select an ADP that gives a reasonable temperature drop and air quantity, and provide enough reheat to make it possible. In other words, dehumidify the air and then reheat it to the proper dry-bulb temperature. For instance, we could swing the $\Delta H_s/\Delta H_t$ line until it intersects the saturation line at, say, 46°. The air quantity will be 4100 cfm. The air must then be heated to:

$$80° - \frac{100,000}{4100 \ (1.08)} = 57.4° \text{ supply air}$$

Fig. 7–13 shows this plotted on the psychrometric chart. This adds an additional load, of course.

$$\text{Btuh} = \frac{4100 \ (60)}{13.85} \ (21 - 18.4)$$

$$\text{Btuh} = 45,500$$

The total load would be:

$$\frac{4100 \ (60)}{13.85} [(31.4 - 18.4) + (21 - 18.4)] = 275,000 \text{ Btuh}$$

Compare this with the calculated load of 180,000 Btuh. It takes 50% more capacity to provide the required condition of 80° and 50% RH.

BYPASSING AIR

Often it is necessary to bypass air around the cooling coil, as is shown in Fig. 7–14. This may be done because a code may require a volume of air in excess of the amount required for cooling and dehumidification. Sometimes air is bypassed around a coil as a means of controlling temperature. Still another reason is to provide reheat when steep $\Delta H_s / \Delta H_t$ lines are encountered.

Fig. 7–15 charts a process with 80° and 50% room air and 55° ADP. The entering conditions are 83° DB and 69° WB, and the leaving conditions 58° DB, 56.5° WB. 8000 cfm of 83° and 69° air is taken through the coil and 2000 cfm of ROOM AIR is bypassed. A bypassing process is the same as the adiabatic mixing of two air streams. The air entering the room lies on a line drawn between point 1 and 4. This is also the $\Delta H_s / \Delta H_t$ line. While we could calculate the mass of the air streams, it is reasonably accurate to use a ratio of the air quantities. In this case, point 5 lies 2000/10,000, or $1/5$ of the distance of point 4 to 1. The air at this point is 63° DB, 58.8° WB.

Notice that if mixed air were bypassed, the mixture would not be on the $\Delta H_s / \Delta H_t$ line. However, if the amount of outside air in the mixture is small, and if the amount of mixed air bypassed is less than $1/3$, the resulting conditions would be close.

But bypassing outside air is a different story. Fig. 7–16 charts the same process, only outside air is bypassed. It can be seen that point 5 now is a significant distance from the $\Delta H_s / \Delta H_t$ line. Read 65.5° DB and 61.5° WB. If outside air must be bypassed, then the air should be cooled to 51° DB, 50.5° WB so that the

resulting mixture lies on the $\Delta H_s / \Delta H_t$ line. This is shown by a dotted line intersecting $\Delta H_s / \Delta H_t$ line at point 5'.

INJECTION OF WATER OR STEAM INTO AIR

We have discussed the case of water, at the same temperature as the entering air wet-bulb, being sprayed into the air stream. But what of the case where the water and air are at different temperatures, or the case of steam? What happens here? If the process is adiabatic, then:

$$mh_1 + m_w h_w = mh_2$$
$$m_w h_w = mh_2 - mh_1$$
$$m_w h_w = m \ (h_2 - h_1) \quad \text{(Equation 7-5)}$$
$$mw_1 + m_w = mw_2$$
$$m_w = mw_2 - mw_1$$
$$m_w = m \ (w_2 - w_1) \quad \text{(Equation 7-6)}$$

Divide Equation 7-5 by 7-6

$$\frac{m_w h_w}{m_w} = \frac{m \ (h_2 - h_1)}{m \ (w_2 - w_1)}$$

$$h_w = \frac{h_2 - h_1}{w_2 - w_1} \quad \text{(Equation 7-7)}$$

where m_w is the mass of water added and h_w is the specific enthalpy of water in any phase—steam, liquid, or solid. On the psychrometric chart h_w is the $\frac{\text{enthalpy}}{\text{humidity ratio}}$ or $\frac{\Delta h}{\Delta w}$.

Example #15 (Fig. 7–17)

One lb of water per min at 150° is injected into 3000 cfm of air at 95° DB and 75° WB. What are the leaving conditions of water and air?

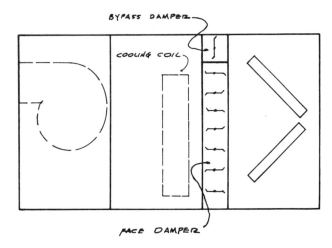

FIG. 7–14 Typical Face And Bypass Damper

From the chart we find that at 95–75 the specific volume $v = 14.3$ and the humidity ratio is 0.0142 lb/lb.

$$m = \frac{cfm}{v} = \frac{3000}{14.3} = 210 \text{ lb/min}$$

The enthalpy of the water at 150° (from Table 7–2) is 117.96 Btu/lb. This is equal to h_w which $= \Delta h / \Delta w$. To determine the leaving-air conditions, first find the leaving humidity ratio:

$$m_w = m (w_2 - w_1)$$
$$\frac{m_w}{m} = w_2 - w_1$$
$$w_2 = \frac{m_w}{m} + w_1$$
$$w_2 = \frac{1}{210} + 0.0142$$
$$w_2 = 0.0188 \text{ lb/lb dry air}$$

Then draw a line through the center of the protractor and $\Delta h / \Delta w = 117.96$. Draw a line parallel to this and through the room conditions. Where this intersects the leaving humidity ratio, read 77° DB and 75.5° WB. If steam at 250° had been used instead of the water, then $h_w = 1164$ Btu/lb, which is the $\Delta h / \Delta w$ ratio. The final humidity ratio is the same. So following the above procedure, we read 96.5° DB and 80.5° WB. Notice that if sufficient steam had been used to raise the humidity ratio to 0.030, the final DB temperature would only be 100°. Obviously then, heating air by injecting steam is not too efficient.

CHEMICAL DEHUMIDIFICATION

There are times when it is desirable or necessary to dehumidify air chemically. This process is the reverse of adiabatic saturation and follows the enthalpy line. The reason that the air is heated while being dehumidified is that the water that condenses must give up its latent heat and this heats up the air. On Fig. 7–18 this process is shown by the solid line from 1 to 2. This is the ideal path. However, in practice, the line is more like 1 to 3 for a variety of reasons, all related to the equipment.

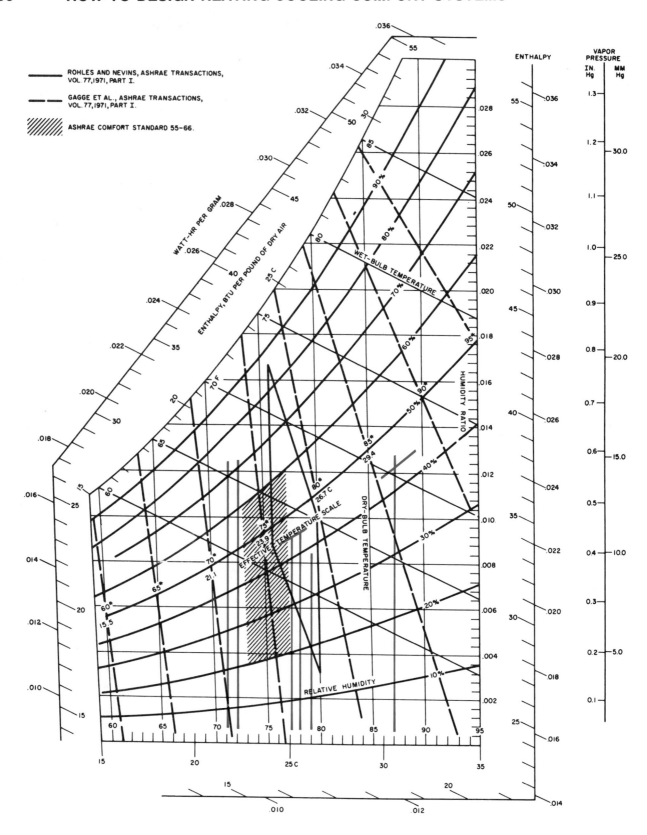

Fig. 7-2 ASHRAE Comfort Chart (Reprinted by permission from ASHRAE)

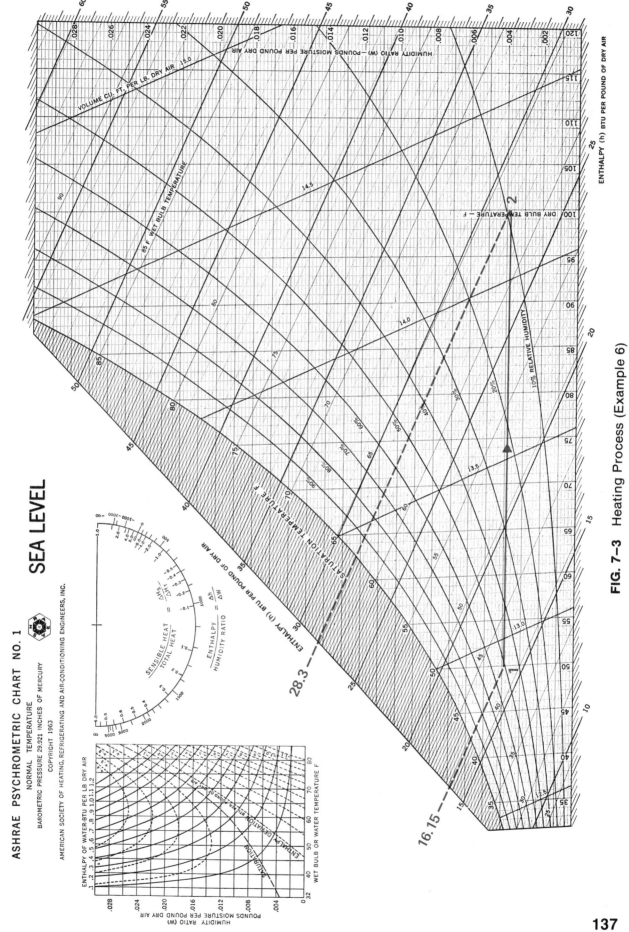

ASHRAE PSYCHROMETRIC CHART NO. 1

NORMAL TEMPERATURE

BAROMETRIC PRESSURE 29.921 INCHES OF MERCURY

COPYRIGHT 1963

AMERICAN SOCIETY OF HEATING, REFRIGERATING AND AIR-CONDITIONING ENGINEERS, INC.

SEA LEVEL

FIG. 7–3 Heating Process (Example 6)

137

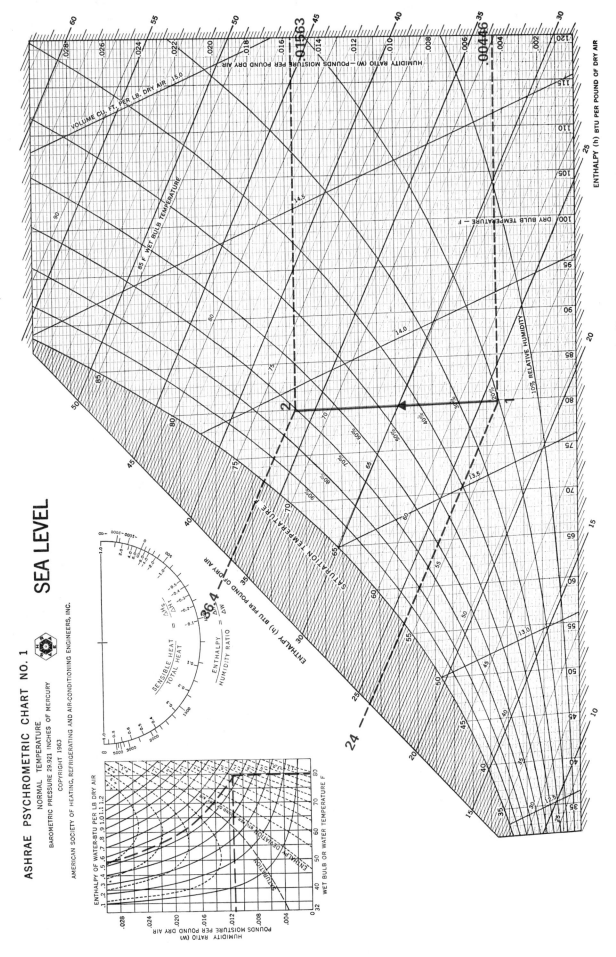

FIG. 7–4 Humidification Process (Example 7)

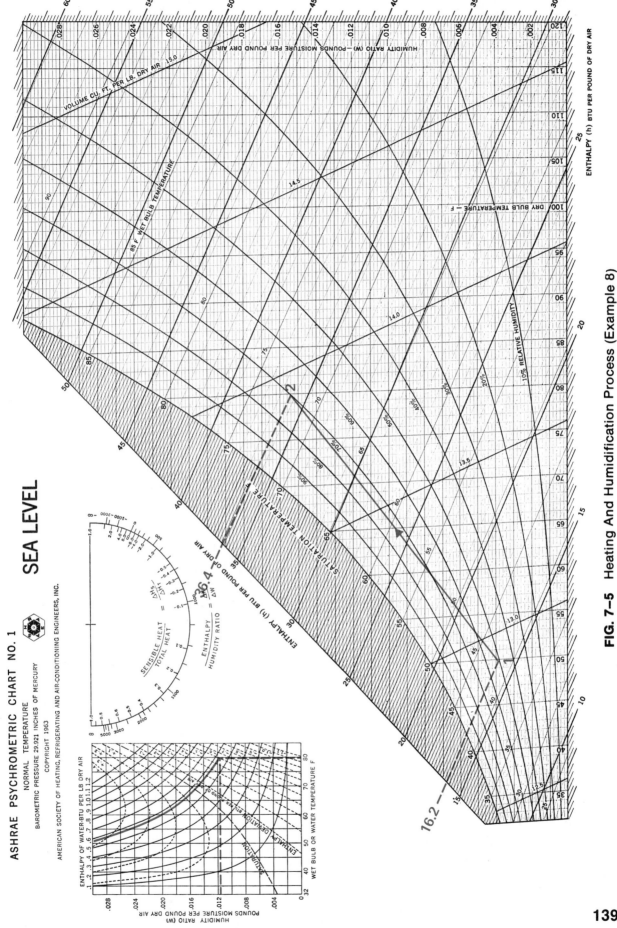

FIG. 7–5 Heating And Humidification Process (Example 8)

139

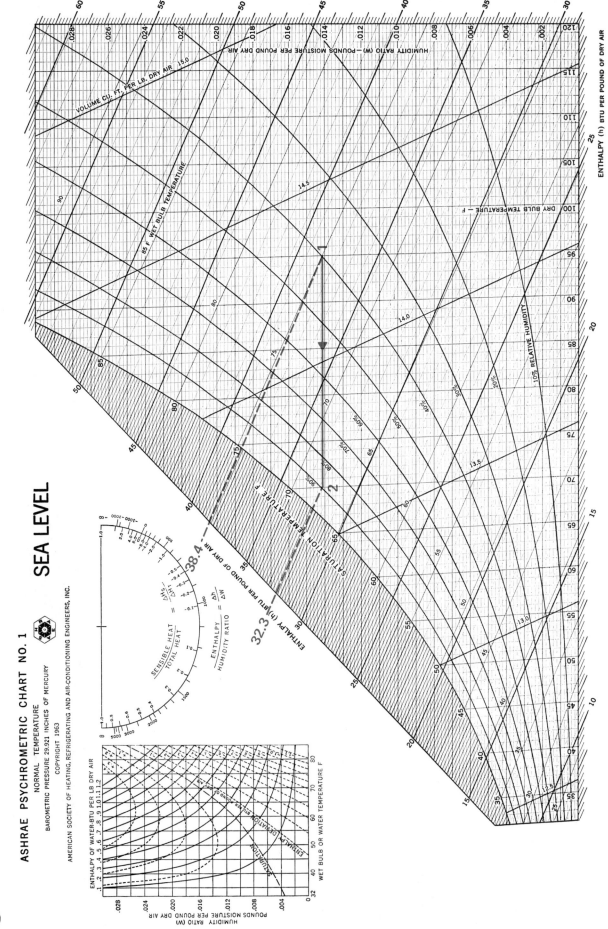

FIG. 7-6 Cooling Process (Example 9)

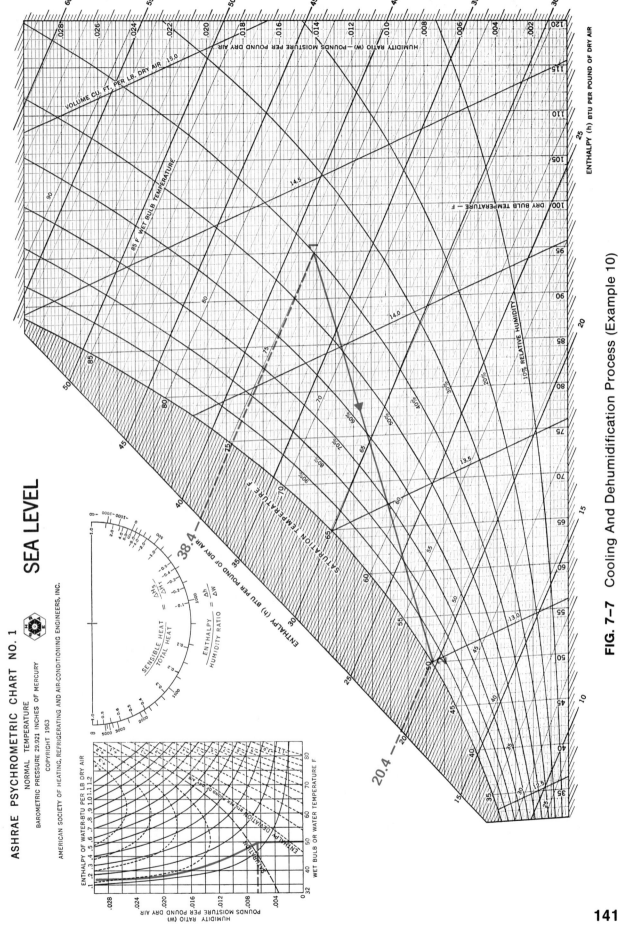

FIG. 7-7 Cooling And Dehumidification Process (Example 10)

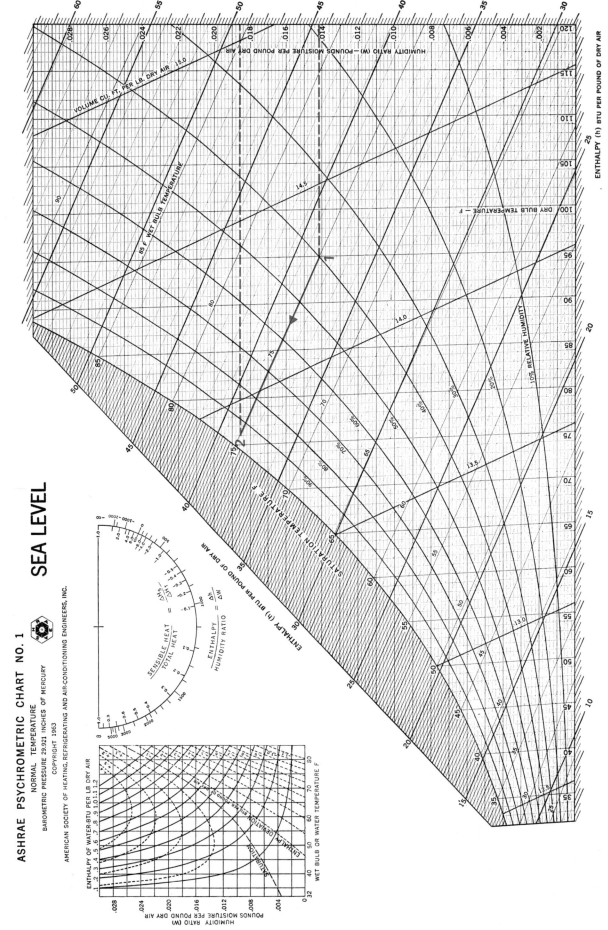

FIG. 7–8 Evaporative Cooling Process (Example 11)

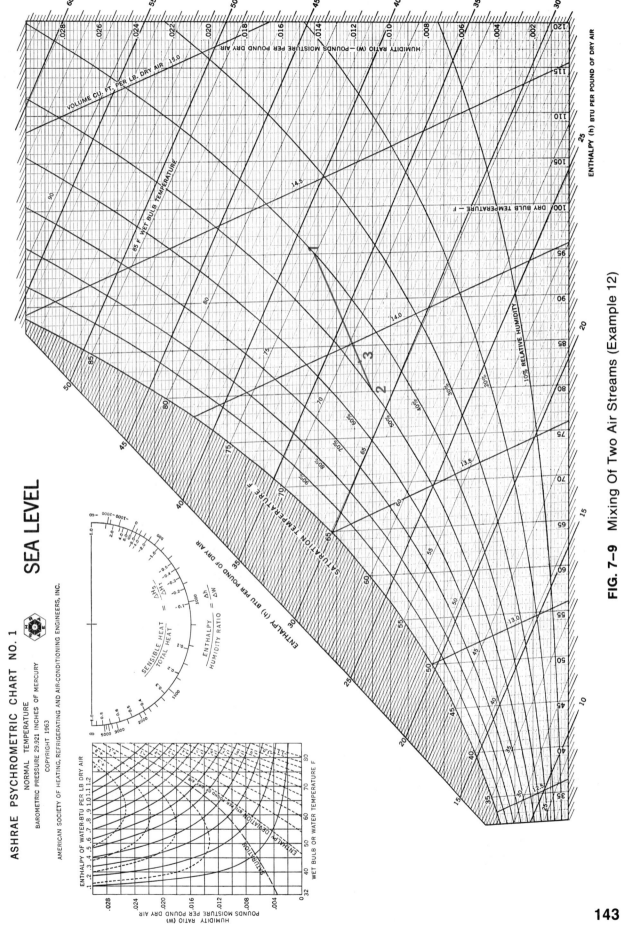

ASHRAE PSYCHROMETRIC CHART NO. 1

NORMAL TEMPERATURE

BAROMETRIC PRESSURE 29.921 INCHES OF MERCURY

COPYRIGHT 1963

AMERICAN SOCIETY OF HEATING, REFRIGERATING AND AIR-CONDITIONING ENGINEERS, INC.

SEA LEVEL

FIG. 7-9 Mixing Of Two Air Streams (Example 12)

143

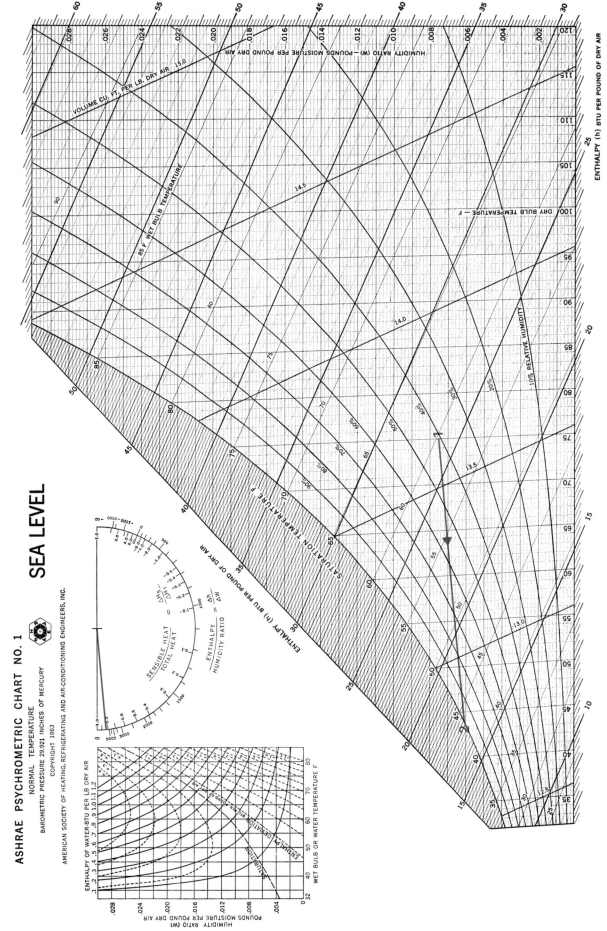

FIG. 7-10 Apparatus Dew Point (Example 13)

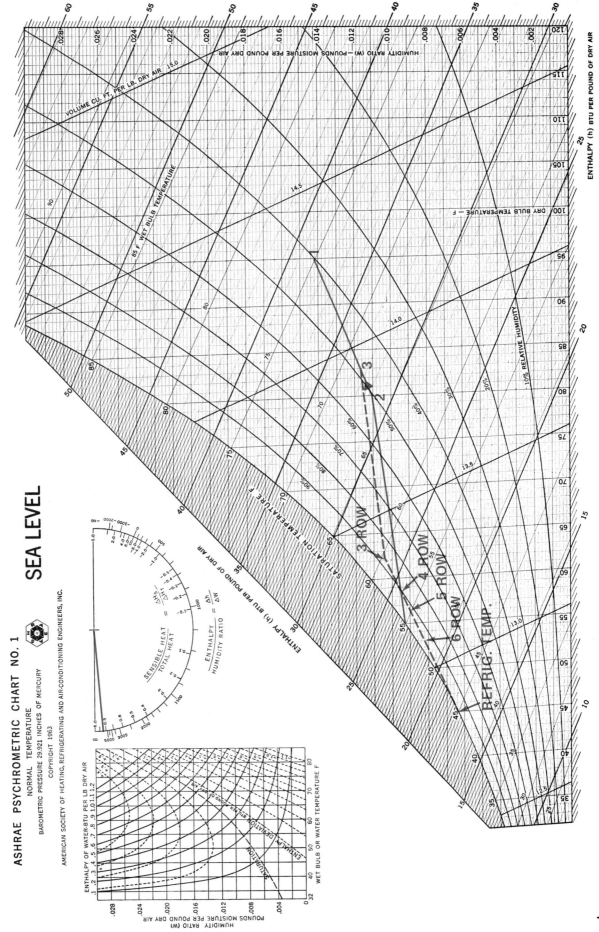

FIG. 7-11 Leaving Conditions for 3, 4, 5, And 6-Row Coils With 83 DB, 68.6 WB Entering Conditions

145

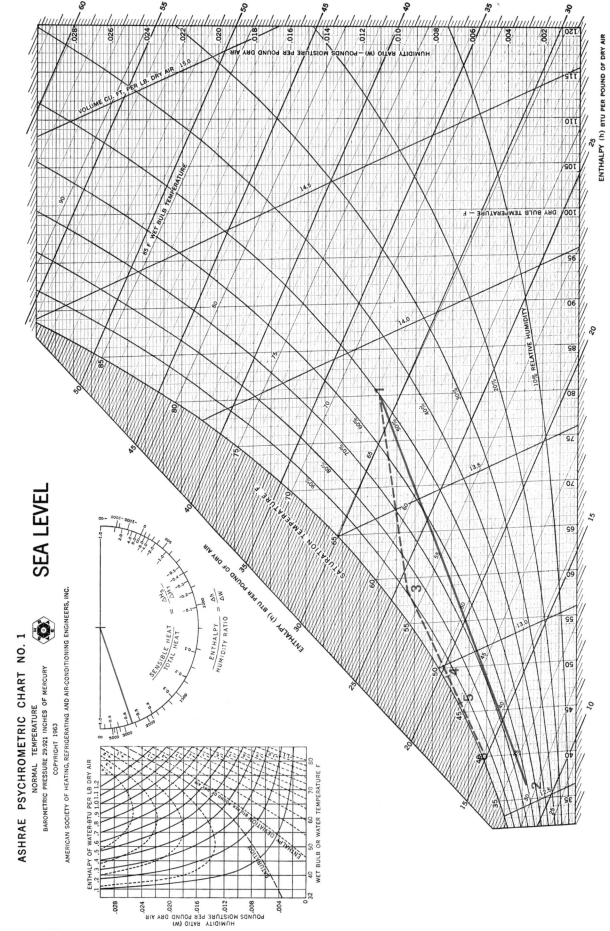

FIG. 7–12 Plot Showing Condition In Which $\Delta H_S/\Delta H_T$ Line Does Not Intercept Saturation Line (Example 14)

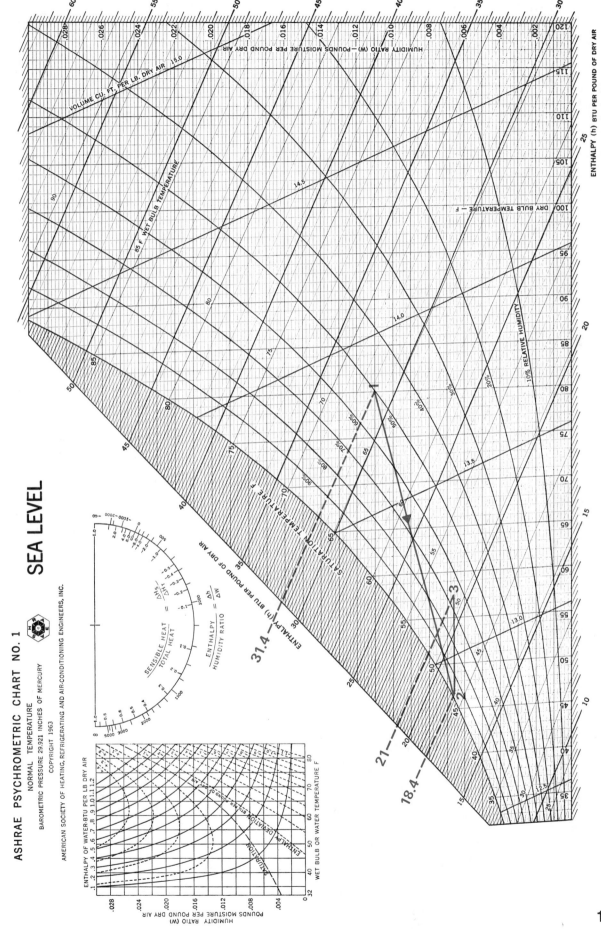

FIG. 7–13 Plot Showing Solution To Condition In Which $\Delta H_S/\Delta H_T$ Line Does Not Intercept Saturation Line (Example 14)

147

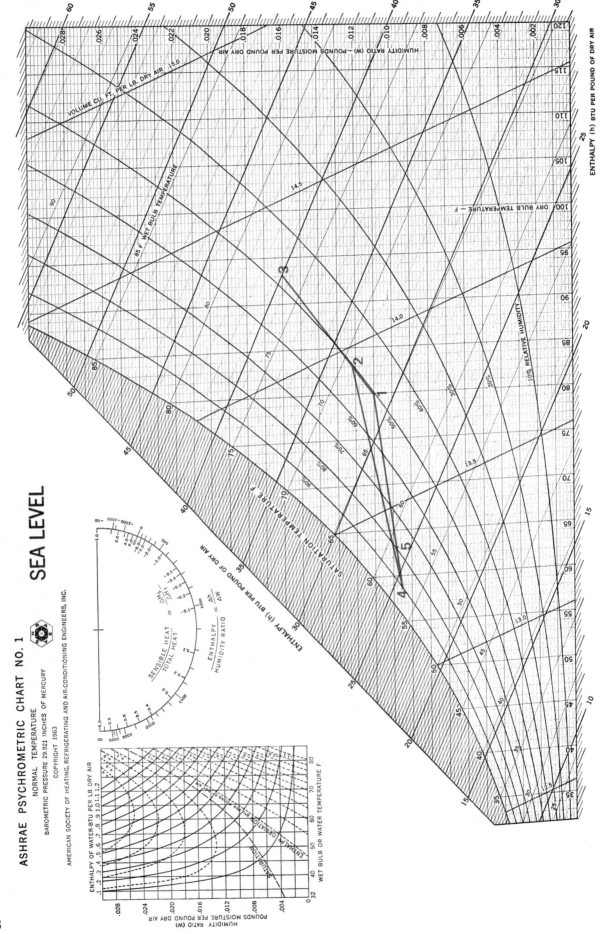

FIG. 7-15 Plot Showing Mixed Air Bypassing Cooling Coil

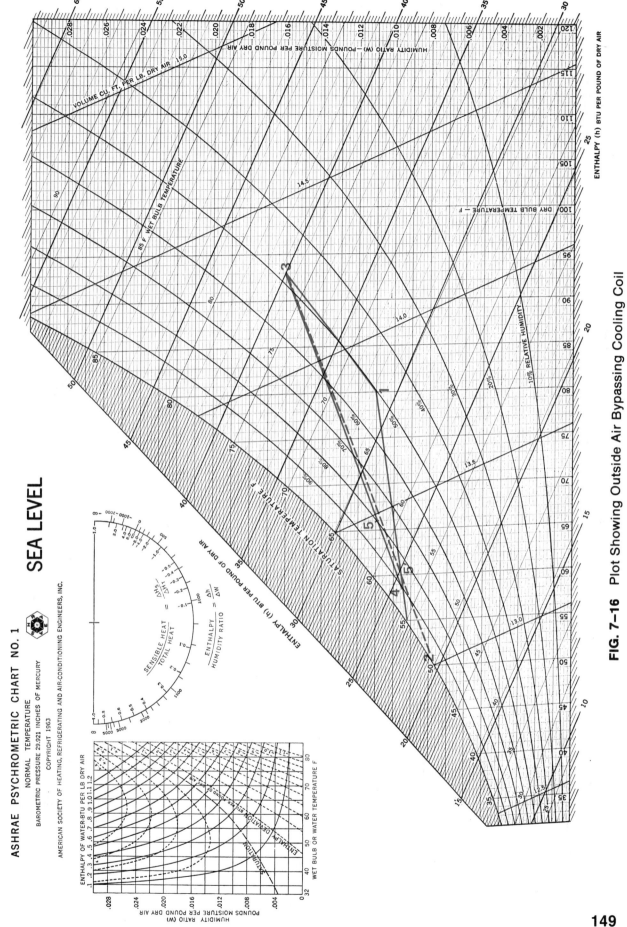

FIG. 7-16 Plot Showing Outside Air Bypassing Cooling Coil

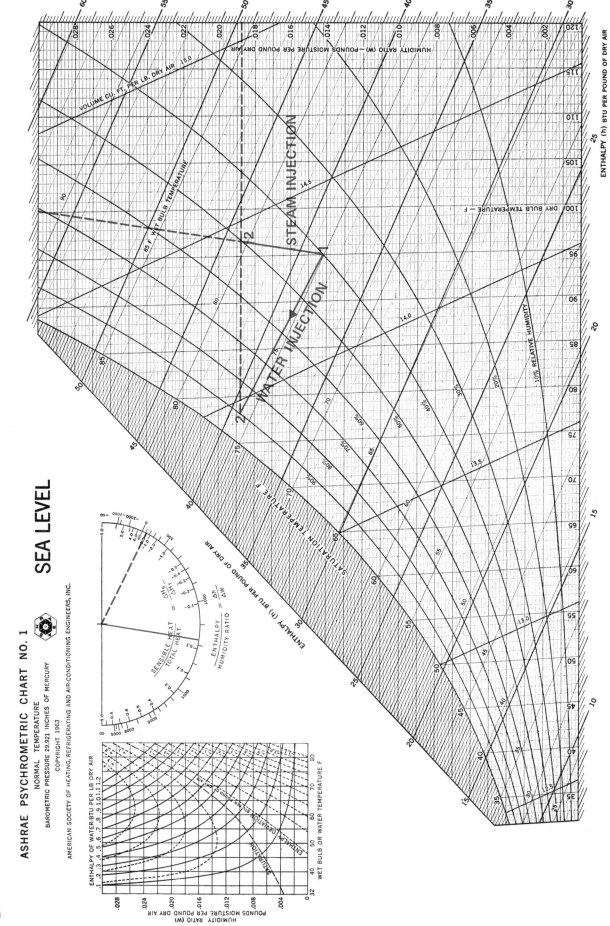

FIG. 7–17 Plot Of Injection Of Water Or Steam Into Air

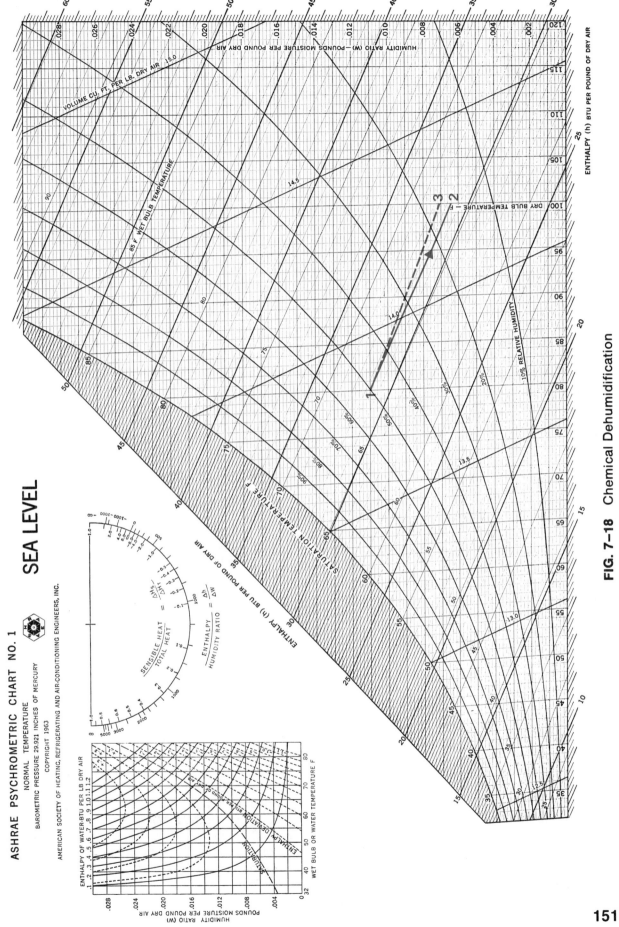

FIG. 7-18 Chemical Dehumidification

CHAPTER 8

Air and How to Move It

Q = AV. This is all it takes to size a duct if you don't care what the friction is.

Q (how much) = A (how big), multiplied by V (how fast). So, if you have 1000 cfm (Q, how much) traveling at a velocity of 1000 fpm (V, how fast), it would require a 1 sq ft (A, how big) duct. Or, to put it in formula:

$$A = \frac{Q}{V}; \; A = \frac{1000 \text{ cu ft/min}}{1000 \text{ ft/min}}; \; A = 1 \text{ sq ft}$$

Had we said 100 ft/min, it would have required 10 sq ft of duct area. At 10,000 fpm it would only take a 0.1 sq ft duct. Ah, but were it this simple. Friction changes with velocity. At 100 fpm, the friction would be less than 0.01 in. of water/100 ft of duct; at 1000 fpm, 0.1 in./100 ft; and, at 10,000 fpm, way over 10 in.

So, you can see that friction does not increase in equal steps, but in increasingly larger steps. This has been expressed in a formula:

$$*H_f = \frac{f\,(L)}{(D)} H_v$$

H_f = head loss due to friction, inches of water
f   = friction factor, which depends on the roughness of the pipe and the Reynolds number.
L = length of duct
D = duct diameter
H_v = velocity pressure $\left(\frac{V}{4005}\right)^2$

$$H_f = \frac{f\,(L)}{(D)} \left(\frac{V}{4005}\right)^2$$

While we could laboriously solve this each time, fortunately, the scientist fellows have arranged ve-

*ASHRAE GUIDE AND DATA BOOK, 1965, page 564.

locity, duct size, cfm, and friction in chart form. These are presented in Figs. 8–1 and 8–2.

Example #1

What size duct would it take to carry 5000 cfm at a velocity of 1000 fpm, 2000 fpm, 4000 fpm, and 8000 fpm? Also, determine the friction loss in inches of water per 100 ft of duct.

Velocity	Duct Size	Friction
1,000 fpm	30-in. dia.	0.04 in.
2,000 fpm	21.5-in. dia.	0.23 in.
4,000 fpm	15-in. dia.	1.25 in.
8,000 fpm	10.7-in. dia.	8.00 in.

You will notice the little note at the bottom of Figs. 8–1 and 8–2 which states that the chart is based on standard air with a density of 0.075 lb/cu ft and clean, round ductwork having approximately 40 joints per 100 ft. Between temperatures of 50° to 90° no corrections need be made. To correct for pipe roughness, use Fig. 8–4.

Example #2

What would be the friction in a medium rough 12-in. dia. pipe carrying 1000 cfm?

First, read Fig. 8–1 and read 0.2 in./100 ft and a velocity of 1275 fpm. Then, go to Fig. 8–4 and find 1275 fpm on the abscissa. Read up to the intersection of 1275 fpm and 12-in. dia. in the medium rough section. Now, read horizontally to 1.37 correction factor. So the corrected friction would be 0.2 in. × 1.37 = 0.274 in./100 ft.

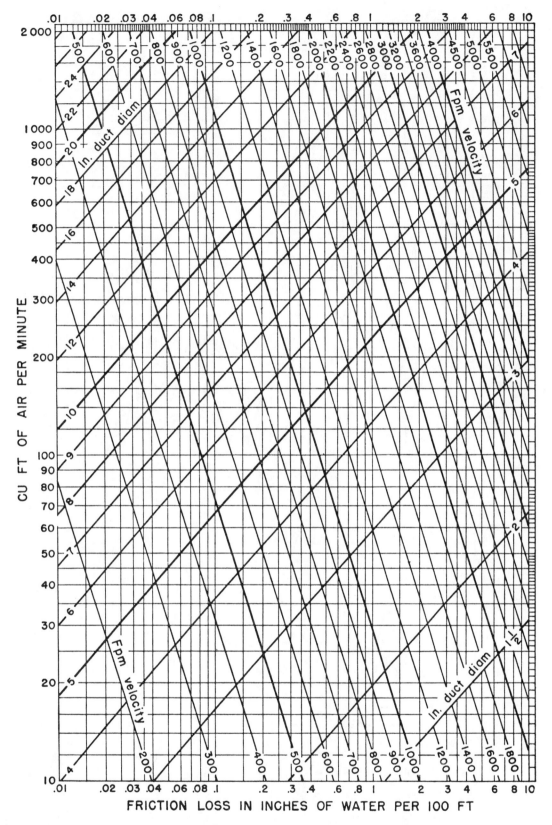

(Based on Standard Air of 0.075 lb per cu ft density flowing through average, clean, round, galvanized metal ducts having approximately 40 joints per 100 ft.) Caution: Do not extrapolate below chart.

FIG. 8–1   Friction Of Air In Straight Ducts For Volumes Of 10 To 2000 CFM (Reprinted by permission from *ASHRAE Guide and Data Book,* 1965)

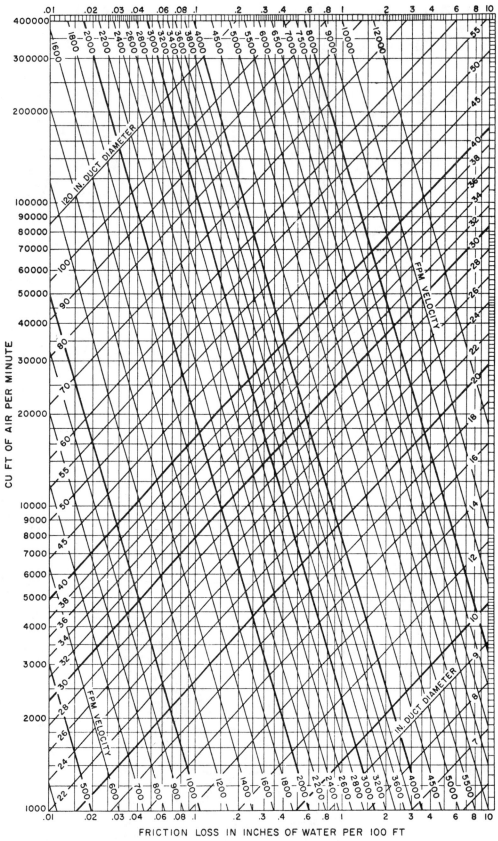

FRICTION LOSS IN INCHES OF WATER PER 100 FT

(Based on Standard Air of 0.075 lb per cu ft density flowing through average, clean, round, galvanized metal ducts having approximately 40 joints per 100 ft.)

FIG. 8–2 Friction Of Air In Straight Ducts For Volumes Of 1000 To 400,000 CFM (Reprinted by permission from *ASHRAE Guide and Data Book,* 1965)

Side Rectangular Duct	4.0	4.5	5.0	5.5	6.0	6.5	7.0	7.5	8.0	9.0	10.0	11.0	12.0	13.0	14.0	15.0	16.0
3.0	3.8	4.0	4.2	4.4	4.6	4.8	4.9	5.1	5.2	5.5	5.7	6.0	6.2	6.4	6.6	6.8	7.0
3.5	4.1	4.3	4.6	4.8	5.0	5.2	5.3	5.5	5.7	6.0	6.3	6.5	6.8	7.0	7.2	7.4	7.6
4.0	4.4	4.6	4.9	5.1	5.3	5.5	5.7	5.9	6.1	6.4	6.8	7.1	7.3	7.6	7.8	8.1	8.3
4.5	4.6	4.9	5.2	5.4	5.6	5.9	6.1	6.3	6.5	6.9	7.2	7.5	7.8	8.1	8.4	8.6	8.9
5.0	4.9	5.2	5.5	5.7	6.0	6.2	6.4	6.7	6.9	7.3	7.6	8.0	8.3	8.6	8.9	9.1	9.4
5.5	5.1	5.4	5.7	6.0	6.3	6.5	6.8	7.0	7.2	7.6	8.0	8.4	8.7	9.0	9.4	9.6	9.8

Side Rectangular Duct	6	7	8	9	10	11	12	13	14	15	16	17	18	19	20	22	24	26	28	30	Side Rectangular Duct
6	6.6																				6
7	7.1	7.7																			7
8	7.5	8.2	8.8																		8
9	8.0	8.6	9.3	9.9																	9
10	8.4	9.1	9.8	10.4	10.9																10
11	8.8	9.5	10.2	10.8	11.4	12.0															11
12	9.1	9.9	10.7	11.3	11.9	12.5	13.1														12
13	9.5	10.3	11.1	11.8	12.4	13.0	13.6	14.2													13
14	9.8	10.7	11.5	12.2	12.9	13.5	14.2	14.7	15.3												14
15	10.1	11.0	11.8	12.6	13.3	14.0	14.6	15.3	15.8	16.4											15
16	10.4	11.4	12.2	13.0	13.7	14.4	15.1	15.7	16.3	16.9	17.5										16
17	10.7	11.7	12.5	13.4	14.1	14.9	15.5	16.1	16.8	17.4	18.0	18.6									17
18	11.0	11.9	12.9	13.7	14.5	15.3	16.0	16.6	17.3	17.9	18.5	19.1	19.7								18
19	11.2	12.2	13.2	14.1	14.9	15.6	16.4	17.1	17.8	18.4	19.0	19.6	20.2	20.8							19
20	11.5	12.5	13.5	14.4	15.2	15.9	16.8	17.5	18.2	18.8	19.5	20.1	20.7	21.3	21.9						20
22	12.0	13.1	14.1	15.0	15.9	16.7	17.6	18.3	19.1	19.7	20.4	21.0	21.7	22.3	22.9	24.1					22
24	12.4	13.6	14.6	15.6	16.6	17.5	18.3	19.1	19.8	20.6	21.3	21.9	22.6	23.2	23.9	25.1	26.2				24
26	12.8	14.1	15.2	16.2	17.2	18.1	19.0	19.8	20.6	21.4	22.1	22.8	23.5	24.1	24.8	26.1	27.2	28.4			26
28	13.2	14.5	15.6	16.7	17.7	18.7	19.6	20.5	21.3	22.1	22.9	23.6	24.4	25.0	25.7	27.1	28.2	29.5	30.6		28
30	13.6	14.9	16.1	17.2	18.3	19.3	20.2	21.1	22.0	22.9	23.7	24.4	25.2	25.9	26.7	28.0	29.3	30.5	31.6	32.8	30
32	14.0	15.3	16.5	17.7	18.8	19.8	20.8	21.8	22.7	23.6	24.4	25.2	26.0	26.7	27.5	28.9	30.1	31.4	32.6	33.8	32
34	14.4	15.7	17.0	18.2	19.3	20.4	21.4	22.4	23.3	24.2	25.1	25.9	26.7	27.5	28.3	29.7	31.0	32.3	33.6	34.8	34
36	14.7	16.1	17.4	18.6	19.8	20.9	21.9	23.0	23.9	24.8	25.8	26.6	27.4	28.3	29.0	30.5	32.0	33.0	34.6	35.8	36
38	15.0	16.4	17.8	19.0	20.3	21.4	22.5	23.5	24.5	25.4	26.4	27.3	28.1	29.0	29.8	31.4	32.8	34.2	35.5	36.7	38
40	15.3	16.8	18.2	19.4	20.7	21.9	23.0	24.0	25.1	26.0	27.0	27.9	28.8	29.7	30.5	32.1	33.6	35.1	36.4	37.6	40
42	15.6	17.1	18.5	19.8	21.1	22.3	23.4	24.5	25.6	26.6	27.6	28.5	29.4	30.4	31.2	32.8	34.4	35.9	37.3	38.6	42
44	15.9	17.5	18.9	20.2	21.5	22.7	23.9	25.0	26.1	27.2	28.2	29.1	30.0	31.0	31.9	33.5	35.2	36.7	38.1	39.5	44
46	16.2	17.8	19.2	20.6	21.9	23.2	24.3	25.5	26.7	27.7	28.7	29.7	30.6	31.6	32.5	34.2	35.9	37.4	38.9	40.3	46
48	16.5	18.1	19.6	20.9	22.3	23.6	24.8	26.0	27.2	28.2	29.2	30.2	31.2	32.2	33.1	34.9	36.6	38.2	39.7	41.2	48
50	16.8	18.4	19.9	21.3	22.7	24.0	25.2	26.4	27.6	28.7	29.8	30.8	31.8	32.8	33.7	35.5	37.3	38.9	40.4	42.0	50
52	17.0	18.7	20.2	21.6	23.1	24.4	25.6	26.8	28.1	29.2	30.3	31.4	32.4	33.4	34.3	36.2	38.0	39.6	41.2	42.8	52
54	17.3	19.0	20.5	22.0	23.4	24.8	26.1	27.3	28.5	29.7	30.8	31.9	32.9	33.9	34.9	36.8	38.7	40.3	42.0	43.6	54
56	17.6	19.3	20.9	22.4	23.8	25.2	26.5	27.7	28.9	30.1	31.2	32.4	33.4	34.5	35.5	37.4	39.3	41.0	42.7	44.3	56
58	17.8	19.5	21.1	22.7	24.2	25.5	26.9	28.2	29.3	30.5	31.7	32.9	33.9	35.0	36.0	38.0	39.8	41.7	43.4	45.0	58
60	18.1	19.8	21.4	23.0	24.5	25.8	27.3	28.7	29.8	31.0	32.2	33.4	34.5	35.5	36.5	38.6	40.4	42.3	44.0	45.8	60
62	18.3	20.1	21.7	23.3	24.8	26.2	27.6	29.0	30.2	31.4	32.6	33.8	35.0	36.0	37.1	39.2	41.0	42.9	44.7	46.5	62
64	18.6	20.3	22.0	23.6	25.2	26.5	27.9	29.3	30.6	31.8	33.1	34.2	35.5	36.5	37.6	39.7	41.6	43.5	45.4	47.2	64
66	18.8	20.6	22.3	23.9	25.5	26.9	28.3	29.7	31.0	32.2	33.5	34.7	35.9	37.0	38.1	40.2	42.2	44.1	46.0	47.8	66
68	19.0	20.8	22.5	24.2	25.8	27.3	28.7	30.1	31.4	32.6	33.9	35.1	36.3	37.5	38.6	40.7	42.8	44.7	46.6	48.4	68
70	19.2	21.	22.8	24.5	26.1	27.6	29.1	30.4	31.8	33.1	34.3	35.6	36.8	37.9	39.1	41.3	43.3	45.3	47.2	49.0	70
72															39.6	41.8	43.8	45.9	47.8	49.7	72
74															40.0	42.3	44.4	46.4	48.4	50.3	74
76															40.5	42.8	44.9	47.0	49.0	50.8	76
78															40.9	43.3	45.5	47.5	49.5	51.5	78
80															41.3	43.8	46.0	48.0	50.1	52.0	80
82															41.8	44.2	46.4	48.6	50.6	52.6	82
84															42.2	44.6	46.9	49.2	51.1	53.2	84
86															42.6	45.0	47.4	49.6	51.6	53.7	86
88															43.0	45.4	47.9	50.1	52.2	54.3	88
90															43.4	45.9	48.3	50.6	52.8	54.8	90
92															43.8	46.3	48.7	51.1	53.4	55.4	92
96															44.6	47.2	49.5	52.0	54.4	56.3	96

Equation for Circular Equivalent of a Rectangular Duct:[5]

$$d_c = 1.30 \frac{(ab)^{0.625}}{(a+b)^{0.250}} = 1.30 \sqrt[8]{\frac{(ab)^5}{(a+b)^2}}$$

where
a = length of one side of rectangular duct, inches.
b = length of adjacent side of rectangular duct, inches.
d_c = circular equivalent of a rectangular duct for equal friction and capacity, inches.

FIG. 8–3 Circular Equivalents Of Rectangular Ducts For Equal Friction And Capacity (Reprinted by permission from *ASHRAE Guide and Data Book*, 1965)

To make corrections for temperatures higher or lower than standard (70°), just remember that increased temperatures cause lower densities. If air is less dense, it expands. So, to determine the new air quantity, multiply the air quantity at 70° by the density at 70° and divide by the density at the operating temperature. Corrections can be made without knowing the densities because densities vary inversely with the absolute temperature. This comes from the perfect gas law $PV = RT$, where P = absolute pressure, V = specific volume cu ft/lb, R = gas constant, T = absolute temperature (460 + t), t = fahrenheit temperature.

$$\frac{P_1 V_1}{T_1} = \frac{P_2 V_2}{T_2}$$

$$P_1 = P_2$$

$$\frac{V_1}{V_2} = \frac{T_1}{T_2}$$

$$V = \frac{1}{\rho} \ (\rho = \text{density})$$

$$\frac{\rho_2}{\rho_1} = \frac{T_1}{T_2}$$

After correcting the air quantity, find the friction H_c from Fig. 8–1 or 8–2. Then find the density and viscosity friction correction factor from Fig. 8–5. Multiply the friction by K, the correction factor, to get the friction loss under operating conditions.

Example #3

5000 cfm std air is flowing through a 24-in. duct. If the air temperature is 240° what is the friction loss?

$$Q_c = 5000 \text{ cfm} \ \frac{(460 + 240)}{(460 + 70)}$$

$Q_c = 6600$ cfm
From Fig. 8–2 6600 cfm & 24-in. dia., H_f
  = 0.22-in./100 ft
From Fig. 8–5 K = 0.79
$H_c = KH_f$
$H_c = 0.79 (0.22)$
$H_c = 0.172$ in./100 ft
Note the H_f for 5000 cfm & 24-in. dia. is 0.13 in./100 ft.

To convert round duct sizes to square and rectangular, use the charts in Fig. 8–3. Notice that the areas of the round duct and rectangular duct are not equal. For example:

Size	Area	Perimeter	Aspect Ratio
12-in. dia.	112 in.	37.8 in.	
11 in. × 11 in.	121 in.	44 in.	1
15.5 in. × 8 in.	124 in.	47 in.	1.94
22 in. × 6 in.	132 in.	56 in.	3.65

Also, notice that as the aspect ratio (ratio of width to height) increases, the required area increases. The key to this is the length of the perimeter. A duct 10.6-in. sq has the same area as a 12-in. dia. duct, but its perimeter is 42.4 in. Friction varies with the amount of surface encountered by the air as it travels down a duct. The more perimeter, the more friction. To compensate for the increased friction, a larger duct is used. What is the ideal duct shape if round ducts cannot be used? A square duct. If you can't use a square duct, stay as close to square as possible. Try to keep the aspect ratio below 3-to-1 and use 4-to-1 as a maximum. The reason for this is cost. There is no engineering reason why you can't go up to 10-to-1, providing you are willing to pay for the additional sheet metal. Not only more metal, but heavier metal. As the span increases you must either increase the gage of the metal or add reinforcing to

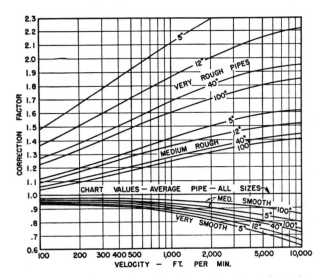

FIG. 8–4 Correction Factors For Pipe Roughness (Reprinted by permission from *ASHRAE Guide and Data Book*, 1965)

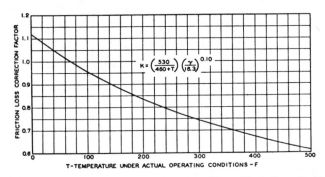

FIG. 8–5 Correction Factors For Density And Viscosity (Reprinted by permission from *ASHRAE Guide and Data Book*, 1965)

keep the metal from buckling. A duct with a 3-to-1 aspect ratio costs 25% more than a square duct, while an 8-to-1 costs twice as much.

Before we go too far, let's stop for a moment and get a couple of definitions out of the way. In the flow of air we are concerned with three pressures: total pressure, static pressure, and velocity pressure. Velocity pressure is the pressure required to get the air up to the desired velocity, while the static pressure is the bursting pressure. When you blow up a balloon it takes velocity pressure to blow the air out of your mouth, and static pressure to inflate the balloon. Static pressure is used to overcome the pressure drop caused by friction and turbulence. Velocity pressure and static pressure are interchangeable. That is, velocity pressure can be converted to static pressure, and vice-versa. The sum of static pressure and velocity pressure is called the total pressure. Later, use will be made of the interconvertability of static and velocity pressure in the static-regain method of duct sizing.

$$\text{Velocity pressure} = \left(\frac{\text{Velocity}}{4005}\right)^2$$

Years ago, rather tight limits were set for velocity and friction. I can remember when we never exceeded 0.1 in./100 ft friction loss or 1000 fpm. If you exceeded these limits, you were a real daredevil. What are our limits today? It is hard to say. The reasons for the low limits of yesterday were to keep down the total friction and noise. Noise is no longer a problem because we have methods to attenuate noise. In addition, duct noises are not as loud as previously believed if good duct installation procedures are followed. The chief causes of noise are sharp edges and fittings that cause turbulence. Friction is a problem because it takes horsepower to overcome friction. Remember, that as you double velocity, the friction increases four times. So, you pay a penalty in horsepower to overcome the increased friction. But this may not be a problem. Is space a problem? Perhaps it is worth the penalty of increased horsepower to save space. You, as the designer, must decide what the limits will be. Maybe it will help if I tell you my guide lines. For conventional or low velocity systems, I use 2000 fpm and 0.2 in./100 ft friction loss. For medium velocity systems, I use 4000 fpm and 1.0 in./100 ft friction loss.

In deciding how high-pressure a system you wish to use, don't lose sight of what increased fan horsepower does to the total refrigeration load. A 20,000 cfm fan, operating at 2-in. SP, requires about 9.5 BHP; at 4-in. SP—16.3 BHP; at 6-in. SP—24.75 BHP; at 8-in.—31.5 BHP. While a fan is raising the pressure

of the air, it is doing work on the air and so raises the temperature of the air. The temperature increase is as follows if the motor is in the air stream:

SP	Temperature Rise	Tonnage
2	1.5°	2.7
4	2.55°	4.6
6	3.9°	7
8	5°	9

If the motor is not in the air stream or air conditioned space, the temperature rise and tonnage are reduced by the inefficiency of the motor. Multiply the above values by 80% when the motor is not in the air stream or conditioned space.

Now, before we forget, let's look into other sources of pressure drop. Ductwork doesn't always go straight. Sometimes ductwork turns. These turns are called elbows. Rarely does ductwork serve one outlet only. In order to serve more than one outlet, a divided flow fitting must be used. Then, too, there are losses caused by area changes. These pressure losses are given in Figs. 8–6, 8–7, and 8–8. I say, "given," but must add, indirectly. They don't just say that a 12-in. dia. elbow with a centerline radius equal to the diameter has a pressure drop of 0.08 in. at 2000 fpm. No, the chart says that it has a "C" of 0.33 or a L/D of 17. What do these mean? The pressure drop of an elbow equals "C" times the velocity pressure. The velocity pressure at 2000 fpm equals 0.25 (from Fig. 8–9), so the pressure drop is 0.33 (0.25) = 0.0825 in. The ratio L/D refers to the equivalent length (L) divided by the diameter. When we speak of equivalent length, we mean the length of straight pipe that has the same pressure drop as the elbow. In this case L/D = 17. Be sure to change the diameter from inches to feet to keep our unit consistent. So change 12 in. to 1 ft. Then L/D = 17; L = 17 D; D = 1; L = 17 (1) = 17 ft.

Now, from Fig. 8–1, a 12-in. dia. duct at 2000 fpm velocity has a pressure drop of 0.48 in./100 ft. Multiply this by the equivalent length, 17 ft, to get the pressure drop of $H_f = 17 (0.48/100) = 0.0816$.

In the case of rectangular ductwork, the ratio L/W is used, where W is the width of the duct or elbow in feet. Notice that while with round ductwork the "C" and "L/D" varied only with the radius ratio, rectangular ductwork has an additional variable, the ratio of height to width (H/W).

Figure 8–11 gives examples of different H/W in the same ductwork system. Always call the side that is turning "W," and so in Fig. 8–11, elbow #1, H/W = 24 in./6 in. = 4 and for elbow #2, H/W = 6 in./24 in. = 0.25. Be sure to remember this and don't arbitrarily call the larger dimension H as some do. At a velocity of 2000 fpm, if the radius in elbow #1 = 6 in. and in

(Additional Equivalent Losses in Excess of Friction to Intersection of Center Lines)

TYPE	ILLUSTRATION	CONDITIONS	C^a	L/D	L/W
N-DEG.	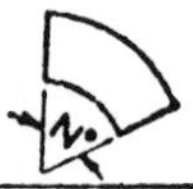	RECTANGULAR OR ROUND; WITH OR WITHOUT VANES	$\frac{N}{90}$ TIMES VALUE FOR SIMILAR 90-DEG. ELBOW		
90-DEG. ROUND SECTION	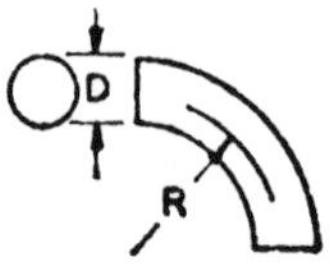	MITER	1.30[b]	65	
		R/D=0.5	0.90		
		0.75	0.45	23	
		1.0	0.33	17	
		1.5	0.24	12	
		2.0	0.19	10	
90-DEG. RECTANGULAR SECTION		H/W 0.25 { MITER	1.25[b]		25
		R/W 0.5	1.25		25
		0.75	0.60		12
		1.0	0.37		7
		1.5	0.19		4
		0.5 { MITER	1.47		49
		0.5	1.10		40
		0.75	0.50		16
		1.0	0.28		9
		1.5	0.13		4
		1.0 { MITER	1.50		75
		0.5	1.00		50
		0.75	0.41		21
		1.0	0.22		11
		1.5	0.09		4.5
		4.0 { MITER	1.38		110
		0.5	0.96		65
		0.75	0.37		43
		1.0	0.19		17
		1.5	0.07		6
90-DEG. SQUARE SECTION WITH SPLITTER VANES	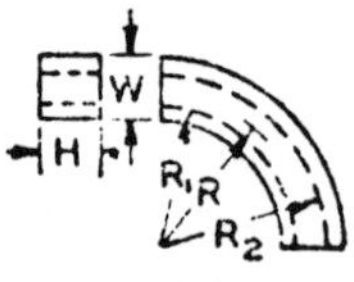	R/W R₁/W R₂/W			
		MITER 0.5			28
		0.5 0.4	0.70		19
		0.7 0.6			12
		1.0 1.0	0.13		7.2
		1.5	0.12		
		MITER 0.3 0.5			22
		0.5 0.2 0.4	0.45		16
		0.75 0.4 0.7	0.12		
		1.0 0.7 1.0	0.10		
		1.5 1.3 1.6	0.15		
MITER WITH TURNING VANES	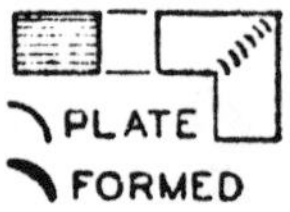 PLATE / FORMED	C = 0.10 TO 0.35 DEPENDING ON MANUFACTURE			
MITER TEE WITH VANES	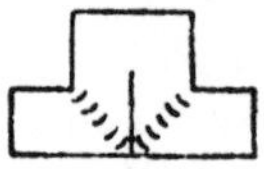	CONSIDER EQUAL TO A SIMILAR ELBOW. BASE LOSS ON ENTERING VELOCITY.			
RADIUS TEE	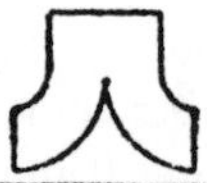				

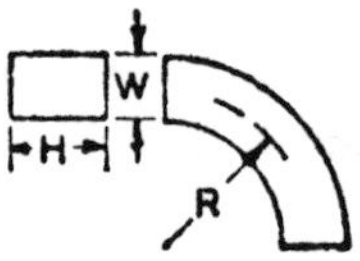

[a] Values based on f values of approximately 0.02.

[b] Values calculated from L/D and L/W values of Reference 6 for $f = 0.02$.

FIG. 8–6   Pressure Losses Due To Elbows (Reprinted by permission from *ASHRAE Guide and Data Book*, 1965)

Left side

TYPE	CONDITIONS		
ABRUPT EXPANSION	A1/A2	C1	C2
	0.1	0.81	81
	0.2	0.64	18
	0.3	0.49	5
	0.4	0.36	2.25
	0.5	0.25	1.00
	0.6	0.16	0.45
	0.7	0.09	0.18
	0.8	0.04	0.06
	0.9	0.01	0.01
GRADUAL EXPANSION	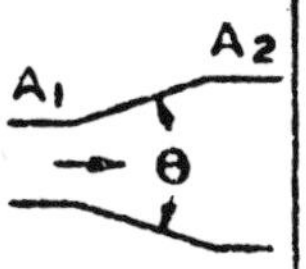 θ	Cr	
	5°	0.17	
	7°	0.22	
	10°	0.28	
	20°	0.45	
	30°	0.59	
	40°	0.73	
ABRUPT EXIT (A2=∞) 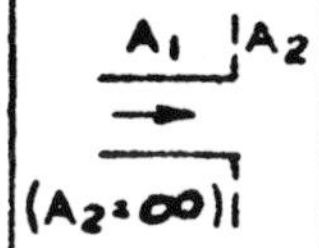	A1/A2=0.0	1.00	
SQUARE EDGE ORIFICE EXIT 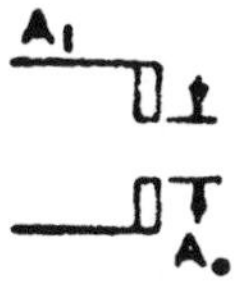	A0/A1	C0	
	0.0	2.50	
	0.2	2.44	
	0.4	2.26	
	0.6	1.98	
	0.8	1.54	
	1.0	1.00	
BAR ACROSS DUCT	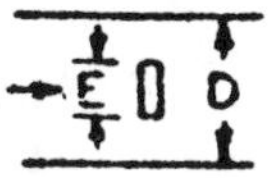 E/D	C	
	0.10	0.7	
	0.25	1.4	
	0.50	4.0	
PIPE ACROSS DUCT	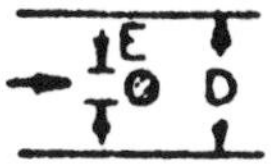 E/D	C	
	0.10	0.20	
	0.25	0.55	
	0.50	2.0	
STREAMLINED STRUT ACROSS DUCT	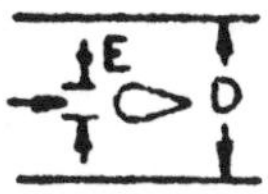 E/D	C	
	0.10	0.07	
	0.25	0.23	
	0.50	0.90	

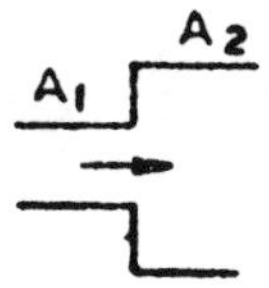

Right side

TYPE	CONDITIONS	LOSS COEFFICIENT
ABRUPT CONTRACTION SQUARE EDGE 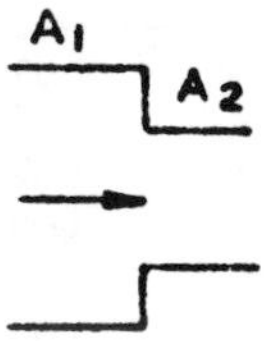	A2/A1	C2
	0.0	0.34
	0.2	0.32
	0.4	0.25
	0.6	0.16
	0.8	0.06
GRADUAL CONTRACTION	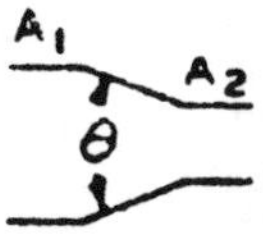 θ	
	30°	0.02
	45°	0.04
	60°	0.07
EQUAL AREA TRANSFORMATION	A1=A2, θ≤14°	C = 0.15
FLANGED ENTRANCE	A=∞	C = 0.34
DUCT ENTRANCE	A=∞	C = 0.85
FORMED ENTRANCE	A=∞	C = 0.03
SQUARE EDGE ORIFICE ENTRANCE 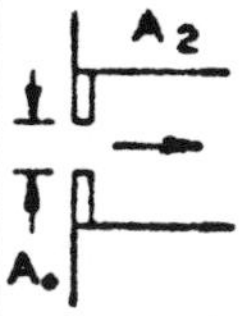	A0/A2	C0
	0.0	2.50
	0.2	1.90
	0.4	1.39
	0.6	0.98
	0.8	0.81
	1.0	0.34
SQUARE EDGE ORIFICE IN DUCT (A1=A2) 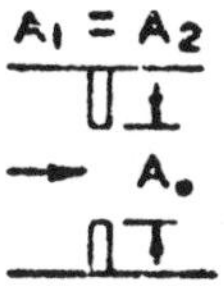	A0/A	C0
	0.0	2.50
	0.2	1.86
	0.4	1.21
	0.6	0.84
	0.8	0.20
	1.0	0.0

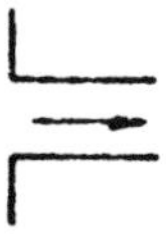

Note 1: Subscript on C indicates cross-section at which velocity is calculated

FIG. 8–7 Pressure Losses Due To Area Changes (Reprinted by permission from *ASHRAE Guide and Data Book*, 1965)

FIGURE 8–8   **Pressure Drop in Divided Flow Fittings**

Main Vel	Branch Vel	Angle of Takeoff			Main Vel	Branch Vel	Angle of Takeoff		
		90°	60°	45°			90°	60°	45°
1000	400	0.01			1700	680	0.04		
	500	0.02				850	0.06		
	600	0.02	0.02			1020	0.08	0.03	
	700	0.03	0.03			1190	0.12	0.05	
	800	0.04	0.04	0.04		1360	0.17	0.09	0.02
	900	0.05	0.05	0.05		1530	0.23	0.11	0.06
	1000	0.06	0.06	0.06		1700	0.26	0.14	0.10
1100	440	0.02			1800	720	0.04		
	550	0.03				900	0.07		
	660	0.03	0.01			1080	0.10	0.03	
	770	0.05	0.02			1260	0.14	0.06	
	880	0.07	0.04	0.01		1440	0.18	0.10	0.02
	990	0.09	0.05	0.03		1620	0.23	0.13	0.07
	1100	0.11	0.06	0.04		1800	0.29	0.16	0.11
1200	480	0.02			1900	760	0.04		
	600	0.03				950	0.08		
	720	0.04	0.02			1140	0.11	0.04	
	840	0.06	0.03			1330	0.15	0.07	
	960	0.08	0.04	0.01		1520	0.21	0.11	0.02
	1080	0.11	0.06	0.03		1710	0.27	0.14	0.07
	1200	0.13	0.07	0.05		1900	0.33	0.17	0.12
1300	520	0.02			2000	800	0.05		
	650	0.04				1000	0.09		
	780	0.05	0.02			1200	0.12	0.04	
	910	0.07	0.03			1400	0.17	0.08	
	1040	0.10	0.05	0.01		1600	0.23	0.12	0.03
	1170	0.12	0.07	0.04		1800	0.29	0.15	0.08
	1300	0.15	0.08	0.06		2000	0.36	0.19	0.14
1400	560	0.02			2100	840	0.05		
	700	0.04				1050	0.10		
	840	0.06	0.02			1260	0.13	0.05	
	980	0.08	0.04			1470	0.19	0.08	
	1120	0.11	0.06	0.02		1680	0.25	0.13	0.03
	1260	0.14	0.08	0.04		1890	0.22	0.17	0.09
	1400	0.17	0.10	0.07		2100	0.40	0.21	0.15
1500	600	0.03			2200	880	0.06		
	750	0.05				1100	0.10		
	900	0.07	0.02			1320	0.14	0.05	
	1050	0.10	0.04			1540	0.21	0.09	
	1200	0.13	0.07	0.02		1760	0.28	0.14	0.03
	1350	0.17	0.09	0.05		1980	0.36	0.18	0.09
	1500	0.20	0.11	0.08		2200	0.44	0.23	0.16
1600	640	0.03			2300	920	0.07		
	800	0.05				1150	0.11		
	960	0.07	0.03			1380	0.16	0.06	
	1120	0.11	0.05			1610	0.23	0.10	
	1280	0.15	0.08	0.02		1840	0.31	0.16	0.03
	1440	0.19	0.10	0.05		2070	0.39	0.20	0.10
	1600	0.23	0.12	0.09		2300	0.48	0.25	0.18

FIGURE 8–8 Pressure Drop in Divided Flow Fittings—*Continued*

Main Vel	Branch Vel	Angle of Takeoff 90°	60°	45°	Main Vel	Branch Vel	Angle of Takeoff 90°	60°	45°
2400	960	0.07			3100	1240	0.12		
	1200	0.12				1550	0.21		
	1440	0.17	0.06			1860	0.28	0.10	
	1680	0.25	0.11			2170	0.41	0.17	
	1920	0.33	0.17	0.03		2480	0.56	0.28	0.05
	2160	0.43	0.23	0.11		2790	0.72	0.37	0.18
	2400	0.53	0.28	0.19		3100	0.88	0.46	0.32
2500	1000	0.08			3200	1280	0.13		
	1250	0.13				1600	0.22		
	1500	0.18	0.06			1920	0.30	0.10	
	1750	0.27	0.12			2240	0.44	0.19	
	2000	0.36	0.19	0.04		2560	0.60	0.30	0.05
	2250	0.46	0.24	0.12		2880	0.77	0.39	0.19
	2500	0.57	0.30	0.21		3200	0.94	0.49	0.34
2600	1040	0.08			3300	1320	0.13		
	1300	0.14				1650	0.23		
	1560	0.20	0.07			1980	0.32	0.11	
	1820	0.29	0.12			2310	0.47	0.21	
	2080	0.39	0.20	0.04		2640	0.63	0.32	0.05
	2340	0.50	0.26	0.13		2970	0.81	0.42	0.20
	2600	0.61	0.32	0.23		3300	1.00	0.52	0.36
2700	1080	0.09			3400	1360	0.14		
	1350	0.16				1700	0.25		
	1620	0.21	0.08			2040	0.34	0.12	
	1890	0.31	0.13			2380	0.50	0.21	
	2160	0.42	0.22	0.04		2720	0.67	0.34	0.06
	2430	0.54	0.29	0.14		3060	0.86	0.44	0.21
	2700	0.67	0.35	0.24		3400	1.06	0.55	0.38
2800	1120	0.10			3500	1400	0.15		
	1400	0.17				1750	0.27		
	1680	0.23	0.08			2100	0.36	0.12	
	1960	0.34	0.14			2450	0.53	0.22	
	2240	0.45	0.23	0.04		2800	0.74	0.36	0.06
	2520	0.58	0.30	0.15		3150	0.93	0.47	0.22
	2800	0.72	0.38	0.26		3500	1.12	0.58	0.40
2900	1160	0.10			3600	1440	0.16		
	1450	0.18				1800	0.28		
	1740	0.25	0.09			2160	0.38	0.13	
	2030	0.36	0.15			2520	0.56	0.24	
	2320	0.49	0.25	0.04		2880	0.76	0.38	0.06
	2610	0.63	0.33	0.16		3240	0.97	0.50	0.23
	2900	0.77	0.40	0.28		3600	1.19	0.62	0.43
3000	1200	0.11			3700	1480	0.17		
	1500	0.19				1850	0.29		
	1800	0.27	0.09			2220	0.41	0.14	
	2100	0.39	0.16			2590	0.59	0.25	
	2400	0.52	0.27	0.04		2960	0.80	0.40	0.06
	2700	0.67	0.34	0.17		3330	1.03	0.52	0.25
	3000	0.83	0.43	0.30		3700	1.26	0.65	0.45

FIGURE 8–8 **Pressure Drop in Divided Flow Fittings—Continued**

Main Vel	Branch Vel	Angle of Takeoff			Main Vel	Branch Vel	Angle of Takeoff		
		90°	60°	45°			90°	60°	45°
3800	1520	0.18			4400	1760	0.24		
	1900	0.31				2200	0.42		
	2280	0.43	0.14			2640	0.57	0.19	
	2660	0.62	0.26			3080	0.84	0.35	
	3040	0.84	0.42	0.07		3520	1.13	0.57	0.08
	3420	1.08	0.55	0.26		3960	1.46	0.74	0.34
	3800	1.33	0.69	0.47		4400	1.79	0.92	0.63
3900	1560	0.19			4500	1800	0.25		
	1950	0.33				2250	0.44		
	2340	0.45	0.15			2700	0.60	0.20	
	2730	0.66	0.27			3150	0.87	0.37	
	3120	0.89	0.45	0.07		3600	1.18	0.59	0.08
	3510	1.14	0.58	0.27		4050	1.52	0.77	0.36
	3900	1.40	0.72	0.50		4500	1.87	0.96	0.66
4000	1600	0.20			4600	1840	0.26		
	2000	0.35				2300	0.46		
	2400	0.47	0.16			2760	0.63	0.21	
	2800	0.70	0.29			3220	0.91	0.38	
	3200	0.93	0.47	0.07		3680	1.24	0.62	0.09
	3600	1.20	0.61	0.28		4140	1.60	0.81	0.38
	4000	1.47	0.76	0.52		4600	1.96	1.00	0.69
4100	1640	0.21			4700	1880	0.27		
	2100	0.39				2350	0.48		
	2460	0.50	0.17			2820	0.66	0.22	
	2870	0.72	0.30			3290	0.95	0.40	
	3280	0.98	0.50	0.07		3760	1.29	0.65	0.09
	3690	1.26	0.65	0.30		4230	1.66	0.85	0.39
	4100	1.55	0.80	0.55		4700	2.04	1.05	0.72
4200	1680	0.22			4800	1920	0.29		
	2100	0.39				2400	0.50		
	2520	0.53	0.18			2880	0.68	0.23	
	2940	0.76	0.32			3360	0.99	0.42	
	3360	1.03	0.52	0.08		3840	1.13	0.67	0.09
	3780	1.33	0.68	0.31		4320	1.63	0.88	0.41
	4200	1.63	0.84	0.58		4800	2.13	1.09	0.75
4300	1720	0.23			4900	1960	0.30		
	2150	0.40				2450	0.52		
	2580	0.55	0.18			2940	0.71	0.24	
	3010	0.80	0.33			3430	1.03	0.43	
	3440	1.08	0.54	0.08		3920	1.41	0.70	0.10
	3870	1.39	0.71	0.33		4410	1.81	0.90	0.42
	4300	1.70	0.88	0.60		4900	2.22	1.11	0.78

#2 = 24 in., the C factors would be as follows: #1 C = 0.07; #2 C = 0.19. The velocity pressure equals 0.25 in. and so the pressure drops are, #1, 0.07 (0.25) = 0.0175, and #2, 0.19 (0.25) = 0.0475.

While Fig. 8–6 gives the pressure drop for various radius ratios including mitered elbows (mitered elbows are square elbows that have no radius), only elbows with a ratio of radius to diameter or radius to width of one or greater should be used. If the R/D or R/W ratio is less than one, then use a mitered elbow with turning vanes. Fig. 8–10 gives the equivalent length of various elbows with various radius ratios. Notice that a 10 × 6 elbow has an equivalent length of 3 ft when the R/W ratio is 1.5, but 31 ft for 0.5 R/W ratio.

FIGURE 8–9
Table of Velocity Pressures

VP	V	VP	V	VP	V	VP	V	VP	V	VP	V
0.01	400	0.43	2626	0.85	3690	1.27	4513	1.69	5206	2.11	5817
0.02	566	0.44	2656	0.86	3709	1.28	4531	1.70	5222	2.12	5831
0.03	694	0.45	2687	0.87	3729	1.29	4549	1.71	5237	2.13	5845
0.04	801	0.46	2716	0.88	3758	1.30	4566	1.72	5253	2.14	5859
0.05	896	0.47	2746	0.89	3779	1.31	4583	1.73	5268	2.15	5872
0.06	981	0.48	2775	0.90	3800	1.32	4601	1.74	5283	2.16	5886
0.07	1060	0.49	2804	0.91	3821	1.33	4619	1.75	5298	2.17	5899
0.08	1133	0.50	2832	0.92	3842	1.34	4636	1.76	5313	2.18	5913
0.09	1201	0.51	2860	0.93	3863	1.35	4653	1.77	5328	2.19	5927
0.10	1266	0.52	2888	0.94	3884	1.36	4671	1.78	5343	2.20	5940
0.11	1328	0.53	2916	0.95	3904	1.37	4688	1.79	5359	2.21	5954
0.12	1387	0.54	2943	0.96	3924	1.38	4705	1.80	5374	2.22	5967
0.13	1444	0.55	2970	0.97	3945	1.39	4722	1.81	5388	2.23	5981
0.14	1498	0.56	2997	0.98	3965	1.40	4739	1.82	5403	2.24	5994
0.15	1551	0.57	3024	0.99	3985	1.41	4756	1.83	5418	2.25	6008
0.16	1602	0.58	3050	1.00	4005	1.42	4773	1.84	5433	2.26	6021
0.17	1651	0.59	3076	1.01	4025	1.43	4790	1.85	5447	2.27	6034
0.18	1699	0.60	3102	1.02	4045	1.44	4806	1.86	5462	2.28	6047
0.19	1746	0.61	3127	1.03	4064	1.45	4823	1.87	5477	2.29	6061
0.20	1791	0.62	3153	1.04	4084	1.46	4840	1.88	5491	2.30	6074
0.21	1835	0.63	3179	1.05	4103	1.47	4856	1.89	5506	2.31	6087
0.22	1879	0.64	3204	1.06	4123	1.48	4873	1.90	5521	2.32	6100
0.23	1921	0.65	3229	1.07	4142	1.49	4889	1.91	5535	2.33	6113
0.24	1962	0.66	3254	1.08	4162	1.50	4905	1.92	5550	2.34	6128
0.25	2003	0.67	3279	1.09	4181	1.51	4921	1.93	5564	2.35	6140
0.26	2042	0.68	3303	1.10	4200	1.52	4938	1.94	5579	2.36	6153
0.27	2081	0.69	3327	1.11	4219	1.53	4954	1.95	5593	2.37	6166
0.28	2119	0.70	3351	1.12	4238	1.54	4970	1.96	5608	2.38	6179
0.29	2157	0.71	3375	1.13	4257	1.55	4986	1.97	5623	2.39	6192
0.30	2193	0.72	3398	1.14	4276	1.56	5002	1.98	5637	2.40	6205
0.31	2230	0.73	3422	1.15	4295	1.57	5018	1.99	5651	2.41	6217
0.32	2260	0.74	3445	1.16	4314	1.58	5034	2.00	5664	2.42	6230
0.33	2301	0.75	3468	1.17	4332	1.59	5050	2.01	5678	2.43	6243
0.34	2335	0.76	3491	1.18	4350	1.60	5066	2.02	5692	2.44	6256
0.35	2369	0.77	3514	1.19	4368	1.61	5082	2.03	5706	2.45	6269
0.36	2403	0.78	3537	1.20	4386	1.62	5098	2.04	5720	2.46	6282
0.37	2436	0.79	3560	1.21	4405	1.63	5114	2.05	5734	2.47	6294
0.38	2469	0.80	3582	1.22	4423	1.64	5129	2.06	5748	2.48	6307
0.39	2501	0.81	3604	1.23	4442	1.65	5144	2.07	5762	2.49	6320
0.40	2533	0.82	3625	1.24	4460	1.66	5160	2.08	5776	2.50	6332
0.41	2563	0.83	3657	1.25	4478	1.67	5175	2.09	5790	2.51	6345
0.42	2595	0.84	3669	1.26	4495	1.68	5191	2.10	5804	2.52	6358

Fig. 8–7 gives losses caused by changes in area or size. The area-change pressure-loss coefficient C has in some cases a subscript such as C_1, C_2, C_0. These are keyed to A_1, A_2, A_0. In the case of an abrupt expansion, C_1 is used with the velocity pressure at area 1 (A_1), and C_2 with the velocity pressure at area 2 (A_2). Let's take a case where a 2-sq ft duct abruptly expands to 4 sq ft. $A_1 = 2$ and $A_2 = 4$, so $A_1/A_2 = 2/4 = 0.5$. $C_1 = 0.25$ and $C_2 = 1.00$. If the velocity at $A_1 =$ 4000 fpm, then the velocity at A_2 will equal 2000 fpm. The velocity pressure at 4000 fpm = 1.00, and at 2000 = 0.25. The pressure drop, if the velocity pressures at A_1 and C_1 factor are used, is 0.25 (1.00) = 0.25 in., and if A_2 and C_2 are used, 1.00 (0.25) = 0.25 in. So you might say you pay your money and take your choice. In the case of an orifice, the velocity pressure corresponding to the orifice velocity and the C_0 orifice loss coefficient is used.

FIGURE 8–10

Equivalent Length of Elbows

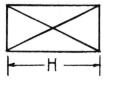

ELEVATION   PLAN

W	H	R/W			
		0.5	1.0	1.5	2.0
6	6	23	5	2	1
8	6	27	7	3	1
10	6	31	8	3	2
12	6	34	9	4	2
14	6	37	10	4	3
16	6	40	11	5	3
18	6	43	12	6	3
20	6	45	13	6	4
22	6	48	14	7	4
24	6	50	15	7	4
8	8	31	7	3	2
10	8	35	8	4	2
12	8	38	9	4	2
14	8	42	11	5	3
16	8	45	12	5	3
18	8	48	13	6	3
20	8	51	14	6	4
22	8	54	15	7	4
24	8	57	16	7	4
26	8	59	17	8	5
28	8	62	18	8	5
30	8	65	19	9	5
32	8	67	20	9	6
10	10	39	9	4	2
12	10	43	10	4	2
14	10	46	11	5	3
16	10	50	12	5	3
18	10	53	13	6	3
20	10	56	15	7	4
22	10	59	16	7	4
24	10	62	17	8	4
26	10	65	18	8	5
28	10	68	19	9	5
30	10	71	20	9	5
32	10	74	21	10	6
34	10	76	22	10	6
36	10	79	22	11	6
38	10	82	23	11	7
40	10	84	24	12	7
12	12	46	11	4	2
14	12	50	12	5	3
16	12	54	13	6	3
18	12	58	14	6	3

FIGURE 8–10 (Continued)

W	H	R/W			
		0.5	1.0	1.5	2.0
20	12	61	15	7	4
22	12	64	16	7	4
24	12	68	17	8	5
26	12	71	18	8	5
28	12	74	20	9	5
30	12	77	21	10	6
32	12	80	22	10	6
34	12	82	23	11	6
36	12	85	24	11	6
38	12	88	24	12	7
40	12	91	25	12	7
42	12	93	26	13	7
44	12	96	27	13	8
46	12	98	28	14	8
48	12	101	29	14	8
14	14	54	12	5	3
16	14	58	14	6	3
18	14	62	15	6	4
20	14	65	16	7	4
22	14	69	17	8	4
24	14	72	18	8	5
26	14	76	19	9	5
28	14	79	20	9	5
30	14	82	21	10	6
32	14	85	22	10	6
34	14	88	23	11	6
36	14	91	24	11	7
38	14	94	25	12	7
40	14	97	26	12	7
42	14	99	27	13	8
44	14	102	28	13	8
46	14	105	29	14	8
48	14	107	30	14	9
50	14	110	31	15	9
52	14	113	32	15	9
54	14	115	33	16	10
56	14	118	34	16	10
16	16	62	14	6	3
18	16	66	15	7	4
20	16	70	17	7	4
22	16	73	18	8	4
24	16	77	19	8	5
26	16	80	20	9	5
28	16	84	21	9	5
30	16	87	22	10	6
32	16	90	23	11	6
34	16	93	24	11	6
36	16	96	25	12	7
38	16	99	26	12	7
40	16	102	27	13	7
42	16	105	28	13	8
44	16	108	29	14	8
46	16	111	30	14	8
48	16	113	31	15	9
50	16	116	32	15	9
52	16	119	33	16	9

FIGURE 8–10 (Continued)

W	H	R/W 0.5	1.0	1.5	2.0
54	16	122	34	16	10
56	16	124	35	17	10
58	16	127	36	17	10
60	16	129	37	18	11
62	16	132	38	18	11
64	16	134	39	19	11
18	18	70	16	7	4
20	18	74	17	7	4
22	18	77	18	8	4
24	18	81	20	8	5
26	18	85	21	9	5
28	18	88	22	10	5
30	18	92	23	10	6
32	18	95	24	11	6
34	18	98	25	11	6
36	18	101	26	12	7
38	18	104	27	12	7
40	18	107	28	13	7
42	18	110	29	13	8
44	18	113	30	14	8
46	18	116	31	15	8
48	18	119	32	15	9
50	18	122	33	16	9
52	18	125	34	16	9
54	18	128	35	17	10
56	18	130	36	17	10
58	18	133	37	18	10
60	18	136	38	18	11
62	18	138	39	19	11
64	18	141	40	19	11
66	18	144	41	20	12
68	18	146	42	20	12
70	18	149	43	21	12
72	18	151	44	21	13
20	20	77	18	7	4
22	20	81	19	8	4
24	20	85	20	9	5
26	20	89	21	9	5
28	20	92	22	10	5
30	20	96	24	10	6
32	20	99	25	11	6
34	20	103	26	12	6
36	20	106	27	12	7
38	20	109	28	13	7
40	20	113	29	13	8
42	20	116	30	14	8
44	20	119	31	14	8
46	20	122	32	15	9
48	20	125	33	15	9
50	20	128	34	16	9
52	20	131	35	16	10
54	20	133	36	17	10
56	20	136	37	17	10
58	20	139	38	18	10
60	20	142	39	18	11
62	20	146	40	19	11

FIGURE 8–10 (Continued)

W	H	R/W 0.5	1.0	1.5	2.0
64	20	147	41	19	11
66	20	150	42	20	12
68	20	153	43	20	12
70	20	155	44	21	12
72	20	158	45	22	13
74	20	160	46	22	13
76	20	163	47	23	13
78	20	166	48	23	14
80	20	168	49	24	14
22	22	85	19	8	4
24	22	89	21	9	5
26	22	93	22	9	5
28	22	97	23	10	6
30	22	100	24	11	6
32	22	104	25	11	6
34	22	107	27	12	7
36	22	111	28	12	7
38	22	114	29	13	7
40	22	117	30	13	8
42	22	121	31	14	8
44	22	124	32	14	8
46	22	127	33	15	9
48	22	130	34	16	9
50	22	133	35	16	9
52	22	136	36	17	10
54	22	139	37	17	10
56	22	142	38	18	10
58	22	145	39	18	11
60	22	148	40	19	11
62	22	151	41	19	11
64	22	153	42	20	12
66	22	156	43	20	12
68	22	159	44	21	12
70	22	162	45	21	13
72	22	164	46	22	13
74	22	167	47	22	13
76	22	170	48	23	14
78	22	172	49	23	14
80	22	175	50	24	14
82	22	177	51	24	14
84	22	180	52	25	15
86	22	182	53	25	15
88	22	185	53	26	15
24	24	93	21	9	5
26	24	97	22	10	5
28	24	101	24	10	6
30	24	104	25	11	6
32	24	108	26	11	6
34	24	112	27	12	7
36	24	115	28	12	7
38	24	119	29	13	7
40	24	122	31	14	8
42	24	125	32	14	8
44	24	129	33	15	8
46	24	132	34	15	9
48	24	135	35	16	9

FIGURE 8–10 (Continued)

W	H	R/W 0.5	1.0	1.5	2.0
50	24	138	36	16	9
52	24	141	37	17	10
54	24	144	38	17	10
56	24	147	39	18	10
58	24	150	40	18	11
60	24	153	41	19	11
62	24	156	42	20	11
64	24	159	43	20	12
66	24	162	44	21	12
68	24	165	45	21	12
70	24	167	46	22	13
72	24	170	47	22	13
74	24	173	48	23	13
76	24	176	49	23	14
78	24	178	50	24	14
80	24	181	51	24	14
82	24	184	52	25	15
84	24	186	53	25	15
86	24	189	54	26	15
88	24	192	55	26	16
90	24	194	56	27	16
92	24	197	56	27	16
94	24	199	57	28	17
96	24	202	58	28	17
26	26	101	23	10	5
28	26	105	24	10	6
30	26	108	25	11	6
32	26	112	27	11	6
34	26	116	28	12	7
36	26	120	29	13	7
38	26	123	30	13	7
40	26	127	31	14	8
42	26	130	32	14	8
44	26	133	33	15	8
46	26	137	35	15	9
48	26	140	36	16	9
50	26	143	37	17	9
52	26	146	38	17	10
54	26	149	39	18	10
56	26	153	40	18	10
58	26	156	41	19	11
60	26	159	42	19	11
62	26	162	43	20	11
64	26	165	44	20	12
66	26	167	45	21	12
68	26	170	46	21	12
70	26	173	47	22	13
72	26	176	48	22	13
74	26	179	49	23	13
76	26	182	50	23	14
78	26	184	51	24	14
80	26	187	52	25	14
82	26	190	53	25	15
84	26	193	54	26	15
86	26	195	55	26	15
88	26	198	56	27	16
90	26	201	57	27	16

FIGURE 8–10 (Continued)

W	H	R/W 0.5	1.0	1.5	2.0
92	26	203	58	28	16
94	26	206	59	28	17
96	26	208	59	29	17
98	26	211	60	29	17
100	26	214	61	30	18
28	28	108	25	10	6
30	28	112	26	11	6
32	28	116	27	12	6
34	28	120	28	12	7
36	28	124	30	13	7
38	28	127	31	13	7
40	28	131	32	14	8
42	28	134	33	15	8
44	28	138	34	15	8
46	28	141	35	16	9
48	28	145	36	16	9
50	28	148	37	17	9
52	28	151	39	17	10
54	28	154	40	18	10
56	28	158	41	18	11
58	28	161	42	19	11
60	28	164	43	20	11
62	28	167	44	20	12
64	28	170	45	21	12
66	28	173	46	21	12
68	28	176	47	22	13
70	28	179	48	22	13
72	28	182	49	23	13
74	28	185	50	23	14
76	28	187	51	24	14
78	28	190	52	24	14
80	28	193	53	25	14
82	28	196	54	25	15
84	28	199	55	26	15
86	28	201	56	26	15
88	28	204	57	27	16
90	28	207	58	27	16
92	28	209	59	28	16
94	28	212	60	28	17
96	28	215	61	29	17
98	28	217	62	29	17
100	28	220	62	30	18
102	28	223	63	30	18
104	28	225	64	31	18
106	28	228	65	31	19
108	28	230	66	32	19
110	28	233	67	32	19
112	28	235	68	33	20
30	30	116	27	11	6
32	30	120	28	12	6
34	30	124	29	12	7
36	30	128	30	13	7
38	30	131	31	14	7
40	30	135	33	14	8
42	30	139	34	15	8
44	30	142	35	15	9

FIGURE 8–10 *(Continued)*

W	H	R/W			
		0.5	1.0	1.5	2.0
46	30	146	36	16	9
48	30	149	37	16	9
50	30	153	38	17	10
52	30	156	39	18	10
54	30	159	40	18	10
56	30	162	41	19	11
58	30	166	43	19	11
60	30	169	44	20	11
62	30	172	45	20	12
64	30	175	46	21	12
66	30	178	47	21	12
68	30	181	48	22	13
70	30	184	49	22	13
72	30	187	50	23	13
74	30	190	51	24	14
76	30	193	52	24	14
78	30	196	53	25	14
80	30	199	54	25	15
82	30	202	55	26	15
84	30	204	56	26	15
86	30	207	57	27	16
88	30	210	58	27	16
90	30	213	59	28	16
92	30	216	60	28	17
94	30	218	61	29	17
96	30	221	62	29	17
98	30	224	63	30	18
100	30	226	64	30	18
102	30	229	64	31	18
104	30	232	65	31	18
106	30	234	66	32	19
108	30	237	67	32	19
110	30	239	68	33	19
112	30	242	69	33	20
114	30	245	70	34	20
116	30	247	71	34	20
118	30	250	72	35	21
120	30	252	73	35	21

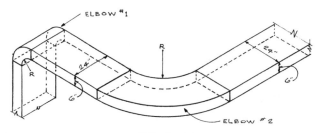

FIG. 8–11

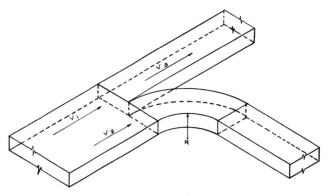

FIG. 8–12 Elbow Type Divided Flow Fitting

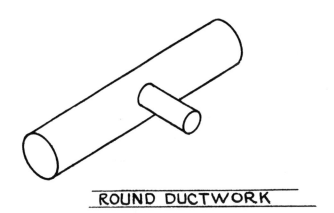

ROUND DUCTWORK

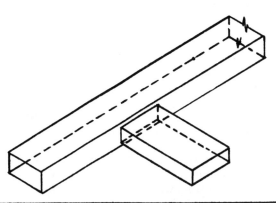

RECTANGLER DUCTWORK

FIG. 8–13 Tap-In Type Divided Flow Fitting

DIVIDED-FLOW FITTINGS

Divided-flow fittings are fittings in which the air is divided into branches. In rectangular ductwork the divided-flow fitting can take the form shown in Fig. 8–12. In this case, the loss would be the same as an elbow. The same standards apply to this type fitting as apply to elbows; if a good radius cannot be used, then use turning vanes. Be sure that $V_1 = V_2 = V_3$. The pressure drop is the same as an elbow. In round ductwork, and sometimes in rectangular ductwork, "tap in" divided-flow fittings are used (Fig. 8–13). To determine the pressure drop in these fittings use the table in Fig. 8–8.

Now, let's put this all together.

Example #4

In Fig. 8–14 we see an open-ended duct carrying 1000 cfm. Size and determine SP.

FIG. 8–14

From Fig. 8–1, 1000 cfm in a 12-in. dia. duct has a velocity of 1270 fpm and a pressure drop of 0.2 in./100 ft.

Duct Loss = 80 ft × 0.2 in./100 ft = 0.16 in.
Abrupt Exit (Fig. 8–7) C = 1.00

$$VP = \left(\frac{1270}{4005}\right)^2$$

$$PD = C \, (VP)$$

$$PD = 1.00 \left(\frac{1270}{4005}\right)^2 = 0.10 \text{ in.}$$

Total Static Pressure = Duct + Abrupt Exit
Total Static Pressure = 0.16 in. + 0.10 in. = 0.26 in.

Example #5

Same as Example #4 but with two 90° elbows (R/D = 1.5).

FIG. 8–15

Duct Loss = 80 ft × 0.2 in./100 =   0.16 in.
Elbows (Fig. 8–6) C = 0.24

$$VP = \left(\frac{1270}{4005}\right)^2$$

$$PD = 0.24 \left(\frac{1270}{4005}\right)^2$$

= 0.024 in. Each Elbow × 2 Elbows = 0.048 in.

| Abrupt Exit | 0.10 in. |
| Total Static Pressure = | 0.308 in. |

Very often, and you might say most often, a duct has more than one opening and/or branch, which makes the duct sizing somewhat complicated. The duct sizing is the same, but adjusting the duct sizing so that the air quantities are balanced is a little more difficult. Air, like people, seeks the path of least resistance.

Suppose a system has two branches, one with a 0.1-in. SP and the other with 0.01-in. SP. Further, each branch requires a 1000 cfm. Is it possible? No! Because again the air will seek the path of least resistance. Actually, more air will go out the 0.01 branch and less out the 0.1 branch until the pressure drops balance at some pressure between 0.1 and 0.01, depending on the fan characteristics. If they balanced at 0.05, then the air quantity in the 0.1 branch air would reduce from 1000 cfm to 630 cfm, and the 0.01 branch air quantity would increase from 1000 cfm to 2000 cfm.

Now, it is possible to install a damper behind the register and add 0.09-in. pressure so that the two branches balance. If you had only a few registers to worry about, then this might be an acceptable solution. But, the rule, generally, is many registers and so, many dampers to adjust, which at times is a formidable task. A better solution is to design the ductwork so that all the branches have approximately equal pressure drops.

Example #6

The first step is to size the common portion of the duct; section AB as shown in Fig. 8–16. Remembering our limits of 0.2 in./100 ft and 2000 fpm, we find in Fig. 8–2 a duct size of 22-in. dia. As shown in the following table, the pressure drop for section AB is 0.10 in. Next, decide which branch to size first. It is generally better to size the shortest run first. In this way, the longer run will be certain to be at a velocity that is less than the limit of 2000 fpm. If the longer run is sized first at a friction as close to 0.2 in./100 ft as possible, then the shorter run will have to be sized at a friction of more than 0.2 in./100 ft. This may put you at a velocity higher than 2000 fpm. For example, if a 100-ft-long, 2000 cfm branch is sized at 0.2 in./100 ft, then to balance a 50-ft-long, 3000 cfm branch, a friction of 0.4 in./100 ft must be used.

$$L_1 H_{f1} = L_2 H_{f2}$$

$$100 \text{ ft} \left(\frac{0.2 \text{ in.}}{100 \text{ ft}}\right) = 50 \text{ ft} \left(\frac{0.4 \text{ in.}}{100 \text{ ft}}\right)$$

$$0.2 \text{ in.} = 0.2 \text{ in.}$$

But, the velocity in a duct carrying 3000 cfm at a friction of 0.4 in./100 ft would be 2200 fpm, which is over our limit.

We size branch BC first because it is the shortest run. As you can see, 0.2 in./100 ft puts us between 18 in. and 20 in. at 18.2 in. An 18-in. duct would be at

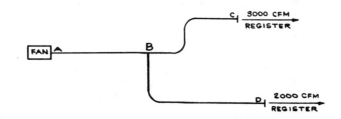

Duct section	CFM	Duct size in.	Length ft	Fittings	Friction per 100 ft	Velocity	Velocity pressure	Pressure drop	Accum. P. D.
AB	5000	22			0.20	1900		0.100	0.10
BC	3000	18	40		0.21			0.082	
				2 – 90° elbows Loss = 2 (.24)VP		1650	0.170	0.082	
				Register				0.050	
								0.214	0.314
BD (1)	2000	18	80		0.10			0.080	
				1 – 90° takeoff (Fig. 8-8)				0.110	
				1 – 90° elbow Loss = 1 (0.24)VP		1160	0.085	0.020	
				Register				0.050	
								0.260	0.360
BD (2)	2000	20	80		0.06			0.048	
				1 – 90° takeoff (Fig. 8-8)				0.080	
				1 – 90° elbow Loss = 1 (0.24)VP		940	0.055	0.013	
				Register				0.050	
								0.191	0.291

Register (From Manufacturer's Catalog)
Total Static Pressure

FIG. 8-16

0.21 in./100 ft and 1650 fpm. It is wiser to exceed the 0.20 slightly then to go to the 20-in. diameter duct.

Section BD must have a ductwork pressure drop of 0.164 in. to balance BC. Now BD is 80-ft long. But, we have fittings which cause a pressure drop. We could assume that the fittings would be equal to 80 ft of straight duct and solve for a trial friction.

$$\frac{0.164 \text{ in.}}{80 \text{ ft} + 80 \text{ ft}} = \frac{0.164 \text{ in.}}{160 \text{ ft}} = \frac{0.10 \text{ in.}}{100 \text{ ft}}$$

The total pressure drop, using 0.10 in/100 ft, was found to be 0.22 in., which is above the desired 0.164 in. The next larger duct, 20 in., is tried. This gives a loss of 0.148 in. — which is close enough to balance. To get the total friction or static pressure, we add the AB loss to BC or BD, whichever is greater, and this yields 0.291 in.

Now, this system works find for branches, but how do we balance a string of registers all installed in a single main?

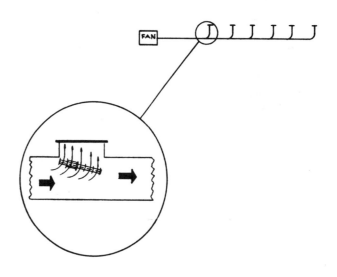

FIG. 8–17

As you can see (Fig. 8–17), there isn't enough room between the register and the main to do any adjusting of the duct size. Here we make use of static regain — the conversion of velocity pressure to static pressure. But, before we do, let's once again discuss total pressure, velocity pressure, and static pressure. In order for air to flow, there must be an increase in pressure. It's like the ball on the track in Fig. 8–18. If the track is flat, the ball won't roll. But, if we elevate one end of the track or give the ball some head, it will roll.

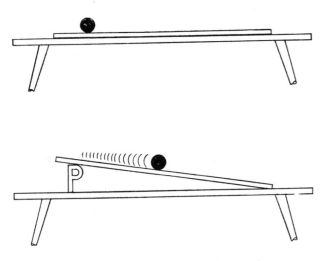

FIG. 8–18 Illustration Of Head

Just as we increased the head above the datum plane (the table), if we increase or decrease the air pressure above or below the datum plane (atmospheric pressure), air will blow. In other words, if we push or pull, the air will flow. This head is called total pressure and has two components, velocity pressure and static pressure. If we have a fan blowing into a duct with an open end, we can plot the pressures along the duct as shown in Fig. 8–19.

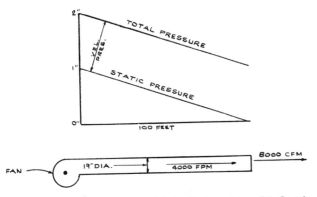

FIG. 8–19 Illustration Of Relationship Of Static Pressure, Velocity Pressure And Total Pressure

At 4000 fpm the velocity pressure is 1 in. Also, a 19-in. dia. duct, carrying 8000 cfm at 4000 fpm, has a friction of 1 in. per 100 ft. At the fan outlet, the total pressure is 2 in., 1-in. velocity pressure and 1-in. static pressure. As the air travels down the duct, the static pressure is used to overcome the friction and at the end of the duct is all consumed, leaving only the 1-in. velocity pressure.

As we stated earlier, the static pressure is the pressure which tries to burst the duct. It is measured by a manometer connected to a carefully drilled and reamed hole in the side of the duct. The velocity pressure is the pressure needed to accelerate the air to the desired velocity. IT CANNOT BE MEASURED DIRECTLY.

At 90° to the air flow only static pressure is measured, as stated earlier. Anywhere else, the pressure measured will have velocity components along with the static. Remember that the static exerts itself in all directions equally, while the velocity pressure varies from 0 to 90° to flow, to a maximum in the direction of flow. To determine velocity pressure, measure total pressure and subtract from it the measured static pressure, as shown in Fig. 8–20.

A pitot tube (Fig. 8–21) is able to sense both total pressure and static pressure. As you can see, it is a tube within a tube. The open end senses total pressure and the holes in the sides sense static pressure. The tube sensing total pressure is connected to one leg of a manometer or inclinded gauge (Fig. 8–22), and the tube sensing static is connected to the other leg. The pressure differential on the gauge or manometer is the velocity pressure. This can be converted to velocity. Some inclined gauges are arranged to read both pressure and velocity.

Now, if the end of the duct is sealed, the total pressure is still two inches, but because there is no flow, there can be no velocity pressure. Therefore, all of the pressure is static pressure. Further, if the ductwork is removed from the fan, there is no static pressure

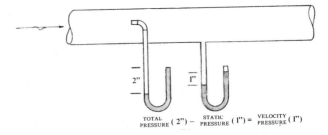

FIG. 8–20 Diagrammatic Illustration Of Measurement Of Total Pressure, Static Pressure And Velocity Pressure

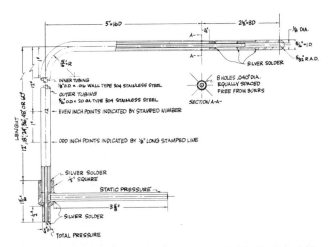

FIG. 8–21 Pitot Tube (Courtesy: F. W. Dwyer Mfg. Co., Inc.)

FIG. 8–22 Manometer Type Air Meter (Courtesy: F. W. Dwyer Mfg. Co., Inc.)

component and the total pressure equals the velocity pressure.

In any two points in a duct system, the total pressure at point 1 equals the total pressure at point 2 + losses from point 1 to 2. For example: A 19-in. dia. duct carrying 8000 cfm at 4000 fpm has a friction of 1.00 in./100 ft. In a 100-ft-long duct, the total pressure at the fan discharge would be 1-in. VP + 1-in. SP = 2-in. TP. Now 10 ft downstream, the total pressure would be equal to 1-in. VP + 0.9-in. SP = 1.9-in. TP. Or, the total pressure has been diminished by the friction loss (0.10 in.) between points 1 and 2.

Now, take Fig. 8–23 and let's see what happens with changes in velocity.

The total pressure, you will remember, is 2 in. at the fan discharge. If point 1–1 is 25 ft from the fan, then the total pressure at this point is less than the total pressure at the fan discharge by the friction to point

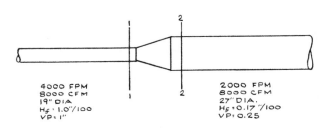

FIG. 8–23 Relationship Of Velocity Changes And Pressure Changes

1–1, or 2 in. − 25 ft $\left(\dfrac{1 \text{ in.}}{100 \text{ ft}}\right)$ = 1.75 in. At this point, VP = 1.00 in. and SP = 0.75 in. At point 2–2, the velocity pressure = $(2000/4000)^2$ = 0.25 in. The total pressure is reduced by the loss in the gradual enlargement. Notice in Fig. 8–7 that the C factor is shown as C_r, which is the fraction of the loss of an abrupt expansion. $A_1/A_2 = 1/2 = 0.5$ and $C_2 = 1.00$. The loss for an abrupt expansion is 1 VP_2 or 1(0.25) = 0.25 in. But the fraction of this, for a gradual enlargement of 30° included angle, is 0.59. So, the loss in the gradual enlargement is 0.59 (0.25 in.) = 0.15 in. And at point 2–2, the total pressure is diminished by 0.15 in. The total pressure is 1.75 in. − 0.15 in. = 1.60 in. Now, the velocity pressure at 2–2 is 0.25 which, when subtracted from the total pressure, leaves 1.60 − 0.25 = 1.35 in. for SP. We have increased our static pressure from 0.75 in. to 1.35 in. We have regained 1.35 in. − 0.75 in. = 0.60 in. to use to offset friction and turbulence.

Duct section	CFM	Duct size, in.	Velocity fpm	H_v	H_f in./100 ft	Length ft	Loss	Gain	Difference	Net
Fan–A	3500	20	1600	0.160	0.170	50	−0.085″	0	−0.085″	−0.085″
AB	3000	20	1380	0.119	0.130	15	−0.020″	+0.020″	0	−0.085″
BC	2500	20	1150	0.082	0.090	15	−0.014″	+0.019″	+0.005″	−0.080″
CD	2000	20	920	0.053	0.060	15	−0.009″	+0.015″	+0.006″	−0.074″
DE	1500	18	850	0.045	0.058	15	−0.009″	+0.004″	−0.005″	−0.079″
EF	1000	16	725	0.033	0.046	15	−0.007″	+0.006″	−0.001	−0.080″
FG	500	12	640	0.026	0.055	15	−0.008″	+0.004″	−0.004″	−0.084″

* A 0.50 regain factor was used.

Fig. 8–25

Now, how can we make use of this to balance our ductwork system? In Fig. 8–24 we take a section out of a ductwork system and size it.

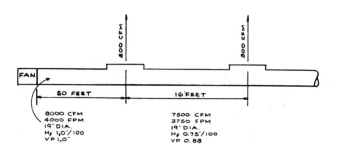

FIG. 8–24 Illustration Of Simple Static Regain Duct Sizing

At point A, 500 cfm leave the system and if the same size duct is used between A and B, the velocity will drop to 3750 fpm which has a VP of 0.88. The total pressure at point A does not change, but the velocity pressure has been reduced by 0.12 in., so the static pressure must increase by 0.12 in. We now have an excess of static pressure to the tune of 0.12 in. What can we use it for? To offset the friction between points A and B, which just happens to be 16 ft × 0.75 in./100 ft = 0.12 in. So, you can see that if the velocity pressure gained equals the static pressure, the two registers are balanced. This can be repeated again and again until the registers are all balanced. Now, we assumed that all of the velocity pressure difference can be recovered but, unfortunately, only a part can be recovered. It is possible to recover 80% of the difference, but it is wise to use 50%.

Now, let's try a complete ductwork system (Figure 8-25). As you can see, the difference in static pressure, between the highest and the lowest, is .011''. This difference is easily handled with dampers, with no noise problems.

Return air duct systems are sized in a slightly different manner than supply systems. Certainly the friction and dynamic losses are the same in both systems, but the relationship between static and velocity pressure has a different impact. You will remember that, when we sized supply duct systems, we converted velocity pressure into static pressure every time we reduced velocity. Nothing was said about what happens when the velocity increases as it most often does in return or exhaust systems. Whether you increase velocity by pushing air, as we do in a supply system, or by pulling air, as we do in a return duct, the results are the same. When the velocity is increased, all the difference in velocity pressure causes an increase in static pressure.

If the velocity increases, the C factor for a fitting can be greater than 1. This is due to the fact that, in addition to dynamic loss, there is an increase in velocity pressure which becomes static pressure.

Return or exhaust systems can be balanced either by the system design or by the use of dampers. Using dampers is the easiest method but, as we stated when discussing the design of supply systems, balancing a large system by adjusting dampers is not easy. In addition, dampers may cause noise problems.

If you intend to use dampers to balance a system, the dampers should not be placed behind the grille, but four feet away from the grille. In addition, the duct should be lined with sound insulation from the grille to the damper.

A balanced return duct system can be designed by varying duct sizes and air quantities, as is done in industrial exhaust systems. Table 8-1 is taken from the Industrial Ventilation Manual published by the American Conference of Industrial Hygienists. We will use this chart, instead of Figure 8-8, for return air systems. Whenever two ducts meet at a junction, the two static pressures are compared. If the difference between the two pressures is 5% or less, no adjustment is made. If the difference is greater than 20%, then one of the ducts is resized. If the difference is greater than 5%, but less than 20%, the air volume is increased using this formula:

$$cfm_2 = cfm_1 \sqrt{\frac{SP_2}{SP_1}}$$

Example #7

Two 14'' diameter branches must carry a minimum of 1400 cfm. Branch A requires a SP of 0.3 to offset friction and dynamic loses, while branch B requires 0.282. Compare the two and make changes as required.

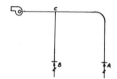

$$\% \text{ Difference} = 100 \frac{(SP_2 - SP_1)}{SP_2}$$

$$\% \text{ Difference} = 100 \frac{(0.3 - 0.282)}{0.3}$$

$$\% \text{ Difference} = 4.33$$

This is less than 5%, so we do nothing. The reason we do nothing is because the increase in air quantity will be the square root of 4.33% or 2.08%. It is unlikely that this increase will be noticeable. In fact, I doubt that you would be able to measure the difference.

Perhaps you are saying to yourself, "But why does the air quantity in branch B increase?" Look at the sketch above. At point C, where the two branches meet, the static pressure is 0.3. Let's assume that the loses from point C to the fan require a static pressure of 0.2, for a total suction static pressure of 0.5. The fan creates a suction pressure of 0.5 at the entrance to the fan. You could prove this to yourself by measuring the static pressure at this point. If you measured the static pressure at point C, you would find that it is 0.3. Please bear in mind that, on the suction side of the fan, all static pressures are below atmospheric and should carry a negative sign. I ignore the sign because our interest is in the absolute value. Just in case you have forgotten, an absolute value is a number without a sign.

Now, we have static pressure of 0.3'' at point C. Branches A and B are parallel, so each has 0.3'' SP available to offset friction and dynamic losses. While branch A needs 0.3'' SP, branch B needs only 0.282. This means we must either add or adjust a damper, to add 0.018'' in dynamic loss, or we must have the air move faster to increase the loss. If the duct size does not change, then the air must move faster. The air can move faster only if the quantity is increased from 1400 to 1431 cfm. Let's look at another example.

Example #8

Two branches each need to carry a minimum of 1400 cfm. Branch A requires a 0.3'' SP, while branch B requires 0.25''.

$$\% \text{ Difference} = 100 \frac{(0.3 - 0.25)}{0.3}$$

$$\% \text{ Difference} = 16.67$$

This is more than 5%, but less than 20%, so we must adjust the air quantity in branch B.

$$cfm_2 = cfm_1 \sqrt{\frac{SP_2}{SP_1}}$$

$$cfm_2 = 1400 \sqrt{\frac{0.3}{0.25}}$$

$$cfm_2 = 1534$$

This means that, from point C to the fan, the duct must carry 2934 cfm, not 2800 cfm, and that the fan must be selected for 2934 cfm.

Now, don't forget to check the new velocity in branch B to be sure that you haven't exceeded whatever velocity limit you have set. Instead of increasing the air quantity in branch B, you could have decreased the air quantity in branch A, as follows:

$$cfm_1 = cfm_2 \sqrt{\frac{SP_1}{SP_2}}$$

$$cfm_1 = 1400 \sqrt{\frac{0.25}{0.3}}$$

$$cfm_1 = 1278$$

Now we will do the complete return air system shown in Table 8-2 . The calculations are shown below the illustration.

We start by sizing the branch furthest from the fan. The loss from point A to B is 0.184.

You may be wondering where we got the grille loss. We got it from the manufacturer who told us that the dynamic loss is 0.02''. However, we must add one velocity pressure to this loss. Bear in mind that the room is still at atmospheric pressure, so its velocity pressure is zero. As we pull this air into the grille and on into the duct, we speed it up to the duct velocity. We must add this as an increase in static pressure. So, the total grille loss in section AB is 0.12.

Next, we size branch GB. This duct joins section AB through a 45° branch fitting. The branch duct connects to the 45° branch fitting with a 45° elbow. The branch loss is 0.18. When we compare branch AB with branch GB, we find that the static pressure difference is 2.2%. Therefore, we make no adjustment in the branch air quantity.

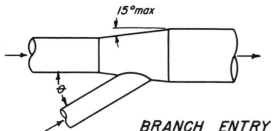

Table 8-1

BRANCH ENTRY LOSSES

Note: Branch entry loss assumed to occur
in branch and is so calculated.

Do not include an enlargement regain
calculation for branch entry enlargements.

Angle θ Degrees	Loss Fraction of VP in Branch
10	0.06
15	0.09
20	0.12
25	0.15
30	0.18
35	0.21
40	0.25
45	0.28
50	0.32
60	0.44
90	1.00

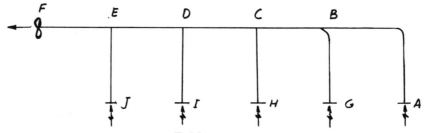

Table 8-2

Duct section	CFM	Duct size in.	Velocity fpm	H_v	H_f in./100 ft	Length ft	Fittings	Losses	Section loss	Accum. loss	Remarks
AB	1000	12	1275	0.10	0.200	20		0.040			
							1 elbow	0.024			
							1 grille	0.120	0.184	0.184	
GB	1000	12	1275	0.10	0.200	10		0.020			2.2% difference. No cfm adjustment required.
							1 45° branch	0.028			
							1 45° elbow	0.012			
							1 grille	0.120	0.180	0.184	
BC	2000	16	1450	0.13	0.175	10		0.018			0.03 SP required. VP increased from 0.10″ to 0.13″.
								0.030	0.048	0.232	
HC	1000	12	1275	0.10	0.200	10	Same as GB		0.180		Over 20% difference, resize.
HC	1000	12	1275	0.10	0.200	10		0.020			2% difference. No cfm adjustment.
							1 90° branch	0.010			
							1 grille	0.120	0.240	0.240	
CD	3000	20	1400	0.12	0.13	10		0.013			
							Regain	0.005	0.008	0.248	
ID	1000	12	1275	0.10	0.200	10	Same as HC		0.240		3.3% difference. No cfm adjustment.
DE	4000	22	1525	0.14	0.14	10		0.014			0.02 SP required. VP up from 0.12″ to 0.14
								0.020	0.034	0.282	
JE	1000	12	1275	0.10	0.200	10	Same as HC		0.240		14.9% difference. Cfm increases to 1084
EF	5084	24	1625	0.16	0.14	10		0.014			0.02 is the SP increase because VP increased from .14 to .16.
								0.020	0.34	0.316	

The loss of 0.048″ in the main section BC is added to the larger branch loss, 0.184″, for a total of 0.232″. Notice that the velocity increased to 1450 fpm with an accompanying velocity pressure of 0.13″. This is 0.03″ greater than the velocity pressure in AB and so it becomes an increase in static pressure.

Branch HC is next. We first use a 12″ duct and, of course, the loss is the same as section BC or 0.18″. When we compare this to main section BC, we find that BC exceeds GB by 22%. This means that the duct must be resized. We could try a 10″ duct, along with a 45° branch connection, and would find that the loss is 0.364″. A better solution is to stay with the 12″ duct and change the branch connection from a 45° to a 90° angle. The 90° branch connection has a loss equal to one velocity pressure, giving us a section loss of 0.24″. This is 2.2% more than the main loss but, because this is less than 5%, we do not need to adjust the cfm. The main static pressure is now 0.24″.

I appear to be using the terms loss and static pressure interchangeably. Remember that static pressure is used to offset friction and dynamic losses, so I use the terms interchangeably.

Main section CD adds another 0.008″ to the main for a new total of 0.248″. Next on the list is is branch ID and we try the same setup as section HC. When we compare this with the main, we find that its static pressure is 3.3% less than the main and that no cfm adjustment is needed.

When we add 1000 cfm to the 3000 cfm in main CD, we get 4000 cfm in main section DE. The loss in this section, including the velocity pressure increase of 0.02″, is 0.034″. Adding this to the loss accumulated to point D, we get a new static pressure of 0.282″.

Branch JE is designed in the same fashion as branch HC. The difference at point E is 14.9%, so we have an increase in branch cfm to 1084 cfm.

Finally, we complete the system by sizing main section EF. The total suction loss is 0.316″. This means that a suction static pressure of 0.316″ is needed to offset the losses. This suction static pressure is added to the supply static pressure to get the total static pressure that the fan must develop.

Now, what about fans and fan laws? What follows is a brief discussion of fans. The reader should read a book like Professor Berry's "Flow and Fan,"* or "Fan Engineering,"† for a complete discussion of the subject.

Fans come in a variety of types. All types fall into either of two categories: centrifugal or axial. All fans are, in effect, buckets or spoons that grab a bucket or spoonful of air and throw it, and so move it along. A simple fan (Fig. 8–26) could consist of a number of spoons mounted on a shaft. The air wouldn't go far. As you can see, it would spill all over. But, we can improve the situation by two methods. First, we can wrap an enclosure around the spoons (Fig. 8–27) so that the air can come out of the spoons at one place only. This will be a simple centrifugal fan.

In Fig. 8–28 we see another way to improve the performance of our spoon machine. Tilt the spoons a bit and the air is pulled through our machine along the axis of the hub, which is similar to the way an airplane propeller moves air. This is called an axial-flow fan.

*C. Harold Berry, *Flow and Fan* (New York, 1963) Industrial Press.
†Richard D. Madison, ed., *Fan Engineering* (Buffalo, N. Y., 1949) Buffalo Forge Co.

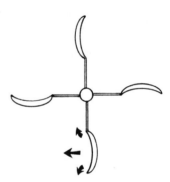

FIG. 8–26 Simple Fan

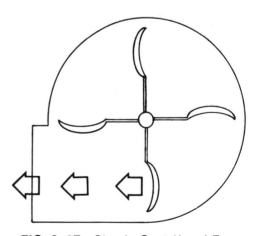

FIG. 8–27 Simple Centrifugal Fan

Early centrifugal fans consisted of a series of flat blades mounted about a shaft. These are still used in material handling. They are called paddle wheel fans. Refinements are the forward-curved and backward-curved type. A forward-curved fan wheel is shown in Fig. 8–29. A recent development is the backward-curved airfoil-type blade, shown in Fig. 8–30. The backward-curved fan has the highest efficiency, is non-overloading, easy to keep clean, and best suited for industrial work. Disadvantages are the higher cost of the backward-curved fan and the large physical size. The forward-curved fan is inexpensive and small in size and so ideally suited for commercial and institutional comfort applications. The big problem is that if the static pressure is overestimated, the fan will compensate by trying to push more air and so overload the fan motor.

The axial-flow fan in its simplest form is a device similar to any airplane propeller mounted in a frame. A good example is the desk-type fan. Variations are the tube-axial (Fig. 8–31), in which a propeller fan is mounted in a short section of round duct, and the vane-axial (Fig. 8–32), which is a propeller fan in a duct with inlet vanes. Some axial-flow fans are built with variable pitch blades. The blade pitch can be varied to meet the system conditions.

FIG. 8–29 Forward Curved Blade (Courtesy: American Standard, Industrial Div.)

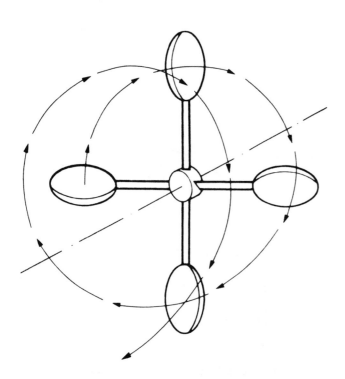

FIG. 8–28 Simple Axial Fan

FIG. 8–30 Backward Curved Air Foil Blade (Courtesy: American Standard, Industrial Div.)

There are three fan laws that are important to the designer of environmental systems. For a fixed fan size, constant air density and constant system:

cfm varies as fan speed

$$\frac{Q_1}{Q_2} = \frac{RPM_1}{RPM_2}$$

Pressure varies as square of fan speed

$$\frac{P_1}{P_2} = \left(\frac{RPM_1}{RPM_2}\right)^2$$

Power varies as cube of fan speed

$$\frac{BHP_1}{BHP_2} = \left(\frac{RPM_1}{RPM_2}\right)^3$$

How are fan curves drawn and what is a system curve? The fan curve is a graphic representation of the performance of a fan under certain conditions. The fan to be tested is mounted on a test stand or dynamometer so that the horsepower can be measured. The fan pulls out of the whole room, and ductwork is added to the fan discharge. Enough ductwork is installed to assure uniform air flow over the complete duct cross section. This is generally a length equal to ten diameters.

The fan is run at a specific speed. The end of the duct is wide open. For all practical purposes, this is a 0 static pressure, or wide-open condition. A pitot tube is used to traverse the duct in two directions, 90° to one another. Ten readings are taken in each direction. The average of these velocities, multiplied by the duct area, gives the air quantity in cfm. Then the horsepower is measured on the dynamometer. We now have cfm, SP, and HP. We now have our first points for plotting. A cfm-SP curve and cfm-HP curve is started.

A plate is placed on the end of the duct (Fig. 8–33). This adds some resistance, giving a reduced flow. Again we measure SP, cfm, and HP and establish a second point on our curve. The plate is moved closer to the duct end, giving additional resistance, and again cfm, HP, and SP are read. This is repeated until the damper completely closes the duct. About ten readings are taken, which are enough to develop a cfm-SP curve, sometimes called Q-H curve. Also, a cfm-HP curve is developed. Fig. 8–34 shows a typical curve developed from a test. Let's say you wished 16,000 cfm at 2-in. SP. Follow 16,000 cfm up from the bottom at 2-in. SP across from the left. At their intersection, read RPM, in this case, 600. Now drop straight down again and where the cfm line intersects the HP @ 600 RPM curve, go horizontally to the right margin and read 6.5 HP. This test can be repeated at various RPM to get a family of curves.

The cfm-SP curve is often called the fan-character-

FIG. 8–31 Tube Axial Fan (Courtesy: Buffalo Forge Co.)

FIG. 8–32 Vane Axial Fan (Courtesy: Buffalo Forge Co.)

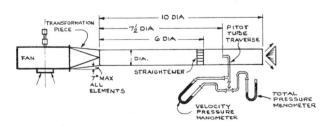

FIG. 8–33 Diagram Of Fan Test Setup

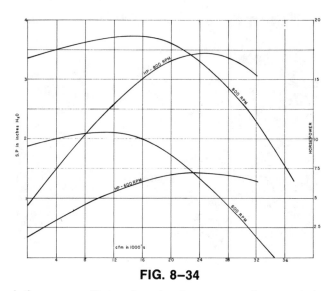

FIG. 8–34

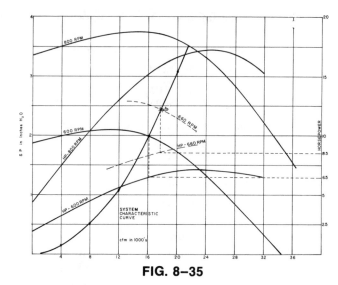

FIG. 8–35

istic curve. But, what is the system-characteristic curve? Any air-handling system has a characteristic curve whose path is determined by the relationships:

$$\frac{SP_2}{SP_1} = \left(\frac{cfm_2}{cfm_1}\right)^2$$

$$SP_2 = SP_1\left(\frac{cfm_2}{cfm_1}\right)^2$$

$$\frac{RPM_2}{RPM_1} = \frac{cfm_2}{cfm_1}$$

$$SP_2 = SP_1\left(\frac{RPM_2}{RPM_1}\right)^2$$

If you have a system in which you wish to move 16,000 cfm and you estimate that the SP is 2 in., the 16,000 cfm and 2-in. SP give you the first point on your curve. To develop your curve, just use the above relationships and find your points. For instance, you could increase the air quantity to 20,000 cfm and calculate the SP required:

$$SP_2 = SP_1\left(\frac{cfm_1}{cfm_2}\right)^2$$

$$SP_2 = 2\left(\frac{20,000}{16,000}\right)^2$$

$$SP_2 = 3.13 \text{ in.}$$

This process could be repeated until you had enough points to plot a curve:

cfm	SP
4,000	.125
8,000	.5
12,000	1.12
16,000	2
20,000	3.13

Plotting these points, we have our system-characteristic curve, which on Fig. 8–35, you will see, intersects the 600 RPM curve at 16,000 cfm, 2-in. SP,

and the 800 RPM curve at 21,350 cfm and 3.55-in. SP.

What does this mean to you? It means that if you speed up or slow down your fan, the capacity of your fan will vary along this curve. As much as we may wish otherwise, this is the only path. Let's say that you are at 16,000 cfm, 2-in. SP, 600 RPM, and 6.5 HP. Further, assume you purchased a 7.5 HP for your fan and you need 10% more air. Just a "stinkin" little 10%. You need only speed up 10% to 660 RPM to get 10% more air, but your static pressure is now 2.42 inches. Boy! Did that jump! But look, your HP is now 8.5 and your 7.5 HP motor is overloaded. It would have been nice if 17,600 cfm would have put us at 2.2 SP and 625 RPM, which would have been 7.5 HP. BUT, NO MATTER HOW WE WISH, WE MUST FOLLOW THE SYSTEM-CHARACTERISTIC CURVE!!!

Probably the idea that we can ignore system-characteristic curves stems from the fact that we generally select fans from a table (Fig. 8–36) and not a curve. From looking at the table you might easily assume that you would get 17,870 cfm at 2 inches. The new speed would be 624 RPM and the new HP would be 7.41. But if we use the fan laws we find:

$$RPM_2 = 600\,\frac{17,870}{16,000}$$

$$RPM_2 = 670$$

$$SP_2 = 2\left(\frac{17,870}{16,000}\right)^2$$

$$SP_2 = 2.5 \text{ in.}$$

$$HP_2 = 6.5\left(\frac{17,870}{16,000}\right)^3$$

$$HP_2 = 9.03$$

FIG. 8–36 Sample Fan Rating Table

Volume	Outlet Velocity	2 in SP		2.25 in SP		2.50 in SP		3.00 in SP	
		RPM	BHP	RPM	BHP	RPM	BHP	RPM	BHP
16,000	1700	600	6.5	622	7.27	647	7.47	694	9.06
16,935	1800	606	6.82	631	7.65	655	8.48	700	10.13
17,870	1900	624	7.41	646	8.18	668	9.03	712	10.73
18,805	2000	636	7.86	659	8.73	681	9.62	723	11.38
19,740	2100	654	8.45	675	9.31	697	10.20	737	12.02
20.675	2200	666	9.03	687	9.93	706	10.81	746	12.65

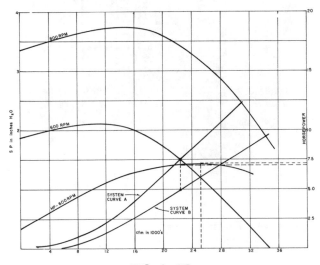

FIG. 8–37

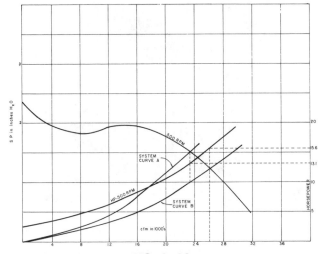

FIG. 8–38

But, what if you protected yourself by overestimating your SP? What would happen? Let's say you want 23,400 cfm and you add .5 to the 1-in. static. On the fan curve (Fig. 8–37), find 23,400 cfm and 1.5-in. static pressure, which hits on the 600 RPM curve. System curve A is drawn through this point. But, where will you actually be? Drop straight down from 23,400 and 1.5 to 23,400 and 1 in. Draw system curve B through this point, and where system curve B intersects 600 RPM, read 25,500 cfm, 1.2-in. SP, and 7.2 HP. The original HP was 7.2, so all that happens with a backward-curved fan is that we handle more air.

What would have happened with a forward-curved fan? Fig. 8–38 shows a typical forward-curved fan characteristic curve. Notice the dip in the curve. This is characteristic of forward-curved fans. Repeating the procedure we used with the backward-curved fan, we find an increase in air to 26,000 cfm. But, notice the HP. It has risen from 13.1 to 15.6. Just a little more and your motor will be overloaded. In this case, it isn't too serious, because we can slow the fan down to 450 RPM.

But, serious problems can occur in systems where the air quantity varies. For example: In a double duct system, the air quantity in the hot and cold ducts can vary from 0 to 100%. The worst case is when each duct carries 50%. With all the air in one duct we could have a condition of 21,500 cfm at 5.4-in. SP. The two-duct portion has a loss, let's say, of 4.4-in. SP while the common portion has 1-in. SP. When the system goes to the 50-50 condition, then the common portion stays at 1 in., but the 4.4 in. becomes:

$$SP = 4.4 \left(\frac{1}{2}\right)^2$$
$$SP = 1.1 \text{ in.}$$

Add 1 in. to 1.1 in. and get 2.1-in. SP.

This situation is diagrammed in Fig. 8–39. Using the same procedure we used previously, we find that we have 29,600 instead of 21,500 cfm. But, look at the horsepower. It has gone from 28 HP to 40 HP. When using a double-duct system you can see that a backward-curved fan is preferred because of its non-overloading characteristic. In either case, damper devices to keep the static constant are very important.

Earlier in our discussion we talked about elbows and divided-flow fittings. We stated that unless an elbow or divided-flow fitting could be built with a radius ratio of at least one, turning vanes should be used. The vanes can be job fabricated to the specifications shown in Fig. 8–40, or can be purchased prefabricated, as shown in Fig. 8–41. In Fig. 8–42 we see the improvement in performance through the use of turning vanes. In the duct without vanes, notice the turbulence and the way the air piles up on one side. The elbow with turning vanes has a smooth even flow.

When rectangular tap-in fittings are used, it is well to use deflecting-type turning vanes of the kind shown in Fig. 8–43. These are important to reduce pressure drop. But, there is an even more important reason. Often, registers or grilles are mounted right on the

FIG. 8–41 Prefabricated Turning Vanes (Courtesy: Barber-Colman Co.)

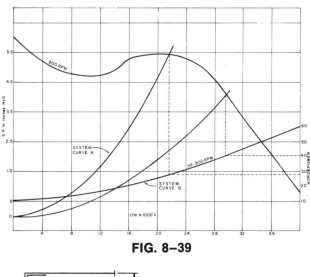

FIG. 8–39

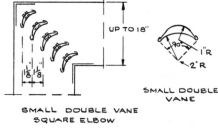

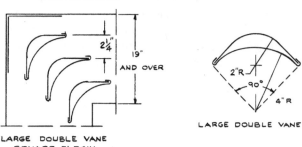

FIG. 8–40 Drawing Of Field Fabricated Turning Vane

FIG. 8–42 Comparison Of Mitered Elbow With And Without Turning Vanes (Courtesy: Barber-Colman Co.)

side of a duct. Without the turning vanes, all of the air piles up on one side of the register, causing uneven distribution and noise. Let's say you selected a register with a face velocity of 600 fpm. At 600 fpm the noise level is 23 DB measured on the A scale. (See Chapter 13.) But, if all the air comes out of 1/2 the register, the velocity rises to 1200 fpm and the noise level is now 49 DB. This is a serious noise increase.

Here we are talking about registers and grilles and I haven't distinguished between them yet. The fan provides the head to move the air, the ductwork carries it, and along the way there are the terminal units through which air is introduced into the space. A grille is the terminal used to introduce the air horizontally from the side of a duct or from a wall. Grilles can also be placed in the floor to introduce the air

vertically up into the space. Typical grilles are shown in Figs. 8–44, 8–45, and 8–46. When is a grille a register? The moment you put a damper behind a grille to adjust the air quantity, you have a register. These dampers can be single-leaf or opposed-blade type (Fig. 8–47). The opposed-blade type is more desirable because of its better throttling characteristics.

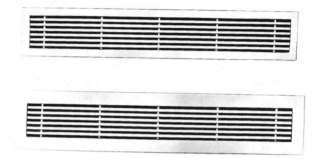

FIG. 8–45 Extruded Aluminum Grille (Courtesy: Carnes Corp.)

FIG. 8–46 Fin-Type Grille (Courtesy: Barber-Colman Co.)

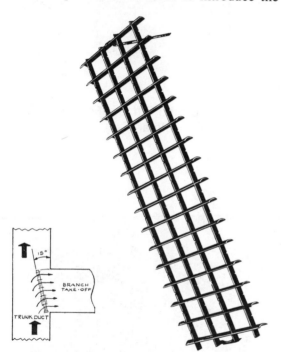

FIG. 8–43 Tap-In Fitting Turning Vanes (Courtesy: Barber-Colman Co.)

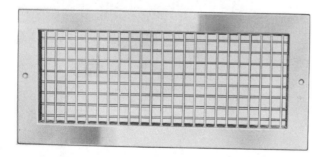

FIG. 8–44 Typical Bar Grille (Courtesy: Carnes Corp.)

FIG. 8–47 Opposed Blade Damper (Courtesy: Barber-Colman Co.)

Grilles can be constructed with horizontal bars only, vertical bars only, or both horizontal and vertical bars. These can be adjustable or fixed. Sheet aluminum or steel is used to manufacture grilles. These are generally only used for wall or duct mounting. Floor grilles are generally constructed of extruded aluminum shapes for greater strength. These can also be used in wall applications. Baseboard diffusers (Fig. 8–48) are also used for low air introduction.

Ceiling outlets are called diffusers. They can be round, square, or rectangular. They can be built as concentric rings, perforated plates, or louvers. These are shown in Figs. 8–49 through 8–56. Diffusers have adjustable air patterns and can introduce air horizontally along the ceiling or can project the air straight down. Normally, they are used with a horizontal throw. Diffusers are built that act as both supply or return outlets. A typical one is shown in Fig. 8–52. A recent development is the combination flourescent light fixture and diffuser, shown in Fig. 8–54. This unit not only acts as a diffuser, but air is returned through the ends. The return air passes through the lamp compartment, picking up the lamp heat. Diffusers should be used with opposed-blade dampers and turning vanes.

Manufacturers of air outlets give, in their literature, rules for applying these outlets. Consult this literature, for there is much to be learned from it. I will present a few rules as a preliminary lesson. In the use of air outlets, we are concerned with the amount of air, the distance it must be thrown, the noise level, and the pressure drop. The noise level is the most difficult problem. Noise is discussed in Chapter 13, but I would urge you to be sure you understand what the manufacturer means by his ratings.

A 10 ft × 30 ft × 8 ft 6-in. room requires 450 cfm and side wall outlets are to be used. How many outlets should we use? If we use a grille with a straight throw, the distance between outlet centerlines should

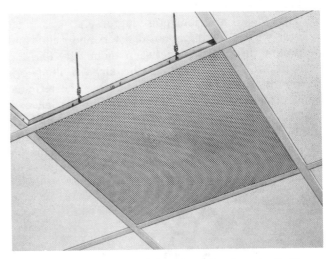

FIG. 8–49   Perforated Diffuser For Lay-In Ceilings (Courtesy: Barber-Colman Co.)

FIG. 8–50   Perforated Diffuser For Plaster Ceiling (Courtesy: Barber-Colman Co.)

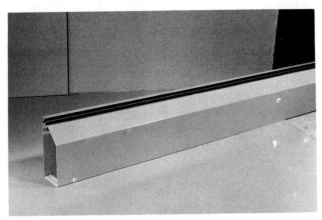

FIG. 8–48   Baseboard Diffuser (Courtesy: Carnes Corp.)

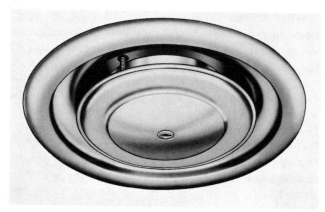

FIG. 8-51 Round Diffuser (Courtesy: Connor Engineering Corp.)

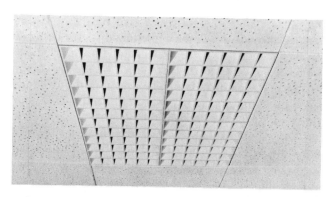

FIG. 8-54 Plastic Modular Diffuser (Courtesy: Carnes Corp.)

FIG. 8-52 Round Supply And Return Diffuser (Courtesy: Connor Engineering Corp.)

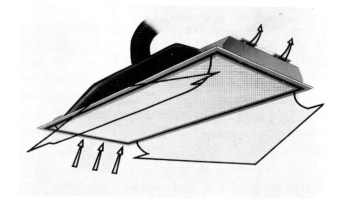

FIG. 8-55 Fluorescent Light Fixture with Diffuser (Courtesy: Sechrist Mfg. Co.)

FIG. 8-53 Round Diffuser With Light Fixture (Courtesy: Connor Engineering Corp.)

FIG. 8-56 Linear Diffuser (Courtesy: Barber-Colman Co.)

be no less than the throw divided by three. If fan-shaped throw is used, the centerline distance between outlets should be no less than the throw. The distance from the wall to the outlet should be ½ the distance between outlets. The throw is the distance from the outlet to the opposite wall. In this case, our throw is 10 ft. If a straight throw is used, the outlets must be 10 ÷ 3 or 3⅓ ft apart, which means we will need 30 ÷ 3⅓ = 9 outlets. If we use a fan-shaped throw, we need 30 ÷ 10 = 3 outlets. We will use fan-shaped outlets and save 6 outlets. Fig. 8–57 is a

OL–A Grille Co.

CFM	Throw in Feet @ 300 FPM Face Velocity				
	8	10	12	14	16
50	10 × 4 F	10 × 4 S			
75	14 × 4 F	14 × 4 S	10 × 5 S	20 × 3 S	
100	26 × 3 F	18 × 4 F	12 × 5 S		
150	40 × 3 F	26 × 4 F	20 × 5 F	16 × 6 S	20 × 5 S

F Indicates Fan-Shaped Throw
S Indicates Straight Throw

FIG. 8–57 Typical Grille Catalog

catalog sheet from a mythical manufacturer. An air quantity of 150 cfm and a 10-ft throw recommends a 26-in. × 4-in. grille with a fan-shaped throw. You will notice that 14-ft and 16-ft throws can only be accomplished with a straight throw. To get a fan shape, we must increase the velocity. The distance air can be thrown is a function of the velocity, the mass, and the shape. The larger a perimeter a mass of air has, the more erosion of the air stream there will be, which reduces the throw. A hose nozzle is a good example. The water goes much further when the water stream is in a straight pattern than when it is in a fan-shaped pattern. Oh, Yes! Mount the outlet so its centerline is at least 1 ft below the ceiling.

If we wished to select diffusers for the same room, our approach would be to try to divide the room into squares. This room can be divided into three 10 × 10 squares. In a procedure similar to the one used for sidewall outlets, we would find that we could use a diffuser with a 4-in. dia. inlet or a 6-in. Both will distribute air over a 5-ft radius, which covers our 10-ft × 10-ft square. But the 4-in. has a sound rating of 36 DBA, while the 6-in. is 30 DBA. We would select the 30 DBA diffuser.

What happens to this air after it leaves an outlet? Remember the hoze nozzle we used as an example? When water is sprayed into air you see no effect on the air around the water. But, if you can rig a hose onto your bathtub faucet you can perform an interesting experiment. Place the nozzle under the water, parallel to the surface. Turn on the water. Notice the effect the stream of water has on the whole body of water. If you can't do this, fill your kitchen sink about half way. Sprinkle some talcum powder on the surface of the water. Now, turn the water on. The straight jet of water falling straight down into the water is like an outlet throwing air into a room. Notice how water from the far corners of the sink is pulled into the stream from the faucet.

This is what happens with air. The forceful entry of air into the room creates a low-pressure area at the outlet. Room air rushes in to fill this low-pressure area. We call this induction. Two to four cfm of room air will be induced into the air stream for every one of supply or primary air. Actually, the whole volume of the room will turn over, shall we say. Now, this is true whether the air is warm or cold. Then why don't we attempt to heat from high sidewall or ceiling grilles? Because, while we can get the heat down within inches from the floor, we cannot stop that cold draft that drops down the cold wall and glass and rolls across the floor. When you heat from overhead your room thermostat will show that the room is warm. But, the room thermostat senses the temperature at the breathing level, not at ankle level. If the room you are heating has an insulated outside wall with 4% or less insulated glass, you can heat from overhead. Otherwise, use a floor or baseboard register and blow up the cold wall and glass and stop the draft.

Now, where should the return outlet be placed? You will hear arguments wax long and loud about where to place the return. The truth is that it doesn't make too much difference where you put the return. If your supply outlet is doing its job of moving and inducing air, then the return won't influence the pattern. The supply outlet is pressurizing the room and the return serves only to relieve the pressure. You would expect the return to cause its own air pattern, but it can't. Take a match and light it. Hold it three inches from your mouth—blow hard. It went out, didn't it? Now, light another match. Try to inhale and put the match out. It can't be done. The same is true of a return outlet. Let's say that in our 10-ft × 30-ft × 8-ft 6-in. room (450 cfm) we used a 14-in. × 12-in. wall return grille.

The grille face velocity is 500 fpm, while one foot from the outlet the velocity is only 50 fpm. The velocity from a supply grille at 500 fpm face velocity would be 500 fpm one foot from the outlet. You can see, therefore, why the return has little effect on air distribution.

While the induced air we have been discussing is quite necessary for comfort, it can cause problems. If this secondary air hits the wall or ceiling near the outlet, it will deposit dirt. You may have seen diffusers with a perfect dirt ring around them. This does note mean dirty filters! Again, this dirt comes from the room and is deposited on the ceiling by the induced air. On some outlets this dirt impinges on the outlet and not the ceiling. On others it will be necessary to use smudge rings. Consult your manufacturer.

Earlier we discussed opposed-blade balancing dampers. Just how do they operate? Well, if you select a register to throw 300 cfm 40 feet, you will need a face velocity of 1000 fpm. To move 300 cfm through the register sized to provide a 1000-fpm face velocity takes 0.135-in. SP. Say this is the first register in a system, and the pressure difference between the first and last register is 0.125. That means that behind the last register you have a SP of 0.135, and behind the first register the pressure is 0.26. So you must "chew up" 0.125.

In a duct with a 1000 fpm velocity, a damper has the following pressure drop for various positions:

Damper Open	SP	Velocity Through Damper
$\frac{2}{3}$	0.125 in.	1500 fpm
$\frac{1}{2}$	0.30 in.	2000 fpm
$\frac{1}{3}$	1.25 in.	3000 fpm

So, to get our 0.125 reduction in pressure, we must close the damper about $\frac{1}{3}$. This will raise the velocity through the damper to 1500 fpm. This can raise the noise level by 8 DBA. But, if it is necessary to dissipate 0.30-in. SP, then the velocity will go up to 2000 fpm. Then our noise level can go up as much as 16 DBA. This can be quite serious.

Dissipating pressure is even more difficult when a low grille or duct velocity is used. If the grille and

the damper are selected at 500 fpm, then at the various positions the SP drop and velocity will be:

Damper Open	SP	Velocity Through Damper
$\frac{2}{3}$	0.05	750
$\frac{1}{2}$	0.125	1000
$\frac{1}{3}$	0.375	1500

At the $\frac{1}{2}$ position the noise level increase is 16 DB, and to dissipate 0.125-in. SP you must put the damper at the $\frac{1}{2}$ open position. So, while balancing a 1000 fpm register by dissipating 0.125, the noise level increase is 8 DB; at 500 fpm the noise level increase is 16 DB; at $\frac{1}{3}$ open, the noise level increase can be 30 DBA or more. So you can see that you might select a register to deliver 1000 cfm at 500 fpm for a quiet room, happy in the thought that the noise level is 25 DB. But then, in balancing, you close the damper $\frac{1}{2}$, and now you have an objectionable noise. The best thing to do is be sure that the difference between the first and last outlet does not exceed 0.05 in. If this is impossible, then do your balancing as far as possible from the register. If the damper is within 10 ft of the grille, then line the duct with 1 in. of sound insulation for 3 ft following the damper.

When high-pressure systems came into being it became necessary to dissipate 1-in. SP or more. The normal damper had to be closed so much to do this that it became quite noisy. In addition, the normal duct construction was not of the caliber needed for the high pressure. So a new type of damper was developed and given a new name—the air valve. The air valve's action is nearly linear, while the damper's is parabolic. These valves are constructed for the pressures at which they operate.

Duct construction can make or break a system. A poorly constructed system can cause extra turbulence and pressure drop, noise, and air leakage. The Sheet Metal and Air Conditioning Contractors' National Association has published two manuals for duct construction—one for low-pressure ductwork and the other for high-pressure. You should acquire both manuals, study them, and be sure that your ductwork is constructed and installed in accordance with these standards.

CHAPTER 9

Water and Environmental Control

THE MOVEMENT of water is similar to the movement of air. Air and water are both fluids. Water is an incompressible fluid, while air is compressible. But, in the field of environmental control, the pressures involved in the movement of air are small, so it, too, is considered incompressible.

Two types of flow are possible in fluid flow, laminar or turbulent. Laminar flow can best be defined by visualizing the fluid as flowing in layers. It kind of looks like a deck of cards moving down a pipe. Turbulent flow is a real mixed-up type of movement. Fasten a rope at one end and begin to snap it up and down like a whip. When the up and down movement is going good, start to rotate so that you get a combination up and down movement and round and round movement. It should resemble a screw thread. This is turbulent flow. In most cases we will be dealing with turbulent flow. In fact, all the tables and charts that we will use to determine pressure drop are based on turbulent flow.

Remember when we were discussing air movement, we used the illustration of a ball on a track. We stated that the ball won't roll unless we elevate the track to give the ball some head. In Fig. 9–1a you will notice a U shaped tube. The water is at the same height in both legs because, as you have so often heard, water seeks its own level. If we measured the pressure at the base of the U, we would find it would be equal to the height of water. This pressure is called the static pressure. It is the pressure caused by the height of a fluid. In order to get flow, we must add water to the higher tube. Now the water again attempts to establish the same height and so water comes spurting out of the shorter leg (Fig. 9–1b). We could accomplish the same thing by adding a piston to the higher leg (Fig. 9–1c). Pushing down on this piston causes water to flow.

Now, what do 'b' and 'c' have in common? In one case, we increased the height of water. In doing this, we added X pounds of water and moved it through a distance Y ft, and so we did XY ft lbs work. By pushing the piston we also did XY ft lbs of work and caused water to flow. So, you can infer from this that in order for water to move, we must do work on it. The work we do must be enough to overcome the friction in the pipe, plus the losses caused by changes in direction in fittings, and the height we are lifting the fluid. The friction is equal to:

$$h = \frac{f(L)\,V^2}{(D)\,2g}$$

which is the same formula (expressed in a different form) we noted on page 143 about friction in ducts.

  h = friction in ft of water
  f = friction factor which depends on the
      Reynolds number
  L = length of pipe in ft
  D = diameter in ft
  V = velocity in ft per second
  g = acceleration due to gravity

Figure 9–2 presents the above formula in chart form for black iron pipe, and Fig. 9–3 for copper. As long as we know any two of the following three—

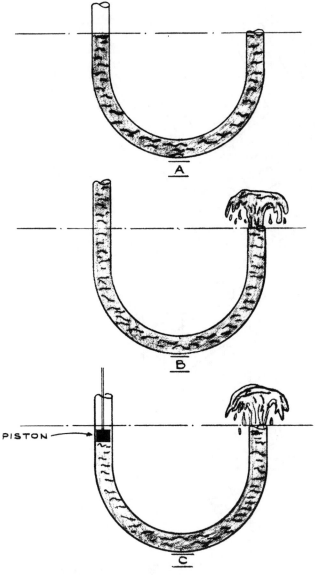

FIG. 9–1 Illustration Of Head

How do you find how much water you need to pump? In some cases, you are told by the manufacturer that the equipment needs X gpm. In most cases, the gpm is a function of heat transfer. You wish to absorb or emit a certain number of Btuh in a finned coil. If you fill this coil with 200° water and blow cold air across the coil, you will heat the air and cool the water. Eventually, unless you replace the cooled water, the water and air will reach the same temperature. If your coil holds 10 gal and you wish to replace the water after it has cooled 10°, you will have to move 10 gal every time the water cools 10°. If the 10° cooling takes place in one minute, you will have to move 10 gpm. How much heat transfer takes place? Just multiply the flow in pounds by the specific heat, temperature drop, and the change in temperature:

$$\frac{lb}{min} = \frac{10\ gal}{min} \times \frac{8.33\ lb}{gal}$$

$$\frac{lb}{min} = 83.3\ \frac{lb}{min}$$

$$\frac{Btu}{min} = 83.3\ \frac{lb}{min} \times \frac{1\ Btu}{lb\ deg} \times 10\ deg$$

$$\frac{Btu}{min} = 833$$

or, if you multiply by 60 min per hour, you get

$$Btuh = 833 \times 60 = 50,000$$

Can you see our formula? To determine the gpm, just divide the heat you wish to transfer per hour by the desired or selected temperature drop, the weight of one gallon of water, the specific heat of water, and 60 minutes per hour.

$$gpm = \frac{Btuh}{\frac{°TD\ (60\ min)}{(hr)}\ \frac{(8.33\ lb)}{(gal)}\ \frac{(1\ Btu)}{(lb\ °TD)}}$$

As you can see, the only variable under normal conditions in determining gpm is the temperature difference. For many years a 20°TD was used practically as a standard in heating systems. In recent years, 40°, 60°, and even 80°TD's have been used. Notice that going from 20° to 40° cuts the gpm in half, 20° to 60° by ²/₃, etc. In low-temperature hot water work, 220° is generally the maximum water temperature entering a coil or radiator. The following average water temperatures result with the various TD's.

TD	AVG
20	210
40	200
60	190
80	180

pipe size, gallons per minute flow (gpm), and friction, we can find the third. For example, if we had a 2-in. pipe carrying 40 gallons per minute we could determine the friction loss. First, find along the top 40 gpm, then read down until you intersect the 2-in. line, and now read horizontally to the left and read 400 milli inch (mi) per foot of pipe. (A milinch equals 0.001 in. or ¹/₁₂,₀₀₀ of a ft.) Back at the intersection of the 40 gpm and 2-in. lines you can also read the velocity of 48 in. per second. So, you can see that if we had 500 ft of 2-in. pipe carrying 40 gpm, our total friction would be:

$$h = 500\ ft\ \frac{400\ mi}{ft}\ \frac{1\ ft}{12,000\ mi}$$

$$h = 16.6\ ft$$

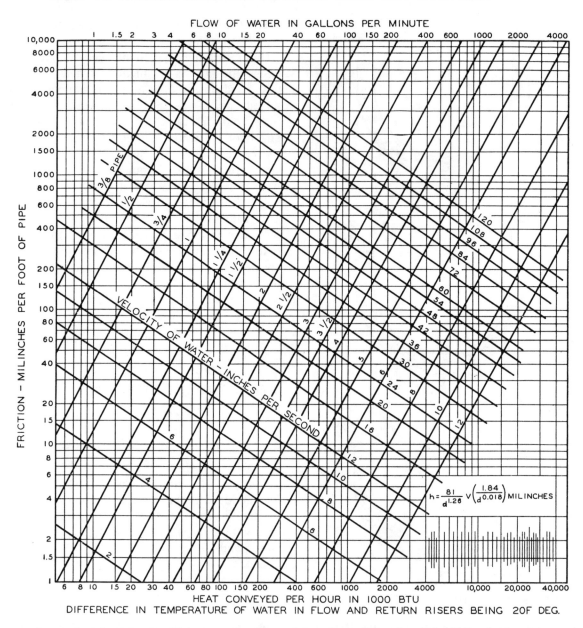

Lower scale of chart is based on 20-deg temperature difference between flow and return risers. To find friction when temperature drop is other than 20 deg, multiply the actual heat conveyed by (20 ÷ actual temp. drop) and read the corresponding friction.

Conversion Ft/(100Ft) to Milinches/Ft

Ft per 100 ft	0.5	1	2	3	4	5
Milinches per ft	60	120	240	360	480	600

FIG. 9–2 Friction Loss Due To Flow Of Water In Iron Pipe (Reprinted by permission from *ASHRAE Guide and Data Book,* 1965)

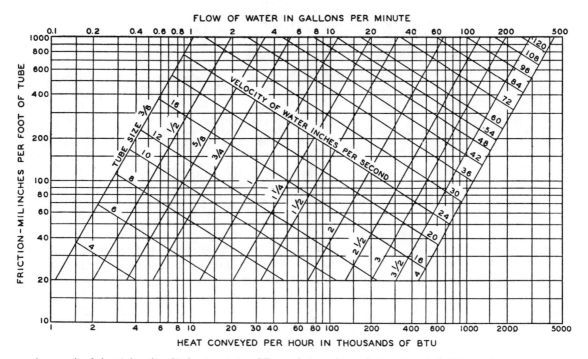

FLOW OF WATER IN GALLONS PER MINUTE

Lower scale of chart is based on 20-deg temperature difference between flow and return risers. To find friction when temperature drop is other than 20 deg, multiply the actual heat conveyed by (20 ÷ actual temp. drop) and read the corresponding friction.

Conversion Ft/(100 Ft) to Milinches/Ft

Ft per 100 ft	0.5	1	2	3	4	5
Milinches per ft	60	120	240	360	480	600

FIG. 9–3 Friction Loss Due To Flow Of Water In Type L Copper Tube (Reprinted by permission from *ASHRAE Guide and Data Book*, 1965)

Vel. Fps	Pipe Size														
	½	¾	1	1¼	1½	2	2½	3	3½	4	5	6	8	10	12
1	1.2	1.7	2.2	3.0	3.5	4.5	5.4	6.7	7.7	8.6	10.5	12.2	15.4	18.7	22.2
2	1.4	1.9	2.5	3.3	3.9	5.1	6.0	7.5	8.6	9.5	11.7	13.7	17.3	20.8	24.8
3	1.5	2.0	2.7	3.6	4.2	5.4	6.4	8.0	9.2	10.2	12.5	14.6	18.4	22.3	26.5
4	1.5	2.1	2.8	3.7	4.4	5.6	6.7	8.3	9.6	10.6	13.1	15.2	19.2	23.2	27.6
5	1.6	2.2	2.9	3.9	4.5	5.9	7.0	8.7	10.0	11.1	13.6	15.8	19.8	24.2	28.8
6	1.7	2.3	3.0	4.0	4.7	6.0	7.2	8.9	10.3	11.4	14.0	16.3	20.5	24.9	29.6
7	1.7	2.3	3.0	4.1	4.8	6.2	7.4	9.1	10.5	11.7	14.3	16.7	21.0	25.5	30.3
8	1.7	2.4	3.1	4.2	4.9	6.3	7.5	9.3	10.8	11.9	14.6	17.1	21.5	26.1	31.0
9	1.8	2.4	3.2	4.3	5.0	6.4	7.7	9.5	11.0	12.2	14.9	17.4	21.9	26.6	31.6
10	1.8	2.5	3.2	4.3	5.1	6.5	7.8	9.7	11.2	12.4	15.2	17.7	22.2	27.0	32.0

FIG. 9–4 Equivalent Length Of Pipe For 90-Degree Elbows (Reprinted by permission from *ASHRAE Guide and Data Book*, 1965)

The emission rate of heat-transfer equipment varies with the average water temperature. For example: A finned radiator's capacity is 1000 Btuh at 180° average, and 1400 Btuh at 210°. So, you can see that while you save pipe by using a higher TD, you must buy more heat-transfer surface to compensate for the lower average temperature. This is significant when gravity heating devices such as convectors, radiators, finned radiators, and baseboards are used. However, when forced-air flow units are used this isn't as significant. A good rule to follow is to use a 20°TD for gravity units, and 40° or 60° on forced-air flow units.

There is another reason for avoiding the large TD's. Safety factors! There is a tendency among men who are unscarred by battle to look down on safety factors and call them ignorance factors. In reality, one need not be ashamed of having a little in reserve for the unforeseen. The construction industry has made great strides in building techniques. However, the possibility of heat losses, in excess of those calculated, still exists. I don't advocate the arbitrary adding of 10 or 20% across the board. I guess I am a little more devious in my methods. I select all my gravity heat emmission units at 200° average with a 20°TD. This means that I have a cushion of 10°. If I used 200° average and a 40°TD I would have no cushion. On forced-air units I use 180° average and a 60°TD, or 190° average and 40°TD.

There is one exception to this rule of mine. When combination heating and cooling coils are used, I lower the water temperature considerably. Sometimes I use as low as 120° water. This can be done because a coil selected for cooling and dehumidification has so much surface that when heating, the leaving-air temperature and entering-water temperature are at times within 5° of each other. No hard and fast rule can be made here. Each case must be examined individually.

But, let's get on with pipe sizing. What do you do about fittings? What are the losses in elbows, tees, valves, etc? Here we are in luck. Fitting losses are easier to determine in water design than in air. Everything is keyed to the elbow. Fig. 9–4 gives the equivalent lengths of 90° elbows for various sizes at various water velocities. As you can see, for a 3-in. elbow at 5 fps the elbow has an equivalent length of 8.7 ft.

Yes, all other fittings are tied to the loss of elbows. All but tees are listed in Fig. 9–5. For instance, an open globe valve has a loss equal to 17 elbows (copper tube). If the velocity is 1 fps, the elbow equivalent

Fitting	Iron Pipe	Copper Tubing
Elbow, 90-deg....................	1.0	1.0
Elbow, 45-deg....................	0.7	0.7
Elbow, 90-deg long turn..........	0.5	0.5
Elbow, welded, 90-deg............	0.5	0.5
Reduced coupling.................	0.4	0.4
Open return bend.................	1.0	1.0
Angle radiator valve..............	2.0	3.0
Radiator or convector...........	3.0	4.0
Boiler or heater..................	3.0	4.0
Open gate valve..................	0.5	0.7
Open globe valve.................	12.0	17.0

ᵃSee Fig. 9–4 for equivalent length of one elbow.

FIG. 9–5 Iron And Copper Elbow Equivalents (Reprinted by permission from *ASHRAE Guide and Data Book*, 1965)

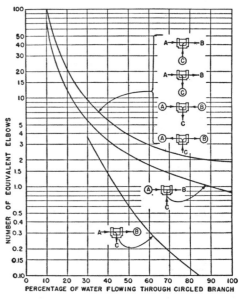

Notes: 1. The chart is based on straight tees, that is, branches A, B, and **C** are the same size.

2. Head loss in desired circuit is obtained by selecting proper curve according to illustrations, determining the flow at the circled branch, and multiplying the head loss for the same size elbow at the flow rate in the circled branch by the equivalent elbows indicated.

3. When the size of an outlet is reduced the equivalent elbows shown in the chart do not apply. The maximum loss for any circuit for any flow will not exceed 2 elbow equivalents at the maximum flow (gpm) occurring in any branch of the tee.

4. The top curve of the chart is the average of 4 curves, one for each of the tee circuits illustrated.

FIG. 9–6 Elbow Equivalents Of Tees At Various Flow Conditions (Reprinted by permission from *ASHRAE Guide and Data Book*, 1965)

length is 4.5 ft. So the equivalent length of a 2-in. open globe valve is 17 × 4.5 = 77 ft.

Figure 9–6 presents the elbow equivalents for tees. Three curves are shown so that all possible branch flow combinations are covered. For example: For the loss through a 2-in. tee, if 25% of the water flows through the branch, find the appropriate curve (the top one), and for 25% to port "C", find 13 elbow equivalents. The equivalent length is 13 × 4.5 = 59 ft. The loss, if all the water goes through the branch, is 2 elbow equivalents. The equivalent length is 2 × 4.5 = 9 ft. Isn't this strange? The less water going through the branch, the higher the equivalent length. The equivalent length is higher, but not the pressure drop. If 10 gpm enter a 1 $\frac{1}{4}$-in. tee, and 2 gpm goes to the branch or C port, what is the pressure drop? The flow to the branch is 2 gpm/10 gpm, or 20%. This gives 20 elbows, or 20 × 3.45 = 69 ft. The loss is 69 ft × 20 mi/ft = 1380 mi. If all of the water goes through the branch or C port, then elbow equivalents are equal to 2. So, the equivalent length is 2 × 3.45 = 6.9 ft. The loss is 6.9 × 300 = 2070 mi. You can see that while the equivalent length is larger as the flow through the branch diminishes, the pressure drop decreases. Unless you wish to be extra accurate, use a loss for tees of 2 elbow equivalents.

The pressure loss through the straight portion of the tee (A-B) can also be estimated from the bottom curve in Fig. 9–6. This is quite complicated as is the branch loss. Many designers ignore this loss because they feel it is not significant. Others figure that the maximum loss is equal to 0.3 elbow equivalents. I go along with those who do not include the loss through the straight flow through a tee. We will let you decide what to do on your projects.

As we discussed in Chapter 4, water piping can be arranged as a one-pipe system, two-pipe direct return, two-pipe reversed return, and the series-loop. Because it is the simplest, let's start with the series-loop. Figure 9–7 shows a typical series-loop system. In practice, you would make your layout and then scale the length. In this case, we will pretend that you have measured and the length is 200 ft. Study Fig. 9–7 and count the fittings. I see 12 elbows, 1 tee with 100% of the water going through the branch, 2 gate valves, and a boiler. The pipe is copper. The heat loss is 100,000 Btuh. Using a 20°TD the flow is:

$$gpm = \frac{100,000}{20° \times 8.3 \times 60}$$

$$gpm = 10$$

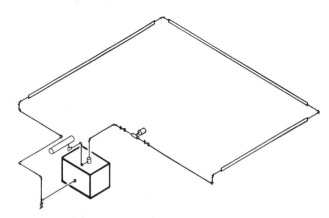

FIG. 9–7 Series Loop System

We are now ready to size our pipe by the use of Fig. 9-3. There are a couple of rules that go with the use of Figs. 9-2 and 9-3. To keep the possibility of noise to a minimum, don't exceed 4 fps for pipes less than 2'', 5 fps for 3'', 6 fps for 4'', 8 fps for 6'' and 10 fps for pipe larger than 6''. But to be sure that entrapped air isn't carried along, don't size pipe at less than 1 fps. The friction is kept between 100 and 600 mi/ft.

When you must choose between going over 600 mi/ft or dropping below 1 fps, always go above 600 mi/ft. This will only happen with small pipe sizes.

For 10 gmp, a 1¼'' pipe is indicated. The velocity is 32 in/second and the friction is 310 mi/ft.

11 elbows	=	11 elbow equivalents
1 tee	=	2 elbow equivalents
2 gate valves = 2(0.7) =		1.4 elbow equivalents
1 boiler	=	4 elbow equivalents
TOTAL	=	18.4 elbow equivalents

One elbow equivalent = 3.5 ft (Fig. 9–4)

So, 18.4 elbow equivalents = 18.4 (3.5) = 65 ft. Our total equivalent length is, therefore, 200 + 65 = 265. The pressure drop is 265 ft × 310 mi = 83,000 mi. Dividing by 12,000 mi per foot, we have 6.9 ft. So, we need a pump that can move 10 gpm, while overcoming 6.9 ft of head.

Now, let's try one that's a little tougher. Figure 9–8 shows a typical two-pipe system using convectors. Each convector has an output of 10,000 Btuh.

$$gpm = \frac{10,000}{20° \,(8.3)\,(60)} = 1$$

Again, you, as the designer, would measure length and count fittings. I like to use a table like the one shown on the next page.

FIG. 9–8 Typical Two-Pipe System

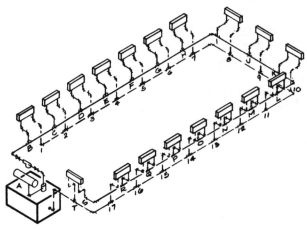

Sect	gpm	Size	mi/ft	Length	Fittings	Total Eq. Lgth	Total mi
					2 gate $2 \times .7 \times 4.4 = 6$		
AB	18	1½	400	20	2 el $2 \times 4.4 = 8.8$	35	14,000
BC	17	1½	375	10		10	3,750
CD	16	1½	310	10		10	3,100
DE	15	1½	290	10		10	2,900
EF	14	1¼	600	10		10	6,000
FG	13	1¼	500	10		10	5,000
GH	12	1¼	400	10		10	4,000
HI	11	1¼	350	20	1 el $1 \times 3.6 = 3.6$	24	8,400
IJ	10	1¼	310	10		10	3,100
JK	9	1¼	280	10		10	2,800
KL	8	1	550	15	1 el $1 \times 2.7 = 2.7$	18	10,000
LM	7	1	420	10		10	4,200
MN	6	1	320	10		10	3,200
NO	5	1	250	10		10	2,500
OP	4	¾	600	10		10	6,000
PQ	3	¾	400	10		10	4,000
QR	2	¾	200	10		10	2,000
RS	1	½	310	20	1 el $1 \times 1.3 = 1.3$	21	6,500
ST	1	½	310	10	7 el $7 \times 1.3 = 9.1$		
					1 tee $2 \times 1.3 = 2.6$	28	8.700
					1 gate $.7 \times 1.3 = .9$		
					1 conv $4 \times 1.3 = 5.2$		
TA	18	1½	400	24	2 el $2 \times 4.4 = 8.8$		
					1 gate $.7 \times 4.4 = 3.0$	54	21,600
					boiler $4 \times 4.4 = 17.6$		110,150 mi
							9.2 ft
1– 2	1	½	310				
2– 3	2	¾	200				
3– 4	3	¾	400				
4– 5	4	¾	600				
5– 6	5	1	250				
6– 7	6	1	320				
7– 8	7	1	420				
8– 9	8	1	550				
9–10	9	1¼	280				
10–11	10	1¼	310				
11–12	11	1¼	350				
12–13	12	1¼	400				
13–14	13	1¼	500				
14–15	14	1¼	600				
15–16	15	1½	290				
16–17	16	1½	310				
17– T	17	1½	375				

As you can see from Fig. 9–8, there are as many possible paths of flow as there are convectors. We must choose one path to determine our system pressure drop. We chose the last convector. We could just as easily have chosen the first and followed a path from the boiler to the first tee (AB), then through the first convector (B1), and followed the return main back to the boiler (1-17–TA). If we wanted to be elegant, we could determine the pressure drop through each possible path and select the highest for our pump. It is sufficient, however, to follow one path, as I did. A word of caution, however. If in the place of one of the convectors we have a cabinet unit heater, take this as your path, because it has the biggest loss. In Fig. 9–9 we show a typical direct-return branch main. Let's say each convector requires 1 gpm. The loss through ABHN is 19,440 mi, and through the longest path ABGMN is 64,490 mi. Obviously, the water doesn't know that you want 1 gpm through each convector. The water will follow the path of least resistance. The last convector will be starved for water. As you can see in Fig. 9-10, by adding a third line, the flow paths are approximately equal so the flow is balanced.

At one time, balancing a direct return system was difficult because, at best, it was a trial and error procedure. Each branch was balanced by adjusting a globe valve.

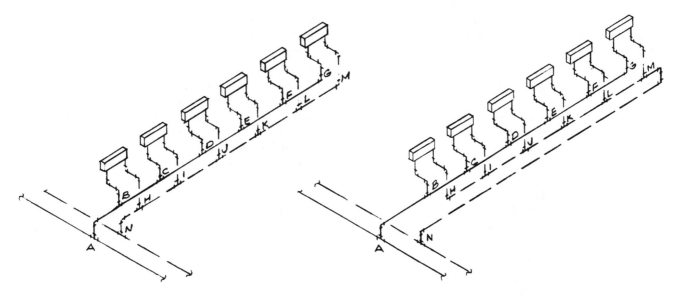

FIG. 9–9 Direct Return Branch Main   **FIG. 9–10** Reversed Return Branch Main

Sect	gpm	Size	mi/ft	Length	Fittings	Total Eq. Lgth	Total mi
AB	6	1	320	10	1 el 1 × 2.5 = 2.5 2 tee 4 × 1.4 = 5.6 6 el 6 × 1.4 = 8.4	13	4,120
BH	1	½	320	10	1 conv 4 × 1.4 = 5.6	30	9,600
HN	6	1	320	15	1 el 1 × 2.5 = 2.5	18	5,420
TOTAL							19,440 mi
AB	6	1	320	10	1 el 1 × 2.5 = 2.5	13	4,120
BC	5	1	250	15		15	3,750
CD	4	1	160	15		15	2,400
DE	3	¾	600	15		15	9,000
EF	2	¾	200	15		15	3,000
FG	1	½	320	15		15	4,800
GM	1	½	320	10	7 el 7 × 1.4 = 9.8 1 tee 2 × 1.4 = 2.8 1 conv 4 × 1.4 = 5.6	28	9,000
ML	1	½	320	15		15	4,800
LK	2	¾	200	15		15	3,000
KJ	3	¾	600	15		15	9,000
JI	4	1	160	15		13	2,400
IH	5	1	250	15		15	3,800
HN	6	1	320	15	1 el 1 × 2.5 = 2.5	18	5,720
TOTAL							64,790

Because the globe valve was only an adjusting device and not an indicator of flow rate, you'd have to constantly be guessing what would happen to flow with each turn of the hand wheel.

Now there is an array of more sophisticated devices that makes balancing a lot easier. There are flow limiting devices that are installed right in the line. You select the device that gives the flow required. Just remember, once you've made the selection that's it. The only way you can change the flow is to remove the device and replace it with another. There is nothing wrong with this when you are sure of your flow requirements, but if you're not certain, you'd better use one of the following devices.

FIG. 9–11a Balancing Device (Courtesy: ITT Bell & Gossett)

FIG. 9-11b. Balancing valve (Courtesy Gerand Engineering Co.)

The balancing device shown in Fig. 9-11a will not only balance but will also give a flow reading and the water temperature. You only have to turn down the adjusting section until you read the desired flow rate. This unit is used for flows less than 10 gmp. For larger flows, there's a similar unit that uses a globe valve or square-head cock to alter the flow.

A more compact balancing valve (Fig. 9-11b) is both an indicator and adjusting valve. A readout meter is connected to two ports, and a ball valve is turned until the desired flow is indicated. This unit is for flows of less than 7 gmp. For larger flows, the indicator is installed as a part of the device.

These devices may be expensive, but they may prove to be considerably cheaper than running the extra pipe required in a reverse return system.

But the simplicity of the self-balancing feature of the reverse return may make it worth the extra cost. My own feeling is that small buildings should be designed with reverse return. Let's say any system in which the main is 1½ in. or smaller should use reverse returns. In systems in which the main is 2 in. or larger, I advise the use of direct return on the mains, with balancing devices, and the use of reverse return on the branch mains.

Now, there is one other way to balance a direct-return system. If the loss in the mains is kept very low and the drop through the convector circuit high, the variation will be small. Let's say you sized the mains for a total pressure drop of 10,000 mi. Then, if the pressure drop through a convector circuit were 90,000 mi, the total pressure drop would be 100,000 mi. The difference between the first and last convector would be 10,000 mi or 10%. Just as in ductwork, the pressure drop varies as the square of the flow.

$$\left(\frac{Q_1}{Q_2}\right)^2 = \left(\frac{P_1}{P_2}\right) \qquad Q_1 = Q_2 \sqrt{\frac{P_1}{P_2}}$$

$$Q_1 = Q_2 \sqrt{\frac{90,000}{100,000}} \quad Q_1 = Q_2 \sqrt{0.9}$$

$$Q_1 = 0.95 \, Q_2$$

So, a difference of 10% in pressure causes a 5% change in flow. If you elect to make use of this principle, don't try to get the large pressure drop by increasing the velocity. You will not only get noise, but you will get pipe erosion. So use orifices or globe valves. The water flow through coils used in fancoil units, air conditioning units, ventilating units, etc., can be circulted to provide a large pressure drop.

The one-pipe system appears to be a paradox. How can you get any flow through the convector shown in Fig. 9–12, when the path of least resistance is the straight one AC? But, if we install an orifice between AC, which has a greater pressure drop than ABC, then water will flow through ABC because it is now the path of least resistance. This is exactly what is done. Special tees or diverting fittings are available which have a greater pressure drop in the straight-through path than through the branch. A typical one is shown in Fig. 9–13. Generally, these diverting fittings are installed in the return only. The manufacturers of these diverting fittings give the pressure drop through them. Fig. 9–14 gives the pressure drop through the Bell and Gossett monoflo

fittings. Actually, these are the maximum branch circuit pressure drops that can be overcome. If the pressure drop through the branch circuit is greater than the drop through a regular tee and a diverting fitting, then either use two diverting fittings or reduce the pressure drop in the convector circuit.

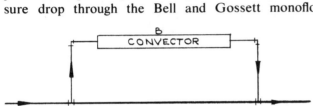

FIG. 9–12 Flow Diagram Of One-Pipe System

FIG. 9–13 Monoflo Fitting For One-Pipe System (Courtesy: ITT Bell & Gossett)

B & G MONOFLO FITTING SELECTION CHART

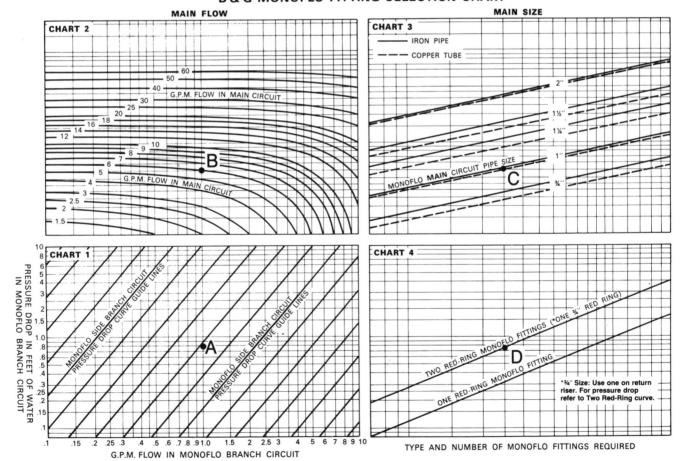

FIG. 9–14 B & G Monoflo Fitting Selection Chart (Courtesy: ITT Bell & Gossett)

A typical one-pipe system is shown in Fig. 9–15. Notice that Fig. 9–15 doesn't only differ from 9–8 because only one-pipe mains are used; we also create two parallel circuits by running the main down to the end of the building and then split into two branch mains and double back, picking up convectors on each branch main. The biggest reason for this is pressure drop. If we used one loop, as we did in Fig. 9–8, we would have 12 diverting fittings in series. If each monoflo fitting has a loss of one foot, then the loss would be 12 feet for the fittings alone. But if we split the system, as shown in Fig. 9–15, we have only six fittings and so a loss of six feet.

Let's size the piping system shown in Fig. 9–15. Assume the convectors each require 1 gpm. This is done on the table below:

To get the pressure drop caused by the diverting fittings and to get the size, we use Fig. 9–14. The first step is to calculate the loss in the convector circuit which is 9920 mi or 0.827 ft. Then start at Chart 1, Fig. 9–14. Find the intersection of 1 gpm and 0.827 ft. Go straight up from this point (A) to Chart 2 and find the intersection of this line and flow in the branch main (6 gpm). This is called point B. Next, go horizontally from point B to the branch main size in Chart 3. This intersection is point C. Now drop vertically to Chart 4 and find the intersection of the line from point C and a horizontal line drawn from A. This point is called D. We are between the lines for two red rings and one red ring. When between two lines, use the higher one which would be two red rings.

FIG. 9–15 Typical One-Pipe System

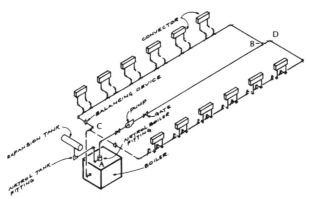

Section	gpm	Size	mi/ft	Length	Fittings	Total Eq. Length	Total mi
AB	12	1¼	400	75	2 gate 1.4 × 3.6 1 elbow 1 × 3.6 3 elbow 3 × 2.6	84	33,600
BC	6	1	320	103	1 Tee 2 × 2.6	116	37,120
BC	6	diverting fittings at 1.2 ft = 7.2 ft					
CA	12	1¼	400	10	2 elbow 2 × 3.6 1 gate .7 × 3.6 1 boiler	44	17,600
Typical Convector Runout	1	½	310	17	4 elbow 4 × 1.3 1 conv. 4 × 1.3 1 rad valve 3 × 1.3	32	9,920
BDC	6	1	320	123	3 elbow 3 × 2.6 1 Tee 2 × 2.6	136	43,520
BDC	6	diverting fittting at 1.2 ft = 7.2 ft					

TOTAL = AB 33,600 mi = 2.8 ft
+ BDC 43,520 mi = 3.6
+ Diverting
   Fittings 7.2
+ CA 17,600 mi = 1.5
             15.1 ft

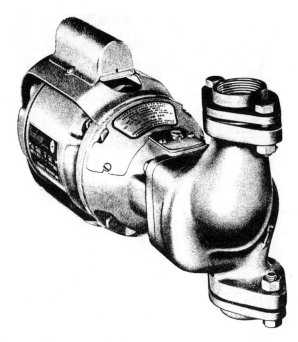

FIG. 9–17 Base Plate Mounted Pump (Courtesy: ITT Bell & Gossett)

FIG. 9–16 In-Line Circulating Pump (Courtesy: ITT Bell & Gossett)

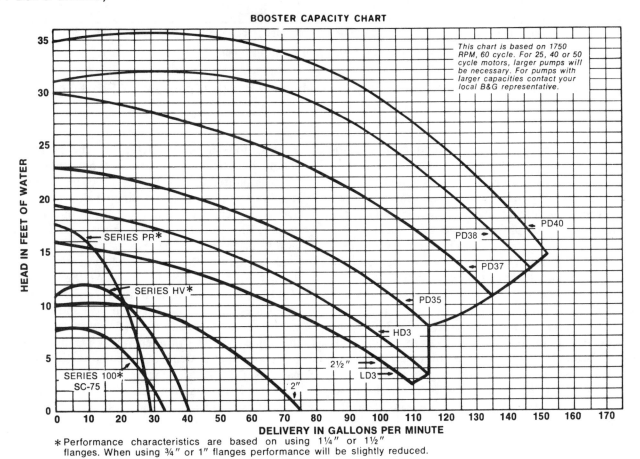

FIG. 9–18 Capacity Curves For In-Line Circulating Pumps (Courtesy: ITT Bell & Gossett)

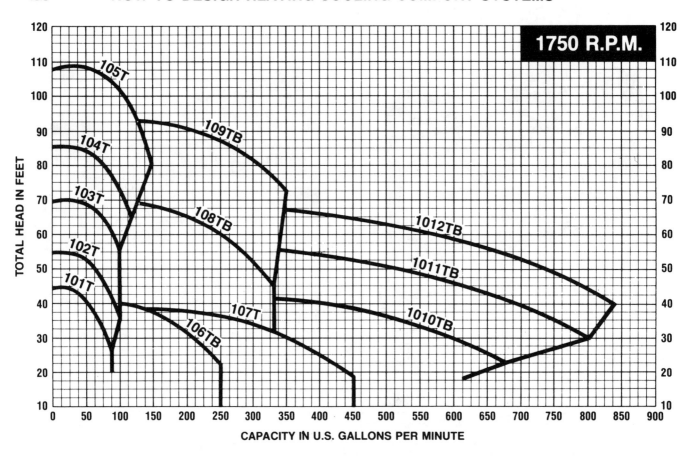

FIG. 9–19   Capacity Curves For Base Plate Mounted Pumps (Courtesy: ITT Bell & Gossett)

Now add the loss of the monoflo fittings to the branch main loss and main loss to get the total pressure drop. Because loop BDC has a larger pressure drop than BC, it is used to determine the total pressure drop. A balancing device is installed, of course, in BC to raise the pressure drop so it equals BDC.

Let's discuss pumps a bit. There are two types— the in-line booster (Fig. 9–16) which fastens right into the line, and the floor-mounted pump (Fig. 9–17). The capacity curves are shown in Figs. 9–18 and 9–19. Our two-pipe system requires a pump that will deliver 18 gpm against a 9.8-ft head. A 2-in. pump will do nicely for us. It will deliver 18 gpm against 10.1-ft head. It would be hard to come any closer. The one-pipe system requires 12 gpm against 8.7-ft head. A 1½-in. pump hits this right on the nose.

As I said earlier, air and water are a lot alike. Remember our discussion of fan-characteristic curves?

Pumps, too, have characteristic curves and they should be used. Fig. 9–20 is a typical pump curve. Flow is plotted against head. The solid, curved lines running from the left and dipping down toward the bottom are curves for different sized impellers. The dashed lines are horsepower curves, and the lines with % on them are efficiency.

Let's suppose that we need 110 gpm against a 47 ft-head. Our selection is good. At the intersection of 110 gpm and 47 ft-head, read 7-in. impeller, 68% efficiency, and we are between 1½ and 2 HP. We will, of course, specify a 2-HP motor. But, we have the same possible problems that we had with air. Let's say that instead of calculating the head you have used a "smear" factor to arrive at your head. And, let's say that the actual head is 25 ft. Come straight down from 110 and 47 to 25. This is a point on your characteristic curve. So, by making use of

$P_2 = P_1 (Q_2/Q_1)^2$ we can get enough points to plot our new characteristic curve. For example: At a flow of 120 gpm, the head would be

$$P_2 = 25 \, (120/110)^2$$
$$P_2 = 30 \text{ ft}$$

The point at which the new characteristic curve, curve B, intersects the 7-in. impeller curve is our point of operation. We will actually deliver 140 gpm, which might cause noise somewhere in the system. But, this isn't all! You are now past the 2-HP motor curve and if you have used a 2-HP motor you are overloaded. So you can see that it is important to calculate the pressure drop accurately.

These system changes don't happen only because of mistakes. Sometimes systems are used that have a variable flow. If you had a system in which the flow were cut in half, it would cut the pressure by four, wouldn't it?

$$P_2 = P_1 \left(\frac{55}{110}\right)^2 \quad P_2 = P_1 \, (\tfrac{1}{2})^2$$
$$P_2 = 47 \, (\tfrac{1}{4}) \quad P_2 = 11.8$$

Following the same procedure as above, we would plot a new system characteristic through 110 gpm and 11.8 ft, and it would intersect the 7-in. impeller curve at about 172 gpm and 30-ft head. Our efficiency, under these conditions, would be very poor, and again we would overload. We have two choices: use a 3-HP motor so we won't overload, or try to find a pump with a steep characteristic curve.

There is one other alternative; install a globe valve or other throttling valve on the pump discharge. This way additional resistance can be added that can move the system characteristic line to the the left and back within range of the 2-HP motor.

As water is heated it expands. The expansion of water can damage the parts of the system. Even if no damage takes place, the relief valve will open to relieve the increased pressure. Then, when the system cools down, there is insufficient water and the make-up valve opens, admitting raw water. This continual introduction of raw water into the system will cause corrosion. An expansion tank is provided to take care of this expansion. Early hot-water systems had open tanks which had to be located at the high point of the system. Today, we use closed tanks that can be located anywhere, but are generally located in the boiler room.

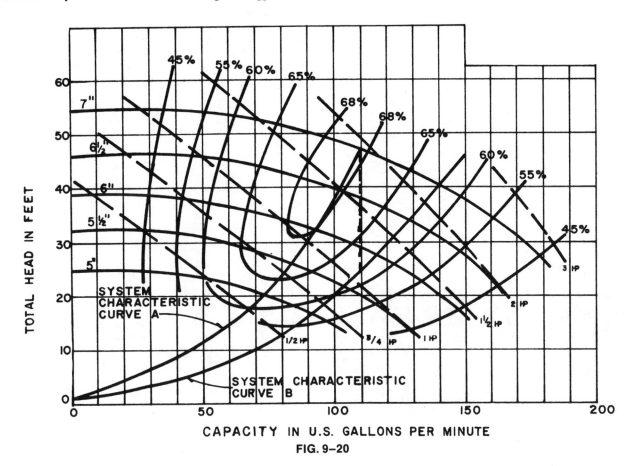

CAPACITY IN U.S. GALLONS PER MINUTE
FIG. 9–20

A second benefit results from the use of closed tanks. As the water expands into the tank it compresses the air. This increases the pressure on the system and, like our pressure cooker at home, also raises the boiling point of the water. If the pressure is raised to 10 psig, the boiling point of water is 240°. So, you can see that with pressurization it is possible to operate above 212°, the boiling point at atmospheric pressure (0 psig). Care must be taken in selecting an expansion tank that it be not too big or too small. If the tank is too small, there won't be room to store the water when it increases. If it is too large, insufficient pressure will build up and we will be unable to operate at the desired elevated temperatures.

The low-temperature, low-pressure hot-water systems have as a limit a pressure of 30 psig and 250°. The system, by code, must have an ASME relief valve set at 30 psig. The initial pressure is equal to the height of the system above the boiler. If the height above the boiler is above 69 feet, then the pressure will exceed 30 psig, the maximum allowable. 1 psi = 2.3 ft, so 30 psi = 69 ft. When the height of the system exceeds 69 ft, either a high-pressure boiler must be used, or an indirect system of water heating, such as a heat exchanger, must be used. In a heat exchanger, the system water is heated in the tubes and hot water or steam from a boiler is circulated in the shell. Fig. 9–21 shows a typical heat exchanger.

The expansion tank is the point of no pressure change in the system. Observation of a system will tell us this. No matter where the pump is located, the water level in the tank remains the same whether the pump runs or not. The pump is where the pressure changes take place. If we locate the tank at the pump discharge, then the system operates at pressures below the static pressure. The pressure at the tank connection to the system equals the static pressure. Let's say that the static pressure equals 30 ft. Let's say that the pump can provide 10 ft of head to the system. Fig. 9–22 shows how the pressure changes take place.

Notice that almost the full 10 ft of head appears as a reduction of pressure at the pump suction. With low head pumps this is not too great a problem. But, had the pump capacity been more than 33 ft, a vacuum condition would have occurred at the pump suction. Air will enter the system at the pump suction, causing corrosion in the system. I have seen boiler tubes completely ruined in a year under these conditions. The vacuum can be great enough to cause the water to flash. Remember that if the pressure at the pump suction reaches 6 ft below atmospheric (5.5 in Hg vacuum), water at 200° and above will boil.

In Fig. 9–23 you can see how the pressure changes occur when the expansion tank is located at the pump suction. Notice that now our pressure to the system is all additive. We enter the pump suction at near the static pressure. Always install the pump and tank so that all pressure changes are positive.

To determine the size of expansion tank required, use the following formula:

$$T = \frac{EV}{\frac{P_a}{P_f} - \frac{P_a}{P_o}}$$

T = tank size
V = volume of water in system
E = net expansion of water (see Fig. 9–24)
P_a = pressure in tank when water enters. Usually atmospheric pressure feet of water absolute
P_f = initial fill pressure feet of water absolute
P_o = maximum operating pressure in tank

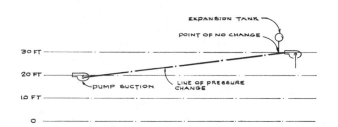

FIG. 9–22 Pressure Changes With Expansion Tank At Pump Discharge

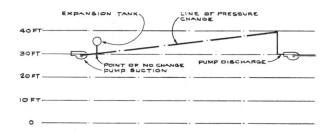

FIG. 9–23 Pressure Changes With Expansion Tank At Pump Suction

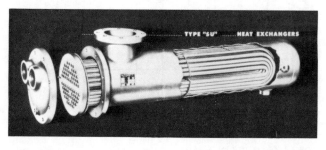

FIG. 9–21 Typical Steam To Hot Water Heat Exchanger (Courtesy: ITT Bell & Gossett)

The P_f is the pressure required to fill the system, plus 4 psi to assure positive venting, and also to prevent boiling. Let's say the highest point of the system is 30 ft above the boiler. If we are going to fill the system we must have enough pressure to raise the water 30 ft. Obviously, 30 ft is required. To this we add 4 psi, or 9.2 (say 9 ft), for a total fill pressure of 39 ft. If automatic makeup is desired, a pressure reducing valve is installed in the city water line. This valve reduces the city water pressure to 39 ft. Then, if the system pressure drops below 39 ft, water enters the system.

The operating pressure P_o is the maximum allowable pressure 30 psig or 69 ft for low-temperature, low-pressure systems. If a high-pressure boiler or a convertor is used, it can be whatever is desired.

Let's size an expansion tank. A system has 1000 gal of water. It will operate at 220°. The highest point in the system is 20 ft above the boiler and a low-pressure boiler is used.

$$T = \frac{E}{\frac{P_a}{P_f} - \frac{P_a}{P_o}}$$

$E = .042$ (Fig. 9–24)
$P_a = 39$ ft absolute
$P_o = 2.3$ (30 psig) + 39 ft = 108 ft absolute
$P_f = 20$ ft + 9 ft + 39 ft = 68 ft absolute

$$T = \frac{.042 \,(1000)}{\frac{39}{68} - \frac{39}{108}}$$

$$T = \frac{42}{.57 - .36}$$

$$T = \frac{42}{.21}$$

$$T = 200 \text{ gal}$$

The size of the expansion tank can be reduced by increasing P_a. If compressed air is added to the tank, P_a is increased. Let's say, in the above problem, we raised the tank pressure to 44 ft, then:

$$T = \frac{42}{\frac{44}{68} - \frac{44}{108}}$$

$$T = \frac{42}{.65 - .41}$$

$$T = 182 \text{ gal}$$

Now, let's try a chilled-water system operating at 40°. When the system is off, the water temperature could reach room temperature of 80°. $P_a = 39$, $P_o = 108$, $P_f = 68$. The system has 1000 gal. From Fig. 9–24, $E = .006$ (use 100° the lowest chart valve)

$$T = \frac{.006 \,(1000)}{\frac{39}{68} - \frac{39}{108}}$$

$$T = 29 \text{ gal}$$

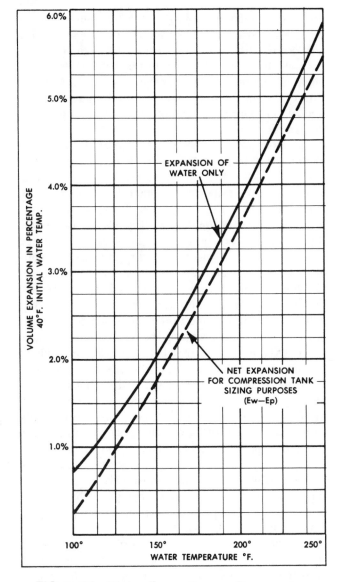

FIG. 9–24 Water Expansion vs. Temperature

Now, let's see. We have discussed pipe sizing, determination of flow (gpm), pump selection, and expansion tank sizing and location. While throughout our discussion we talked in terms of heating, you will notice all pipe sizing was based on gpm flow. In this way, the lessons learned can be used on chilled-water systems.

Two things remain to be discussed. Air control and equipment-piping diagrams. The amount of air that water can absorb varies with water temperature. The colder water is, the more air it can hold. The hotter water is, the less air it can hold. As water is heated it gives up air. This air causes corrosion, but also can cause "air binding," which stops circulation and so prevents heating or cooling. In the past, automatic air vents were used which permitted air to escape. We now know that while air escapes through the vent, air can also enter the system through the air vent. Now, it is recommended that manual air vents be used. These are opened on initial start-up to purge the system of air. During operation, the air is removed at the source.

An air-control system is installed at the boiler or chiller. The system consists of a boiler (or chiller) unit to remove air. This is generally a device that slows down the water flow, so that the air is separated from the water. Figs. 9–25A and 9–25B show two types of air separation devices: Fig. 9–25A is the type that fits in the boiler, while Fig. 9–25B is an in-line type. A tank unit is also needed to prevent the interchange of boiler and tank water, while permitting air to enter the tank (Fig. 9–26). Fig. 9–27 shows a typical boiler-piping diagram. Notice that the make-up water is introduced in the line to the expansion tank, which is the point of no pressure change. A double check valve assembly is installed in the make-up line to prevent any contamination of the city water system. Pressure reducing valves have built-in back-flo preventers, but extra protection is a good idea and is required in some communities. Notice, on the supply main, the flo-control valve. This is a weighted check valve that prevents gravity circulation of the water. Some systems are controlled by starting and stopping the pump. Without the flo-control valve it is possible for water to circulate through the system by convection, causing overheating. A heat exchanger-piping diagram is shown in Fig. 9–28.

Let's take a quick look at anti-freeze systems. In snow-melting systems, skating rinks and heat pumps, anti-freeze solutions are used. In general, the glycols are used. The glycol should have a corrosion inhibiter and the solution should be checked yearly to be sure of a neutral PH. If you look back, you will notice that each time we calculated the gpm, I always used TD x 8.3 x 60 and did not say (TD x 500). My reason was to try to have you keep in mind that the Btuh is divided by the TD, the specific heat of the fluid and the weight of fluid. If you were using a glycol solution and divided the Btuh by (TD x 500) you would circulate insufficient fluid. The specific heat and weight per gallon of a 50-50 glycol solution are 0.682 and 9.2 at −20°, and 0.745 and 9. at 40°. If you wished to transfer 100,000 Btuh, you would need to circulate at −20°.

$$gpm = \frac{100,000}{20° (0.682) (9.2) (60)}$$

$$gpm = 13$$

FIG. 9–25A Boiler Fitting For Airtrol System (Courtesy: ITT Bell & Gossett)

FIG. 9–25B In-Line Airtrol Fitting (Courtesy: ITT Bell & Gossett)

If 20° x 500 had been used, the flow would only have been 10 gpm resulting in a 26° TD. In addition, the viscosity of a glycol-water mixture at 40° is 4.5 times that of water. At −20° the pump efficiency is reduced by 20%, which may mean a larger motor. At 40° the change is insignificant. Pressure drops should be multiplied by 2 at −20° and 1.50 at 40°.

Selecting pumps for cooling towers is the same as for the supply systems, with one exception. Cooling towers are open systems and so the static lift must be considered. We actually consider it in a closed system, but it cancels itself. In a closed system you lift the water to the top floor but it drops down the same distance, so the lift is cancelled by the drop. But, in a cooling tower, you lift water to the top of the tower and pick it back up at the bottom, leaving a gap. This distance, or gap, must be added to your friction in order to get the total head (Fig. 9–29). Let's assume that the tower shown in Fig. 9–29 is part of a system that is pumping 300 gpm through the condenser of a water chiller. There are 200 ft of pipe, 20 elbows, 4 gate valves, a condenser with 20-ft pressure drop, and a static lift of 6 ft. Use Fig. 9–2 and select a 5-in. pipe. The friction is 175 mi and the velocity 57 in/sec, say 5 fps. The elbows from Fig. 9–5 have an equivalent length of 13.6 ft. So, let's put it together:

200 ft × 175 mi/ft =	35,000 mi
4 elbows × 13.6 ft × 175 =	9,500
4 gate × 0.5 × 13.6 × 175 =	4,550
	49,050 mi
Total for Piping	4.1 ft
Condenser	20.0 ft
Static Lift	6.0 ft
	30.1 ft

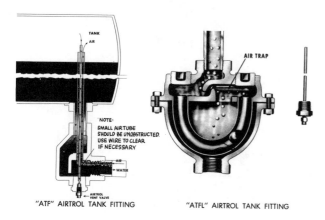

FIG. 9–26 Tank Fittings For Airtrol System (Courtesy: ITT Bell & Gossett)

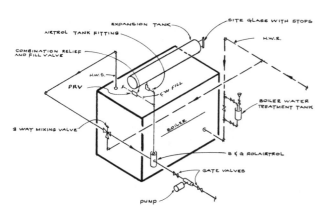

FIG. 9–27 Boiler Piping Diagram

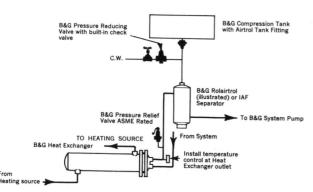

FIG. 9–28 Heat Exchanger Piping Diagram (Courtesy: ITT Bell & Gossett)

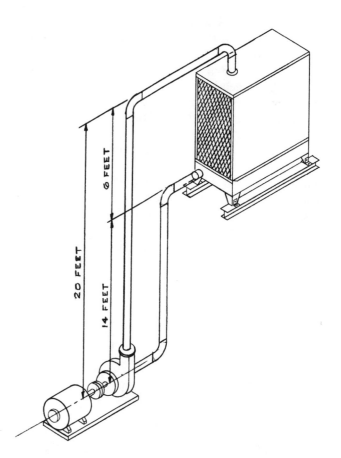

FIG. 9–29 Cooling Tower Piping Diagram

So, you would need a pump that can pump 300 gpm against a head of 30.1 ft. A word of caution about cooling towers. Sometimes, towers are placed on grade. The pump MUST be lower than the tower so that the pump suction has a positive head on it. You must be certain that the head on the pump suction is also enough to overcome the friction in the suction main. For example, if the above tower had been on grade 70 equivalent feet from the pump, the suction friction would be $\frac{70 \times 175}{12,000} = 1$ ft. The pump suction should be located $1 + 1 = 2$ ft below the cooling tower to be sure of a positive head on the suction. Also, do not install globe valves, strainers, or other high-pressure devices in the pump suction. A pump can pass $\frac{1}{4}$-in. objects, so put the strainer in the pump discharge line.

A few final words on pipe sizing. Do not overestimate the pressure drop because of the shift in the pump curve. Try to get actual pressure drop data if possible. This is particularly important on equipment that has large pressure drops, such as cooler sections of water chillers, coils on air conditioning units, temperature control valves, etc. Don't guess. Find out! Be sure to allow enough head for temperature control valves, which work best when selected with a high pressure drop.

Primary-secondary systems are a modern addition to hot-water heating. Fig. 9–30 shows a typical primary secondary system. Notice that we have a primary pump which pumps through the boiler and primary piping. Secondary pumps pump to each secondary zone. One of the biggest advantages of primary-secondary systems is that the primary mains can be sized for a large TD to achieve minimum pipe sizes and pump horsepowers. At the same time, the secondary system can be sized for conventional TD's to have an optimum heat transfer at the terminal units. In addition the system is self-balancing, eliminating the cost of the reverse return and the difficulty of balancing a direct return.

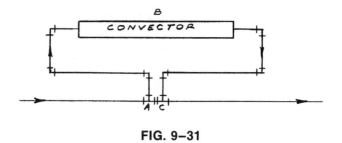

FIG. 9–31

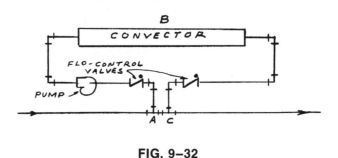

FIG. 9–32

How does the system work? Remember the monoflo system, Fig. 9–12? We stated that the natural tendency is for the water to go through AC because this path has a much smaller pressure drop. If we move points A and C closer together, as shown in Fig. 9–31, the pressure drop between AC is so small by comparison to ABC that all of the water flows through AC. Remember that the monoflo system only works because the monoflo tees introduce a large pressure drop to the straight-through or AC circuit. The flow through ABC is a function of the pressure drop between AC.

Now if there is no pressure drop between AC, there will be no flow through ABC. To get flow we need a secondary pump, as shown in Fig. 9–32. To prevent gravity flow, flo-control valves are installed as shown. A most important use of the primary-secondary system is in hot-water, chilled-water systems, as shown in Fig. 9–33. Notice that in addition to the secondary pumps we have three-way mixing valves.

The secondary circuits circulate in a closed loop. Let's say we are heating. If the primary-loop water temperature is satisfied, then the water in the primary loop goes right on by the secondary connections. At the same time, the secondary pump circulates through the boiler to the diverting valve and back to the boiler. Let's say the primary loop needs some heat. An aquastat in the primary line would sense this and open the diverting valve, permitting some of the primary water to enter the secondary circuit. As soon as this happens,

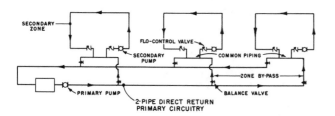

FIG. 9–30 Typical Primary-Secondary System (Courtesy: ITT Bell & Gossett)

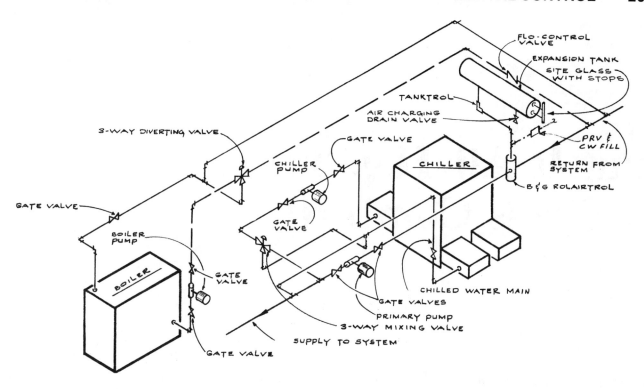

FIG. 9–33 Hot-Water Chilled-Water System Using Primary-Secondary Principle

hot water from the secondary circuit is injected into the primary circuit. The water temperature in the primary loop is reset by an outdoor thermostat.

When cooling is desired, or required, the chiller circuit acts in a manner similar to the boiler circuit to maintain desired primary chilled-water temperature. The advantage of primary-secondary for chillers is that we have a constant flow through the chiller. Also, on switchover from heating to cooling, primary water is slowly permitted to flow into the secondary circuit so that no slugs of hot water enter the chiller to cause possible damage.

What we have presented here is an introduction to primary-secondary systems. The Bell and Gosset Company has several fine publications that present a complete discussion of primary-secondary system. I urge you to read these if you are interested in learning more about these systems.

CHAPTER 10

Put It All Together-
It Spells Comfort

AT THIS POINT, you should be able to calculate the heat loss or gain of a building, and also size ductwork and piping. But this still isn't enough background to allow you to design building environmental systems. Additional factors must be considered prior to designing. Remember,

    C is for the first cost of the system
    O is for the cost of operation
    M is for the money that is needed
    F is for the function of the building
    ORT is that is ort to be compatible with the architecture

Put them all together, they spell comfort. And that should be your goal. I'll admit my attempt at verse is terrible, but it does point out the facts that must be considered in selecting the proper system. Never approach a building with the attitude that this or that system is the vogue and should be used. Be logical and consider the points I've mapped out for you.

The first cost is of considerable importance and is the first or coarse screen in sorting the many available systems. Why consider a system with wall-to-wall control when your budget permits only man-to-manual control?

Then, of course, comes operating cost. There are times when you can spend a few additional dollars in first cost and get it back in operating-cost savings within two years. Then again, the converse can be true. You may save enough in first cost to permit a higher than normal operating cost.

The function of the building is too often neglected when selecting systems. What do I mean by function? Well consider a church. Its principle function is to house a large audience of people in worship for several hours once a week. The large number of people means we need plenty of ventilation. (Refer to Chapter 5) So you must select a system that will provide top-drawer ventilation, even if you have to sacrifice ideal heating.

Integration of the system with the architecture is probably the most important of our four rules of procedure. Years ago, there was a series of cartoons that showed the airplane as seen by the people who designed the various parts. The engine designer had a plane that had a giant engine, stubby wings, midget body and no room for the pilot. Each of the other designers made his part the dominant one and ignored the other. Similar sins are committed in the design of buildings and their environmental systems. We can all pause to curse and vilify the architectural designer who sees the building only as a piece of sculpture with no regard to function or comfort. But aren't we equally at fault if we take a chisel to gouge and scratch this sculpture to fit in a heating system? The environmental control system must mold itself into the structure and provide nearly invisible comfort. Of course, you can't do this alone. It's basically a team effort, but try to go the extra yard.

First costs are determined in many engineering offices by keeping an up-to-date file of costs. In this way, a trip to the files will tell you approximately what a certain system will cost. Normally, this is good enough, but when entering the arena to be judge and jury in the battle of fuels, accurate first costs are necessary. To get these costs, preliminary drawings of each system, which is under consideration, is prepared and priced-out.

Operating costs consist of owning costs, fuel costs, cost of operating auxiliaries (such as fans and pumps), cost of supplies, maintenance costs and salaries for operators where they're required. Sometimes, when electricity is used as a fuel, a lower overall power rate results. This may yield a savings in the cost of all electricity used, so be sure to include the total electrical consumption from all uses.

Fuel consumption can be determined in several ways. One of the most popular methods is the degree-day method developed by the American Gas Association (AGA) and adapted for use with all fuels. Noting that fuel consumption varies directly with the outdoor temperature, they established the *degree-day*. There are as many degree-days in a day as the difference between the day's mean temperature and the base of 65°. So if the average or mean temperature for a day was 30°, then that day would have (65 − 30) = 35 degree-days.

The formula used is:

$$F = \frac{N \times 24 \times DD}{DTD \times U \times E}$$

F = Fuel consumption

N = Heat loss in 1000 Btuh

24 = Hours/day

DD = Degree-day (Table 10-1)

DTD = Design Temperature Difference

E = System yearly efficiency

U = Btu/unit of fuel
    Gas = 100 MBtu/Therm (1 Therm = 100,000 Btu)
    Oil = 140 MBtu/Gal (#2 Fuel Oil)
    Coal = 12 MBtu/lb

Example #1

Determine the amount of gas used in a residence in Alpena, Michigan, that has an hourly heat loss of 100,000 Btuh. Outdoor design condition is − 10°. Efficiency is assumed to be 80%.

N = 100,000 ÷ 1,000 = 100

DD = 8073

E = 0.80

$$F = \frac{N \times 24 \times DD}{DTD \times U \times E}$$

$$F = \frac{100 \times 24 \times 8073}{(70 - (10)) \times (100) \times (.80)}$$

F = 3,027 Therms (302,700 ft³)

Certainly this is a simple method of determining fuel consumption.

TABLE 10-1

Yearly Number of Degree Days

State	City	Average Winter Temp.*	Yearly Total
Ala	Birmingham	51.8	2780
	Mobile	58.9	1529
Ariz	Flagstaff	35.9	7525
	Phoenix	59.5	1698
Ark	Bentonville		4036
	Little Rock		2982
Calif	Los Angeles	59.3	1451
	Sacramento	53.0	2600
	San Francisco	54.2	3069
Colo	Denver	37.0	6132
Conn ...	Hartford		6139
D.C......	Washington	43.4	4258
Fla.......	Jacksonville	60.6	1243
	Miami	71.4	173
Ga......	Atlanta		2826
	Savannah	56.7	1710
Idaho...	Boise	39.8	5890
Ill.........	Cairo	46.4	3756
	Chicago	35.1	6310
	Springfield	37.7	5693
Ind	Evansville	45.1	4360
	Fort Wayne	37.6	6287
Iowa	Des Moines	35.4	6446
	Dubuque	34.6	7271
Kan	Topeka	42.1	4919
	Wichita		4571
Ky	Louisville	45.1	4279
La	New Orleans	60.6	1317
Me......	Portland	33.0	7681
Md	Baltimore	44.1	4787
Mass....	Boston		5791
Mich....	Alpena	29.6	8073
	Detroit (City Airport)		6404
	Sault Ste. Marie	26.0	9475
Minn....	Duluth		9937
	Minneapolis		7853
Miss	Meridian		2333
Mo	Kansas City		4888
	Saint Louis	42.3	4699
Mont....	Billings	34.9	7106
Neb	Lincoln	35.6	6104
Nev......	Reno		6036

TABLE 10-1 (Continued)

Yearly Number of Degree Days

State	City	Average Winter Temp.*	Yearly Total
N.H......	Concord		7612
N.J	Newark		5252
N.M	Santa Fe		6123
N.Y......	Albany		6962
	Buffalo		6838
	New York Central Park Obs.		4965
N.C......	Charlotte		3205
	Hatteras		2392
N.D......	Bismarck	22.9	9033
Ohio	Cincinnati		5195
	Cleveland	37.2	5717
	Columbus	38.2	5615
Okla	Oklahoma City	47.9	3519
Ore......	Baker	35.2	7087
Pa	Erie	37.3	6116
	Philadelphia	41.4	4866
	Pittsburgh		
	Greater Pittsburgh	38.7	5905
R.I.......	Providence	37.5	5607
S.C......	Charleston	55.0	1973
	Greenville	49.2	3060
S.D......	Rapid City		7535
Tenn....	Chattanooga	47.8	3384
Texas...	Dallas		2272
	Houston	60.1	1388
Utah	Salt Lake City	38.3	5866
Vt........	Burlington		7865
Va	Norfolk		3454
	Richmond	47.0	3955
Wash ...	Seattle-Tacoma	45.1	5275
W.Va....	Elkins	39.4	5773
Wis......	Madison	31.4	7300
	Milwaukee	29.0	7205
Wyo.....	Cheyenne		7562

*Average winter temperature is from September 1 to May 31 inclusive.

Extracted by permission from *ASHRAE Guide and Data Book*, 1968.

The only problem comes when it is used in buildings with large internal heat gains, such as lights, people or machinery. Another problem is outside air used for ventilation. It is part of the heat loss but may not be on 24 hours a day, seven days a week. So if a building is a residence, warehouse or other building type with little or no internal heat or ventilation load, use the degree-day method as presented.

For buildings with large internal heat gains and ventilation loads, the degree-day method must be modified as follows:

$$N = (N_{HL} + N_V) - N_{HG}\left(\frac{DTD}{ATD}\right)$$

Where

N_{HL} = Heat loss in 1,000 Btuh

N_V = Ventilation load in 1,000 Btuh

N_{HG} = Heat gain load in 1,000 Btuh

DTD = Design temperature difference (Room Temp − Outside Temp)

ATD = Average temperature difference (65° − Avg. Winter Temp)

Example #2

Determine the oil consumption for a factory with an hourly heat loss of 1,000,000 Btuh, a ventilation heat load of 500,000 Btuh, and a heat gain from lights, people and machinery of 300,000 Btuh. The factory is located in Denver, Colorado, and the design temperature is −12°. Assume 70% efficiency.

$$N_{HL} = \frac{1,000,000}{1,000} = 1,000 \text{ (For heat loss)}$$

$$N_V = \frac{500,000}{1,000} = 500 \text{ (For outside air)}$$

$$N_{HG} = \frac{300,000}{1,000} = 300 \text{ (For heat gain)}$$

U = 140 MBtu/gal #2 fuel oil

DD = 6,132

DTD = (70 − (−10)) = 80

ATD = 65 − 37 = 28

E = 0.70

$$F = \frac{N \times 24 \times DD}{(DTD) \times 21 \times E}$$

$$N = (N_{HL} + N_V) - N_{HG}\left(\frac{DTD}{ATD}\right)$$

$$N = (1,000 + 500) - 300\left(\frac{80}{28}\right)$$

N = 640

$$F = \frac{(640 \times 24 \times 6132)}{(80) \times (140) \times (0.7)}$$

F = 12,014 gallons

If the internal heat gains had not been considered, the fuel consumption estimate would have been 28,000 gallons.

Example #3

If the factory in Example 2 only worked 80 hours a week, how much fuel would be consumed? We can safely assume that the ventilation and heat gain would be present only during working hours. Not only must we correct for 80 hours of operation/week, but we must also add the infiltration load to the heat loss during the non-working hours.

$$N = N_{HL} = N_V\left(\frac{HO}{TO}\right) = N_1\left(\frac{TO - HO}{TO}\right) - N_{HG}\left(\frac{DTD}{ATD}\right)\left(\frac{HO}{TO}\right)$$

HO = Hours of operation

TO = Total hours

$$N = N_{HL} + N_V\left(\frac{80}{168}\right) + N_1\left(\frac{88}{168}\right) - N_{HG}\left(\frac{DTD}{ATD}\right)\left(\frac{80}{168}\right)$$

$$N = 1000 + 500\left(\frac{80}{168}\right) + 200\left(\frac{88}{168}\right) - 300\left(\frac{80}{28}\right)\left(\frac{80}{168}\right)$$

$$N = 1000 + 238 + 105 - 300\left(\frac{80}{28}\right)\left(\frac{80}{168}\right)$$

N   = 935

$$F = \frac{935 \times 24 \times 6132}{80 \times 140 \times (0.70)}$$

F   = 17,551 gallons

Now, there will be times when the degree-day method cannot be used. Let's consider an air-source heat pump. In an air-source heat pump the Btu output varies as the temperature of the heat source. In this case, a rational method must be used. Ideally, we should work in increments of one degree. However, 10° increments may safely be used. In this method, the heat loss is calculated for 10° increments from the room temperature to the design temperature. From the heat loss, you subtract the heat gain and add the ventilation load. You will then have net heat loads, loss or gain in increments of 10°. From manufacturers' literature, the output can be determined for the same increments and the kW input determined. The hours of occurance of each increment can be obtained from the Weather Bureau.

Example #4

What is the cost of heating an office building with a heat pump? The heat loss is 200,000 Btuh, the ventila-

OCCUPIED LOAD*

Outside Temp.	Heat Loss	Heat Gain	Net Load	Btuh/KW	KW	Hrs	Total KW Hrs
60	40,000	60,000	+ 20,000	15,000	0	1200	0
50	80,000	60,000	− 20,000	13,600	1.5	1284	1,900
40	120,000	60,000	− 60,000	12,000	5	1380	6,900
30	160,000	60,000	−100,000	10,000	10	1710	17,100
20	200,000	60,000	−140,000	9,000	15.6	786	12,200
10	240,000	60,000	−180,000	7,500	24	132	3,150
0	280,000	60,000	−220,000	6,800	33	6	200
							41,450

*Heat loss includes ventilation load.

UNOCCUPIED LOAD*

Outside Temp.	Heat Loss			Btuh/KW	KW	Hrs	Total KW Hrs
60	35,000			15,000	2.3	1200	2,750
50	70,000			13,600	5.15	1284	6,600
40	105,000			12,000	8.8	1380	12,200
30	140,000			10,000	14	1710	24,000
20	175,000			9,000	19.5	786	15,400
10	210,000			7,500	28.	132	3,700
0	245,000			6,800	36.6	6	220
							64,870

*Heat Loss includes infiltration load

Days— 41,450 × 50/168 = 12,300
Nights—64,870 × 118/168 = 45,700
                      58,000 KW Hrs

tion load is 80,000 Btuh, the infiltration load is 45,000 Btuh and the heat gain is 60,000. Assume 50 hours/week occupancy. The building is located in Detroit, Michigan.

There are other operating costs to be considered. The first is amortization. To me, amortization is the cost of owning. If you borrow money for an environmental system, amortization is how you repay that money/year, including interest. If you pay cash, it represents the cost/year, plus the interest you would have been paid if you had invested this money in stocks, bonds and/or savings.

To arrive at amortization, you can consult a handbook of amortization schedules. But it's just as easy to calculate your multiplier on an electronic calculator.

$$M = \frac{I(1 + I)^t}{(1 + I)^t - 1}$$

M = Amortization multiplier

I = Interest rate (yearly or monthly)

t = Time in years or months depending on I

To calculate $(1 + I)^t$ with your calculator, use the X^Y or Y^X button on most calculators. Let's assume there's a 15% yearly interest rate for a ten year period. To determine $(1 + .15)^{10}$, press 1.15, X^Y, then 10 and read 4.0455577, which can be rounded out to 4.046. The amortization multiplier will be:

$$M = \frac{0.15 (1 + 0.15)^{10}}{(1 + 0.15)^{10} - 1}$$

$$M = \frac{0.15 (4.046)}{(4.046) - 1}$$

$$M = 0.199$$

To arrive at the yearly owning cost, multiply the first cost by the amortization multiplier. In general, the amortization period for environmental control systems can be taken as 20 years.

In addition to amortization, the cost of insurance, taxes, power for auxiliaries, water and maintenance must be included. Insurance and taxes should be discussed with the owner. The owner or his staff are the most knowledgeable when it comes to determining the approximate cost of insurance and taxes.

To determine the cost of power for auxiliaries, multiply the kW of the various motors, such as fans and pumps, by the hours of usage. In general, water pumps run continuously during the heating season. To determine the hours of operation of boiler fan motors, multiply the hours of operation of the heating system (Table 10-2) by

40%. Room terminal-unit fans and control-system fans generally run continuously during occupied periods and intermittently during unoccupied periods. Therefore, the total operating hours of these fans is equal to the percentage of the occupied period, multiplied by the total heating hours (Table 10-2), plus the percentage of the unoccupied hours, multiplied first by the total hours and then multiplied by 40% usage factor.

TABLE 10-2
Hours of Operation of Heating Systems

Degree Days	Hours of Operation
8000	7200
6000	6600
5000	6000
4000	5400
3000	4800
2000	4200
1000	3600

Example #5

What are the kWh for a hot water heating plant in a 5000 degree-day area? The building is occupied 50 hours/week. The twenty rooms each have a ⅙-HP unit ventilator. The boiler burner unit has ½-HP forced-draft fan, and the pump has a 1-HP motor.

```
Boiler Burner Unit
  ½-HP Forced-Draft Fan  ½ × 6,000 hrs × .40 = 1,200
  Pump 1-HP              1 × 6,000 hrs       = 6,000
Unit Ventilators
  Occupied    20 × ⅙ × 6,000 × 50/168       = 6,000
  Unoccupied  20 × ⅙ × 6,000 × 118/168 × .40 = 5,600
                                            18,800
```

You will notice that I used a HP as equal to a kW which is not completely correct. But for our purposes, it's close enough.

When estimating the cost of operation of a heating plant, we generally do not include the cost of operating lights. But when making a cost comparison between an electric heating system and a gas-oil-coal heating plant, the cost of lighting must be included. Many times the use of an electric plant qualifies the owner for a special rate. This special rate often applies towards the power consumed for lighting as well. So the cost of lighting must be included to get an accurate picture. The easiest way to estimate the lighting load is to multiply the kW of lighting by the hours of operation. The operating hours are generally equal to the hours the building is occupied or in use. If the building, such as a factory, uses a lot of non-lighting power, include an estimate of this load in your comparison. There are no easy rules here. The owner can give you an idea of how much power is used or can make a close approximation which you can then use for your calculations.

Maintenance costs can vary tremendously. In the case of the owner who does little or no preventive maintenance, the cost can be very small for the first few years and very high in later years when he pays for his neglect. Over the life span of the system, the owner who properly cares for his system will have a lower average cost than the negligent owner. Maintenance costs include supplies such as oil and grease, filters and fan belts.

The problem of fitting an environmental system into the architectural design is truly a formidable one. As we discussed earlier, in an area where the outside temperature is consistently below 45°, the heating must be introduced along the outside wall and at the floor. This is sometimes easier said than done.

Let's look at some examples of the problems that can arise. Take a room 20'L × 10'W × 8'H. The 20' long wall has two 3' wide by 5' high windows. An architect looks

at a wall as a unit. He doesn't see the wall, the windows and the heating units separately. Rather, he sees them as one integral unit. So what can be used? You can use convectors under the windows, finned tube radiation, baseboard radiation, air register or air-baseboard register.

In Fig. 10-1 you can see the poor application of a convector. The convector looks as if it was *stuck* on the wall, and it does not integrate. Figure 10-2 shows a better application. The convector is as wide as the window, it's recessed and inlet/outlet grilles are used. A poor application of finned-tube radiation is displayed in Fig. 10-3. The heating application is fine, but the use of a 12" high cover acts as a divider to the eye. A 32" high cover is used in Fig. 10-4 and again integrates. In Fig. 10-5 you can see two pieces of baseboard radiation *stuck* on the wall, and as a result, they tend to stand out. But in Fig. 10-6, the enclosure runs wall-to-wall along the baseboard and integrates well.

When an air-baseboard register is used, the same rules apply as in the use of baseboard radiation. A poor place-

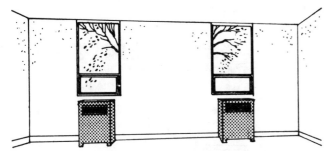

FIG. 10-1 Poor Placement Of A Convector

FIG. 10-2 Good Placement Of A Convector

FIG. 10-3 Poor Placement Of Finned Radiator

FIG. 10-4 Better Placement Of Finned Radiator

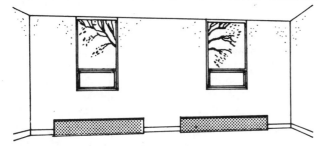

FIG. 10-5 Poor Arrangement Of Baseboard Radiator

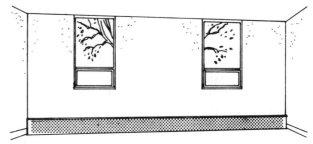

FIG. 10-6 Wall-To-Wall Baseboard Radiator Enclosure

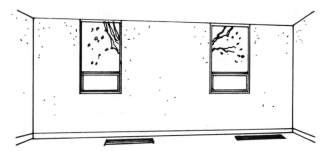

FIG. 10–7 Poor Placement Of Floor Register

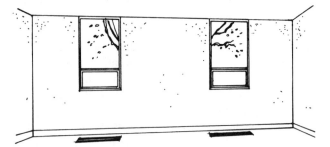

FIG. 10–8 Balanced Placement Of Floor Register

FIG. 10–9 Registers Placed For Horizontal Air Delivery

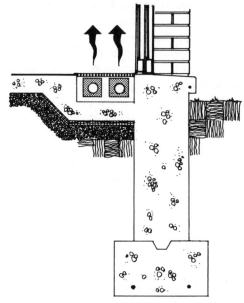

FIG. 10–10 Radiation Recessed In Floor For Use With Floor-To-Ceiling Glass

FIG. 10–11 Use of Wall Radiant Panels To Offset Cold Wall Effect

ment of floor registers is shown in Fig. 10-7. They appear to stand out, but notice in Fig. 10-8 that the eye follows the window to the floor register. Another good rule of thumb is to use long, narrow registers in preference to short, wider ones. Size the register to make it as long as the window is wide if possible.

Try to avoid the placement of registers as shown in Fig. 10-9. While they integrate well, comfort problems can result. When the air is discharged horizontally, it can strike occupants in the room and prove to be uncomfortable because of the draft-like action it creates. Registers should be installed in the window sills for the best in comfort and integration. They can be placed in the sill if the wall is deep enough to allow the installation of the duct. Front outlet convectors can also be used because the air motion is a warm, gentle gravity type.

When terminal units such as two-duct boxes, induction units or fancoil units are employed, follow rules similar to those used with convector applications.

However, because the air discharge is forced, it can feel cold or cool if it strikes someone directly. It is wise to use a top discharge to avoid striking the occupants of a room.

Sliding glass doors and floor-to-ceiling glass present problems if hot-water units are to be used. Convectors cannot be used, but finned radiation can be recessed in the floor, as shown in Fig. 10-10. The capacity of the radiation is considerably reduced when applied in this manner. More radiation must then be used raising the cost of the installation.

One solution is the use of a floor-radiant or ceiling-radiant panel. Another excellent solution is the use of wall radiant panels on both sides of the windows, as shown in Fig. 10-11. Fancoil units, shown in Fig. 10-12, installed perpendicular to the outside wall discharge air along that outside wall. The units can be free standing, as shown, or recessed into the wall. These fancoil units work, but at the same time, they can also cause a draft problem.

To solve this problem, the units can be positioned above the ceiling as illustrated in Fig. 10-13. Notice the unit is connected to a strip diffuser. The diffuser is split by a damper, which is positioned by a discharge thermostat, and deflects air through one section of the diffuser vertically, at high velocity when heating is required. But when cooling is necessary, the damper deflects air horizontally along the ceiling.

Locating ceiling outlets requires the same type of criteria that is applied to wall unit placement. Plan your light and diffuser layout so the components appear harmonious. Be geometric and disciplined in your thinking. The diffusers shown in Fig. 10-14 are on the quarter

points of the room, as recommended in Chapter 8. But notice how they clash with the lighting. In Fig. 10-15, the diffusers have been equally spaced between the lights for a more pleasing appearance. A popular ceiling is the 24" × 48" lay-in type, using inverted T bars. Again watch your pattern. Instead of using round diffusers, use square or rectangular diffusers. Now, at the same time, don't attempt to use a 12" square diffuser. It's wiser to use a diffuser with one dimension of 24". Preferably limit yourself to using diffusers with dimensions of 24" × 12", 24" × 24" or 24" × 48". Figure 10-16 shows an integrated pattern. The diffusers are 24" × 24" with three-sided discharge.

FIG. 10–12 Use Of Fancoil Units To Heat Glass Wall

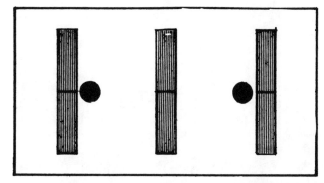

FIG. 10–14 Poor Architectural Placement Of Diffusers

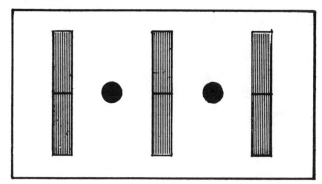

FIG. 10–15 Better Diffuser Placement For An Integrated Appearance

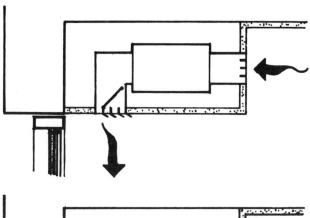

FIG. 10–13 Ceiling Fancoil Units With Variable Discharge

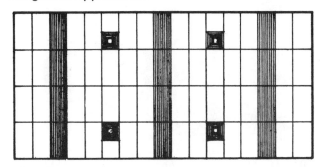

FIG. 10–16 Good Diffuser Placement For An Integrated Appearance

Rather than using diffusers to introduce the air into a room, one alternative that's become popular is the installation of breathing ceilings. An example is depicted in Fig. 10-17a. Figure 10-17b shows a diagrammatic section through a typical breathing ceiling. Still another method of integrating is the use of diffuser light fixtures where the diffuser is actually built into the light fixture. (See Fig. 10-18)

Space for the transmission system often times poses yet another problem. There may be several reasons why an architect would want to keep the space between floors to a minimum. Using a piping system instead of a ductwork system can save space. But remember that combination hot-chilled water systems should only be used as perimeter systems unless combined with a primary or ventilating-air ductwork system for the interior area.

Suppose an architect wants a building with no suspended ceiling space. Consequently, there's no room between floors for ducts or piping. What can you do? You can use through-the-wall, self-contained units or fancoil units for the perimeter by running the pipe in vertical chases as shown in Fig. 10-19. An induction system can also be used. Again, the primary air duct and secondary water piping can be run in vertical chases.

If a suspended ceiling can be used above the corridor for return air, then a double duct system can be used as illustrated in Fig. 10-20. The hot duct is run down one column, and the cold duct down another. Runouts are channeled into the mixing box. The return-air duct is run above the corridor ceiling.

This appears to be a good solution for the exterior, but what can you do for interior areas? Figure 10-21 shows a solution I've used. Run a high-velocity duct down one side of an interior column. Just above the floor, install an air-valve, reheat mixing box and connect this to a duct

FIG. 10–17A Perforated Ceiling For Air Distribution (Courtesy: Armstrong Cork Co.)

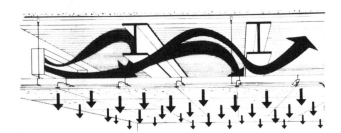

FIG. 10–17B Diagrammatic Section of Air Distribution Ceiling (Courtesy: Armstrong Cork Co.)

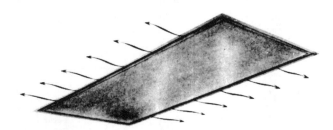

FIG. 10–18 Combination Light Fixture Diffuser (Courtesy: Lightcraft Co. and Connor Engineering Corp.)

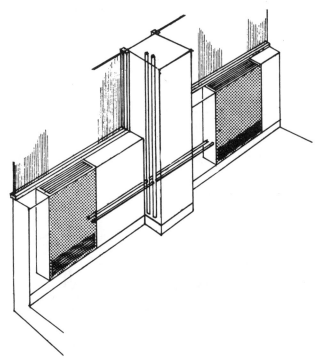

FIG. 10–19 Perimeter Units And Pipe Chases For Use In No Suspended Ceiling Installations

that is wrapped around three sides of the column. Grilles are then installed on these three sides. For a five story building, this assembly measures about 24'' × 24'' and includes the column.

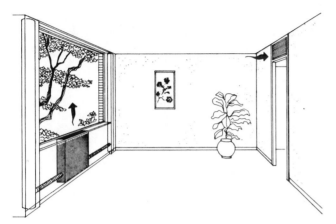

FIG. 10–20 Double Duct System For Use In No Suspended Ceiling Installations

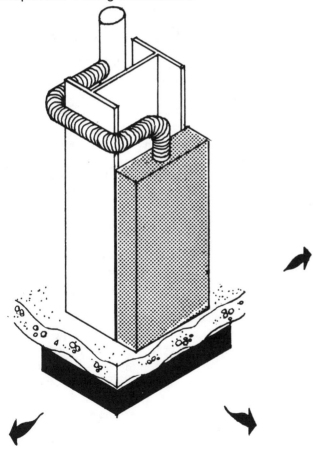

FIG. 10–21 High Velocity Installation In Interior Areas

DESIGNING COMFORT SYSTEMS

Now let's design comfort systems. The first step in a designing comfort system is to assemble a library of manufacturers' catalogs. This is done by calling or writing manufacturers' representatives, or company offices, and requesting catalogs. I've assembled a small library at the back of the chapter. And to save space, I've condensed the catalogs.

If you recall, back in Chapter 3 we did a heat loss and heat gain for a clinic. Using that same building, we will design the following systems:

- Series loop baseboard system
- Air conditioning system using self-contained, through-the-wall units
- Hot water-chilled water fancoil system
- Multizone overhead air system
- Variable air volume system (VAV)

The series loop baseboard system heat loss summary is comparable but not exactly the same as that used in Chapter 3. The heat loss figures used for the other systems do not include infiltration, except at the entrances. All of our air conditioning systems introduce treated outside air which pressurizes the building, offsetting infiltration. So instead of using the TETD method, the cooling loads are obtained by applying the CLTD method because it's more accurate and more nearly simulates actual building loads.

SERIES LOOP BASEBOARD SYSTEM

We start by summarizing the heat losses in Fig. 10-22. Next, we select the amount of radiation needed in each room. To do this, we go to our library and select a baseboard radiation catalog. Figure 10-53 is a sheet out of a Sterling Company catalog. Ratings are usually given for two flow rates and various average water temperatures. The output is greater for the greater flow rate because an increase in velocity increases heat transfer. At this point we are not sure of our flow rate, so we use the lower one. I like to use 200°F average water, so at 1 gpm flow rate and 200°F, read 910 Btuh/ft of radiation.

Dividing the room heat loss by 910 gives the required footage. Notice in the third column of Fig. 10-22 that the required footage comes out in odd lengths such as 2.4 and 5.4. Baseboard radiation is purchased to the nearest foot, so we must *round out* our required footages. To *round*, we go to the smaller value if the amount is 0.3 or less, and to the larger amount if it is greater than 0.3. This is what's represented in the fourth column.

In Fig. 10-22 we see that 12' of baseboard radiation is required for the reception area. However, if you go back to the front elevation in Chapter 3, you will see that

Room Name	Heat Loss Btuh	Baseboard Radiation-ft	Corrected Footage	Btuh	
Records	2,057	2.26	2	1,820	
Shop	3,334	3.66	4	3,640	
Employees Lounge	1,663	1.83	2	1,820	
Exam Rm 1 to 6	3,360	3.69	4	3,640	
Exam Rm 7	4,918	5.40	6	5,460	
Entry #4	4,958	5.44	6	5,460	
Hall	3,264	3.59	4	3,640	
Office	4,918	5.40	6	5,460	
Exam Rms 8, 9	3,360	3.69	4	3,640	
Drop Room	9,015	9.9	10	9,100	
Visual Fields	3,683	4.05	4	3,640	
Doctors Office	5,591	6.14	6	5,460	
Entry #2	6,055	6.65	7	6,370	
Womens Toilet	1,518	1.67	2	1,820	
Mens Toilet	1,518	1.67	2	1,820	
Storage	3,197	3.5	4	3,640	
Entry #3	12,601	13.84	14	12,740	
Hall	969	1.06	1	910	
Office	1,417	1.56	2	1,820	
Work Rm	4,398	4.83	5	4,550	
Stock	541	0.59			
Fitting	12,567	13.8	14	12,740	
Reception	10,592		10	9,100	} 11,260
		(Type A)	6	2,160	
Entry #1	12,601	13.85	14	12,740	
Waiting Rm	10,720	11.78	12	10,820	
General Office	4,304	4.73	5	4,550	
	153,279			161,410	

FIGURE 10-22   Summary of Heat Loss

the window on the east wall goes right down to the floor. It's best to put the radiation under the window, but how can it be done? Figure 10-10 shows how to recess radiation in the floor. In Fig. 10-24 we call this *type A* radiation. The manufacturer indicates that the radiation has a capacity of 360 Btuh when installed in this manner. The 6' gives 2,160 Btuh leaving 8,760 to be supplied by conventional baseboard. Ten feet of baseboard is installed on the north wall at a capacity of 9,100 Btuh.

In the series loop system, the tube in the radiation is used as the main. As the water travels down the tube, it loses temperature. Consequently, the last radiation gives off less heat than the first. In large systems, adjustments must be made to supply more radiation in the last rooms than in the first. In buildings of this size, and those that are smaller, it should not be a problem.

The next step is to spot the baseboard on the plans as shown in Figs. 10-23, 10-24 and 10-25. Notice that the baseboard units are placed on the outside wall if possible. This is done for two reasons. First, because it is the source of greatest heat loss, and secondly, because it reduces drafts. In the entries and halls the baseboard is run on the hall walls. Notice that entry 1, entry 2 and

the connecting hall all merge into one room when we started laying out the baseboard.

After the baseboard radiation is drawn, a piping system is laid out. Type L copper tubing is commonly used in this type and size of project, so that's what we will use. There are no hardfast rules for piping other than to use common sense. Avoid buried or underground piping as much as possible. Rise from the boiler room to the ceiling space of the first floor. As you examine the building, some zoning is suggested. First, the exam-room wing breaks down into an east and west zone. In this way, the winter sun shining on one side will not upset the system. As you can see, if the sun was shining on the west side of the building, some of the heat loss would be offset. Now, if the thermostat was on the east side, it wouldn't sense the sun effect and the west would overheat. If the thermostat was on the west, it would shut down the heat and the east would be cold. Two thermostats and two zones make good sense.

The north end of the building is another area that obviously lends itself to zoning — not only for orientation, but because of the different space usage. This is where glasses are fitted and manufactured while the other wing

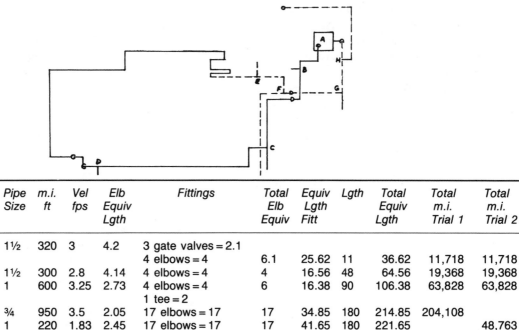

Pipe Sect	Btuh	Gpm	Pipe Size	m.i. ft	Vel fps	Elb Equiv Lgth	Fittings	Total Elb Equiv	Equiv Lgth Fitt	Lgth	Total Equiv Lgth	Total m.i. Trial 1	Total m.i. Trial 2
AB	157,770	15.8	1½	320	3	4.2	3 gate valves = 2.1 4 elbows = 4	6.1	25.62	11	36.62	11,718	11,718
BC	148,670	14.9	1½	300	2.8	4.14	4 elbows = 4	4	16.56	48	64.56	19,368	19,368
CD	85,540	8.5	1	600	3.25	2.73	4 elbows = 4 1 tee = 2	6	16.38	90	106.38	63,828	63,828
DE	44,590	4.5	¾	950	3.5	2.05	17 elbows = 17	17	34.85	180	214.85	204,108	
DE		4.5	1	220	1.83	2.45	17 elbows = 17	17	41.65	180	221.65		48,763
EF	85,540	8.2	1	600	3.25	2.73	1 elbow = 1 1 tee = 2	3	8.19	10	18.19	10,914	10,914
FG	123,190	12.3	1¼	350	2.75	3.53		0	0	4	4	1,400	1,400
GH	132,290	13.2	1¼	500	3.9	3.7	2 elbows = 2 1 tee = 2	4	14.8	33	47.8	23,900	23,900
HA	157,770	15.8	1½	320	3.0	4.2	1 Gate valve = 0.7 1 tee = 2 2 elbows = 2 1 boiler = 4	8.7	36.54	28	64.54	20,653	20,653

TOTAL					200,544
Diverting Tee	4500 × 40%				1,800
Three way mixing valve	3 ft allowance (36,000 mi)				36,000
Zone valve	3 ft allowance (36,000 mi)				36,000
					274,344
					22.86

FIGURE 10-26   Pipe Sizing

millinch. This yields a total pressure drop of 35.6' which means that a PD40 pump must be used. A PD40 has a 1 ½-HP motor. Keeping in mind energy costs, we should use a 1'' pipe that gives only 51,700 millinch. This establishes a total pressure requirement of 22.1'. Now we can use a PD35 with a ½-HP motor. As you can see, this reduction in horsepower should provide savings. At seven cents per kWh, using the ½-HP motor can save us as much as $460 per year. Let's re-examine our velocity. The lower limit of 2 fps is conservative and it is permissible to drop to 1.5 fps. Section DE is then sized at 1''.

The head is 22 ft. Notice that an allowance was included for a 3 ft pressure drop for the three-way mixing valve and the zone valve. In the specifications, make a note of this so that the temperature-control contractor will select a valve within your limit. Turning to Fig. 10-55,

a Bell and Gossett catalog, we find that a PD35 will give 15.9 gpm against a 22.2 ft head. What is our flow if we use a 20° drop? You say 15.3 quick as a flash because you remember to divide 153,280 by 10,000. But you forgot the piece of radiation that is not part of the loop. Recall the piece in the hall that's on a monoflo system. Subtract this amount and you get 15.0 gpm. What you are really doing is increasing the system temperature drop slightly by using the monoflo system for this one piece of radiation. That branch unit is carrying 8.43 gpm so the temperature drop will be:

$$\frac{3,960 \text{ (rating of radiation at } 210°)}{8.43 \, (8.33)(60)} = 0.94°$$

And as you can see, we have reduced the water in the main from 210° to 209.06°.

About all that remains to complete the system is the expansion tank. (See Fig. 10-27). As you can see, we require a 7.9 gallon tank. The smallest standard tank available is 15 gallons, and its dimensions are: 34½'' long with a 13'' diameter.

Notice on the boiler piping diagram (Fig. 10-25) that we show a three-way mixing valve. This valve blends boiler water and return water to maintain a reset water temperature. What do we mean by reset? This means an outdoor thermostat will change or reset the water temperature according to the outdoor temperature. At $-10°$ outdoors, we will have 210° water; while at 30° outdoors, 145°; and at 70° outdoors, 80° water. Our individual zone valves then only *trim out* to get the final desired room conditions.

Pipe Size	Length	Gal H_2O /ft Pipe	Gal
1½	90	.0925	8.33
1	650	.0442	28.73
¾	80	.0250	2.00
½	150	.0121	1.82
			40.88
Boiler (from mfg)			16.80
			57.68

E = expansion factor = 0.038
V = system water capacity = 57.68 gal
Pa = 34 ft absolute = pressure in tank when water enters
Po = 103 ft absolute = minimum operting pressure
Pf = 15 ft (approximate distance above boiler)
  9 ft (added for positive pressure)
  34 ft (atmospheric pressure)
  58 ft (fill pressure absolute)

$$T = \frac{0.035(57.68)}{\frac{34}{58} - \frac{34}{103}} = \frac{2.019}{0.256} = 7.9 \text{ gal}$$

FIGURE 10-27   Expansion Tank Sizing

As pipe is heated, it expands. Unless expansion joints or loops are used, serious damage will result to both piping and building. Copper pipe will expand 2.0''/100' between 40° and 210°. An expansion compensator, similar to Flexonics Model L (shown in Fig. 10-56), will handle ¾'' of motion. Install one every 30' in copper lines. The Model L is well suited to baseboard radiation because it will fit under the cover. The Model HB will handle 1¾'' but, because of its size, can only be used on mains above the ceiling or in tunnels.

HOT WATER-CHILLED WATER FANCOIL SYSTEM

Our first step is to select the heating and air conditioning units or fancoil units. Consult a fancoil catalog such as American Air Filter shown in Fig. 10-57. The units depicted are called cabinet air conditioners.

Back in Chapter 3 we calculated the heat gain for this building using indoor conditions of 78° DB and 45% RH. These are the conditions specified in many energy codes. The comfort chart on page 7 shows that this is just within the comfort zone. If we use 10% outside air, then the entering air conditions will be:

$$.10 \times 87° = 8.7° \qquad .10 \times 72° = 7.20°$$
$$.90 \times 78° = \underline{70.2°} \qquad .90 \times 63.5° = \underline{57.15°}$$
$$78.9° \text{ DB} \qquad\qquad 64.35° \text{ WB}$$

The entering air conditions are 78.9° DB and 64.35° WB. To make the selection charts in Fig. 10-57 a bit easier to use, we can assume that the entering air conditions will be 79° DB and 64° WB. The resulting error is minor and not significant for the purposes of this example.

The catalog gives instructions for the use of the capacity nomographs. The Fancoil Unit Capacity Table provides the data in tabular form.

To be correct, we should make a selection at 78° DB and 63.5° WB for the units that do not have an outside air connection. However, it is reasonably accurate to use the calculated values for room conditions given above. Figure 10-28 tabulates our heat gain, heat loss and unit selection.

Notice that the information tabulated from the nomographs is for total heat which includes both the room loads and the outside air loads. To arrive at the room loads it's necessary to subtract the outside air loads from the total loads. The outside air loads were calculated as follows:

Unit cfm (10%) (87 − 78) (1.10) = outside air sensible load
Unit cfm (10%) (35.6 − 29.4) (60 ÷ 14.1) = outside air total load

10%	= amount of outside air
87°	= outdoor design dry bulb temperature
78°	= room design dry bulb temperature
35.6	= outdoor design enthalpy
29.4	= room design enthalpy
14.1	= specific volume of outside air
60	= minutes/hour

As you scan Fig. 10-28 you may notice that the stock and records rooms were omitted. They are rarely occupied and do not warrant the cost of a unit. The stock room heat loss is so small that, if the door is left open, its temperature will be very close to the temperature in the adjoining space. The basement records room may be a little cool in winter. However, if it is going to be used for any length of time, a portable electric heater can be used to warm it.

Units were not installed in the hallways. Rather, the entry units were selected to cover both the entry and adjacent hallway. When you compare the total cfm of 8,200

FANCOIL UNIT CAPACITY TABLE

Unit No.	cfm	Total Sens. Heat	Total Heat	Room Sens.	Room Total	gpm	Heating 100° Total	Capacity Water-Air TD Room
2,000	200	4,500	5,700	4,300	5,172	1.14	12,000	10,504
3,000	300	6,900	8,000	6,600	7,209	1.6	19,000	16,756
4,000	400	8,900	10,500	8,500	9,445	2.1	25,000	22,000
5,000	500	11,100	13,100	10,604	11,781	2.62	32,000	28,260
6,000	600	14,200	15,800	13,600	14,217	3.16	37,000	32,512

to the amount used in the outside air calculations, you may feel that an error has been made. If you look at Fig. 10-30 you will see that the entry units and the general office unit do not have outside air connections. When these are subtracted from 8,200 cfm, 6,600 cfm results. We then multiply 6,600 by 10% which yields 660 cfm.

The heating capacities listed in the Fancoil Unit Capacity Table are for a 100° temperature difference between the entering water and the entering air. As you scan Fig. 10-28, notice that in each case the unit has the potential to provide much more heat than is necessary. The cor-

rection chart in the manufacturer's literature only goes down to an 80° TD. Therefore, a little experimenting on the job may be necessary. You will probably be able to operate with water as low as 120° at 4°F outdoors. Start with this temperature and raise it or lower it as necessary to achieve comfort.

The heat loss summary in Fig. 10-28 shows lower values than those in Fig. 10-22. The reason, the fancoil units introduce outside air, and in the process, pressurize the room and prevent infiltration. Therefore, the infiltration was subtracted from the loss except in the entries.

Room Name	Heat Loss Btuh	Heat Gain			Unit		
		Rm. Sens. Btuh	Rm. Lat. Btuh	Rm. Total Btuh	Model No.	cfm	Gpm each
Shop	3,180	4,658	510	5,168	3,000	300	1.6
Lounge	1,509	3,349	1,020	4,369	2,000	200	1.14
Exam 1 to 6	3,272	3,018	255	3,272	2,000	200	1.14
Exam 7	4,830	3,125	255	3,380	2,000	200	1.14
Entry 4 and Hall	8,222	5,313	642	5,955	3,000	300	1.6
Office	4,840	4,148	510	4,658	2,000	200	1.14
Exam 8, 3	3,272	4,041	255	4,296	2,000	200	1.14
Drop Room	8,839	13,511	5,100	18,611	2@4,000	400	2.1
Visual Fields	3,683	2,865	510	3,375	2,000	200	1.14
Drs. Office	5,409	6,870	510	7,380	4,000	400	2.1
Entry 2	6,055	3,181	642	3,823	2,000	200	1.14
Gen. Office	4,304	7,384	1,275	8,659	4,000	400	2.1
Women	1,518	1,021	0	1,021	2,000	200	1.14
Men	1,518	1,021	0	1,021	2,000	200	1.14
Storage	3,197	911	0	911	2,000	200	1.14
Entry 3 and Hall	13,086	6,146	2,309	8,455	4,000	400	2.1
Office	1,417	1,098	255	1,353	2,000	200	1.14
Work Room	4,398	5,112	765	5,877	3,000	300	1.5
Fitting	12,216	12,521	1,275	13,796	2@3,000	300	1.5
Reception	10,592	7,375	255	7,630	2@2,000	200	1.14
Entry 1 and Hall	13,086	4,058	2,309	6,367	3,000	300	1.6
Waiting Room	10,544	8,912	1,020	9,932	2@3,000	300	1.6
Room Totals	148,619	128,769	21,202	148,969		8,200	44.96
Outside Air 660 cfm	49,368			31,174			
Grand Totals	197,987			180,143			

FIGURE 10-28

Heat Loss — Heat Gain; Summary — Unit Selection

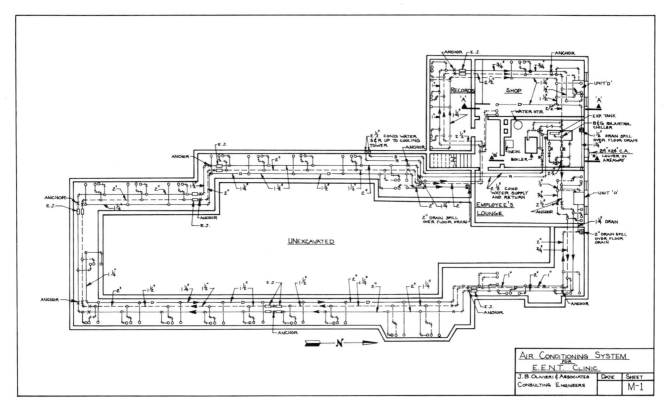

FIG. 10-29 Basement And Tunnel Plan

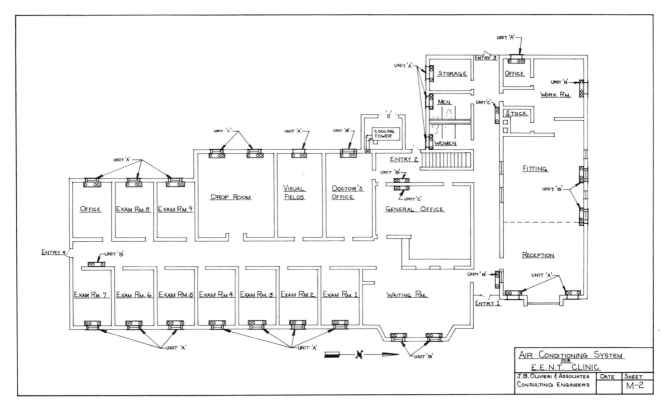

FIG. 10-30 First Floor Plan

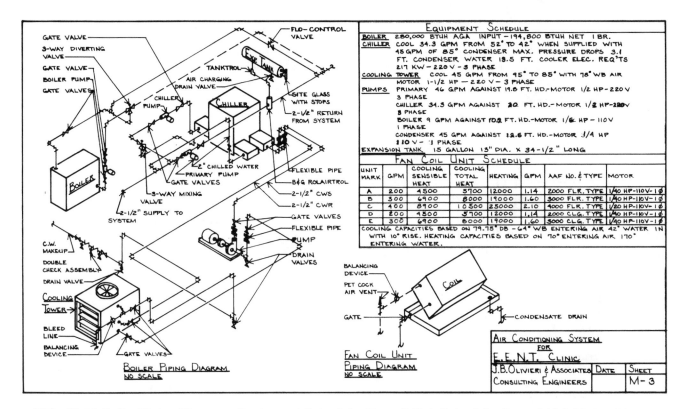

FIG. 10-31 Boiler Piping Diagram; Equipment Schedule; Fancoil Unit Schedule; Fancoil Unit Piping Diagram

After selecting the units, we spot them on the plan as shown in Figs. 10-29, 10-30 and 10-31. Notice that in the basement we use horizontal-discharge ceiling units. Now to run a piping system. To serve the first floor units from overhead would be difficult, so it's advisable to use a tunnel. While a walk-through tunnel would be ideal, our purpose is just as well served by a 4' deep crawl-type tunnel. Utilizing a shallower tunnel in this case can also prove to be less expensive.

For the boiler, we turn once again to Fig. 10-54. Our system requires 197,987 Btuh. We'll select a GEM-300A-WC that has a net IBR output of 208,700. The chimney called for requires a height of 20' with a 9'' diameter.

A McQuay chiller is selected from Fig. 10-58. A nominal 15 ton chiller, WHR 015C, supplies 16.2 tons at 42° leaving chilled water and a 10° rise. This chiller will cool 38.9 gpm form 52° to 42°. We require 44.96 gpm. There is a temptation to select a larger chiller to be sure that all of our 44.96 gpm will be cooled from 52° to 42°. But you must remember that each unit has capacity in excess of our requirements. Certainly, if every fan-coil unit operated wide open, we would need 228,880 Btuh. However, the room thermostats will permit each unit to take only its fair share of the total capacity so don't be concerned. The chiller has a pressure drop of 7.7 ft while the condenser has a 12.1 ft drop at a flow of 44.96 gpm.

When the same mains are used for hot water and chilled water, the primary-secondary piping system is the best method to use. In this system, a primary pump circulates the hot or chilled water to the room units. A secondary pump is provided for a boiler loop and another for a chiller loop. These are piped as shown in the system-piping diagram in Fig. 10-31 where we use larger temperature drops in the secondary circuits than in the primary. The primary circuit uses a small temperature drop and a high flow to achieve optimum heat transfer in the room-unit coils. We don't need the small TD in the secondary, so we use a large TD and save on piping and pumps.

Briefly, the principle in primary-secondary piping circuits is that the pressure drop between the points where the secondary circuits connect to the primary main is so small that no flow will occur in the secondary circuit. In order to have flow in the secondary circuit, we need a secondary pump. So a secondary pump is installed in the boiler secondary loop and the chiller secondary loop. In addition, three-way diverting valves are installed in the secondary circuits. The secondary circuits circulate in a closed loop. Let's say we are heating. If the primary loop

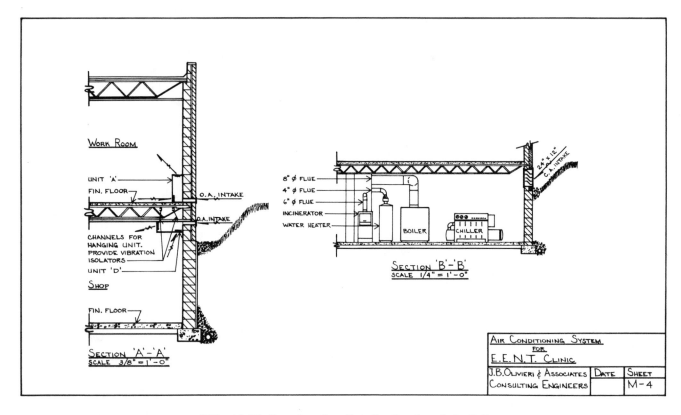

FIG. 10-32 Construction Details Section A-A, B-B

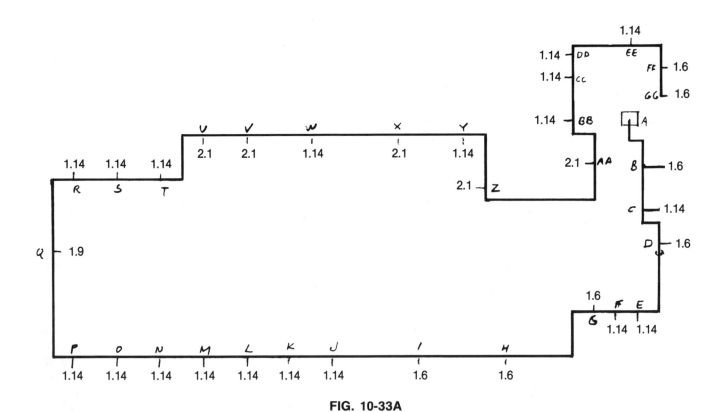

FIG. 10-33A

Section	Gpm	Size	mi/ft	Length	Fittings	Equiv. Length	Total mi.
AB	44.96	2½	180	10	2el = 12.8	22.8	4,104
BC	43.36	2½	160	4		4	640
CD	42.22	2½	150	12	2el = 12.8	24.8	3,720
DE	40.62	2	440	14	3el = 16.8	30.8	13,552
EF	39.48	2	430	14		14	6,020
FG	38.34	2	400	6		6	2,400
GH	36.74	2	360	20	2el = 10.8	30.8	11,088
HI	35.14	2	340	10		10	3,400
IJ	33.54	2	300	14		14	4,200
JK	32.4	2	290	9		9	2,610
KL	31.26	2	270	10		10	2,700
LM	30.12	2	250	10		10	2,500
MN	28.98	2	240	10		10	2,400
NO	27.84	2	220	10		10	2,200
OP	26.7	2	200	10		10	2,000
PQ	25.56	2	195	18	1el = 5.1	23.1	4,505
QR	23.96	2	180	22		22	3,960
RS	22.82	2	155	10		10	1,550
ST	21.68	2	140	10		10	1,400
TU	21.54	2	120	17	2el = 10.2	27.2	3,264
UV	18.44	1½	450	10		10	4,500
VW	16.34	1½	330	10		10	3,300
WX	15.2	1½	300	10		10	3,000
XY	13.1	1½	230	12		12	2,760
YZ	11.96	1½	195	8	1el = 3.9	11.9	2,321
ZAA	9.86	1¼	320	23	2el = 7.2	30.2	9,664
AABB	7.76	1¼	200	13	2el = 6.6	19.6	3,920
BBCC	6.62	1	400	12		12	4,800
CCDD	5.48	1	280	4		4	1,120
DDEE	4.34	¾	800	20	1el = 2	22	17,600
EEFF	3.2	¾	450	11	1el = 1.9	12.9	5,805
FFGG	1.6	½	800	15	12el + 1tee = 16.8	31.8	13,440
GGA	44.96	2½	180	9		9	1,620
Unit from Mfg							24,000
Diverting Valve							36,000
							212,063
							17.6754 ft

FIG. 10-33B Pipe Sizing

1. Primary pump: PD35, 46 gpm against 19.8 ft.

2. Boiler circuit pump: 9 gpm, 1", 760 m.i./ft
40 ft + 3 tees + 8 els + check valve + 3 gate valves + boiler
40 + 16.8 + 22.4 + 20 + 4.2 + 3 = 106.4

$$\frac{106.5 \times 760}{12,000} = 6.7 \text{ ft}$$

B & G 2" supplies 9 gpm against 10.2 head

3. Chiller circuit pump: 38.8 gpm, 2", 380 m.i./ft
40 ft + 2 tees + 6 els + 3 gate valves + chiller
40 + 22.4 + 33.6 + 11.8 = 107.8

$$\frac{107.8 \times 380}{12,000} = 3.4 \text{ ft} + 13.5 \text{ ft (chiller)} + 3 \text{ ft}$$

(diverting valve) = 19.9 ft
B&G PD35 supplies 40 gpm against 20 ft head

4. Condenser water pump: 48.6 gpm, 2½", 200 m.i./ft
120 ft + 3 tees + 15 els + 4 gate valves
120 + 38 + 96 + 18 = 272

$$\frac{272 \times 200}{12,000} = 4.5 \text{ ft.} + 5 \text{ ft (static lift)} + 2 \text{ ft}$$

(condenser) = 11.5 ft
B&G 2½" supplies 50 gpm against 12 ft head.

NOTE: Chiller and condenser pressure drops are taken from manufacturer's literature. Ours are found in figure 10-58.

FIG. 10-33C Pump Selection

water temperature is satisfied, the water in the primary loop goes right on by the secondary connections. At the same time, the secondary pump circulates water through the boiler to the diverting valve and back to the boiler. Let's say the primary loop needs some heat. An aquastat in the primary line would sense this and open the diverting valve, permitting some of the primary water to enter the secondary circuit. As soon as this happens, hot water from the secondary circuit is injected into the primary circuit. The water temperature in the primary loop is reset by an outdoor thermostat.

When cooling is desired, or required, the chiller circuit acts in a manner similar to the boiler circuit to maintain desired primary chilled-water temperature. The advantage of primary-secondary for chillers is that we have a constant flow through the chiller. Also, on switch over from heating to cooling, primary water is slowly permitted to flow into the secondary circuit so no slugs of hot water can enter the chiller to cause damage.

A detailed discussion of primary-secondary systems is available from the Bell and Gossett Company.

The chiller selected is water cooled and should have a cooling tower to conserve water. A Havens Company catalog is included in our library (Fig. 10-59). We can match our chiller with a CBC-15 which will cool 45 gpm from 95° to 85° when supplied with 78° WB air. When we did the heat gain, we stated that we were designing for 72° WB. But when selecting a tower, adding a few degrees is good insurance. A 72° WB design does not mean that the wet-bulb temperature will not exceed 72°. It means that it does only for a small number of hours. Because the capacity of a tower is a function of the wet-bulb temperature, we want to be able to work even during these few hours, so we select at 78°.

The piping was sized and the pumps selected as shown in the work sheets in Fig. 10-33. Notice that when calculating the pressure drop for the cooling tower pump, we added something called static lift. A cooling tower circuit is not a closed circuit. The water is pumped to the top and flows down to the sump by gravity. The water must be lifted this distance between the bottom and top of the tower. This is called the static lift.

The cooling tower presents an architectural problem. The appearance of this low building could be ruined by the installation of this 8' high piece of machinery on the roof. Notice in Fig. 10-30 that it is placed next to Entrance 2 on the outside of the building. This unit is enclosed by a perforated or open brick grillage. Also note the double doors at the front of the tower to facilitate servicing. As shown in Fig. 10-31, a ¾" CW make-up must be run to the tower.

If you examine the plans, you'll see that a drain line is run from each fancoil unit. The water removed in the dehumidification process must be disposed of. It is collected in a drain line and spilled over a floor sink in the boiler room. Remember to keep an air gap equal to two diameters between the drain line and the floor sink. This is a code requirement.

How do you size the drain line? A unit will discharge from 0.25 to 1 gph depending on its size. Consult your local plumbing code for proper sizing of the drain line. If this is not covered in the code then use the following chart:

DRAINAGE SIZES FOR FANCOIL UNITS

Pipe Size	1¼	1½	2	2½	3	4
# of Units	5	15	30	45	90	450

These sizes are based on pitching the pipe ⅛" to the foot.

Finally, an expansion tank selection was made in Fig. 10-34. A combustion air louver is needed. Providing 1 in²/1,000 Btuh input means a 300 in² louver. A 24" × 24" louver should suffice.

Type HB expansion joints and alignment guides (Fig. 10-56) were used to compensate for pipe expansion.

Pipe Size	Ft	Gal Ft.	Gal
2½	200	0.247	49.4
2	400	0.161	64.4
1½	100	0.0925	9.25
1¼	72	0.0655	4.72
1	32	0.0442	1.4
¾	62	0.025	1.6
½	510	0.0121	6.17
Units 32 at 1.5			48
Boiler from manufac.			27.3
Chiller from manufac.			20.4
			232.64

$$T = \frac{0.0175\ (232.64)}{34/48 - 34/103} = \frac{4.07}{0.38} = 10.7\ \text{gal}$$

Use 15 gallon tank 13" dia. × 34½" long

FIGURE 10-34
Expansion Tank for
Hot-Water Chilled-Water System

SELF-CONTAINED, THROUGH-THE-WALL UNITS

Again we start at the beginning and select units as tabulated in Fig. 10-35. From the McQuay catalogs (Fig. 10-60) we select our equipment. In the basement we use the Type J because the condenser intake is high and can be installed above the grade. It also has a horizontal discharge that is necessary because the unit is installed near the ceiling. The first floor units are Type K which

Room Name	Heat Loss Btuh	Heat Gain			Unit Model No.	Gpm
		Rm. Sens. Btuh	Rm. Lat. Btuh	Rm. Total Btuh		
Shop	3,180	4,658	510	5,168	J 9	1.27
Lounge	1,509	3,349	1,020	4,369	J 9	1.27
Exam 1 to 6	3,272	3,018	255	3,273	K 6	1.25
Exam 7	4,830	3,125	255	3,380	K 6	1.25
Entry 4 and Hall	8,222	5,313	642	5,455	CAB HTR 2,000	1.14
Office	4,840	4,148	510	4,658	K 10	1.25
Exam 8, 9	3,272	4,041	255	4,296	K 9	1.25
Drop Room	8,839	13,511	5,100	18,611	2@ K 14	1.33
Visual Fields	3,683	2,865	510	3,375	K 6	1.25
Drs. Office	5,409	6,870	510	7,380	K 14	1.25
Entry 2	6,055	3,181	642	3,823	CAB HTR 2,000	1.14
Gen. Office	4,304	7,384	1,275	8,659	RK 11	1.33
Women	1,518	1,021	0	1,021	2 FT BB RAD	.20
Men	1,518	1,021	0	1,021	2 FT BB RAD	.20
Storage	3,197	911	0	911	2 FT BB RAD	.20
Entry 3 and Hall	13,086	6,146	2,309	8,455	CAB HTR 3,000	1.60
Office	1,417	1,098	255	1,353	K 6	1.25
Work Room	4,398	5,112	765	5,877	K 11	1.25
Fitting	12,216	12,521	1,275	13,796	2@ K 11	1.25
Reception	10,592	7,375	255	7,630	2@ K 6	1.25
Entry 1 and Hall	13,086	4,058	2,309	6,367	CAB HTR 3,000	1.60
Waiting Room	10,544	8,912	1,020	9,932	2@ K 9	1.25
Outside Air 1,300 cfm	95,472					
	244,091					

FIGURE 10-35   Heat Loss — Heat Gain Summary — Unit Selection

have the vertical discharge desired for floor-mounted units. The entries, halls and the general office present a problem because they either do not have an outside wall or enough outside wall to install a unit.

In the case of the general office, we use a Type RK which is a water-cooled unit available with a hot water coil. At the entries, we install cabinet unit heaters that are similar to the fancoil units used in the previous example. Ideally, we should use air conditioners in the halls and entries. Still it is not improper to not use air conditioners. Enough cooling will work its way into the halls to keep them reasonably cool. You will notice that here we install no cooling in the toilets. There is not enough room and, in addition, the heat gain is very small. The exhaust fans required by code will keep the toilets at a reasonable temperature for short occupancies.

Notice for the type S unit the catalog indicates that the stipulated cooling capacity has been determined with the outside air damper closed. If the damper is open (as in our case), you must deduct 10% from the ratings. Yet the ratio of sensible heat to total heat is not indicated

anywhere in the catalog. You must call the manufacturer's representative again, who will in turn call the factory. He will report back that the ratio is 70%. Based on that figure, you deduct 10% from the ratings to get the room-total heat. Multiply this by 70% to get room-sensible heat.

The type K catalog states that the rated conditions are based on 95° DB, 75° WB outside and 80° DB, 67° WB room conditions. Once again we consult the factory representative and inform him of our outdoor and room conditions.

Remember that our outdoor design conditions are 87° DB, 72° DB while the room conditions are 78° DB, 63.5° WB. Therefore, assuming 20% outside air, the entering air conditions are 80° DB and 65.2° WB.

The next step is to select the boiler. Add the outside air load to our room heat loss of 148,619 Btuh. The total outside air is 1,300 cfm and the total heating load is 244,091 Btuh.

From another Peerless boiler catalog, not shown, select a G-1061 with a 360,000 Btuh input and a net output of

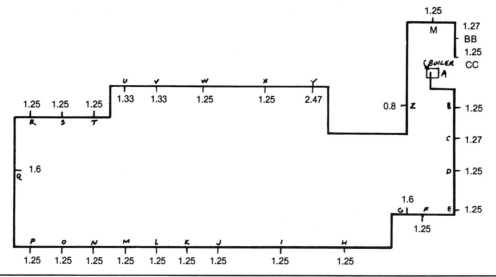

Section	Gpm	Size	m.i./ft	Length	Fittings	Equiv. Length	Total m.i.
AB	30.67	2	350	20	3 el + 2 gate = 25	45	15,750
BC	35.42	2	340	9		9	3,060
CD	34.15	2	330	8		8	2,640
DE	32.90	2	300	14		14	4,200
EF	31.65	2	285	14	1 el = 5.6	20	5,700
FG	30.40	2	270	6		6	1,620
GH	28.80	2	240	20	2 el = 11	31	7,450
HI	27.55	2	225	10		10	2,250
IJ	26.30	2	210	14		14	2,940
JK	25.05	2	195	9		9	1,760
KL	23.80	2	175	10		10	1,750
LM	22.25	2	160	10		10	1,600
MN	21.30	2	150	10		10	1,500
NO	20.05	1½	460	10		10	4,600
OP	18.80	1½	400	10		10	4,000
PQ	17.55	1½	390	18	1 el = 4.2	22	8,580
QR	15.95	1½	310	22	1 el = 4.2	26	8,060
RS	14.70	1½	270	10		10	2,700
ST	13.45	1½	240	10		10	2,400
TU	12.20	1¼	400	17	2 el = 7	24	9,600
UV	10.87	1¼	350	10		10	3,500
VW	9.45	1¼	300	10		10	3,000
WX	8.29	1¼	250	10		10	2,500
XY	7.04	1¼	410	12		12	4,020
YZ	4.57	1	200	32	3 el = 7.5	40	8,000
ZAA	3.77	¾	410	13		13	5,330
AABB	2.52	¾	260	12	1 el = 1.3	14	3,640
BBCC	1.25	½	460	13	1 el = 1.4	14	6,440
CCDD	1.25	½	460	*15	12 el + 1 Tee = 20	35	16,100
DDA	36.67	2	350	16	3 el + 1 gate = 21	37	12,950
							157,230

Unit
Coil   Received verbally from manufacturer   12,000

                                                                          169,230
                                                                          14.1 ft

*Includes distance and fittings through unit but does not include unit.

FIGURE 10-36 Pipe Sizing

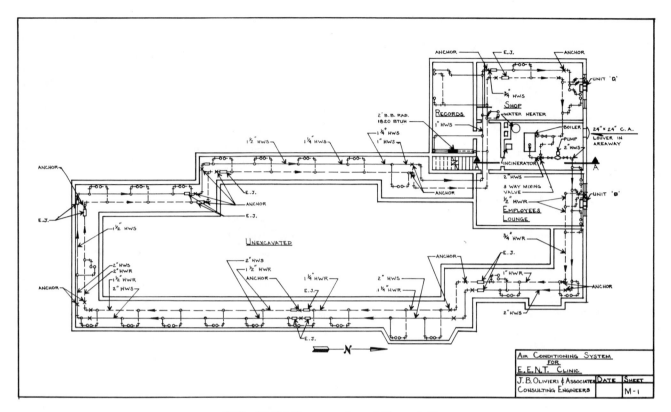

FIG. 10-37 Basement and Tunnel Plan

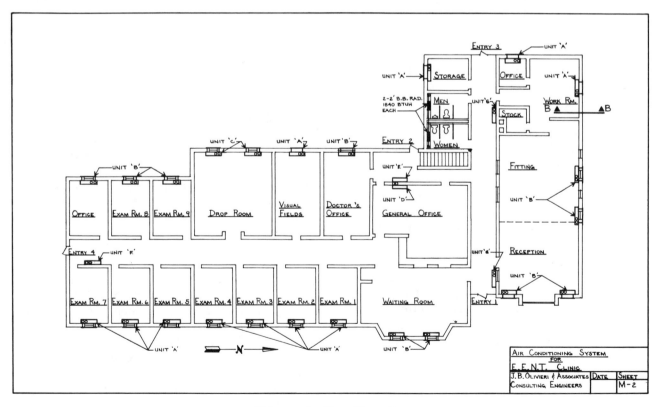

FIG. 10-38 First Floor Plan

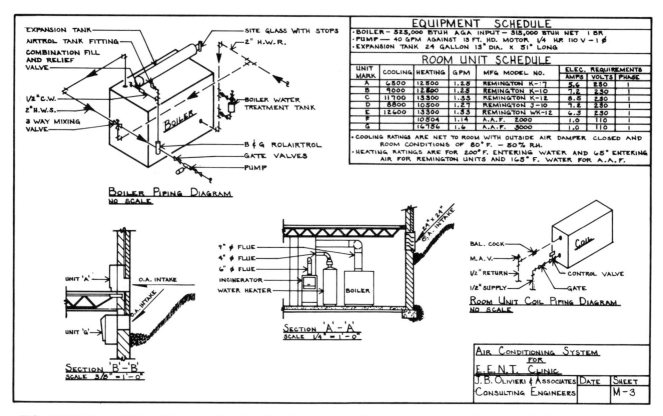

FIG. 10-39 Boiler Piping Diagram, Section B-B; Section A-A; Equipment Schedule; Room Unit Schedule; Room Unit Coil Piping Diagram

250,400 Btuh.

Next, run the piping. As in the last example, it might be wise to use a tunnel for the piping. Notice that cold water and a drain must be run to the type EW unit in the general office. After the piping is drawn, the piping is sketched (Fig. 10-36) and the pipe is sized. These sizes are transferred to the plans and the details drawn (Figs. 10-37, 10-38, 10-39).

The pressure drop is 14.1 ft as determined in Fig. 10-36. We must select a pump that will deliver 36.67 gpm against a 14.1 ft head. The Bell & Gossett catalog shows a HD3 pump that delivers 36.67 gpm against 16.6 ft of head. The expansion tank, from Fig. 10-40, is 30 gallons.

The temperature control system should include a three-way mixing valve to vary the water temperature with outdoor temperature. Each unit, except the heating-only units, should have a modulating valve controlled by a room thermostat — either unit or wall mounted. The fan will run continuously. If a room continues to overheat after the coil valve closes, then the mixture of outside air and return will cool the room (in winter). The outside air damper can be manually controlled or motorized. If desired, these motorized dampers can be controlled by a programming clock which will open them during the day and close them at night. The cooling controls come with the unit.

MULTIZONE AIR SYSTEM

To design an air system it is first necessary to determine the amount of air that must be circulated. Initially, we must calculate the room-sensible heat/room-total heat ratio (Fig. 10-41). Then using a psychrometric chart (Fig. 10-42), we can determine the Apparatus Dew Point (ADP). Using the following formula,

$$\text{cfm} = \frac{\text{Btuh (sensible)}}{1.08 \, (\text{air supply temp} - \text{room temp})}$$

we can find the air quality for each room. If we use 10% outside air, the entering air conditions will be 78.9° DB and 64.35° WB. Our ADPs vary from 44° to 54° which presents a bit of a problem because we can only have one ADP. Let's assume that we are going to select 52°, a good median selection. On Fig. 10-42 we draw our ADP line for the worst case 44°. For the moment, we will assume a bypass factor of 0.10 which places us at 54.69 DB and 53.3 WB for 52°ADP. Now draw a line parallel to the

44°ADP line but through 54.69 DB, 53.3 WB. Next read the room condition of about 50% R.H. at 78° which is still within the comfort zone. As a result, we will use 52° ADP for all rooms to determine required cfm. Now that we have our cfm (Fig. 10-41) we can select our diffusers, air handler, heating coil and cooling coil. And as indicated, we'll use 52° ADP for all rooms to determine the required cfm. The diffusers are selected from the Titus catalog page D2 (Fig. 10-61).

Pipe Size	Ft Pipe	Gal /Ft	Gal
2	400	.161	64.4
1.5	160	.0925	14.8
1.25	50	.0655	3.3
1	90	.0422	4.0
.75	25	.025	.6
.5	40	.0121	.5
29 units @ 1.5			43.5
Boiler			63
TOTAL			194.1

$$T = \frac{0.035\,(194)}{\frac{34}{58} - \frac{34}{103}} = \frac{6.8}{.25} = 27 \text{ gal}$$

Use 30 gal tank 13″ × 61½″

FIGURE 10-40   Expansion Tank Sizing

When selecting condensing units and cooling coils, care must be taken not to pyramid peaks. If you add up the peak loads for each room, you may get an artificially high load. Because we did our room heat gains for 8:00 A.M., 10:00 A.M., Noon, 4:00 P.M. and 6:00 P.M., we can get a total for each time period. In this building, most of the rooms peak at 4:00 P.M., so the 4:00 P.M. total of individual room peaks are very close. Yet this may not always be true.

We are using a condensing unit and not a water chiller. Water chillers are normally used when you have more than one air handling unit or are using a VAV system. Condensing units can be water-cooled or air-cooled. If a water-cooled unit is used, a cooling tower is also used in order to conserve water. Water-cooled units use less electricity than air-cooled but require more maintenance. For a small system like this, an air-cooled unit is a better choice. If all the air goes through the cooling coil, the load will be 13.6 tons. From the McQuay condensing unit catalog (Fig. 10-62), we find that at a 95°F outdoor temperature and 40°F saturated suction temperature, an ST-013 unit has a capacity of 13.5 tons. This is a little confining and does not allow for line losses. Rather, we should then select a ST-015 that has a slightly larger 15.2 ton capacity.

From the McQuay air handler catalog (Fig. 10-63), we select a 108 blow-through or multizone unit with a range of 2,200 to 5,500 cfm. The cooling coil is available with 8.5 ft² of face area and the heating coil with 4.5 ft². Dimensions are shown on page 42 of the catalog. The unit does not have a combination filter and mixing box so we turn to page 51. Without the mixing box, the unit has overall dimensions of 53⅛″W × 80⅜″L × 56″H. If we add the mixing box on the back of the unit, the overall length increases by approximately 38″. We can conserve space by mounting the mixing box on top of the fan section. The cooling coil dimensions are 30″ × 40½″ and the heating coil is 18″ × 35½″.

Next, we turn to the coil catalog Fig. 10-64. Coil capacities and leaving conditions vary depending on the following factors: entering air dry bulb and wet bulb temperatures, refrigerant temperature, rows of tubes, face velocity and fin spacing (also referred to as fin series which is equivalent to the number of fins/inch).

We are not making a final coil selection. Rather, that will be done by the successful bidder. We are only interested in a preliminary selection so that we can calculate the coil pressure drop in order to determine our maximum motor HP. Our electrical engineer needs to know this information also so he can proceed with his drawings.

First, calculate the entering conditions. Once again, we will assume 10% outside air.

Ent DB = (87 × .1) + (78 × .9) = 78.9°
Ent WB = (72 × .1) + (63 × .9) = 64.35°

The entering dry bulb temperature of 78.9 is not to be found so we must interpolate between 78° and 80°. The charts don't list 65 WB, so another interpolation is necessary. However, we are only estimating, so let's not interpolate and merely use 80° DB and 64° WB for our entering air conditions.

If all of the air goes through the cooling coil, the face velocity will be equal to the cfm divided by the coil face area. Dividing 4,752 cfm by 8.5 ft² gives a face velocity of 559 fpm. Again, we will not interpolate but instead use an estimate of 600 fpm. Remember our coil leaving dry bulb temperature needs to be 53°F or less. The following is a partial list of possible selections.

Refrig Temp	Rows	Fin Series	DB	WB*
40	4	12	51.9	51.4D
41	4	12	52.5	51.9D
42	4	12	53.1	52.5D
40	5	08	52.6	51.5E
41	5	10	51.6	51.1D
42	5	10	52.2	51.7D
43	5	10	52.8	52.3D
44	5	12	52.5	52.2D

*The letters indicate the degree of wetness and are used later in determining the air side pressure drop.

We could select the 4 row, 12 fin series unit that is practically on the button when operating at 42 °F. We could, but this might be cutting it too close. Keep in mind these ratings are for a perfect coil and ideal conditions just don't exist in an installation. Coils can get dirty or fins slightly bent. We are in luck though. Our condensing unit has a capacity of 15.2 tons so we can operate down to about 40.5 °F suction temperature. This provides an adequate safety factor, particularly when you consider that we assumed higher inlet temperatures and face velocities to avoid having to interpolate.

The air side pressure drop is found by using the nomograph on page 10 of the McQuay coil catalog (Fig. 10-65). Enter the bottom at 559 face velocity. At the intersection of 559 and D, rise vertically until you reach 12 fin series. Now move horizontally to the right and read, under 4 rows, 0.65.

We now come to the heating. There are two choices: we can try to heat from overhead or combine overhead heating with baseboard heating. In Fig. 10-41, notice that we have calculated the heating temperature differences. They range from 11 °F to 87 °F. The cfm in rooms with a temperature difference in excess of 50 °F were recalculated for a 50 ° temperature difference and the new air quantities listed in the last column, HTNG CFM ADJ. If the heating temperature differences had been below 25 °F, it would be safe to heat from overhead. In this case it would be wise to add a baseboard radiation system. The baseboard system will provide 75% of the heating requirements, except in the entries where all heating will be done from overhead.

The heating coil leaving air temperature will be set at $72 + 50 = 122$ °F to satisfy the entries. If we assume that all 4,752 cfm must be heated to 122 °F, we will have a grossly oversized coil. To determine the amount of air going through the heating coil, first multiply the heat loss (exclude entries) by 25%. The heating coil heating load will be $108,170 \times .25 + 40,449 = 67,541.5$. Dividing 67,541.5 by 50 °F and 1.08 gives 1250 cfm. The coil heating load is equal to $1,250 (1.08)(122 - 65.2)$ or 76,680 Btuh.

The coil face velocity is 1,250 cfm/4.5 ft^2 or 278 fpm. If we use the same 20 °F water temperature difference that we used for the radiation, our GPM will be,

$$GPM = \frac{76,680}{500 (20)} = 7.7$$

The next step is to determine the number of feeds. Assuming 4 gpm for feed, we will need $7.7 \div 4$ or 1.925 feeds. Let's use two feeds or 3.85 gpm/feed.

We have to determine the heat transfer value M_t. To do this we find two factors, factor a and factor b. These two factors and the nomograph on page 22 (Fig. 10-65) of the catalog will give us M_t.

equation for M_t

$$a = \frac{\text{Air Temp Rise}}{\text{Ent Water Temp}} - \text{Ent Air Temp}$$

$$a = \frac{(122 - 65.2)}{(210 - 65.2)}$$

$$a = 0.39$$

$$b = \frac{\text{Water Temp Drop}}{\text{Air Temp Rise}}$$

$$b = \frac{20}{(122 - 65.2)}$$

$$b = 0.35$$

$$M_t = 0.58$$

Now determine the heat transfer value or R_{ft}. $R_{ft} = R_{f1} + R_{f2}$. Locate R_{f1} and R_{f2} on page 21 (Fig. 10-65) of the catalog. At 3.85 gpm/feed and 200 °F average water, read $R_{f1} = 0.1$. Although our face velocity is 278 fpm, we must use 300 fpm because that is how low the chart reads. At 300 fpm and 06 fin series (first trial), we read $R_{f2} = 0.88$.

$R_{ft} = 0.1 + 0.88$ or 0.98. We can now determine how many rows deep the coil must be.

$$\text{Rows Deep} = R_{ft}(M_t)(\text{Face Velocity} \div 100)$$

$$\text{Rows Deep} = 0.98(0.58)(278 \div 100)$$

$$\text{Rows Deep} = 1.58 \text{ rows}$$

Let's see if we can get down to 1 row by using a higher fin series. We will work backwards and solve for R_{f2} since it is the only factor influenced by fin series.

$$\text{Rows Deep} = (R_{f1} + R_{f2})(M_t)(\text{Face Velocity} \div 100)$$

$$1 = (0.1 + R_{f2})(0.58)(278 \div 100)$$

$$R_{f2} = 0.52$$

From the chart we see that a 10 fin series will give us an R_{f2} of less than 0.52.

Using the same chart as we used for the cooling coil we determine that at 278 fpm and series 10 the air side pressure drop will be 0.038.

As I stated earlier, we are using a baseboard system similar to that used in the first example (Figs. 10-23,

Room Name	*Heat Loss Btuh	Room Heat Gain Btuh			SHF	ADP	CFM	ADJ CFM	HTNG Temp Rise	HTNG CFM ADJ
		Sens.	Latent	Total						
Shop	3,180	4,658	510	5,168	.90	52	173	170	17	
Lounge	1,509	3,349	1,020	4,369	.77	48	124	125	11	
Exam 1 to 6	3,272	3,818	255	3,273	.92	52	112	110	28	
Exam 7	4,830	3,125	255	3,380	.92	52	116	115	39	
Entry 4 and Hall	8,222	5,313	642	5,955	.89	52	89	90	85	150
Office	4,840	4,148	510	4,658	.89	52	154	155	29	
Exam 8 and 9	3,272	4,041	255	4,296	.94	53	150	150	20	
Drop Room	8,839	13,511	5,100	18,611	.73	44	500	500	16	
Visual Fields	3,683	2,865	510	3,375	.85	51	106	105	32	
Drs. Office	5,409	6,870	510	7,380	.93	52	254	255	20	
Entry 2	6,055	3,181	642	3,823	.83	51	119	120	45	
Gen. Office	4,304	7,384	1,275	8,659	.85	51	274	275	14	
Women	1,518	1,021	0	1,021	1.00	54	38	40	35	
Men	1,518	1,021	0	1,021	1.00	54	38	40	35	
Storage	3,197	911	0	911	1.00	54	24	25	87	45
Entry 3 and Hall	13,086	6,146	2,309	8,455	.73	44	232	230	54	242
Office	1,417	1,098	255	1,353	.81	50	41	40	33	
Work Room	4,398	5,112	765	5,877	.87	51.5	187	190	21	
Fitting	12,216	12,521	1,275	13,796	.91	52	464	465	23	
Reception	10,592	7,375	255	7,630	.97	54	273	275	36	
Entry 1 and Hall	13,086	4,058	2,309	6,367	.64	32	155	155	79	245
Waiting Room	10,544	8,912	1,020	9,932	.90	52	330	330	30	
	148,619	128,769	21,202	149,971				4,660		4,752
10% outside air	34,899									
	183,518									

*Heat loss does not include infiltration except at entries.

FIGURE 10-41 Heat Loss — Heat Gain Cfm Summary

10-24, 10-25). We have not shown this on the drawings for the sake of brevity. Similarly, we will not go through a complete pipe sizing, pump or expansion tank selection. Our pressure drop in our piping will be no larger than our baseboard system, as shown in Fig. 10-47. Our flow will be greater due to the increased load, but the pump head will not be any greater. We need 108,170 Btuh or 10.8 gpm for the baseboard plus 7.7 gpm for the coils yielding a total of 18.5 gpm. A B&G HD3 will deliver 18.5 gpm against 18 ft of head. We require 20 ft of head. Our natural inclination is to go to the next larger pump, a PD35. But is it really necessary? Using the relationship,

$$Q_2 = Q_1 \times \sqrt{\frac{P_2}{P_1}}$$

$$Q_2 = 18.5 \text{ gpm} \times \sqrt{\frac{18 \text{ ft}}{20 \text{ ft}}}$$

$$Q_2 = 17.6 \text{ gpm}$$

we reduce our flow by 0.9 gmp, but more importantly, this only results in a 1 °F rise in TD. Instead of operating at 20°TD, we are at 21°TD which is an insignificant difference. The coil has less than 2 gallons of water, so we can use the same expansion tank as used in Example 1.

Now that we have assembled most of our data, let's start drawing. First, select diffusers from the Titus catalog (Fig. 10-61) and spot them on your drawing. Notice that the noise criteria table doesn't list examination rooms, so we must make a judgement. Examination rooms should be somewhat similar to private hospital rooms, so let's use NC 25 to 35, yet staying as close to NC 25 as possible. The offices will be NC 30 to 40, and the remaining rooms can be treated as general offices, NC 35 to 45. As a rule of thumb, let's select all diffusers to be below NC 30.

The next step is to zone the building for exposure and occupancy (Figs. 10-43, 10-44, 10-45). Examination rooms 1 through 7 are alike and all face east, so a natural zone presents itself. Similarly, examination 8 and 9, office, visual fields and doctors' offices should comprise zone 2. The drop room is better as a separate zone (zone 3) because of the variation in occupancy. The waiting room should be zone 4 for the same reason. The general office and reception room are completely interior and will require cooling even while other zones are heating, so they should be designated as zone 5. Unassigned, office, work-

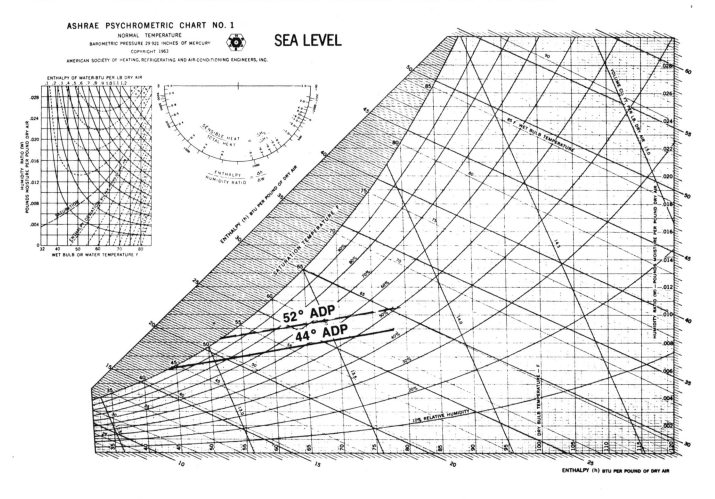

FIG. 10-42 ASHRAE Psychrometric Chart (Reprinted by permission from *ASHRAE Guide and Data Book, 1963*)

room, fitting area and reception should constitute zone 6. Entries 1 and 3 make up zone 7; and the basement, zone 8. Entry 4, which has little usage, can either be added to zone 1 or 2. The hall should be part of zone 5. Entry 2 should be part of zone 3.

I must point out that this is a lot of zoning for a building of this size. Still, the only better solution would be to have individual room control. In that way we could reduce the zones to three on the first floor and one in the basement. But would we save much? Not really. Once you've committed yourself to zoning, going from four to eight zones increases the cost by only one or two percent.

The next step is to lay-out the boiler room. As you can see, there isn't enough room for the multizone unit in the area designated as the boiler room (Fig. 10-45). In addition, many communities will not permit an air-handling unit in a boiler room. So this takes some architectural juggling. By switching the boiler room and the em-

ployee's lounge, we'll allocate a larger space to house the multizone unit. We will divide that newly designated area into two rooms, a boiler room and a fan room.

At this time, it's a good idea to rough out the main duct sizes and decide just how to get them upstairs.

Zone	cfm	Duct Size
1	925	16 × 8
2	815	12 × 8
3	620	10 × 8
4	330	8 × 6
5	275	6 × 6
6	1015	14 × 8
7	475	8 × 6
8	290	6'' dia.

Notice that the space next to the chimney makes a natural duct space. In addition to the supply ducts, we will need a return-air duct which will be 44'' × 10''. These all fit rather neatly, as shown, in a duct shaft with inside

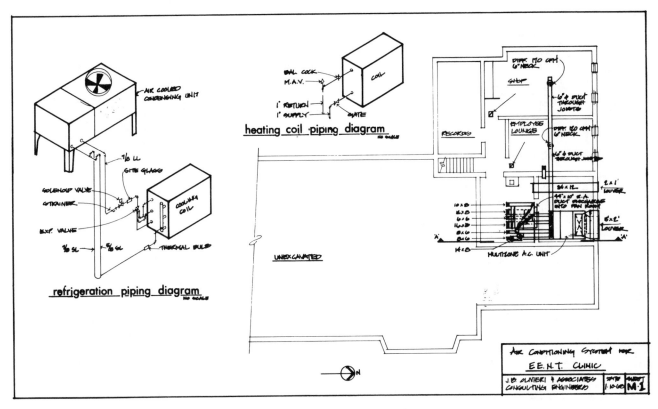

FIG. 10-43 Basement Plan

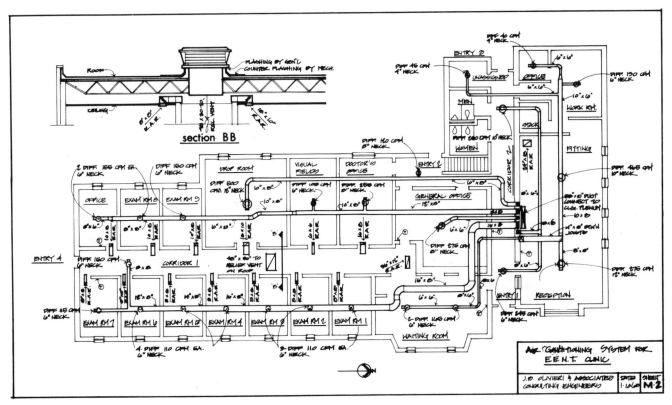

FIG. 10-44 First Floor Plan

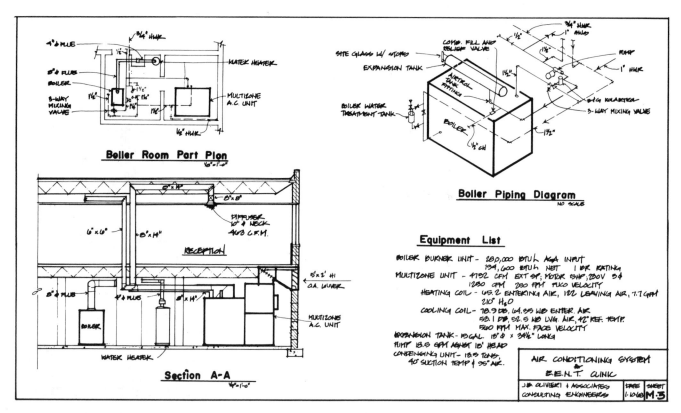

FIG. 10-45 Boiler Room Part Plan; Section A-A; Boiler Piping Diagram; Equipment Schedule

dimensions of 10' × 2'.

The architectural plans show the ceiling tight to joists. In order to get our ductwork in, it will be necessary for the architect to raise the roof by at least 10'' to provide sufficient suspended ceiling space for the ductwork.

The ductwork is sized using the static regain and equal friction method. The duct sizing is shown in Fig. 10-46. Notice how we tried to keep the pressure drop in each duct nearly equal for ease of balancing. As the pieces begin to fall into place, note that the fan room serves as a plenum and the return air damper of the mixing box draws out of the room. The outside air is connected to the outside-air connection. A louver is installed in the areaway on the outside wall. Louvers are sized to have maximum face velocities as follows:

12'' high	350 fpm
24'' high	400 fpm
36'' high	500 fpm
60'' high	500 fpm

This will assure you of a maximum net face velocity of 800 fpm and a pressure drop of about 0.06''. This means that we can have a 14' × 1', 6' × 2' or a 3' × 3' louver. We must also provide a combustion-air louver and duct into the boiler room. The louver is the same size as in the previous example, 2' × 1'. A duct, the full size of

the louver, is run into the boiler room. As long as the duct is the same size as the louver, and no longer than 20' with no elbows, no combustion-air fan is necessary. With atmospheric burners, just remember that the maximum permissible pressure drop in the combustion-air louver and duct is 0.01''.

The return air system consists of return-air registers with short lengths of lined ductwork discharging into the ceiling space. This is a good way of handling return air for several reasons. First, it saves ductwork. Secondly, it ventilates the ceiling or attic space and, finally, we pick up part of the roof heat gain with the return air.

When using a ceiling-plenum system, be sure that any walls that run up to the roof deck are sleeved to permit the air to get into the duct shaft. In some communities, it may be necessary to install a fire damper at each register or where you pierce a wall. Check your code! But code or not, it is still a good idea to install a fire damper in the main return air duct where it connects to the suspended ceiling and where it discharges into the fan room. Also install a smoke detector in the return air duct. When the detector senses smoke, the outside air damper opens fully and the return is closed. This dilutes smoke and makes it easier for occupants to evacuate the building. It also makes fighting a fire easier.

The next step is to determine the total external static

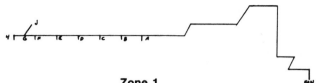

Zone 1

Duct Sect	Cfm	Vel	Size	$\frac{H_f}{100}$	H_v	Lgth	Fittings	Equiv Lgth	Loss	Gain	Diff	Acum Loss
Fan A	925	1,040	16 × 8	0.165	0.067	80	9 elbows	140	− 0.231	0	− 0.231	− 0.231
AB	815	917	16 × 8	0.135	0.052	10		10	− 0.014	+ 0.008	− 0.006	− 0.237
BC	705	793	16 × 8	0.10	0.039	10		10	− 0.010	+ 0.007	− 0.003	− 0.240
CD	595	669	16 × 8	0.054	0.028	10		10	− 0.005	+ 0.006	+ 0.001	− 0.239
DE	485	546	16 × 8	0.047	0.019	10		10	− 0.005	+ 0.005	0	− 0.239
EF	375	480	14 × 8	0.036	0.014	10		10	− 0.004	+ 0.003	− 0.001	− 0.240
FG	265	398	12 × 8	0.029	0.010	5		5	− 0.001	+ 0.002	+ 0.001	− 0.239
GH	115	256	8 × 8	0.017	0.004	5		5	− 0.002	+ 0.003	+ 0.001	− 0.238
GJ	150	270	10 × 8	0.015	0.005	10	Takeoff = 1 elbow	14	− 0.002	+ 0.003	+ 0.001	− 0.238

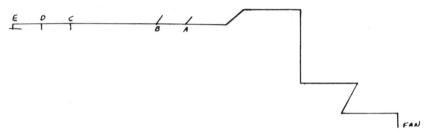

Zone 2

Duct Sect	Cfm	Vel	Size	$\frac{H_f}{100}$	H_v	Lgth	Fittings	Equiv Lgth	Loss	Gain	Diff	Acum Loss
Fan A	815	1,222	12 × 8	0.24	0.093	60	5 el + 2 − 45°el	90	− 0.216	0	− 0.216	− 0.216
AB	560	1,008	10 × 8	0.16	0.069	13		13	− 0.021	+ 0.012	− 0.009	− 0.225
BC	455	819	10 × 8	0.125	0.042	30		30	− 0.038	+ 0.027	− 0.011	− 0.236
CD	305	686	8 × 8	0.015	0.029	10		10	− 0.011	+ 0.007	− 0.004	− 0.240
DE	155	465	8 × 6	0.060	0.013	10		10	− 0.006	+ 0.008	+ 0.002	− 0.238

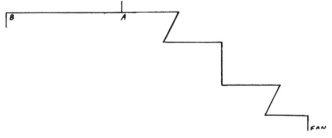

Zone 3

Duct Sect	Cfm	Vel	Size	$\frac{H_f}{100}$	H_v	Lgth	Fittings	Equiv Lgth	Loss	Gain	Diff	Acum Loss
Fan A	620	1,116	10 × 8	0.230	0.078	55	7 elbows	83	− 0.191	0	− 0.191	− 0.191
AB	500	900	10 × 8	0.150	0.050	45		45	− 0.068	+ 0.014	− 0.054	− 0.245

FIG. 10-46 Duct Sizing

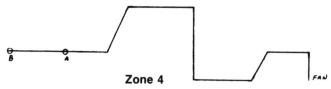

Zone 4

Duct Sect	Cfm	Vel	Size	$\frac{H_f}{100}$	H_v	Lgth	Fittings	Equiv Lgth	Loss	Gain	Diff	Acum Loss
Fan A	330	990	8 × 6	0.250	0.061	55	7 elbows	71	−0.178	0	−0.178	−0.178
AB	105	660	6 × 6	0.140	0.027	12		12	−0.017	+0.017	0	−0.178

Zone 5

Duct Sect	Cfm	Vel	Size	$\frac{H_f}{100}$	H_v	Lgth	Fittings	Equiv Lgth	Loss	Gain	Diff	Acum Loss
Fan A	275	1,100	6 × 6	0.360	0.075	45	5 elbows 1 − 30°el	61	−0.220	0	−0.220	−0.220

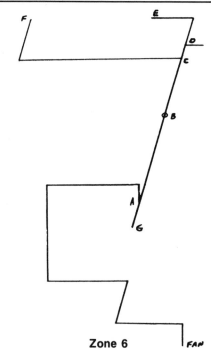

Zone 6

Duct Sect	Cfm	Vel	Size	$\frac{H_f}{100}$	H_v	Lgth	Fittings	Equiv Lgth	Loss	Gain	Diff	Acum Loss
Fan A	1,015	1,305	14 × 8	0.26	0.106	35	6 elbows	70	−0.182	0	−0.182	−0.182
AB	740	1,110	12 × 8	0.15	0.077	15	1 elbow	19	−0.029	+0.015	−0.014	−0.196
BC	275	619	8 × 8	0.085	0.024	20		20	−0.017	+0.027	+0.010	−0.186
CD	230	517	8 × 8	0.060	0.017	1		1	−0.001	+0.004	+0.003	−0.183
DE	40	160	6 × 6	0.010	0.002	10	1 elbow	13	−0.001	+0.008	+0.007	−0.176
CF	45	180	6 × 6	0.010	0.002	26	1 elbow 1 Takeoff	32	−0.003	+0.008	+0.005	−0.171
AG	540	1,215	8 × 8	0.310	0.092	6	1 elbow	9	−0.028	+0.007	−0.021	−0.192

FIG. 10-46 *(Continued)*

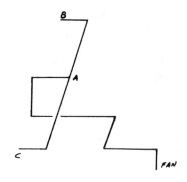

Zone 7

Duct Sect	Cfm	Vel	Size	$\frac{H_f}{100}$	H_v	Lgth	Fittings	Equiv Lgth	Loss	Gain	Diff	Acum Loss
Fan A	475	1,425	8 × 6	0.500	0.127	30	5 elbows	50	− 0.250	0	− 0.250	− 0.250
AC	245	735	8 × 6	0.140	0.034	35	1 elbow 1 Takeoff	43	− 0.060	+ 0.047	− 0.013	− 0.263
AB	230	690	8 × 6	0.125	0.030	16	1 elbow 1 Takeoff	24	− 0.030	+ 0.047	+ 0.017	− 0.246

Zone 8

Duct Sect	Cfm	Vel	Size	$\frac{H_f}{100}$	H_v	Lgth	Fittings	Equiv Lgth	Loss	Gain	Diff	Acum Loss
Fan A	290	1,500	6″ DIA	0.650	0.140	20	1 elbow	26	− 0.169	0	− 0.169	− 0.169
AB	170	870	6″ DIA	0.240	0.047	15		15	− 0.036	+ 0.047	+ 0.011	− 0.158

Return Air

15 ft ductwork + elbow	= 20 ft.
Loss = 20 ft. × 0.2/100	= − 0.04
Register (From Manufacturer)	= − 0.055
Entrance Loss (clg plenum to duct)	= − 0.15
Entrance Loss (from room to unit)	= − 0.056
Total losses return air	− 0.301
Supply main	= − 0.263
Diffuser (From Manufacturer)	= − 0.11
Diffuser connection to duct	
= Elbow or turning vane	= − 0.015
Total losses supply air	− 0.388
Total external static pressure required	= − 0.689
Coil, filters, Cabinet	= − 1.26
Mixing box	= − 0.10
Total system static pressure required	− 2.049

58″×10″ RA   .16″/100′   1800 fpm

FIG. 10-46 *(Continued)*

pressure, also shown in Fig. 10-46. Generally, you specify the external static pressure, not the total. The manufacturer's representative determines the loss through his unit and adds this to your external to get the total. The internal losses vary from manufacturer to manufacturer, so it is wise to specify only the losses external to the unit.

Coil Pressure Drop

From page 12, Airtemp Multizone Catalog, we find that a 1N unit heating coil has 14 tubes per row and is 47 in. long. Coil is circuited single serpentine, so:

$$\text{Water velocity} = \frac{24 \text{ gpm}}{14 \text{ circuits}} \times 1.24$$

Water velocity = 2.1 fps

On page 54, table 35, by interpolation find a pressure drop of 3 ft for a 6 pass coil 50 in. long. Single serpentine coil has 2 passes (table 36). So we subtract 4 (0.26) to get 1.96 ft.

Piping = 30 ft + 6 els + 2 tees

Pipe size = 1½ in., 600 m.i./ft

Total equiv lgth = 30 + 6 (4.4) + 2 (4.4) = 65 ft

$$\text{Piping} \quad \frac{65 \text{ ft} \times 600 \text{ m.i.}}{12,000 \text{ m.i./ft}} = 3.25 \text{ ft}$$

Coil	1.96 ft
	5.21 ft

FIGURE 10-47 Pump Selection

Room Name	Heat Loss Btuh	Cfm	Htn'g Temp. Rise	Mixing Box Model No
Records	1700	000	00°	
Shop	2685	220	12°	L-6C
Employees Lounge	1705	135	13°	L-5C
Exam Rm 1 to 6	3886	130	27°	6/24W
Exam Rm 7	5750	130	41°	6/24W
Entry 4	7456	200	35°	6/24W
Hall	3900	190	19°	6/24W
Office	5750	170	32°	6/24W
Exam Rm 8, 9	3886	165	22°	6/24W
Drop Rm	10290	540	18°	8/36W
Visual Fields	4285	75	53°	6/24W
Drs. office	6442	300	20°	8/36W
Entry 2	8478	255	31°	8/36W
Gen. Office & Reception	4800	275	16°	8/36W
Women	1700	40	40°	6/24W
Men	1700	40	40°	6/24W
Unassigned	2918	45	60°	6/24W
Entry 3	16866	415	38°	8/36W
Hall	1680	80	20°	6/24W
Office	2177	40	50°	6/24W
Work Rm	5930	190	29°	6/24W
Stock	635	35	17°	6/24W
Fitting	13650	400	32°	8/36W
Reception	13780	540	24°	8/36W
Entry 1	16866	245	65°	8/36W
Waiting Rm	12310	435	26°	8/36W

FIGURE 10-48 Summary of Loads

Now the size of the fan and motor can be determined (Fig. 10-66). McQuay manufacturers both forward curved and backward curved airfoil fans. For industrial buildings, where the air may be dirty, it is considered good practice to use backward inclined fans. Backward inclined fans should also be used in double duct and VAV systems. In our case, a clean building using a multizone system, a forward curved fan can be used. This will save us a little in first or initial cost. A 15" fan, page 20 in the McQuay catalog, requires just over 3 BHP for 4,750 cfm at 2.049 in. static pressure. So we will specify a 5-HP motor as the minimum acceptable size.

VAV SYSTEM

The design of the VAV System is similar to the design of the multizone system. The same room air quantities are used as were used in the multizone system. These are shown in Fig. 10-41. The VAV system will make every room a zone. A damper is placed in the branch duct that runs from the main to the diffuser. This damper is motor operated and is controlled by a room thermostat. The ductwork system is used for cooling in the summer and as a ventilating system in the winter. This VAV System is really a split system because the heating is provided by radiation, in this case baseboard. There are two reasons why we don't use the ductwork system for heating. First of all, in a building that's located in a northern climate, we don't want to heat from overhead. The other reason is that there may be rooms that will simultaneously need to be cooled during the heating season. A good example is the general office. Here the only heat loss takes place through the roof because it is an interior room. The lights alone will offset the roof load down to 30°F outdoors.

First we'll select a 108 draw-through unit from our McQuay air handler catalog (Fig. 10-63). This unit has a range of 2,200 to 5,500 cfm. We select the large area cooling coil which is 30" × 40.5" and has a 8.5 ft² face area. We spot this on our basement plan.

We have our air quantities in Fig. 10-41, so we can select our diffusers and air valves (dampers) from the Titus catalog, pages D19 and G8 (Fig. 10-61). When selecting diffusers for a VAV System, check to see what the performance will be at the expected minimum cfm. What's the minimum air flow? This is not easy to predict. We can estimate what it will be by doing cooling loads at various outdoor conditions. Our cooling load assumed a bright sunshine day. Even so, the 8:00 A.M. cooling load is only 739 Btuh, less than a fifth of the peak load. This means that at 8:00 A.M., the room will receive 27 cfm of 52°F air. If this was all outside air or air that has been filtered through activated charcoal, it would be enough for the two people expected to be in that room.

Duct Sect	cfm	Rect Duct	Equiv Dia	Vel FPM	H_v	$\dfrac{H_f}{100}$	Lgth	Fittings	Loss	Gain	Diff.	Net
Fan A	4,055	38×10	20.3	1,537	0.144	0.200	12	1 elbow H/W = 1 c = 0.09 1 trans c = 0.15 Loss = 0.035	−0.059	0	−0.059	−0.059
AB	3,760	38×10	20.3	1,425	0.127	0.180	10	1 elbow H/W = 3.8 c = 0.071 Loss = −0.0009	−0.027	+0.009	−0.018	−0.077
BC	2,680	38×10	20.3	1,016	0.064	0.090	34	1 elbow H/W = 3.8 c = 0.071 2 elbow H/W = .26 c = .19 Loss = −0.017	−0.048	+0.032	−0.016	−0.093
CD	2,080	36×10	19.8	832	0.043	0.070	17		−0.012	+0.013	+0.001	−0.092
DE	1,720	36×10	19.8	688	0.030	0.048	13		−0.006	+0.007	+0.001	−0.091
EF	1,510	34×10	19.3	640	0.025	0.045	10		−0.005	+0.003	−0.002	−0.093
FG	1,400	34×10	19.3	593	0.022	0.042	10		−0.004	+0.002	−0.002	−0.095
GH	900	26×10	17.2	498	0.015	0.030	10		−0.003	+0.004	+0.001	−0.094
HI	790	26×10	17.2	438	0.012	0.024	10		−0.002	+0.002	0	−0.094
IJ	530	20×10	15.2	382	0.009	0.022	10		−0.002	+0.002	0	−0.094
JK	270	12×10	11.9	324	0.007	0.017	10		−0.002	+0.001	−0.001	−0.095
Fan A												−0.059
AB												−0.077
BL	1,080	14×10	12.9	1,111	0.077	0.16	16	1 elbow H/W = 1.4 c = .087 1 elbow H/W = 0.7 c = .110 Loss = 0.015	−0.041	+0.025	−0.016	−0.093
LM	330	6×10	8.4	792	0.039	0.15	10		−0.015	+0.019	+0.004	−0.089
LN	750		10	1,376	0.118	0.28	4	Branch loss = 1VP = −0.118 + VAV damper = −0.06 + diff in VP (0.077 to 0.118) = −0.041	−0.189	−0.041	−0.230	−0.323
NO	750		10				11	1 elbow c = 0.24 Loss = −0.046	−0.059	0	−0.059	−0.382
OP	500		9	1,150	0.082	0.22	13		−0.029	+0.018	−0.011	−0.393
PQ	250		8	720	0.032	0.11	13		−0.014	+0.025	+0.011	−0.382

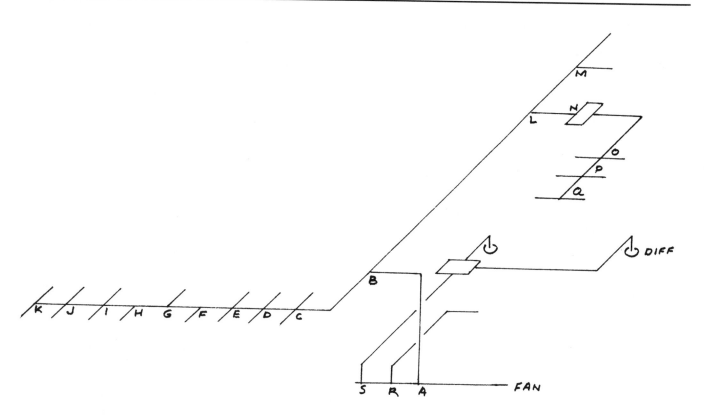

FIGURE 10-52 Duct Sizing

Because smoking is not permitted, we can go down to 14 cfm and still have good air quality. If we assume these circumstances, then the diffuser must be able to introduce 52 °F air at rates of 150 cfm and 14 cfm without causing a draft.

Drop or droop is always cited as a problem at low air flows. However, the reverse is true according to Harold Straub, Director of Research, Titus Products. During his presentation at an ASHRAE meeting, Straub stated: as the drop decreases, the air quality decreases.

In the text: *Air Diffusion Dynamics*, by the late Ralph G. Nevins, drop is addressed in more detail on pages 15 – 17. Drop can be predicted from the following

$$\frac{T_{50}}{D_{50}} = 0.95(N_B)^{0.27}$$

T_{50} = Throw to a point where the jet velocity = 50 fpm
D_{50} = Vertical drop at T_{50}

$$N_B = \frac{V_o^2}{\beta(\Delta^t_o) \, g \, \sqrt{A_c}}$$

V_o = Outlet velocity fps

$\beta = \dfrac{1}{T_a}$

T_a = Room temp degrees Rankine

Δ^t_o = Difference, room air − supply air temp °F

g = Acceleration due to gravity 32.2 fps

A_c = Core area

If we use a 12" × 12" perforated diffuser with a 5" neck and a 4-way throw, we find (from Fig. 10-61, page D19 of the Titus catalog) that the throw for 150 cfm is 9.5'. The neck velocity is 1,100 fpm. At the lowest cfm listed, 40 cfm, the throw is 4' and the neck velocity is 300 fpm. You may be wondering where I got the throw since three throws are listed. The first throw is for a terminal velocity of 150 fpm, the second 100 fpm and the third 50 fpm. To get the outlet or jet, divide the cfm by the core or face area of the diffuser. The core area of a 12" × 12" perforated-face diffuser is 0.85 ft². This gives us a core velocity of 176 fpm when delivering 150 cfm (remember Q = AV).

$$N_B = \frac{V_o^2}{\beta \, \Delta \, t_o g \sqrt{A_c}}$$

$$N_B = \frac{(176 \div 60)^2}{\dfrac{20(32.2)}{532} \, \sqrt{.85}}$$

$$N_B = \frac{8.6}{1.12} = 7.68$$

$$\frac{T_{50}}{D_{50}} = 0.95 \, (N_B)^{0.27}$$

$$\frac{9.5}{D_{50}} = 0.95 \, (7.68)^{0.27}$$

$$D_{50} = 5.76'$$

For 40 cfm V_o = 47 fpm or 0.78 fps

$$N_B = \frac{(0.78)^2}{\dfrac{20(32.2)}{532} \, \sqrt{.85}} = \frac{.61}{1.12} = 0.545$$

$$\frac{T_{50}}{D_{50}} = 0.95 \, (0.545)^{0.27}$$

$$\frac{4}{D_{50}} = 0.806$$

$$D_{50} = 4.96'$$

The drop is even less at 10 cfm, on the order of 3 ft., so it appears that we won't have to worry about drop.

Now that we have selected our diffusers, we spot them on our plans. Next we draw a single line duct layout on a print of the plans. From this, we sketch an isometric diagram and then size the ducts (Fig. 10-52).

Let's check and see how close our duct system is to being balanced. At point A in the main, the loss is equal to 0.059". The highest loss in the main is 0.095" at point K. This is only a difference of .036" which is really quite close. Of course, this is only the main loss. To get the complete loss, we must add the branch losses that consist of divided flow fittings, ductwork, dampers and diffusers. It's advisable to do this for every branch, but it's nearly an impossible task to accomplish manually. If we had a computer and a duct sizing program, it wouldn't be too difficult. Because we don't, we'll use our engineering judgement and check the basement branch — which has the smallest main loss, the branch serving the north end and the branch at point G that has the highest main loss.

Notice that we come off the unit at 21⅛" × 19⅜", the fan outlet size. After we make our turn and before we leave the fan room, we transform to 38" × 10". The elbow H/W is so close to one that we can consider it one. Its C factor is 0.09. The transformation C factor is 0.15. The loss for these two fittings is 0.144 (0.09 + 0.15) = 0.035. To this, we add the duct loss of 0.024 for a total of 0.059. The 6" × 10" duct, A to S, has a negligible loss because it is such a short piece. We will check out the run to the work room because it is the longest run and will have the highest pressure drop. The velocity in the 6" × 10" main is 408 fpm (H_v = 0.046) while the 6" diameter duct will have a velocity of 860 fpm. We can't find this in Fig. 8-8, so the safest strategy is to assume a loss equal to one velocity pressure plus the difference in velocity pressures. The loss in the elbow must also be added. Assuming R/D = 1.5, the elbow loss is 0.24 VP. The total for the two is 1.24(0.046) + 0.0351 = 0.092. Next, we have 22' of straight duct that has a loss of 22(0.215 ÷ 100) = 0.047. We find that the VAV damper has a pressure drop or loss of 0.044 by interpolating data found in the Titus catalog. The box inlet has a 5'' diameter, so we need to transform from 6'' to 5''.

To calculate the loss coefficient for the 60° transformation, we first find the loss coefficient for an abrupt contraction (Fig. 8-7) and then multiply by the correction factor Cr for a 60° transformation. The ratio of the area of a 5'' duct to a 6'' duct is 0.69, so by interpolation, Cr is equal to 0.11. Multiply this by the correction factor, 0.07, and the velocity pressure, 0.046, and get the transformation loss, 0.0004 which is negligible.

The run from the VAV damper assembly to the farthest diffuser is 16' of 5'' diameter duct and includes an elbow. To connect the branch duct to the diffuser, we have another elbow. So the loss is 16(0.15 ÷ 100) + 2(0.24) (620 ÷ 4005)² = 0.036. The diffuser loss is 0.014. The total loss from the fan to the diffuser is as follows:

Fan A	0.059
Takeoff plus elbow	0.092
Branch duct	0.047
VAV damper unit	0.044
Duct to diffuser	0.036
Diffuser	0.014
Total	0.292

At point B, the duct splits. The section going west is B-L. The loss to this point is 0.093. At L, a 10'' duct runs to the VAV damper assembly. As you can see in Fig. 10-52, the loss to point P is 0.393. To this, we add the takeoff plus 4' of ductwork, the elbow to the diffuser and the diffuser.

Fan P	0.393
Takeoff main 1,150 fpm	
branch 640 fpm	
Loss	0.035
4'6'' dia. duct	
4(.13 ÷ 100)	0.005
Elbow .24(640 ÷ 4005)²	0.006
Diffuser	0.033
Total	0.472

As you can see, there is a significant difference between these two, so we go back to the basement and see what happens if we run a 5'' instead of a 6'' diameter duct. The takeoff plus elbow increases to 0.204 while the duct

loss becomes 0.132. The total is now 0.487. For all practical purposes, these two are balanced.

Finally let's see what happens at point G where we branch to the drop room. The main loss to this point is 0.095. When we add all the other losses we get a total of about 0.42. This puts us within 0.07 of the worst case, so we can consider it balanced.

The return air duct loss is as follows:

Register (from manufacturer)		0.055
Elbow	.24(250 ÷ 4005)²	0.001
Entrance loss, plenum to duct, 1VP		0.132
25' ductwork	25(0.2 ÷ 100)	0.050
2 elbows H/W = 4	2(.24)(0.132)	0.063
1 elbow H/W = 0.25	1(.19)(0.132)	0.025
Connection to unit, 1 VP		0.132
Total R.A.		0.458

As you can see from the drawing, the connection to the unit R.A. part of the mixing box does not appear on our dynamic loss chart so I arbitrarily used 1 VP.

Now that we have calculated our supply and return air static pressure requirements, we can proceed with our cooling coil selection so that we can calculate the coil pressure drop in order to determine our maximum motor HP. Our electrical engineer needs this information so that he can proceed with his drawings.

First, calculate the entering conditions. Once again, we will assume 10% outside air.

$$Ent\ DB = (87 \times .1) = (78 \times .9) = 78.9°$$
$$Ent\ WB = (72 \times .1) + (63 \times .9) = 64.35°$$

The entering dry bulb temperature of 78.9° is not readily available, so we must interpolate between 78° and 80°. However, the charts don't list 65 WB, so another interpolation is required. Remember, we are only estimating, so for our purposes it will suffice to use 80° DB and 64° WB for our entering air conditions.

The face area of our 27'' × 36'' coil is equal to 6.75 ft² (27 × (36 ÷ 144) = 6.75). Dividing the face area of 6.75 into the air quantity of 4,055 cfm yeilds a face velocity of 600 fpm. The actual face velocity is 4,055 ÷ 8.5 = 477 fpm. So which one should we use? For our estimating selection, the 477 fpm is what we will use. Examine the coil table in Fig. 10-67 and you should notice that as a coil gets longer, the leaving air temperatures get lower. So if we elect to use 477 fpm with a 36'' length, we should be safe. Since we're making concessions for the convenience of estimating, let's just use 500 fpm face velocity and thereby save ourselves another interpolation.

We select a 5 row coil, ½ serpentine, 36" finned length and 500 fpm and find that our leaving conditions are 53.1 DB, 52.1 WB for 10° temperature rise and 8 fins/in. At 10 fins/in. and 10° temperature rise, we read 51.2 DB and 50.7 WB. At 12° temperature rise, we read 52.3 DB and 51.7 WB. While we based our cfm on 53.1 DB, it's best to build-in a little safety factor. Remember that these ratings were determined under theoretically perfect conditions. In practice, coils get dirty and fins get damaged, so it's wise to protect yourself by allowing a safety margin. In this instance, I would use the 5 row coil, 12° rise and 10 fins/in. By the way, notice the C after wet bulb. This indicates the degree of wetness, which affects pressure drop, so keep this in mind because we will refer to it later.

Now that we've chosen the coil, let's check out the air and water side pressure drops. But before we can determine the water pressure drop, we must determine the gpm and temperature rise. We predetermined a 12° rise, so that's pretty much settled. Recall the gpm equation from Chapter 9:

$$gpm = \frac{Btuh}{500(TR)}$$

$$Btuh = \frac{cfm(\Delta H)60}{v}$$

Entering air enthalpy = 29.0 Btu/lb (78.9° DB, 63.35° WB)
Leaving air enthalpy = 21.3 Btu/lb (52.3° DB, 51.7° WB)
Specific volume (v) = 13.8 ft³/lb (78.9° DB, 63.35° WB)

$$Btuh = \frac{4055(29 - 21.3)60}{13.8}$$

$$Btuh = 135,754 \text{ (11.3 tons)}$$

$$gpm = \frac{135,754}{500(12)} = 22.6$$

(We will use the same chiller as we used in the fan coil system.)

Remember that a coil with a ½ serpentine piping arrangement was selected previously. Why was that arrangement selected, and exactly what does it mean? In single serpentine, the coil has a supply and return header and the coil tubes are fed in parallel. However, the water flows in series through the coil. Look at it this way, suppose you have an 8 row coil (that is 8 rows of tubes in direction of air flow), that's 18 tubes high. In a single serpentine coil, the header feeds each tube, and the water flows through 8 tubes in parallel. In ½ serpentine, the header feeds every other tube and the water flows through 16 tubes. For double serpentine, the header feeds two parallel tubes at a time, and as a result, the water only makes 4 passes. What this does is determine the quantity of water flowing through each tube and also the velocity

of the water. The higher the velocity, the better heat transfer because we get a scrubbing action along the inside surface of the tube. Compare flow for 36 gpm:

Half serpentine	4 gpm/tube
Single serpentine	2 gpm/tube
Double serpentine	1 gpm/tube

While a faster velocity is better for heat transfer, it also results in higher pressure drops. Our catalog gives the following equation for water velocity in feet/second (fps) for a coil with ⅝" tubes.

$$WV = 1.07 \text{ gpm} \div \text{no. of tubes fed}$$

A 17" high coil has 18 tubes and ½ serpentine, 9 tubes are fed

$$WV = 1.07(22.6) \div 9 = 2.68$$

The velocity falls into the recommended range of 2-6 fps. Figure 10-68 has a nomograph for determining our pressure drop. Our coil is 5 WH. Enter at the bottom at 22 gpm and just follow the example. You should read 3.25' for the tube and less than the smallest value shown for the header, so we will assume zero or negligible header loss. Consequently, our water side pressure drop is 3.25' of water.

The air side pressure drop is found by using the other nomograph in Fig. 10-68. We enter the bottom at 477 fpm and follow the example. Remember that our coil has a C wetness factor, so follow the 477 fpm line to the intersection of the C horizontal line and then go vertically and stop at the 10 fin line. Now read to the right, to the 5 row column, and the pressure drop is close to 0.42" of water.

Let's total up our static pressure requirements. The ductwork total was 0.37". We must also figure in the coil, filter and cabinet pressure drops.

Supply ductwork	0.49
Return ductwork	0.46
Coil	0.42
Filters	0.10
Mixing box	0.10
Cabinet	0.25
Total	1.90

The next step is to examine the fan curve for the unit. Our unit comes with a 14⅞" airfoil fan or a 15" forward curved fan. We will use the backward curved airfoil fan because it is best suited for a variable air volume system. In Fig. 10-66 find the intersection of 4,055 cfm and 1.90 sp. You should read 2,200 rpm and 2.50 bhp.

We will specify a minimum 3-HP motor. We must next check to see what we can have as a minimum air flow. This is done by the use of:

$$CFM_{min} = CFM_D \times \sqrt{\frac{SP_c}{SP_1\left(\frac{CFM_D}{CFM_1}\right) + SP_C - SP_D}}$$

CFM_{min} = Minimum cfm

CFm_D = Design cfm = 4,055

SP_C = Control static pressure needed to operate VAV terminals = 0.57

SP_1 = Static pressure at point where rpm curve enters *do not select* area = 4"

CFM_1 = Cfm corresponding to SP_1 = 1,250

SP_D = Design static pressure = 1.90

$$CFM_{min} = 4,055 \times \sqrt{\frac{0.67}{4\left(\frac{4055}{1250}\right) + 0.67 - 1.90}}$$

CFM_{min} = 968

CHILLED WATER PIPE SIZING AND PUMP SELECTION

Our flow requirement is 22.6 gpm. We'll elect to use copper pipe. From Fig. 9-3, we find that we can use either a 1 ½" pipe at 525 mi/ft and 48 IPS velocity or a 2" pipe at 150 mi/ft and 24 IPS velocity. We are right on our high limit for velocity, so we'll use the 2" pipe. The total length of 2" pipe is 40' and the fittings are as follows; 6 gate valves, 9 elbows and 2 tees. These add as follows: 9 + (6 × 0.7) + (2 × 2) = 17.2 elbow equivalents. One elbow = 5.1' (Fig. 9-4). From manufacturer information, we learn that the chiller pressure drop at 22.6 gpm is 4.3 ft. Let's calculate the total.

Piping 40' × 150 mi/ft	= 6000 mi
Fittings 17.2" × 5.1 × 150 mi/ft	= 13158 mi
Total pipe and fittings	= 19,158 mi = 1.6'
Chiller	= 4.3'
Total	= 5.9'

Our pump requirements are 22.6 gpm against 5.9 ft head.

PUT IT ALL TOGETHER 247

EXPANSION TANK

Coil	6.0 gallons	(from manufacturer)
Piping	6.4 gallons	(40' × 2" pipe × 0.16 gal/ft)
Chiller	6.3 gallons	(from manufacturer)
Total	18.7 gallons	

Water will rise from 40°F to room temperature when shut off. Figure 9-24 does not go below 100°F, so we will use 100°F (where E = 0.25%).

$$T = \frac{EV}{\frac{P_a}{P_f} - \frac{P_a}{P_o}}$$

E = 0.25% = 0.0025

V = 18.7 gallons

P_a = 34' absolute

P_o = 2.3 (30 psig) + 34' = 103' absolute

P_f = 10' + 9' + 34' = 53' absolute

$$T = \frac{18.7(0.0025)}{\frac{34}{53} - \frac{34}{103}}$$

T = 0.150 gallons

As you can see, this is a mighty small tank so we will use the smallest stock size tank available which holds 3 gallons, has a 9" diameter and measures 14" long.

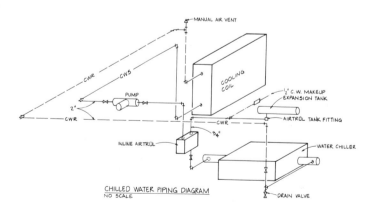

CHILLED WATER PIPING DIAGRAM
NO SCALE

THE COST OF AIR CONDITIONING

Now that we've completed the design of our systems, let's examine how the costs break-out.

COST SUMMARY

System	Cost	Cost ft²*	Cost ton**
Baseboard series loop	21,472	2.90	
Fan coil	79,805	10.78	5,320
Through-wall	80,993	10.95	5,396
Multizone	72,198	9.76	4,813
Variable air volume	82,440	11.14	5,496

* Gross area of 1st floor and basement
**Cooling load is a nominal 15 tons

BASEBOARD SYSTEM

Item	Equipment	Labor	Total
Boiler	1,701.00	796.95	2,497.95
Breeching			847.37
Pumps	518.70		518.70
Hot Water Specialties	99.23		99.23
Baseboard Radiation	2,538.54	3,320.62	5,859.16
Pipe and Fittings	2,902.50	1,992.37	4,894.87
Insulation			1,682.63
Toilet Exhausts			726.32
Temperature Controls			3,982.63
Electrical			363.16
			21,472.01

FAN COIL SYSTEM

Item	Equipment	Labor	Total
Boiler	2,254.50	796.95	3,051.45
Breeching			847.37
Chiller	8,532.00	531.30	9,063.30
Pumps	1,339.20		1,339.20
Cooling Tower	1,134.00	531.30	1,665.30
Hot Water Specialties	162.00		162.00
Fancoil Units	7,992.00	8,500.79	16,492.79
Expansion Joints	2,199.15		2,199.15
Pipe and Fittings	14,542.20	4,117.57	18,659.77
Insulation			3,091.68
Toilet Exhausts			726.32
Temperature Controls			15,010.53
Electrical			3,842.30
Tunnel			4,842.11
			80,993.26

THROUGH-WALL SYSTEM

Item	Equipment	Labor	Total
Boiler	1,984.50	1,062.60	3,047.10
Breeching			847.37
Pumps	268.49		268.49
Hot Water Specialties	116.78		116.78
Baseboard Radiation			266.32
Self Contained Units	26,676.00	9,780.74	36,456.74
Expansion Joints	2,199.15		2,199.15
Pipe and Fittings	5,902.20	2,258.02	8,160.22
Insulation			2,597.79
Toilet Exhausts			726.32
Entrance Units			861.89
Temperature Controls			1,283.16
Electrical			10,525.53
Tunnel			4,842.11
			72,198.94

MULTIZONE SYSTEM

Item	Equipment	Labor	Total
Boiler	1,984.50	796.95	2,781.45
Breching			847.37
Pumps	788.70		788.70
Hot Water Specialties	99.23		99.23
Baseboard Radiation	2,538.68	3,320.62	5,859.30
Multizone Unit	6,210.00	796.95	7,006.95
Condensing Unit	8,910.00	265.65	9,175.65
Expansion Joints	908.09		908.09
Pipe and Fittings	2,902.50	2,258.02	5,160.52
Insulation			2,222.53
Toilet Exhausts			726.32
Temperature Controls			7,820.00
Electrical			4,701.68
Ductwork w/lining			21,617.58
Grilles, Registers, Diffusers	1,188.00	912.87	2,100.87
Refrigeration Piping			1,936.84
Extra Height 1st Floor			6,042.63
			79,805.69

VARIABLE VOLUME SYSTEM

Item	Equipment	Labor	Total
Boiler	1,701.00	796.95	2,497.95
Breeching			847.37
Chiller	8,532.00	531.30	9,063.30
Pumps	1,339.20		1,339.20
Cooling Tower	1,134.00	531.30	1,665.30
Hot Water Specialties	99.23		99.23
Baseboard Radiation			266.32
HVAC Unit	6,210.00	796.95	7,006.95
Expansion Joints	908.09		908.09
Pipe and Fittings	2,902.50	2,258.02	5,160.52
Insulation	1,125.90		1,125.90
Toilet Exhausts			726.32
Entrance Heaters			861.89
Temperature Controls			9,502.63
Electrical			4,701.68
Ductwork w/lining			24,133.05
Grilles, Registers, Diffusers	1,188.00	912.87	2,100.87
VAV Boxes	5,845.50	4,588.49	10,433.99
			82,440.56

HI-PAK I·B·R RATINGS

REG. U.S. PAT. OFF.

Element 2¾ x 3²³⁄₃₂ Fin	*Water Rate G.P.M.	Av. Temp. °F-Forced Hot Water					Pressure Drop Per Foot In Milinches
		180	190	200	210	220	
¾ Tube	1	750	840	920	1000	1090	47
50 Fins (.011)/Ft	4	790	890	970	1060	1150	525
1 Tube	1	750	830	910	990	1070	13
50 Fins (.011)/Ft	4	790	880	960	1050	1130	145

Ratings base on 65° entering air temperature.

The above ratings are condensate ratings plus 15% for heating effect. Ratings are based on finned length. Finned length is 4″ shorter than element length. The use of ratings at 4 G.P.M. is limited to installations (usually loop) where the flow rate is 4 G.P.M. or greater. When the flow rate is not known the standard flow rate of 1 G.P.M. must be used.

If the calculated water flow rate through a baseboard unit in a completely designed hot water heating system is greater than the standard flow rate (500 lb/hr), the rating of that unit may be increased by multiplying the standard water rating at 500 lb/hr by the factor shown for the calculated flow rate.

*WATER FLOW RATE CORRECTION FACTORS

Flow LB/Hr	Flow G.P.M.	"F"	Pressure Drop Milinches Per Ft. of Copper Tube	
			¾	1
500	1.00	1.00	47	13
750	1.50	1.016	96	26
1000	2.00	1.028	157	43
1250	2.50	1.038	230	63
1500	3.00	1.045	320	87
1750	3.50	1.051	420	114
2000	4.00	1.057	525	145
2250	4.50	1.062	650	178
2500	5.00	1.067	775	216
3000	6.00	1.074	1060	290

HI-PAK INSTALLATION DATA

Place brackets on back-plate **before attaching to wall.** Follow instructions in package for bracket location.

Install back-plate on wall. Fasten firmly at top and bottom. Fastening through brackets makes a sturdy installation.

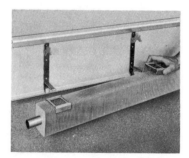

Place slide shoes on element in line with brackets.

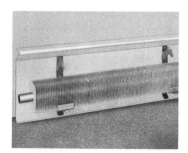

Place element on brackets. A slide shoe should rest on each bracket.

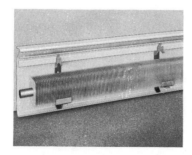

Slide the damper carriers into place on vane-damper. Use one at each bracket. Position vane-damper.

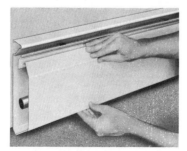

Hook enclosure front over bottom bracket. Push enclosure top over top bracket into place.

FIG. 10-53 Sterling Radiator Company Inc. Catalog

Castings Mfg. by the Eastern Foundry Co., a division of Peerless Industries, Inc., Boyertown, Pa.

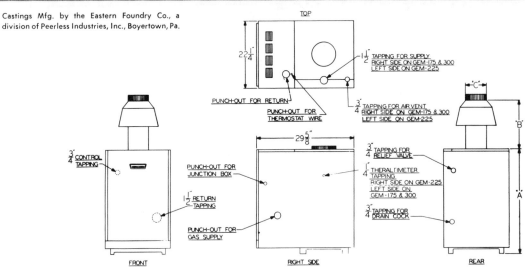

NATURAL & PROPANE GASES

MODEL NUMBER	A.G.A. INPUT BTU/HR.	A.G.A. OUTPUT BTU/HR.	*NET I.B.R. RATING BTU/HR.	**NET RATING SQ. FT.	SUPPLY TAPPING	RETURN TAPPING	NATURAL GAS VALVE SIZE	PROPANE GAS VALVE SIZE	HEIGHT "A" DIM.	DRAFT HOOD "B" DIM.	FLUE OUTLET "C" DIM.
GEM-175A-WC	175,000	140,000	121,700	810	1½"	1½"	½" x ¾"	½" x ¾"	31"	23"	7"
GEM-225A-WC	225,000	180,000	156,500	1,045	1½"	1½"	¾" x ¾"	½" x ¾"	34½"	23"	7"
GEM-300A-WC	300,000	240,000	208,700	1,390	1½"	1½"	¾" x ¾"	½" x ¾"	39½"	29¾"	9"

*The Net I-B-R Ratings shown include allowance for normal piping and pick-up load, in accordance with the Testing and Rating code for Low Pressure Cast Iron Heating Boilers of the Hydronic Institute. Water ratings are based on a piping and pick-up allowance factor of 1.15 steam ratings based on graduated factors in accordance with the I-B-R Code.

The Peerless Heater Company should be consulted before selecting a boiler for gravity hot water installations and installations having unusual piping and pick-up requirements such as exposed piping, night shut-down, etc.

**Net ratings based on Net I-B-R Ratings in BTU per hr. @ 170 boiler water temperature with heat emission of 150 BTU per sq. ft.

STANDARD EQUIPMENT
Forced Hot Water (WC Models)

1. Complete Boiler Factory Assembled, Prewired and Shipped in Dust Proof Carton.
2. Deluxe Extended Steel Jacket, Insulated-High Grade Green Baked Enamel Finish
3. Cast Iron Water Tube Sections. Extra Heavy Duty Finned Surfaces.
4. Special Built-in Air Eliminator.
5. Factory tested for 50# Working Pressure.
6. Slotted Port Main Burners.
7. Honeywell Combination Gas Valves (24 Volt or Milli. Volt Self Generating).
8. 100% Pilot Shut-off.
9. Aquastat Relay (High Limit & Circulator Control).
10. Low Voltage Thermostat.
11. 1¼" Circulator and Fittings.
12. Theraltimeter.

SHIPPED IN ONE SEPARATE CARTON

1. Drain Cock
2. Pressure Relief Valve (ASME) 30 lb.
3. Draft Hood
4. Flue Brush

OPTIONAL ACCESSORIES

Floor Pan—Part No. Z-2004 for Installation on Combustible Floor. Peerless Comfort Manifold with two (2) or more zone valves (Packaged). Diaphram Expansion Tank, Vent, Fill System (Packaged).

CARTON DIMENSIONS AND SHIPPING WEIGHTS

Model No.	Length	Width	Height	Approx. Shipping Weight
GEM-175A-WC	33"	26"	34"	511#
GEM-225A-WC	33"	26"	37"	586#
GEM-300A-WC	33"	26"	41½"	675#

THE RIGHT IS RESERVED BY THE MANUFACTURER TO MAKE CHANGES AT ANY TIME WITHOUT NOTICE

FIG. 10-54 Peerless Industries, Peerless Heater Co. Division, Boiler Catalog

IRON AND BRONZE BOOSTER PUMP

*Performance characteristics are based on using 1¼″ or 1½″ flanges. When using ¾″ or 1″ flanges performance will be slightly reduced.

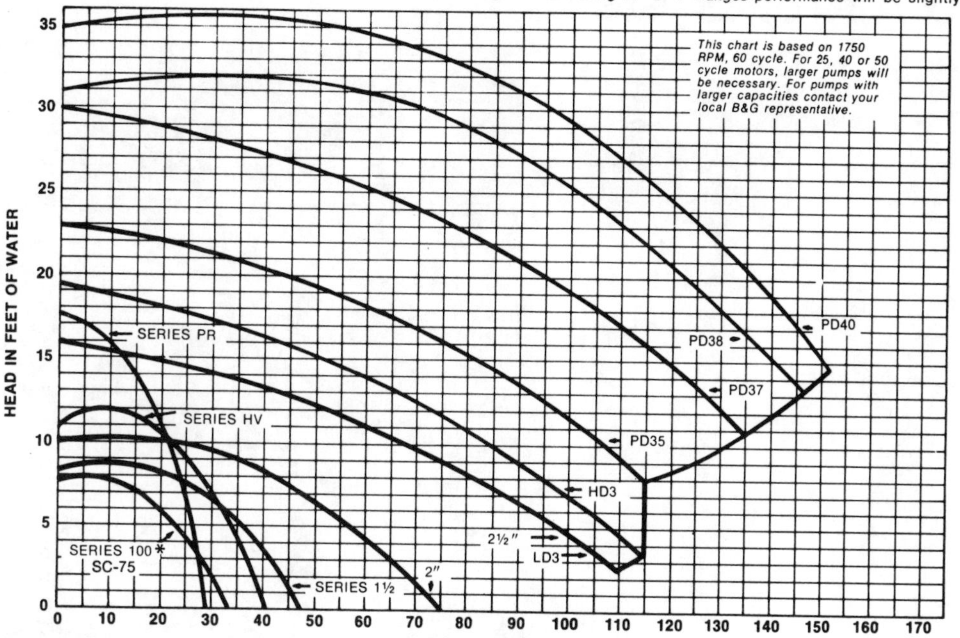
This chart is based on 1750 RPM, 60 cycle. For 25, 40 or 50 cycle motors, larger pumps will be necessary. For pumps with larger capacities contact your local B&G representative.
HEAD IN FEET OF WATER
DELIVERY IN GALLONS PER MINUTE
0
5
10
15
20
25
30
35
0 10 20 30 40 50 60 70 80 90 100 110 120 130 140 150 160 170
SERIES PR
SERIES HV
SERIES 100*
SC-75
SERIES 1½
2″
2½″
LD3
HD3
PD35
PD37
PD38
PD40

DIMENSIONS & WEIGHTS

MODEL NO.	FLANGE SIZE NPT INCHES (specify size)	DIMENSIONS IN INCHES (open drip-proof) A	B	C	D	E	*ELEC. BOX ARNGMT.	APPROX. SHPG. WT. LBS. IRON BODY	BRONZE
	¾				9/16	—			
SERIES 100	1 & 1¼	15	6⅜		¾	—		26	26
	1½			12⅞	15/16	—			
SC-75	SWEAT		7⅜		—	—		—	
	¾				9/16	—			
SERIES PR	1 & 1¼	16¼		13⅝	¾	—			
	1½				15/16	—	1	40	40
SERIES HV	1		8½		⅝	—			
	1¼ & 1½	16⅛			¾	—			
SERIES 1½	1			13¾	⅝	—			
	1¼ & 1½				¾	—		36	39
2	2	16⅜			13/16	—		45	49
2½	2½	17⅜				—		60	66
LD3			10	14⅛		—	2*		
HD3		18⅜		15⅛	1 1/16	—		75	78
PD35-S		20¾		17⅜				78	84
PD35-T		20¼	12	16⅞		1⅛	3		
PD37-S	3	21¼		17⅞				82	91
PD37-T		20¾		17⅜					
PD38-S		24		19½		—		130	147
PD38-T		24¼	14½	19¾	1¼	—	4	132	149
PD40-S		24⅝		20⅛		—		140	153
PD40-T		25⅛		20⅝		—		145	158

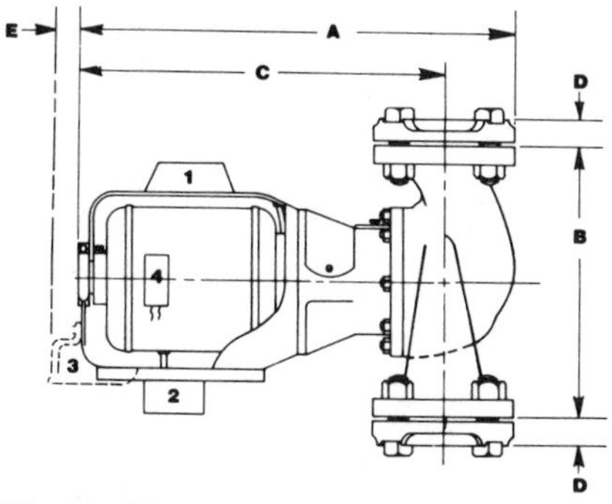
E
A
C
D
B
D
1
2
3
4

*Fractional 3Ø motors are available with electrical box arrangement #3.

TYPICAL SPECIFICATION

The Contractor shall furnish and install In-The-Line Pumps as illustrated on the plans and in accordance with the following specifications:

1. The pumps shall be of the horizontal, oil-lubricated type, specifically designed and guaranteed for quiet operation. Suitable for 125# working pressure.

2. The pumps shall have a ground and polished steel shaft with integral thrust collar. The shaft shall be supported by two horizontal sleeve bearings designed to circulate oil. The pumps are to be equipped with a water-tight seal to prevent leakage. Mechanical seal faces to be carbon on cast iron or ceramic. The motor shall be non-overloading at any point on pump curve.

3. The motor shall be of the open, drip-proof, sleeve-bearing, quiet-operating, rubber-mounted construction. Motors shall have built-in thermal overload protectors. (Exception—PD models with 3-phase motors, see paragraph 4.)

4. For PD models with 3-phase motors, add the following: The Contractor shall furnish and install a magnetic starter for each booster pump, with at least two thermal overload protectors. The starter shall be equipped with manual reset buttons.

The pump shall be Bell & Gossett Model No. ____________, or approved equal with a capacity of ____________ GPM at ____________ Ft. head when directly driven through a self-aligning flexible coupling by an oil-lubricated motor, ________ volts ________ cycle ________ phase(Ø).

PRINTED IN U.S.A. 8-75

BELL & GOSSETT ITT
8200 N. AUSTIN AVE. • MORTON GROVE, ILL 60053
INTERNATIONAL TELEPHONE AND TELEGRAPH CORPORATION

FIG. 10-55 Bell & Gossett Company Catalog

B&G Airtrol® System

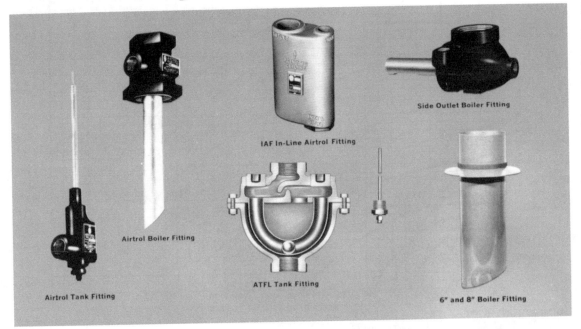

IAF In-Line Airtrol Fitting

Side Outlet Boiler Fitting

Airtrol Boiler Fitting

ATFL Tank Fitting

Airtrol Tank Fitting

6" and 8" Boiler Fitting

Efficient circulating water (or anti-freeze) systems must be completely free of air for proper operation. Air properly controlled, can offer both the expansion space and pressurization needs of a closed system.

The Bell & Gossett Airtrol System is based upon a proven concept of air control for closed hydronic systems of all sizes. The Airtrol System consists of three basic components:

1. The air separator, which may be an Airtrol Boiler Fitting, IAF In-Line Airtrol, or the Rolairtrol, serves to separate air bubbles from the system water before they can enter the system.

2. The Airtrol Tank Fitting serves to help confine air in the compression tank, reduces gravity circulation which, through lower tank temperature, reduces tank size and helps establish the correct initial air content in the tank when filling a system.

3. The guaranteed air-tight compression tank or tanks, where all free air should be confined.

The Airtrol System has been in use on hydronic systems for over twenty years and as a result of its complete success on all types and sizes of systems, is offered with both a product and a performance guarantee when installed in accordance with the published Bell & Gossett design data. Every hydronic system requires air control—guaranteed with Airtrol.

ASME Compression Tanks

The compression tank on a closed hydronic system serves a very important function in providing adequate pressurization under all operating conditions. Above all, a compression tank must be absolutely air tight. Tanks are available nationally in both standard and ASME construction, are of carbon steel construction, thoroughly tested and guaranteed leakproof. Tanks are rust-proof coated on the exterior only. Gauge glass tappings furnished on all ASME tanks as standard equipment. See B&G Air Control Design Manual for compression tank selection procedure. Always use B&G Airtrol Tank Fittings for minimum tank size and effective air control.

AIRTROL SYSTEM PERFORMANCE GUARANTEE

B&G Airtrol System consists of B&G Compression Tank(s) with Airtrol Tank Fitting(s) and Airtrol Boiler Fitting or In-Line Airtrol or Rolairtrol Air Separators.

The Airtrol System is guaranteed to prevent the accumulation of air in heating and cooling units and prevent noises caused by entrained air in piping. In case of failure of any B&G Airtrol System (within the USA) to operate correctly, when installed and operated in accordance with our published instructions on an air tight system, we will provide, free of charge, the services of a factory trained engineer who will supervise steps necessary to provide satisfactory results.

AIRTROL PRODUCT GUARANTEE

The Airtrol Tank Fitting, Airtrol Boiler Fitting, In-Line Airtrol and Rolairtrol Air Separators are guaranteed to last the entire life of the heating system in which they are installed. Labor charges for replacement are not allowed.

ASME CONSTRUCTION

Model No. and Tank Capacity in Gallons	Capacity Sq. Ft. Radiation	Tank Dimensions	Gauge Glass Tappings	
			Size	Center to Center
15		13" x 34½"		7½"
24		13" x 51"		
30		13" x 61½"		
40		16¼" x 53"		10"
60	Write Factory For Tank Sizing Selection Table	16¼" x 76½"	½"	
80		20¼" x 68"		14"
100		20¼" x 82"		
120		24¼" x 71½"		18"
144		24¼" x 83½"		
163		30" x 60"		
202		30" x 72"		24"
238		30" x 84"		
270		30" x 96"		
306		30" x 108"		30"
337		36" x 84"		
388		36" x 96"		

All welding to conform to Part UW, rules for construction of unfired pressure vessels, Section VIII of the ASME Boiler Construction Code.
125 P.S.I.G. Working Pressure—350°F Maximum Operating Temperature

FIG. 10-55 *(Continued)*

INSTALLATION PROCEDURES

Pipe lines must be aligned and properly guided to assure maximum life of the Flexonics Compensators. Under these conditions, the Compensator ends will easily fit into place. Positive anti-torque device assures proper alignment of bellows element in Flexonics Model HB compensators.

Pipe guides at proper spacing and anchors at both ends of pipe runs should be installed to insure proper alignment. Install Flexonics Expansion Compensator with installation clip in place, then remove clip to allow bellows to absorb pipe travel. (Installation instructions are also on label.)

Refer to the table and chart below for pipe movement anticipated. In general, one Flexonics Model HB Expansion Compensator should be used for each 1½" to 1¾" of expansion anticipated.

CHART I
THERMAL EXPANSION OF COPPER TUBING IN INCHES PER 100 FEET

Saturated Steam (Vacuum In. Hg below 212°F.; Pressure p.s.i. above 212°F.)		Temp. Degrees Fahr.	Copper
		−20	−0.210
		0	0
		20	0.238
		32	0.366
		40	0.451
Vacuum in inches of Hg	29.39	60	0.684
	28.89	80	0.896
	27.99	100	1.134
	26.48	120	1.336
	24.04	140	1.590
	20.27	160	1.804
	14.63	180	2.051
	6.45	200	2.296
	0	212	2.428
PSIG	2.5	220	2.516
	10.3	240	2.756
	20.7	260	2.985
	34.5	280	3.218
	52.3	300	3.461
	74.9	320	3.696
	103.3	340	3.941
	138.3	360	4.176

The Model HB Expansion Compensator is designed for a total of 2" axial motion. In the preset position (clip in place), 1¾" of movement is for thermal growth. ¼" is for contraction to protect against system damage during cold water testing or temperature drops during the construction period.

The figures shown in Chart I show the actual expansion of a 100 foot length of copper tubing at different temperatures. Chart II has been compiled to show the temperature to pipe length relationship which will produce 1¾" of thermal growth. The dotted line represents the maximum motion that the Model HB Expansion Compensator can absorb. The solid line allows ¼" safety factor to take care of unforeseen circumstances.

> **EXAMPLE:** It is recommended for a piping system operating at a temperature of 300°F. that one Model HB be used for every 46 feet of copper tubing, including ¼" safety factor.

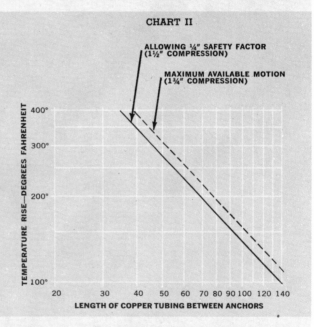

CHART II

The Model HB Expansion Compensator is designed for a total of 2" axial motion.

model **HB** expansion compensators

FLEXONICS Division of CALUMET & HECLA, Inc., Bartlett, Illinois

FIG. 10-56 Flextronics, Div. of Calumet & Hecla Corp. Catalog

SELECTION PROCEDURE — CHILLED WATER COOLING

Unit Total and Sensible Capacity is determined graphically by using the chart on this page for 200 CFM, Series 201, NELSON/aire units, and pages 11 and 12 for High and Low Speed, respectively, on all other models.

An example using the 200 CFM, Series 201 Model chart on this page will demonstrate use of the charts.

EXAMPLE — With the following design conditions, find the Total and Sensible Cooling of a 200 CFM, Series 201 NELSON/aire unit.

$$\begin{array}{ll} \text{Entering Water Temperature} & = 45° \\ \text{Desired Water Temperature Rise} & = 10° \\ \text{Entering Dry Bulb Temperature} & = 80° \\ \text{Entering Wet Bulb Temperature} & = 67° \end{array}$$

1. Using the chart at the bottom of this page, drop vertically from a Water Temperature Rise of 10° to an Entering Water Temperature of 45°. Project horizontally to the Sensible and Total Cooling Reference Lines.

2. From the point of intersection with the Total Cooling Reference Line, project a line parallel with the Total Cooling Direction Line downward to a point near the capacity scale. Project a line vertically from 67° on the Entering Wet Bulb Temperature scale to the Wet Bulb Reference Line, then horizontally to where it intersects with the Total Cooling Projection found in 1. From this intersection, project down to the capacity scale to find a Total Capacity of 5.4 MBH.

3. To determine Sensible Capacity, project a line parallel to the Sensible Cooling Direction Line from the intersection found in 1 to a point near the capacity scale. Enter the Entering Dry Bulb Temperature scale at 80° and project vertically to the Dry Bulb Reference Line, then horizontally to the Sensible Cooling Projection. From this intersection, drop vertically to the capacity scale to find a Sensible Capacity of 4.3 MBH.

4. To determine the Flow Rate required at this set of design conditions, enter Chart on page 13 at a Total Capacity of 5.4 MBH. By projecting horizontally to the 10° Water Temperature Rise line and vertically down to the Flow Rate scale, it is seen that a flow of 1.1 GPM is required.

COOLING CAPACITY
200 CFM — High and Low Speed

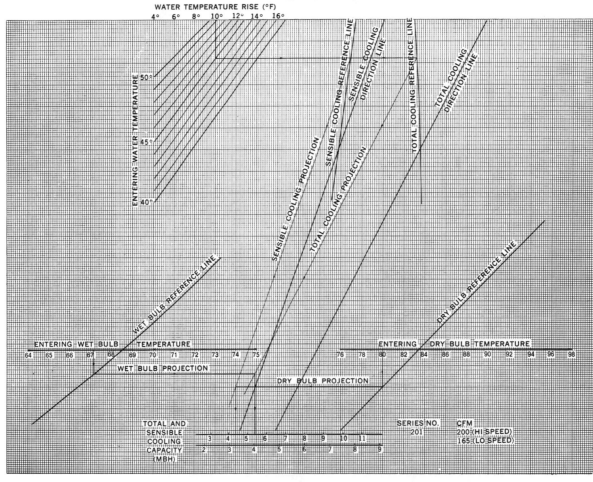

FIG. 10-57 American Air Filter Catalog

COOLING CAPACITY — 300 through 1250 CFM — HI-SPEED

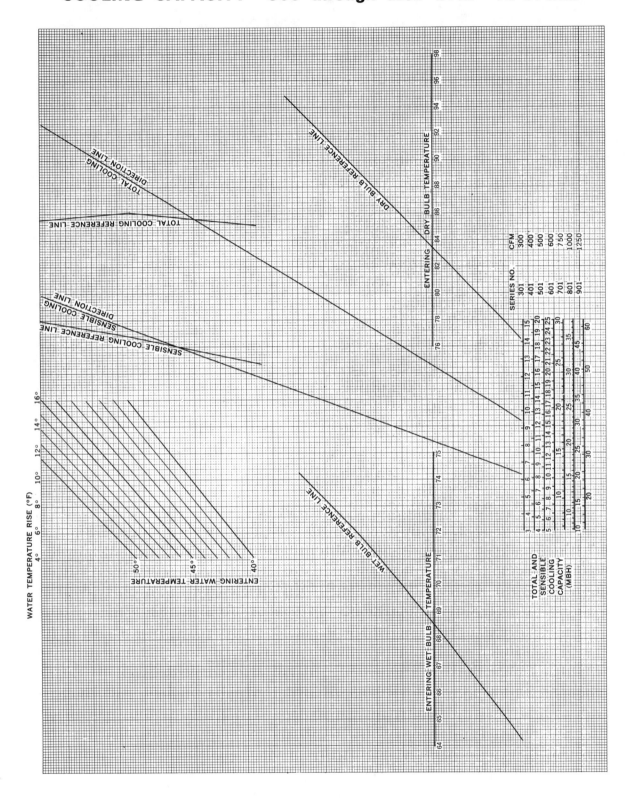

FIG. 10-57 *(Continued)*

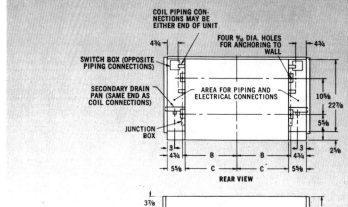

FLOOR MODEL DIMENSIONS

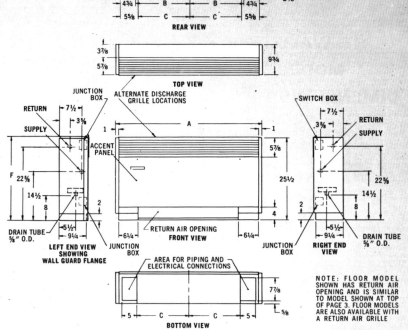

NOTE: FLOOR MODEL SHOWN HAS RETURN AIR OPENING AND IS SIMILAR TO MODEL SHOWN AT TOP OF PAGE 3. FLOOR MODELS ARE ALSO AVAILABLE WITH A RETURN AIR GRILLE.

TABLE 6

Series No.	CFM Std. Air	No. Fans	A	B	C	Wall Opening Height F
2010 3010	200 300	1	36	12¼	11⅜	25⅞
4010 5010	400 500	2	47	17¾	16⅞	25⅞
6010 7010	600 750	3	58	23¼	22⅜	25⅞
8010	1000	4	69	28¾	27⅞	25⅞
9010	1250	5	80	34¼	33⅜	25⅞

All dimensions in inches

NOTES:

1. Key-operated two-speed switch accessible through discharge grille. Optional remote wall-mounted two-speed switch also available. Unit mounted push-button switch also available.

2. Front panel removable for filter access or filter removable through return air opening if specified.

3. Electrical junction box is always in end of unit opposite coil piping connections.

4. Supply and return connections are always on same end of coil.

5. Coil piping connections are ¾" male sweat (⅞" O.D.).

6. Wall opening length dimension for recessed units is A dimension + ⅜".

7. Width of Wall Guard Flange is 1³⁄₁₆".

NOTE: ALL DIMENSIONS ARE APPROXIMATE ONLY AND ARE SUBJECT TO CHANGE WITHOUT NOTICE. REQUEST CERTIFIED PRINTS FOR ROUGH-IN DETAILS AND CONSTRUCTION PURPOSES.

FIG. 10-57 *(Continued)*

Capacity data

Table 3 — Water cooled ratings

UNIT TYPE & SIZE	LVG. COOLER WATER TEMP. (°F)	ENTERING CONDENSER WATER TEMPERATURE (10° ΔT)														
		75F			80F			85F			90F			95F		
		TONS	KW	EER	TONS	KW	EER	TONS	KW	EER	TONS	KW	EER	TONS	KW	EER
WHR 008C	42	8.1	6.5	15.0	7.8	6.7	13.9	7.6	7.0	13.1	7.3	7.2	12.1	6.9	7.5	11.0
	44	8.4	6.5	15.5	8.1	6.8	14.3	7.9	7.1	13.3	7.5	7.3	12.4	7.1	7.6	11.2
	45	8.6	6.6	15.6	8.3	6.8	14.6	8.0	7.1	13.6	7.7	7.4	12.5	7.4	7.7	11.5
	46	8.7	6.6	15.8	8.4	6.9	14.6	8.2	7.2	13.7	7.8	7.4	12.6	7.5	7.7	11.6
	48	9.1	6.7	16.3	8.8	6.9	15.4	8.5	7.3	13.9	8.1	7.5	13.0	7.8	7.8	12.0
	50	9.5	6.8	16.8	9.1	7.0	15.6	8.8	7.4	14.3	8.4	7.6	13.3	8.1	8.0	12.1
WHR 010C	42	10.9	9.3	14.0	10.6	9.6	13.2	10.3	9.8	12.6	9.9	10.1	11.8	9.6	10.3	11.2
	44	11.3	9.4	14.4	11.0	9.7	13.6	10.6	10.0	12.7	10.3	10.3	12.0	10.0	10.5	11.4
	45	11.6	9.5	14.6	11.2	9.8	13.7	10.8	10.1	12.8	10.5	10.4	12.1	10.2	10.6	11.5
	46	11.8	9.6	14.8	11.3	9.9	13.7	11.0	10.2	13.0	10.7	10.5	12.2	10.4	10.7	11.6
	48	12.1	9.8	14.8	11.8	10.1	14.0	11.5	10.3	13.4	11.0	10.6	12.5	10.7	10.8	11.9
	50	12.5	10.0	15.0	12.2	10.3	14.2	11.9	10.5	13.6	11.5	10.7	12.9	11.1	11.0	12.1
WHR 015C	42	17.4	13.3	15.7	16.8	13.9	14.5	16.2	14.4	13.6	15.5	14.9	12.5	14.8	15.4	11.5
	44	18.1	13.5	16.1	17.4	14.1	14.8	16.8	14.5	13.9	16.2	15.1	12.8	15.4	15.6	11.9
	45	18.5	13.5	16.4	17.9	14.2	15.1	17.2	14.6	14.2	16.5	15.2	13.1	15.8	15.8	12.0
	46	18.8	13.6	16.6	18.2	14.2	15.4	17.5	14.7	14.3	16.8	15.3	13.2	16.1	15.9	12.1
	48	19.5	13.8	16.9	18.9	14.4	15.7	18.3	14.9	14.8	17.5	15.5	13.6	16.8	16.1	12.5
	50	20.3	13.9	17.5	19.5	14.5	16.1	18.9	15.0	15.1	18.2	15.7	13.9	17.4	16.3	12.8
WHR 020C	42	22.0	18.1	14.6	21.3	18.5	13.8	20.6	19.1	13.0	19.9	19.7	12.1	19.1	20.2	11.4
	44	22.9	18.2	15.1	22.3	18.8	14.3	21.4	19.4	13.2	20.6	20.1	12.2	19.9	20.6	11.6
	45	23.3	18.3	15.2	22.7	18.9	14.4	21.8	19.5	13.4	21.0	20.2	12.5	20.3	20.8	11.8
	46	23.7	18.4	15.5	23.1	19.1	14.5	22.3	19.7	13.6	21.4	20.3	12.6	20.7	21.0	11.9
	48	24.7	18.6	16.0	23.9	19.3	14.9	23.1	20.0	13.9	22.3	20.6	13.0	21.5	21.3	12.1
	50	25.6	18.9	16.2	24.9	19.6	15.2	23.9	20.2	14.2	23.1	20.9	13.3	22.4	21.6	12.5
WHR 025C	42	26.9	20.2	16.0	25.9	21.0	14.8	25.0	21.8	13.8	24.2	22.5	13.0	23.2	23.2	12.0
	44	27.9	20.5	16.3	27.0	21.3	15.2	26.0	22.0	14.2	25.1	22.8	13.2	24.2	23.6	12.4
	45	28.5	20.6	16.6	27.5	21.4	15.5	26.5	22.2	14.3	25.5	23.0	13.3	24.6	23.8	12.4
	46	29.0	20.8	16.7	28.0	21.6	15.6	27.0	22.4	14.5	26.0	23.2	13.4	25.0	24.0	12.5
	48	30.1	21.1	17.2	29.1	21.9	16.0	28.0	22.7	14.8	27.0	23.6	13.7	25.9	24.4	12.7
	50	31.2	21.4	17.5	30.1	22.2	16.3	29.1	23.0	15.2	27.9	23.9	14.0	26.9	24.8	13.0
WHR 030C	42	33.4	26.6	15.1	32.3	27.6	14.0	31.3	28.5	13.2	30.2	29.4	12.4	29.1	30.3	11.5
	44	34.8	26.9	15.5	33.7	27.8	14.5	32.6	28.8	13.6	31.4	29.7	12.7	30.2	30.7	11.8
	45	35.4	27.1	15.7	34.3	28.0	14.8	33.3	29.1	13.7	32.1	30.0	12.8	31.0	31.0	12.0
	46	36.1	27.3	15.8	35.0	28.2	14.9	33.9	29.3	13.9	32.7	30.2	13.0	31.5	31.2	12.1
	48	37.6	27.7	16.3	36.3	28.6	15.2	35.3	29.6	14.3	33.9	30.6	13.3	32.8	31.6	12.5
	50	39.1	27.9	16.8	37.7	28.9	15.6	36.5	29.9	14.6	35.3	31.0	13.7	34.1	32.0	12.8
WHR 040C	42	43.5	35.9	14.5	42.2	37.0	13.7	40.5	38.0	12.8	39.1	39.1	12.0	37.6	40.2	11.3
	44	45.4	36.5	14.9	44.1	37.5	14.2	42.2	38.7	13.1	40.7	39.8	12.2	39.2	41.0	11.5
	45	46.2	36.7	15.1	44.9	37.8	14.3	43.1	39.0	13.3	41.6	40.1	12.5	39.9	41.2	11.6
	46	47.3	36.9	15.4	45.7	38.0	14.8	43.9	39.2	13.4	42.4	40.3	12.6	40.7	41.5	11.8
	48	49.0	37.3	15.7	47.5	38.6	14.8	45.6	39.8	13.8	44.1	41.0	13.0	42.4	42.3	12.0
	50	51.0	37.7	16.2	49.4	39.1	15.1	47.5	40.4	14.2	45.8	41.6	13.2	44.1	42.9	12.4
WHR 050C	42	54.5	44.4	14.8	52.8	45.8	13.8	50.8	47.3	12.8	48.9	48.7	12.0	47.3	50.2	11.3
	44	56.7	44.9	15.1	54.9	46.4	14.2	52.9	48.0	13.2	51.0	49.4	12.4	49.1	50.9	11.5
	45	57.9	45.2	15.4	56.0	46.7	14.4	54.0	48.3	13.4	52.1	49.8	12.6	50.1	51.3	11.5
	46	58.9	45.4	15.6	57.0	46.9	14.6	55.0	48.6	13.6	53.1	50.2	12.7	51.1	51.7	11.9
	48	61.2	46.0	16.0	59.2	47.6	14.9	57.2	49.2	13.9	55.2	50.8	13.1	53.2	52.4	12.2
	50	63.5	46.6	16.3	61.4	48.3	15.2	59.4	49.9	14.3	57.3	51.5	13.3	55.3	53.2	12.5

FIG. 10-58 McQuay-Perfex Inc., Water Chiller Catalog

Dimensional data

Figure 3   WHR-008C – 015C (Arrangement 1)

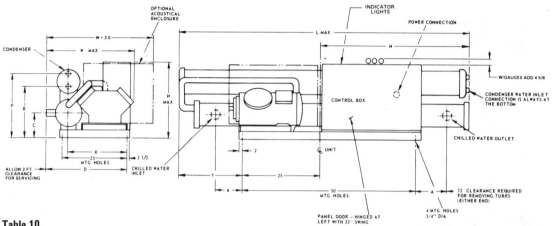

Table 10

MODEL NO.	MAXIMUM OVERALL DIMENSIONS			CHILLED WATER CONNECTIONS				CONDENSER WATER CONNECTIONS					
	L	W	H	SIZE NPT (EXT)	A	C	D	SIZE NPT (INT)	IN E	OUT F	K	N	T
WHR-008C	70¾	33	32½	2	1½	10½	30¼	1½	17¾	21¼	21½	32	13¾
WHR-010C	72	33	32½	2	1½	10½	30¼	1½	17¾	21¼	21½	32	15
WHR-015C	102	38	32½	2	1½	11½	31¼	2	21¼	25½	21½	51	26

WHR-020C – 030C (Arrangement 1)

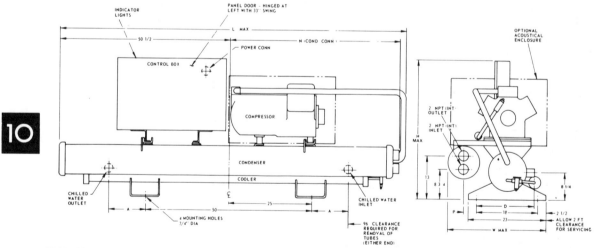

Table 11

MODEL NO.	MAXIMUM OVERALL DIMENSIONS			CHILLED WATER CONNECTIONS			CONDENSER WATER CONNECTIONS		
	L	W	H	SIZE	A	D	N	SIZE	P
WHR-020C	104¼	26¼	46½	2 NPT (EXT)	13½	17¾	51	2 NPT (INT)	1½
WHR-025C	101½	29½	47½	2 NPT (EXT)	7½	18¾	51	2 NPT (INT)	1¾
WHR-030C	106	29½	46½	2 NPT (EXT)	13½	18¾	51	2 NPT (INT)	1¾

FIG. 10-58 *(Continued)*

Water pressure drop

Chart 1 – Cooler

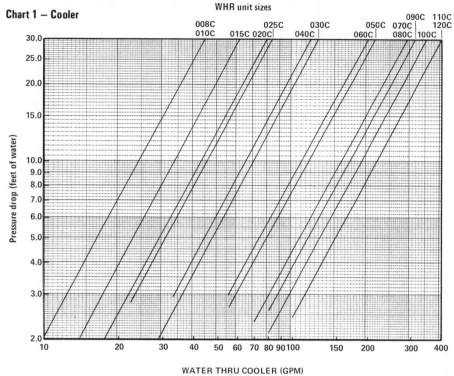

WATER THRU COOLER (GPM)

Chart 2 – Condenser

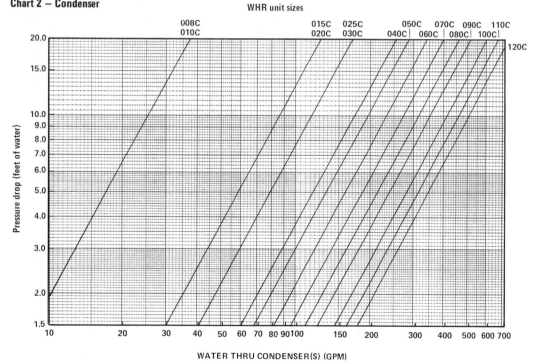

WATER THRU CONDENSER(S) (GPM)

FIG. 10-58 *(Continued)*

Havens CBC SERIES COOLING TOWERS

ENGINEERING DATA

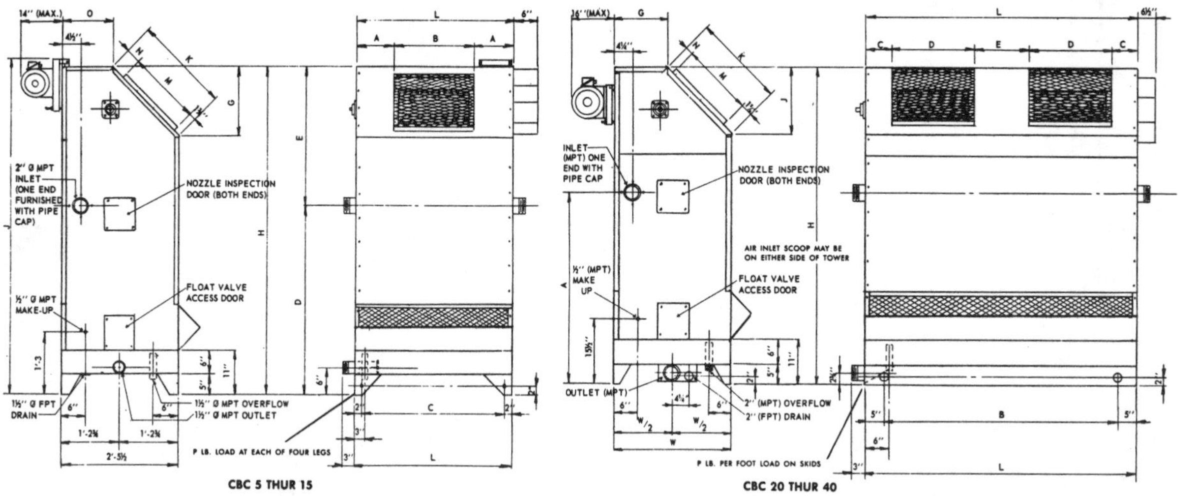

DIMENSIONAL DATA

NOT CERTIFIED FOR CONSTRUCTION

TOWER MODEL	DIMENSIONS (IN INCHES ±⅛")														WEIGHTS (IN POUNDS)			MOTOR H.P.	PIPE SIZES	
	A	B	C	D	E	G	H	J	K	L	M	N	O	W	P	SHIP.	OPER.		INLET	OUTLET
CBC-5	6⅞	12¾	22½	46½	22½	11	68½	70½	16	26½	12½	1¾	18		130	365	520	½	SEE DRAWING	SEE DRAWING
CBC-8	6⅞	12¾	22½	46½	22½	11	68½	70½	16	26½	12½	1¾	18		132.5	370	530	¾		
CBC-10	12¾	16½	38	46½	31	15¾	77	79	22½	42	16½	4¼	13¼		185	500	740	1		
CBC-15	10⅝	20¾	38	46½	31	15¾	77	79	22½	42	18¾	2	13¼		191.25	520	765	1½		
CBC-20	46½	59	9	16½	18	13½	77½	15½	22¼	69	16½	4⅛		29	121	990	1390	2	2½	3
CBC-25	46½	59	6⅞	20¾	13¾	13½	77½	15½	22¼	69	18¾	2⅛		29	126	1050	1450	3	2½	3
CBC-30	60	62	7⅝	20¾	15¼	18	95½	17½	24½	72	20¾	2⅛		35½	208	1350	2490	3	3	4
CBC-40	60	62	7⅝	20¾	15¼	18	95½	17½	24½	72	20¾	2⅛		35½	212	1400	2550	5	3	4

PERFORMANCE DATA

CAPACITIES IN TONS

G. P. M./Ton		3															4										MOTOR H. P.		
Hot Water		90	87	90	95	92	95	96	95	97	95	96	97	96	96	97	82.5	85	89	89.5	92.5	93.5	95	95	95	97			
Cold Water		80	77	80	85	82	85	86	85	87	85	86	87	86	86	87	75	77.5	81.5	82	85	86	87.5	87.5	87.5	89.5	0"		
Wet Bulb		65	70	70	70	72	72	73	75	75	78	78	78	79	80	80	65	70	70	72	75	78	78	79	80	80	S.P.	R.P.M.	Drive
	CBC-5	7.2	4.0	5.6	7.9	5.7	7.3	7.5	6.3	7.3	5.0	5.6	6.2	5.2	4.8	5.3	4.4	3.9	5.7	5.2	5.7	5.0	5.9	5.5	5.0	6.3	½	1750	V-Belt
Model	CBC-8	10.8	6.1	8.4	11.9	8.7	11.0	11.3	9.4	11.0	7.5	8.4	9.3	7.8	7.1	7.9	6.5	5.8	8.5	7.8	8.5	7.5	8.8	8.2	7.5	9.4	1	1750	V-Belt
	CBC-10	14.4	8.1	11.1	15.9	11.5	14.7	15.0	12.5	14.7	10.0	11.1	12.3	10.4	9.5	10.6	8.8	7.8	11.4	10.4	11.4	10.1	11.7	11.0	10.1	12.6	1	1750	V-Belt
Number	CBC-15	21.5	12.2	16.7	23.9	17.3	22.1	22.5	18.7	22.0	15.0	16.7	18.5	15.6	14.2	15.9	13.1	11.6	17.1	15.6	17.1	15.1	17.5	16.4	15.1	18.8	1½	1750	V-Belt
	CBC-20	28.7	16.2	22.2	31.8	23.0	29.4	30.0	24.9	29.3	20.0	22.2	24.6	20.8	18.9	21.2	17.5	15.5	22.8	20.8	22.8	20.1	23.3	21.9	20.1	25.1	2	1750	V-Belt
	CBC-25	35.9	20.3	27.8	39.8	28.8	36.8	37.5	31.1	36.6	25.0	27.8	30.8	26.0	23.6	26.5	21.9	19.4	28.5	26.0	28.5	25.1	29.1	27.4	25.1	31.4	3	1750	V-Belt
	CBC-30	43.1	24.3	33.3	47.7	34.5	44.1	45.0	37.4	44.0	30.0	33.3	36.9	31.2	28.4	31.8	26.3	23.3	34.2	31.2	34.2	30.2	35.0	32.9	30.2	37.7	3	1750	V-Belt
	CBC-40	57.4	32.4	44.4	63.6	46.0	58.8	60.0	49.8	58.6	40.0	44.4	49.2	41.6	37.8	42.4	35.0	31.0	45.6	41.6	45.6	40.2	46.6	43.8	40.2	50.2	5	1750	V-Belt

NOTES

1. Wet operating weight includes normal operating depths of water in both distribution pipe and basin but no allowance for wind load. Data for designing foundation requirements will be furnished on request.
3. Capacities are based on 250 BTU per minute per ton.
4. Specifications subject to change without notice.

Havens COOLING TOWERS

7219 EAST 17th STREET, Kansas City, Missouri 64126

CBC-104-66

FIG. 10-59 Havens Cooling Towers Catalog

Mechanical, Performance and Electrical Specifications, Type K/EK

Nominal cooling capacities are at standard ASHRAE rating conditions; outside temperatures of 95°F DB and 75°F WB; indoor temperatures of 80°F DB and 67°F WB. Nominal heating capacities are at 65°F DB return air temperature.

Note that all published specifications and dimensions for Incremental® equipment are subject to change without notice; we therefore urge the engineer to obtain confirming information from the factory through his representative and to use only certified prints for design and construction. It is the responsibility of the specifier to determine the applicability of codes and to specify any necessary equipment changes.

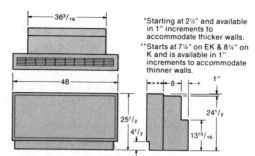

*Starting at 2½" and available in 1" increments to accommodate thicker walls.

**Starts at 7¼" on EK & 8¼" on K and is available in 1" increments to accommodate thinner walls.

Performance and Electrical Data, Type K

Model		K 6L			K 9L			K 11L			K 14L		
Cooling	BTUH(1)	6200			8000			10,500			13,400		
		208V	230V	265V	208V	230V	265V	208V	230V	265V	208V	230V	265V
	Full Load Amps	4.5	3.9	3.7	5.2	4.7	4.1	7.7	6.7	5.9	10.5	9.9	8.3
	Watts	810			1060			1500			2030		
	Power Factor — %	97	98	98	98	98	98	97	97	97	93	93	92
	EER-BTUH/Watt	7.7			7.5			7.0			6.5		
Heating	BTUH(2)												
	Hot Water	13,300			13,300			13,300			15,500		
	Steam	16,500			16,500			16,500			19,200		
Air-CFM	Cooling	250			290			290			350		
	Heating	210			210			210			270		
	Ventilation-Cooling	50			60			60			70		
Minimum Circuit Amps		5.2	4.7	4.0	6.2	5.6	4.9	8.9	8.0	7.0	12.8	10.4	10.0
Delay Fuse — Max. Amps		15	15	15	15	15	15	15	15	15	15	15	15
Net Shipping Wt., Lbs.		305			305			320			320		

(1) Based on ASHRAE and ARI Test Conditions of 95°F DB/75°F WB Outdoors, 80°F DB/67°F WB Inside.
(2) Based on 200°F E.W.T., 180°F L.W.T., 2 psig Steam — 65°F E.A.T.

Performance and Electrical Data, Type EK

Model		EK 6L			EK 9L			EK 11L			EK 14L		
Cooling	BTUH(1)	6200			8000			10,500			13,400		
		208V	230V	265V	208V	230V	265V	208V	230V	265V	208V	230V	265V
	Full Load Amps	4.5	3.9	3.7	5.2	4.7	4.1	7.7	6.7	5.9	10.5	9.9	8.3
	Watts	810			1060			1500			2030		
	Power Factor — %	97	98	98	98	98	98	97	97	97	93	93	92
	EER — BTUH/Watt	7.7			7.5			7.0			6.5		
Electric	"S" Heater — Amps(2)	12.9	10.8	9.8	12.9	10.8	9.8	12.9	10.8	9.8	—	—	—
	Watts	2600	2400	2600	2600	2400	2600	2600	2400	2600	—	—	—
	BTUH	8900	8200	8900	8900	8200	8900	8900	8200	8900	—	—	—
	"M" Heater — Amps(2)	16.8	14.3	12.4	16.8	14.3	12.4	16.8	14.3	12.4	16.8	14.4	12.5
	Watts	3400	3200	3200	3400	3200	3200	3400	3200	3200	3400	3200	3200
	BTUH	11,600	10,900	10,900	11,600	10,900	10,900	11,600	10,900	10,900	11,600	10,900	10,900
	"L" Heater — Amps(2)	—	—	—	—	—	—	20.2	18.7	16.2	20.3	18.8	16.3
	Watts	—	—	—	—	—	—	4200	4200	4200	4200	4200	4200
	BTUH	—	—	—	—	—	—	14,340	14,340	14,340	14,340	14,340	14,340
Air-CFM	Cooling	250			290			290			350		
	Heating	210			210			210			270		
	Ventilation	50			60			60			70		
Minimum Circuit Amps	"S" Heater	16.0	13.4	12.5	16.0	13.4	12.6	16.1	13.5	12.6	—	—	—
	"M" Heater	20.8	17.8	15.8	20.8	17.8	15.8	20.9	17.8	15.8	20.9	17.9	15.9
	"L" Heater	—	—	—	—	—	—	25.7	23.7	20.6	25.7	23.9	20.6
Delay Fuse Maximum Amps	"S" Heater	20	15	15	20	15	15	20	15	15	—	—	—
	"M" Heater	25	20	20	25	20	20	25	20	20	25	20	20
	"L" Heater	—	—	—	—	—	—	30	25	25	30	25	25
Shipping Wt., Lbs.		305			305			320			335		

(1) Based on ASHRAE and ARI Test Conditions of 95°F DB/75°F WB Outdoors, 80°F DB/67°F WB Inside.
(2) Includes Fan Motor.

Specifications subject to change without notice.

Agency listing is indicated by the label affixed to the equipment.

PRINTED IN U.S.A. FLP
3705 8/85 CYM 2006

FIG. 10-60 McQuay, Snyder General Copr., Through-The-Wall Diffuser Catalog

McQuay Type J/EJ Dimensional & Performance Data
Incremental® Comfort Conditioners for through-the-wall or over-the-sill Installation

Type EJ Performance and Electrical Data

Model	BTUH[1]	Cooling					Electric Heaters						Air-CFM			Min. Circuit Amps		Delay Fuse Max. Amps		Net Shipping Weight (lbs.)
		Total Cooling Load	Full Load Amps	Watts	Power Factor %	EER BTUH/ Watt	"S" Heater			"M" Heater			Cooling	Heating	Ventilation (Cooling)	"S" Heater	"M" Heater	"S" Heater	"M" Heater	
							Amps[2]	Watts	BTUH	Amps[2]	Watts	BTUH								
EJ6C	6,000	208V 4.2 / 230V 3.8 / 265V 3.3		850	97	7.1	12.9 / 10.8 / 9.4	2600 / 2400 / 2400	8900 / 8200 / 8200	16.8 / 14.3 / 12.4	3400 / 3200 / 3200	11600 / 10900 / 10900	260	200	50	16.1 / 13.5 / 12.6	20.9 / 17.9 / 15.5	20 / 15 / 15	25 / 20 / 20	280
EJ9C	7,900	208V 5.0 / 230V 4.5 / 265V 3.9		1000	96	7.9	13.1 / 10.9 / 9.5	2600 / 2400 / 2400	8900 / 8200 / 8200	17.0 / 14.4 / 12.5	3400 / 3200 / 3200	11600 / 10900 / 10900	260	200	60	16.2 / 13.7 / 12.8	21.0 / 18.0 / 15.6	20 / 15 / 15	25 / 20 / 20	280
EJ11C	11,000	208V 8.2 / 230V 7.4 / 265V 6.3		1550	91	7.1	13.0 / 10.8 / 9.4	2600 / 2400 / 2400	8900 / 8200 / 8200	16.9 / 14.3 / 12.4	3400 / 3200 / 3200	11600 / 10900 / 10900	290	230	60	16.0 / 13.5 / 12.6	20.8 / 17.9 / 15.5	20 / 15 / 15	25 / 20 / 20	300

Type J Performance and Electrical Data

Model	BTUH[1]	Cooling					Heating BTUH[3]		ARI-CFM			Minimum Circuit Amps	Delay Fuse Amps	Net Shipping Weight (lbs.)
		Total Cooling Load	Full Load Amps	Watts	Power Factor %	EER BTUH/ Watt	Hot Water	Steam	Cooling	Heating	Ventilation (Cooling)			
J6C	6,000	208V 4.2 / 230V 3.8 / 265V 3.3		850	97	7.1	10,700	17,500	260	200	50	4.9 / 4.5 / 4.1	15 / 15 / 15	280
J9C	7,900	208V 5.0 / 230V 4.5 / 265V 3.9		1000	96	7.9	10,700	17,500	260	200	60	5.9 / 5.3 / 4.6	15 / 15 / 15	280
J11C	11,000	208V 8.2 / 230V 7.4 / 265V 6.3		1550	91	7.1	11,600	19,100	290	230	60	8.4 / 7.7 / 6.6	15 / 15 / 15	300

[1] Based on ASHRAE & ARI Test Conditions of 95° F DB/75° F WB Outside, 80° F DB/67° F Inside.
[2] Includes Fan Motor.
[3] Based on 200° F E.W.T., 180° F L.W.T., 2 PSIG Steam @ 65° F E.A.T.
  Specifications subject to change without notice.

Type JC Unit Dimensions — Higher EER   Type EJC Unit Dimensions — Higher EER

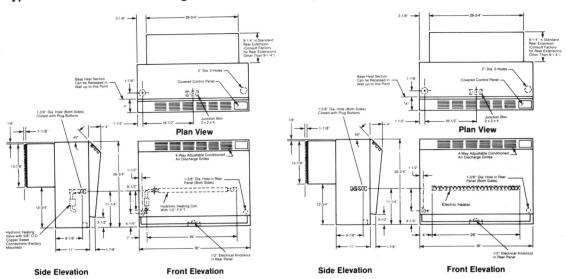

FIG. 10-60 (Continued)

ENELCO®
TITUS®

Round Adjustable Ceiling Diffusers

TM 3-Position
TMA Rotating
TMA-1 Sliding

TITUS TM, TMA and TMA-1 adjustable ceiling diffusers have wide application in heating, cooling and ventilating. All have four cones with the center three movable for adjustment. This not only produces optimum performance, but also assures uniformity of appearance when different sizes are used in the same area.

Features:

TM: Sizes 6 through 36 in. with 3 positions of adjustment available by removing inner cones and repositioning six screws.

TMA: Sizes 6 through 36 in. with adjustment from full horizontal to full vertical by turning the small center cone.

TMA-1: Sizes 6 through 12 in. with full air pattern adjustment by sliding the three center cones up and down.

Standard Outer Cone: Presents an unusually attractive appearance, and is used in the majority of TM Series installations. Available in all sizes.

Style "B" Anti-Smudge Outer Cone: Deep, specially curved contour reduces ceiling smudging. Also used where plenum height or opening size is limited (see drawing at right). Available in 6" through 24" sizes.

Spring Lock: Allows easy removal and replacement of three inner cones.

Material:

Heavy gauge steel.

Finishes:

Exclusive baked ENVIRO-THERM® coating with choice of #1 Aluminum or #25 Off-White.

Other finishes available. See inside back cover or your TITUS Products representative for Color Selection Guide AA-O.

Accessories:

A wide variety is available. See pages D63 through D70.

D1

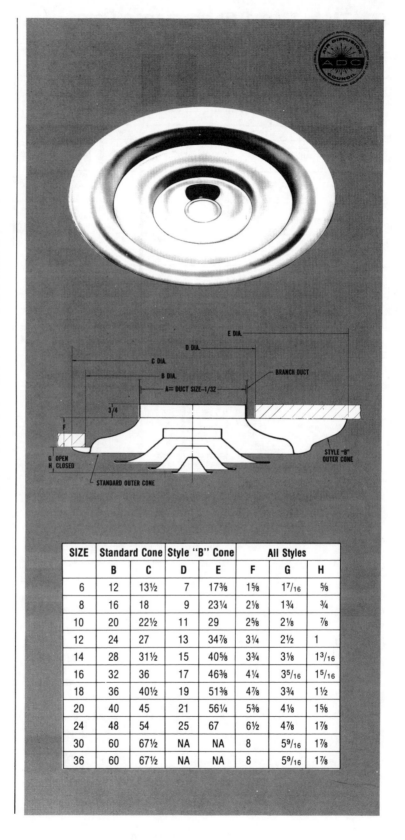

SIZE	Standard Cone		Style "B" Cone		All Styles		
	B	C	D	E	F	G	H
6	12	13½	7	17⅜	1⅝	1⁷/₁₆	⅝
8	16	18	9	23¼	2⅛	1¾	¾
10	20	22½	11	29	2⅝	2⅛	⅞
12	24	27	13	34⅞	3¼	2½	1
14	28	31½	15	40⅝	3¾	3⅛	1³/₁₆
16	32	36	17	46⅜	4¼	3⁵/₁₆	1⁵/₁₆
18	36	40½	19	51⅜	4⅞	3¾	1½
20	40	45	21	56¼	5⅜	4⅛	1⅝
24	48	54	25	67	6½	4⅞	1⅞
30	60	67½	NA	NA	8	5⁹/₁₆	1⅞
36	60	67½	NA	NA	8	5⁹/₁₆	1⅞

FIG. 10-61 Titus Products Div., Philips Industries Inc., Diffuser Catalog

TM, TMA, TMA-1 Performance Data

Size			400	500	600	700 (NC 20)	800	900	1000 (NC 30)	1200	1400 (NC 40)	1600
	Neck Velocity, fpm		400	500	600	700	800	900	1000	1200	1400	1600
	Vel. Press. in. W.G.		.010	.016	.023	.031	.040	.051	.063	.090	.122	.160
	Total Press.	Horizontal	.021	.034	.048	.065	.084	.107	.132	.189	.256	.346
		Vertical	.027	.044	.063	.085	.109	.139	.172	.246	.333	.437
6″	Flow Rate, cfm		80	100	120	140	160	180	200	235	275	315
	Radius of Diff., ft.		1-2-3	2-3-4	2-3-5	2-4-6	3-4-7	3-5-7	4-5-8	4-6-10	5-7-11	6-8-13
	NC		—	—	15	20	24	27	31	36	41	45
8″	Flow Rate, cfm		140	175	210	245	280	315	350	420	490	560
	Radius of Diff., ft.		2-3-4	2-3-5	3-4-7	3-5-8	4-5-9	4-6-10	5-7-11	5-8-13	6-9-15	7-11-17
	NC		—	—	16	21	26	29	32	38	43	47
10″	Flow Rate, cfm		220	270	330	380	435	490	545	655	765	870
	Radius of Diff., ft.		2-3-5	3-4-7	3-5-8	4-6-9	4-7-11	5-8-12	6-8-14	7-10-16	8-12-19	9-13-22
	NC		—	—	17	22	27	30	33	39	44	48
12″	Flow Rate, cfm		315	390	470	550	630	705	785	940	1100	1255
	Radius of Diff., ft.		3-4-7	3-5-8	4-6-10	5-7-11	5-8-13	6-9-15	7-10-16	8-12-19	9-14-23	11-16-26
	NC		—	—	18	23	27	31	34	40	45	50
14″	Flow Rate, cfm		425	530	635	745	850	955	1060	1270	1490	1695
	Radius of Diff., ft.		3-5-8	4-6-9	5-7-11	5-8-13	6-9-15	7-11-17	8-12-19	9-14-22	11-16-26	13-19-30
	NC		—	13	19	24	28	32	35	41	46	50
16″	Flow Rate, cfm		560	700	840	980	1120	1260	1400	1680	1960	2240
	Radius of Diff., ft.		4-5-9	5-7-11	5-8-13	6-9-15	7-11-17	8-12-20	9-14-22	11-16-26	13-19-30	14-22-35
	NC		—	14	19	25	29	32	36	41	46	51
18″	Flow Rate, cfm		710	885	1060	1240	1420	1590	1770	2120	2480	2830
	Radius of Diff., ft.		4-6-10	5-8-12	6-9-15	7-11-17	8-12-20	9-14-22	10-15-24	12-18-29	14-21-34	16-24-39
	NC		—	15	20	26	30	33	36	42	47	52
20″	Flow Rate, cfm		875	1100	1310	1530	1750	1970	2190	2610	3060	3500
	Radius of Diff., ft.		4-7-11	6-9-14	7-10-16	8-12-19	9-14-22	10-15-24	11-17-27	13-19-32	16-24-38	18-27-43
	NC		—	15	21	26	30	34	37	43	48	52
24″	Flow Rate, cfm		1260	1570	1880	2200	2510	2820	3140	3770	4400	5020
	Radius of Diff., ft.		5-8-13	7-10-16	8-12-19	9-14-23	11-16-26	12-18-29	14-20-32	16-24-39	19-28-45	22-32-52
	NC		—	16	22	27	31	35	38	44	49	53
30″	Flow Rate, cfm		1960	2450	2940	3430	3920	4410	4900	5880	6860	7840
	Radius of Diff., ft.		7-10-16	8-13-20	10-15-24	12-18-28	13-20-32	15-23-36	17-25-41	20-30-49	24-35-57	27-40-65
	NC		—	17	23	27	32	36	39	45	50	54
36″ **	Flow Rate, cfm		2820	3520	4230	4930	5630	6340	7040	8450	9850	11,260
	Radius of Diff., ft.		8-12-20	10-15-24	12-18-29	14-21-34	16-24-39	18-27-44	20-30-49	24-36-58	28-42-68	32-48-78
	NC		—	18	24	28	33	37	40	46	51	55
Downward Projection ** of Heated Air, ft.												
	10 Deg. Differential		6-6-3	8-8-6	10-12-11	13-15-16	15-19-24	17-23-28	19-25-33	21-32-42	25-38-52	27-40-60
	20 Deg. Differential		4-4-2	6-7-5	7-8-7	9-11-11	10-14-16	12-16-20	13-18-24	15-17-30	17-25-36	19-29-42
	30 Deg. Differential		3-3-2	5-5-4	6-7-6	7-9-9	9-11-13	10-13-16	11-15-19	13-18-25	14-20-30	15-24-35
	40 Deg. Differential		3-2-2	4-4-3	5-6-6	7-8-9	8-10-12	9-12-15	10-13-17	11-16-22	12-18-27	14-20-31

Downward Projection values represent the distance to a total air velocity of essentially zero.
The three values are for a 6″, 12″ and 24″ Diffuser respectively.

- Units are tested in accordance with the Air Diffusion Council (ADC) Code 1062R4 and ASHRAE Standard 36-72, and ratings are certified by the ADC.
- All pressures are in inches of water. To obtain static pressure, subtract the velocity of pressure at the head of the column from the total pressure.
- Minimum radii of diffusion are to a terminal velocity (VT) of 150 fpm, middle to 100 fpm and maximum to 50 fpm. If the diffuser is mounted on an exposed duct, multiply the radii of diffusion shown by 0.70.
- The NC values are based on a room absorption of 10 dB, re 10^{-12} watts. Values shown are for a horizontal pattern. Add 1 dB for a vertical pattern.

Selection Aids:

To help in selecting the diffuser size, consider the three criteria below. Some designers may prefer to use just one or a combination of these methods.

Comfort by Radius of Diffusion (throw)*:

For optimum comfort, select the radius of diffusion at 50 fpm terminal velocity in the range shown for the space length, L. See page A8 for detailed discussion.

L (Space Length)	5	10	20	30
Radius of Diffusion at 50 fpm	3-6	6-12	12-24	18-36

*Based on ADPI (Air Distribution Performance Ratio)

Sound:

The NC data in the performance table are only for outlets. You can find further discussion on selection by noise criteria on pages A9 and A10. To determine the acoustical result of adding a damper see page A17.

Mapping:

This method identifies the most probable portion of a space to be uncomfortable by mapping the jet performance. For a complete discussion see page A9.

Product improvement is a continuing endeavor at TITUS Products. Therefore, product descriptions are subject to change without notice. Contact your TITUS Products representative to verify details.

D2

FIG. 10-61 *(Continued)*

Performance Data TXS & TXR

12 x 12 Module Size

Duct Size		Neck Velocity, FPM	300	400	500	600	700	800	1000	1200	1400
		Vel. Press., In. W.G.	.006	.010	.016	.023	.031	.040	.063	.090	.123
5" RD		Tot. Press., In. W.G.	.011	.019	.030	.044	.059	.076	.120	.171	.234
		Flow Rate, CFM	40	55	70	80	95	110	135	165	190
		NC	—	—	15	20	24	28	34	39	43
	Throw, Ft.	4-WAY	1-2-4	2-2-5	2-3-6	2-4-7	3-5-7	3-6-8	5-6-9	6-7-10	6-7-10
		3-WAY	1-2-4	2-3-6	2-3-7	2-4-8	3-5-9	4-6-10	5-7-10	6-8-12	6-9-13
		2-WAY	1-2-5	2-3-6	2-4-8	3-5-10	4-6-10	4-7-12	6-8-13	7-10-14	7-10-15
		1-WAY	2-3-6	2-4-8	3-5-9	4-6-10	5-7-11	6-8-13	6-9-13	8-10-14	9-10-15
6" RD		Tot. Press., In. W.G.	.015	.025	.040	.058	.078	.100	.158	.225	.308
		Flow Rate, CFM	60	80	100	120	140	160	195	235	275
		NC	—	—	17	22	26	30	36	41	45
	Throw, Ft.	4-WAY	1-2-4	2-3-5	3-3-7	3-4-8	3-5-8	3-6-9	5-7-10	6-8-10	7-8-11
		3-WAY	1-2-4	2-3-6	3-3-8	3-4-9	3-5-10	4-6-10	5-8-11	6-9-13	7-10-14
		2-WAY	1-2-5	2-3-7	3-4-9	3-5-10	4-6-11	4-7-12	6-9-14	7-10-15	8-11-17
		1-WAY	2-3-6	3-4-9	3-5-11	4-6-11	5-8-12	6-9-13	7-10-15	9-10-16	10-12-17
7" RD		Tot. Press., In. W.G.	.022	.037	.059	.084	.113	.146	.230	.330	.450
		Flow Rate, CFM	80	105	135	160	190	215	270	320	375
		NC	—	15	21	26	30	34	40	45	49
	Throw, Ft.	4-WAY	1-2-5	2-3-6	3-4-9	3-5-10	4-6-10	4-8-11	6-9-12	8-10-13	9-10-14
		3-WAY	1-2-5	2-4-7	3-4-10	3-5-11	4-6-12	5-8-13	6-10-14	7-11-16	9-12-17
		2-WAY	1-2-6	2-4-9	3-5-11	4-6-12	5-8-14	5-9-15	7-11-17	9-13-18	10-14-20
		1-WAY	2-4-8	3-5-11	4-6-12	5-8-13	6-10-14	8-11-15	9-12-17	11-13-18	12-14-20
6 x 6		Tot. Press., In. W.G.	.018	.032	.051	.073	.099	.130	.200	.292	.395
		Flow Rate, CFM	75	100	125	150	175	200	250	300	350
		NC	—	13	19	24	28	32	38	43	47
	Throw, Ft.	4-WAY	1-2-5	2-3-6	3-4-8	3-5-9	4-6-9	4-7-10	6-8-11	7-9-12	8-9-13
		3-WAY	1-2-5	2-4-7	3-4-9	3-5-10	4-6-11	5-7-12	6-9-13	7-10-15	8-11-16
		2-WAY	1-2-6	2-4-8	3-5-10	4-6-12	5-7-13	5-8-14	7-10-16	8-12-17	9-13-19
		1-WAY	2-4-7	3-5-10	4-6-11	5-7-12	6-9-13	7-10-14	8-11-16	10-12-17	11-13-19
10 x 10		Neg.Stat.Press., In.W.G.	.030	.050	.075	.110	.155	.200	.310	.450	.615	Return
		Flow Rate, CFM	210	280	345	420	485	555	695	830	970
		NC	—	17	24	30	35	39	46	53	58

12 x 24 Module Size

Duct Size		Neck Velocity, FPM	300	400	500	600	700	800	1000	1200	1400
		Vel. Press., In. W.G.	.006	.010	.016	.023	.031	.040	.063	.090	.123
5" RD		Tot. Press., In. W.G.	.010	.018	.028	.040	.054	.070	.110	.157	.215
		Flow Rate, CFM	40	55	70	80	95	110	135	165	190
		NC	—	—	14	19	23	27	33	38	42
	Throw, Ft.	4-WAY	1-2-4	2-2-5	2-3-6	2-4-7	3-5-7	3-6-8	5-6-9	6-7-10	6-7-10
		3-WAY	1-2-4	2-3-6	2-3-7	2-4-8	3-5-9	4-6-10	5-7-10	6-8-12	6-9-13
		2-WAY	1-2-5	2-3-6	2-4-8	3-5-10	4-6-10	4-7-12	6-8-13	7-10-14	7-10-15
		1-WAY	2-3-6	2-4-8	3-5-9	4-6-10	5-7-11	6-8-13	6-9-13	8-10-14	9-10-15
6" RD		Tot. Press., In. W.G.	.013	.021	.034	.048	.065	.084	.132	.189	.258
		Flow Rate, CFM	60	80	100	120	140	160	195	235	275
		NC	—	—	17	22	26	30	36	41	45
	Throw, Ft.	4-WAY	1-2-4	2-3-5	3-3-7	3-4-8	3-5-8	3-6-9	5-7-10	6-8-10	7-8-11
		3-WAY	1-2-4	2-3-6	3-3-8	3-4-9	3-5-10	4-6-10	5-8-11	6-9-13	7-10-14
		2-WAY	1-2-5	2-3-7	3-4-9	3-5-10	4-6-11	4-7-12	6-9-14	7-10-15	8-11-17
		1-WAY	2-3-6	3-4-9	3-5-11	4-6-11	5-8-12	6-9-13	7-10-15	9-10-16	10-12-17
7" RD or 6 x 6		Tot. Press., In. W.G.	.015	.025	.039	.057	.076	.098	.155	.221	.302
		Flow Rate, CFM	80	105	135	160	190	215	270	320	375
		NC	—	14	20	25	29	33	39	44	48
	Throw, Ft.	4-WAY	1-2-5	2-3-6	3-4-9	3-5-10	4-6-10	4-8-11	6-9-12	8-10-13	9-10-14
		3-WAY	1-2-5	2-4-7	3-4-10	3-5-11	4-6-12	5-8-13	6-10-14	7-11-16	9-12-17
		2-WAY	1-2-6	2-4-9	3-5-11	4-6-12	5-8-14	5-9-15	7-11-17	9-13-18	10-14-20
		1-WAY	2-4-8	3-5-11	4-6-12	5-8-13	6-10-14	8-11-15	9-12-17	11-13-18	12-14-20
6 x 6		Tot. Press., In. W.G.	.017	.030	.048	.069	.094	.123	.189	.276	.374
		Flow Rate, CFM	75	100	125	150	175	200	250	300	350
		NC	—	13	19	24	28	32	38	43	47
	Throw, Ft.	4-WAY	1-2-5	2-3-6	3-4-8	3-5-9	4-6-9	4-7-10	6-8-11	7-9-12	8-9-13
		3-WAY	1-2-5	2-4-7	3-4-9	3-5-10	4-6-11	5-7-12	6-9-13	7-10-15	8-11-16
		2-WAY	1-2-6	2-4-8	3-5-10	4-6-12	5-7-13	5-8-14	7-10-16	8-12-17	9-13-19
		1-WAY	2-4-7	3-5-10	4-6-11	5-7-12	6-9-13	7-10-14	8-11-16	10-12-17	11-13-19
6 x 18		Tot. Press., In. W.G.	.041	.068	.109	.157	.211	.273	.430	.613	.840
		Flow Rate, CFM	225	300	375	450	525	600	750	900	1050
		NC	17	25	31	36	40	44	50	55	59
	Throw, Ft.	4-WAY	5-7-15	6-10-17	8-12-19	10-15-21	11-16-22	13-17-24	16-19-27	17-21-30	19-23-32
		3-WAY	5-7-15	7-10-17	8-13-19	10-15-21	12-16-22	13-17-24	16-19-27	17-21-30	19-23-32
		2-WAY	5-8-15	7-11-17	9-13-19	11-15-21	13-16-22	14-17-24	16-19-27	17-21-30	19-23-32
		1-WAY	8-12-21	10-15-24	13-19-27	15-21-30	18-23-32	20-24-34	22-28-39	24-30-42	27-32-46
10 x 22		Neg.Stat.Press., In.W.G.	.030	.050	.075	.110	.155	.200	.310	.450	.615	Return
		Flow Rate, CFM	460	610	770	920	1080	1225	1530	1830	2140
		NC	—	19	25	31	36	41	48	55	60

D19

FIG. 10-61 (Continued)

Enviro-Master® Application Data

Inlet See Note 3	CFM	Minimum ΔP$_s$ Basic Ass'y Only	With Atten.	Basic Ass'y Plus Coil 1 Row	2 Row	Minimum T.P. Basic Ass'y Only	@ Min. ΔP$_s$ Basic Ass'y Only	With Atten.	0.5 In. ΔP$_s$ Basic Ass'y Only	With Atten.	Radiated	1.0 In ΔP$_s$ Basic Ass'y Only	With Atten.	Radiated	3.0 In ΔP$_s$ Basic Ass'y Only	With Atten.	Radiated
4" Rd. 100%	75	.02	.04	.04	.06	.07	—	—	29	—		34	—		39	—	
	100	.04	.08	.08	.12	.12	—	—	33	—	17	37	—	22	44	24	28
	150	.08	.17	.16	.24	.26	27	—	37	—	19	41	21	24	47	27	32
	200	.15	.31	.30	.46	.47	34	—	40	20	20	44	24	25	50	30	33
5" Rd. 100%	100	.02	.04	.04	.06	.05	—	—	31	—	16	35	—	21	43	23	28
	200	.06	.13	.12	.18	.18	22	—	36	—	19	41	21	24	48	28	31
	300	.13	.27	.26	.40	.40	31	—	40	20	20	45	25	25	52	32	33
	350	.18	.38	.36	.55	.55	34	—	41	21	21	46	26	26	53	33	34
6" Rd. 50%	200	.03	.06	.06	.09	.09	—	—	28	—	17	35	—	22	44	24	30
	300	.07	.15	.14	.21	.19	—	—	32	—	19	38	—	24	47	27	32
	400	.12	.25	.24	.37	.32	22	—	34	—	20	40	20	25	49	29	33
	450	.15	.31	.30	.46	.41	25	—	35	—	21	41	21	26	50	30	33
7" Rd. 50%	350	.05	.14	.13	.20	.14	—	—	28	—	19	34	—	23	43	29	31
	450	.08	.22	.21	.32	.22	—	—	31	—	20	37	19	25	46	30	32
	550	.12	.32	.31	.48	.32	21	—	33	—	21	39	20	26	48	31	33
	650	.16	.43	.42	.64	.45	25	—	34	—	21	40	21	26	49	32	34
8" Rd. 40%	400	.04	.11	.10	.16	.09	—	—	27	—	18	33	19	23	42	29	31
	600	.08	.22	.21	.32	.19	—	—	32	—	20	37	20	25	46	32	33
	700	.11	.30	.28	.44	.26	21	—	33	—	21	39	21	26	48	33	33
	800	.14	.38	.36	.56	.33	24	—	34	—	21	40	22	26	49	34	34
9" Rd. 40%	450	.03	.08	.08	.12	.07	—	—	28	—	18	33	—	23	42	31	31
	650	.07	.19	.18	.28	.17	—	—	31	—	20	37	22	25	46	33	32
	850	.12	.32	.31	.48	.29	22	—	34	—	21	40	24	26	49	35	33
	1050	.18	.49	.47	.72	.43	27	19	36	21	22	43	25	27	51	36	34
10" Rd. 30%	750	.06	.16	.16	.24	.13	—	—	32	—	20	38	24	24	47	35	32
	950	.10	.27	.26	.40	.21	21	—	34	—	21	40	25	25	49	37	33
	1150	.15	.40	.39	.60	.31	26	—	36	19	21	42	26	26	51	38	34
	1350	.21	.58	.55	.84	.43	30	22	38		22	43	27	27	53	39	35
12" Rd. 20%	900	.05	.14	.13	.20	.09	—	—	29	—	19	35	23	24	44	35	32
	1300	.10	.27	.26	.40	.19	20	—	33	—	21	39	25	26	48	37	33
	1700	.17	.46	.44	.68	.33	27	20	36	20	22	41	27	27	50	39	34
	2100	.26	.70	.68	1.0	.51	32	25	38		23	43	28	28	52	40	35
14" Rd. 15%	1700	.10	.27	.26	.40	.19	20	—	33	20	21	39	27	26	48	39	33
	2200	.17	.46	.44	.68	.31	27	20	36	22	22	42	29	27	51	41	34
	2700	.26	.70	.68	1.0	.48	32	26	38		23	44	30	28	53	42	35
	3200	.36	.97	.94	1.4	.66	37	31	40		23	45	31	28	54	43	36
16" Rd. 10%	2000	.08	.22	.21	.32	.15	—	—	32	20	21	37	27	25	47	39	33
	3000	.13	.49	.47	.72	.35	27	21	36	23	22	42	30	27	51	41	35
	3500	.25	.68	.65	1.0	.46	31	25	37		23	43	31	28	52	42	35
	4000	.32	.86	.83	1.3	.59	35	29	39		23	44	32	28	53	43	36
24x16 5%	2500	.03	.08	.08	.12	.07	—	—	28	21	22	34	28	26	43	40	34
	4000	.08	.22	.21	.32	.18	18	—	33	23	24	39	30	28	48	42	36
	6000	.18	.49	.47	.72	.40	29	—	37	25	26	43	32	30	52	44	38
	8000	.33	.89	.86	1.3	.69	36	23	40		27	46	34	32	55	46	40

Room Noise Criterion (NC)

1. These data have been obtained in the manner outlined in the Air Diffusion Council Test Code 1062R4 and American Society of Heating and Air Conditioning Engineers Standard 36-72.
2. For sound power spectra for octave bands 2 through 6, please refer to pages G31 and G32.
3. NC index values are based on one outlet handling percent total air flow shown in Inlet Size column.

FIG. 10-61 (Continued)

G8

Quick selection table

ST UNIT SIZE	SAT. SUCTION TEMP. °F	CONDENSER ENTERING AIR TEMPERATURE									
		75F		85F		95F		105F		115F	
		CAP. TONS	KW INPUT	CAP. TONS	KW INPUT	CAP. TONS	KW INPUT	CAP. TONS	KW INPUT	CAP. TONS	KW INPUT
008	35	8.6	7.8	8.0	8.2	7.4	8.6	6.7	8.8	5.9	9.0
	40	9.4	8.2	8.7	8.8	8.0	9.1	7.4	9.4	6.6	9.7
	45	10.3	8.7	9.4	9.3	8.8	9.7	7.9	10.0	7.3	10.3
	50	11.2	9.2	10.4	9.8	9.4	10.3	8.7	10.6	7.9	11.0
010	35	11.2	10.6	10.3	11.1	9.4	11.5	8.6	11.8	7.8	12.0
	40	12.2	11.1	11.4	11.7	10.5	12.5	9.4	12.6	8.6	12.9
	45	13.3	11.8	12.5	12.4	11.5	13.0	10.4	13.4	9.4	13.8
	50	14.4	12.4	13.5	13.1	12.5	13.7	11.4	14.3	10.1	14.9
013	35	14.5	13.8	13.4	14.6	12.3	15.2	11.3	15.7	10.4	16.1
	40	15.8	14.5	14.7	15.4	13.5	16.0	12.5	16.6	11.4	17.2
	45	17.1	15.4	15.9	16.1	14.6	16.9	13.5	17.6	12.4	18.3
	50	18.5	16.0	17.3	16.9	15.9	17.8	14.7	18.6	13.3	19.4
015	35	16.3	16.1	15.0	16.9	13.9	17.4	12.6	17.8	11.4	18.2
	40	17.7	17.1	16.5	17.9	15.2	18.5	13.9	19.0	12.6	19.5
	45	19.3	18.1	17.9	18.9	16.5	19.7	15.2	20.6	13.9	20.9
	50	20.9	19.1	19.4	20.0	17.9	20.8	16.5	21.7	15.2	22.4
020	35	22.6	22.7	20.4	23.8	19.1	24.8	17.5	25.7	15.8	26.3
	40	24.6	24.0	22.7	25.3	21.1	26.4	19.1	27.3	17.3	28.1
	45	26.5	25.6	24.6	26.8	22.7	28.0	20.8	29.1	19.0	29.9
	50	28.6	27.2	26.5	28.6	24.5	29.7	22.6	30.9	20.6	31.9
025	35	28.2	28.0	25.6	29.5	23.9	30.8	22.0	31.8	20.1	32.4
	40	30.7	29.8	28.3	31.2	26.1	32.5	24.1	33.9	22.0	34.5
	45	33.2	31.3	30.7	32.8	28.4	34.4	26.1	35.6	24.1	37.0
	50	35.8	32.7	33.2	34.4	30.7	36.3	28.2	27.8	26.1	39.2
030	35	33.1	31.8	30.7	33.4	28.2	34.6	25.7	35.6	23.3	36.3
	40	36.3	33.9	33.6	35.4	31.0	36.8	28.2	37.8	25.8	38.7
	45	39.3	35.8	36.8	37.5	33.8	39.0	31.1	40.5	28.2	41.7
	50	42.8	37.7	39.9	39.6	36.8	41.3	33.8	43.1	30.8	44.5

NOTE: Tabular and curve values do not include condenser fan motor KW. For total fan KW, see Table 4, page 11.

FIG. 10-62 McQuay Air Conditioning, Condensing Unit Catalog

Physical data

Table 5.

Description	103	104	106	206	108	209	111	114	117
CFM Range By Unit Type									
LYF—Ventilating	700—2000	1000—3000	1750—5000	1800—5400	2200—7000	2600—7800	3000—10,000	4000—13,000	5000—15,000
LHD—Heating and Ventilating	700—2000	1000—3000	1750—5000	1800—5400	2200—7000	2600—7800	3000—10,000	4000—13,000	5000—15,000
LSL—Low Pressure Draw-through	700—1800	1000—2700	1750—4000		2200—5500		3000—	4000— 9600	5000—11,600
LML—Low Pressure Blow-through			1750—4000		2200—5500		3000— 7500	4000— 9600	5000—11,600
MSL—Medium Pressure Draw-through					2200—5500		3000— 7500	4000— 9600	5000—11,600
MMM—Medium Pressure Blow-through					2200—5500		3000— 7500	4000— 9600	5000—11,600
HSH—High Pressure Draw-through									5000—10,900
HMH—High Pressure Blow-through									5000—10,900
FAN DATA									
LYF/LHD/LSL/LML/MSL/MMM — Standard FC — Diameter (in.)	9 1/2	12	12 1/4	(2) 9	15	(2) 12 1/4	16 1/2	18 1/4	18 1/4
Standard FC — Outlet Area (sq.ft.)	.84*	1.14*	1.88	1.87	2.82	3.76	3.45	4.2	4.2
Standard FC — Shaft & Bearing Size (in.)	1	1	1 3/16	1 3/16	1 3/16	1 3/16	1 7/16	1 7/16	1 7/16
Standard AF — Diameter (in.)			13 7/32		14 9/16		16 3/16	19 11/16	19 11/16
Standard AF — Outlet Area (sq.ft.)			1.88		3.83		3.45	4.79	4.79
Standard AF — Shaft & Bearing Size (in.)			1 3/16		1 7/16		1 7/16	1 15/16	1 15/16
Optional FC — Diameter (in.)								20	20
Optional FC — Outlet Area (sq.ft.)								5.19	5.19
Optional FC — Shaft & Bearing Size (in.)								1 7/16	1 7/16
Optional AF — Diameter (in.)								21 9/16	21 9/16
Optional AF — Outlet Area (sq.ft.)								5.93	5.93
Optional AF — Shaft & Bearing Size (in.)								1 15/16	1 15/16
MSL/MMM 108**/111 FC — Diameter (in.)					13 1/2		15		
MSL/MMM FC — Outlet Area (sq.ft.)					2.82		3.45		
MSL/MMM FC — Shaft & Bearing Size (in.)					1 11/16		1 15/16		
HSH/HMH Standard AF — Diameter (in.)									19 11/16
HSH/HMH Standard AF — Outlet Area (sq.ft.)									3.45
HSH/HMH Standard AF — Shaft & Bearing Size (in.)									1 15/16
COIL DATA									
EXTRA LARGE FACE AREA — Number/Size (in.)	1-15x26.5	1-18x32.5	1-21x42.5		1-30x40.5		1-30x55.5	1-30x70.5	1-30x85.5
EXTRA LARGE FACE AREA — Face Area (sq.ft.)	2.8	4.1	6.2		8.5		11.6	14.7	17.8
LARGE FACE AREA — Number/Size (in.)	1-15x21.5	1-18x27.5	1-21x37.5	1-15x56.5	1-30x35.5	1-18x70.5	1-30x50.5	1-30x65.5	1-30x80.5
LARGE FACE AREA — Face Area (sq.ft.)	2.3	3.5	5.5	5.9	7.4	8.8	10.6	13.7	16.8
SMALL FACE AREA — Number/Size (in.)	1-12x21.5	1-15x27.5	1-15x37.5	1-12x56.5	1-24x35.5	1-15x70.5	1-24x50.5	1-24x65.5	1-24x80.5
SMALL FACE AREA — Face Area (sq.ft.)	1.8	2.9	3.9	4.9	5.9	7.3	8.4	10.9	13.4
BLOW-THROUGH HEATING HOT DECK — Number/Size (in.)			1-12x37.5		1-18x35.5		1-18x50.5	1-18x65.5	1-18x80.5
BLOW-THROUGH HEATING HOT DECK — Face Area (sq.ft.)			3.1		4.5		6.3	8.2	10.1
MAXIMUM ZONES ON MULTIZONE				7	7		9	12	14
FILTER DATA									
FLAT FILTER SECTION — Number/Size (in.)	2-16x20x2	2-20x20x2	3-16x25x2	3-20x20x2	2-16x20x2 2-16x25x2	4-20x20x2	6-16x20x2	6-16x25x2	4-16-20x2 4-16x25x2
FLAT FILTER SECTION — Filter Area (sq.ft.)	4.4	5.6	8.4	8.3	10.0	11.1	13.3	16.7	20.0
ANGULAR FILTER SECTION — Number/Size (in.)	2-16x25x2	4-16x20x2	6-16x20x2	6-16x20x2	6-16x25x2	8-16x20x2	6-20x25x2	4-16x25x2 4-20x25x2	8-20x25x2
ANGULAR FILTER SECTION — Filter Area (sq.ft.)	5.6	8.9	13.3	13.3	16.7	17.8	20.8	24.9	27.8
HEAVY-DUTY FILTER SECTION — Number/Size (in.)						8-20x20x2	9-20x20x2	9-20x25x2	9-20x20x2 6-16x20x2
HEAVY-DUTY FILTER SECTION — Filter Area (sq.ft.)						22.2	25.0	31.2	38.3
METAL GAUGES									
Blower Section — Structural Frame — All	14,16	14,16	14,16	14,16	12,14,16	12,14,16	12,14,16	12	12
Blower Section — Discharge Panel — Low & Medium Pressure	18	18	16,18	16	16	16	16	16	16,14
Blower Section — Discharge Panel — High Pressure									16
Blower Section — Removable Panels — Low & Medium Pressure	20	20	20	18	20,18	18	18	18	18,16
Blower Section — Removable Panels — High Pressure									18
Draw-Through Coil Section — Structural Frame — All	14,16	14,16	14,16		14,16		14,16	14,16	14,16
Draw-Through Coil Section — Bottom Panel — All	16	16	16		16		16	16	16
Draw-Through Coil Section — Drain Pan — Horizontal	18	18	18		18		18	18	18
Draw-Through Coil Section — Drain Pan — Vertical	18	18	18		16		16	16	16
Draw-Through Coil Section — Removable Panels — All	20	20	20		20		18	18	18
Blow-Through Coil Section — Structural Frame — Low & Medium Pressure			16,14		16,14		16,14	16,14	16,14
Blow-Through Coil Section — Structural Frame — High Pressure									12
Blow-Through Coil Section — Bottom Panel — Low & Medium Pressure			14		14		14	14	14
Blow-Through Coil Section — Drain Pan — Low & Medium Pressure			14		14		14	14	14
Blow-Through Coil Section — Drain Pan — High Pressure									12
Blow-Through Coil Section — Removable Panels — Low & Medium Pressure			16		16		16	16	16
Blow-Through Coil Section — Removable Panels — High Pressure									16
LAC AIRCON DATA									
Condenser Coil AIRCON Unit — Face Area (sq.ft.)					10.1		13.5	17.0	20.4
Condenser Coil AIRCON Unit — Rows					6		6	6	6
Condenser Coil AIRCON Unit — Tubes in Face					22		22	22	22
Refrigerant Charge (Lbs.) Summer to 60°F					9		11	14	17
Connections (Standard Unit) — Hot Gas					1 3/8		1 5/8	1 5/8	2 1/8
Connections (Standard Unit) — Liquid					1 1/8		1 3/8	1 3/8	1 5/8

☐ NOT AVAILABLE

*Not Available in MSL, LML or MMM. ** MSL/MMM-108 & 111 only.

FIG. 10-63 McQuay Commercial Air Conditioning, Air Handler Catalog

TYPE LML & MMM
LOW & MEDIUM PRESSURE
BLOW-THROUGH UNITS
Unit Sizes 106C through 111C

Figure 25a.

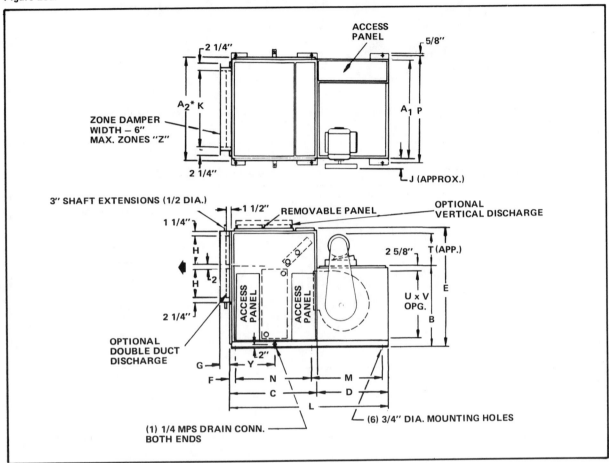

Figure 25b. **AIR INTAKE AND DISCHARGE ARRANGEMENT**

NOTE: With I air intake motor located as shown in solid circle.
With H air intake motor located as shown in phantom circle.

Table 28.   ALL DIMENSIONS ARE APPROXIMATE. CERTIFIED DRAWINGS AVAILABLE ON REQUEST.

Unit Size	A₁	A₂*	B	C	D	E**	F	G	H	J	K	L	M	N	P	T	U	V	Y	Z
106C	50	51 1/8	29¼	35 5/8	28	44	4 7/8	5 5/8	10	6½	42	63 5/8	26 5/8	26½	51¾	16	23¾	45 7/8	18 1/8	7
108C	48	49 1/8	38	43 5/8	36¾	56	4 7/8	5 5/8	14	8	42	80 3/8	35 3/8	34½	49¾	18	32½	43 7/8	22 1/8	7
111C	63	64 1/8	38	43 5/8	36¾	56	4 7/8	5 5/8	14	8	54	80 3/8	35 3/8	34½	64¾	18	32½	58 7/8	22 1/8	9

*Add 4 inches to "A₂" dimension for extra large face area coils.

FIG. 10-63 (Continued)

Figure 37. COMBINATION ANGULAR FILTER AND MIXING BOX

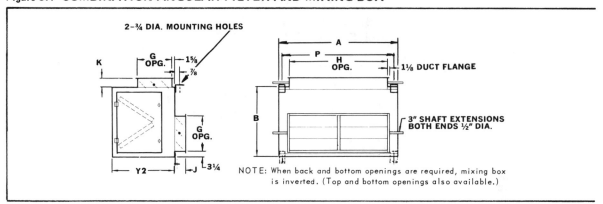

NOTE: When back and bottom openings are required, mixing box is inverted. (Top and bottom openings also available.)

Figure 38. BASE SECTIONS

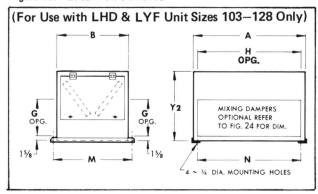

(For Use with LHD & LYF Unit Sizes 103–128 Only)

MIXING DAMPERS OPTIONAL REFER TO FIG. 24 FOR DIM.

Figure 39. MIXING DAMPERS

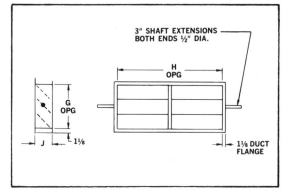

Table 39. ALL DIMENSIONS IN THIS CATALOG ARE IN INCHES UNLESS OTHERWISE INDICATED.

Unit Size	A	B	C	H	J	K	M	N	P_1	P_2	T*	Y_2
103	34	21 5/8	12 3/8	25 3/8	6¼	4 5/8	23 5/8	31 3/8	28 1/2	35 3/4	25 1/2	22½
104	40	24 5/8	12 3/8	31 3/8	6¼	4 5/8	26 5/8	37 3/8	34 1/2	41 3/4	20 1/2	23
106	50	27 7/8	12 3/8	41 2/8	6¼	4 5/8	29 7/8	47 3/8	44 1/2	51 3/4	22 1/4	27¼
206	69	21 5/8	12 3/8	60 3/8	6¼	4 5/8	23 5/8	66 3/8	63 1/2	—	16 1/2	22½
108	48	36 5/8	18 3/8	39 3/8	6¼	4 5/8	38 5/8	45 3/8	42 1/2	49 3/4	26	33½
209	83	24 5/8	12 3/8	74 3/8	6¼	4 5/8	26 5/8	80 3/8	77 1/2	—	20 1/2	23
111	63	36 5/8	18 3/8	54 3/8	6¼	4 5/8	38 5/8	60 3/8	57 1/2	64 3/4	26	33½
114	78	36 5/8	18 3/8	69 3/8	6¼	4 5/8	38 5/8	75 3/8	72 1/2	79 3/4	26	33½
117	93	36 5/8	18 3/8	84 3/8	6¼	4 5/8	38 5/8	90 3/8	87 1/2	94 3/4	26	33½
122	97	43 5/8	24 1/8	88 3/8	7¼	5 5/8	45 5/8	94 3/8	91 1/2	98 3/4	23	32½
128	120	43 5/8	24 1/8	111 3/8	7¼	5 5/8	45 5/8	117 3/8	114 1/2	121 3/4	23	32½
137	119 3/8	55 1/8	31 7/8	112 3/8	7¼	5 5/8	—	—	117 7/8	117 7/8	25 1/2	39 5/8
141	119 3/8	63 3/8	39 5/8	112 3/8	7¼	5 5/8	—	—	117 7/8	117 7/8	21 1/8	47½
150	119 3/8	73 1/8	39 5/8	112 3/8	7¼	5 5/8	—	—	117 7/8	117 7/8	25 1/2	47½
164	119 3/8	91 1/8	47 5/8	112 3/8	7¼	5 5/8	—	—	117 7/8	117 7/8	27 1/8	55½
172	119 3/8	91 1/8	47 5/8	112 3/8	7¼	5 5/8	—	—	217 7/8	117 7/8	27 1/8	55½

*Clearance required for filter removal.
P_1 is used with horizontal units. P_2 is used with vertical units.

FIG. 10-63 *(Continued)*

McQuay Inc.

REFRIGERANT R-22

EVAPORATOR COILS CAPACITY DATA

ENTERING AIR { 80F db / 64F wb

ROWS			3			4			5			6			8		
SUCT. TEMP.	FACE VEL.	FIN SERIES	TOT. (MBH)	LVG. D.B.	LVG. W.B.	TOT. (MBH)	LVG. D.B.	LVG. W.B.	TOT. (MBH)	LVG. D.B.	LVG. W.B.	TOT. (MBH)	LVG. D.B.	LVG. W.B.	TOT. (MBH)	LVG. D.B.	LVG. W.B.
42F	300 FPM	06	8.7	57.3	54.2E	10.4	53.9	52.1E	11.7	51.4	50.4E	12.7	49.6	49.0E	14.2	47.2	47.0E
		08	9.9	54.5	52.8E	11.5	51.4	50.6E	12.8	49.4	49.0E	13.7	47.9	47.7E	14.9	46.1	45.9E
		10	10.7	52.7	51.8E	12.3	50.0	49.6E	13.5	48.2	48.0E	14.3	47.1	46.9E	15.4	45.5	45.3E
		12	11.3	51.4	50.9E	12.9	49.0	48.8E	14.0	47.5	47.3E	14.7	46.5	46.3E	15.7	45.1	44.9E
		14	11.8	50.5	50.2D	13.3	48.4	48.2E	14.3	47.0	46.8E	15.0	46.0	45.8E	15.9	44.8	44.6E
	400 FPM	06	10.2	59.3	55.5E	12.3	55.9	53.6E	14.1	53.3	51.9E	15.5	51.4	50.5E	17.6	48.7	48.4E
		08	11.6	56.5	54.2D	13.8	53.4	52.2E	15.5	51.1	50.5E	16.8	49.5	49.2E	18.7	47.4	47.2E
		10	12.7	54.6	53.2D	14.8	51.7	51.2E	16.5	49.8	49.5E	17.8	48.4	48.2E	19.5	46.6	46.4E
		12	13.5	53.2	52.4D	15.7	50.6	50.3D	17.2	49.0	48.8E	18.4	47.8	47.6E	20.0	46.1	45.9E
		14	14.2	52.2	51.8D	16.3	49.9	49.7D	17.8	48.4	48.2E	18.9	47.3	47.1E	20.3	45.7	45.5E
	500 FPM	06	11.4	60.8	56.4D	14.0	57.4	54.6E	16.0	54.9	53.1E	17.9	52.9	51.7	20.6	50.0	49.6E
		08	13.1	58.0	55.3D	15.7	54.9	53.3D	17.8	52.6	51.7D	19.6	50.8	50.4E	22.1	48.5	48.3E
		10	14.4	56.1	54.3D	17.1	53.1	52.3D	19.2	51.1	50.7D	20.8	49.6	49.4D	23.1	47.6	47.4E
		12	15.4	54.7	53.5D	18.1	52.0	51.5D	20.1	50.1	49.9D	21.7	48.9	48.7D	23.9	47.0	46.8E
		14	16.3	53.6	52.9D	18.9	51.2	50.9D	20.8	49.5	49.3D	22.3	48.3	48.1D	24.4	46.6	46.4E
	600 FPM	06	12.4	62.0	57.2C	15.4	58.7	55.5D	17.8	56.1	54.0D	19.9	54.1	52.7E	23.2	51.2	50.5E
		08	14.3	59.4	56.1C	17.4	56.1	54.2D	19.9	53.8	52.7D	21.9	52.0	51.4D	25.2	49.4	49.2E
		10	15.8	57.4	55.2C	19.0	54.3	53.2D	21.4	52.2	51.7D	23.5	50.6	50.4D	26.5	48.5	48.3D
		12	17.0	55.9	54.5C	20.2	53.1	52.5D	22.6	51.2	50.9D	24.6	49.8	49.6D	27.4	47.9	47.7D
		14	18.0	54.8	53.9C	21.1	52.2	51.9D	23.5	50.6	50.4D	25.4	49.3	49.1D	28.1	47.4	47.2D
	700 FPM	06	13.4	63.1	57.7C	16.6	59.8	56.1C	19.4	57.2	54.7D	21.7	55.2	53.5D	25.6	52.1	51.4D
		08	15.5	60.5	56.7C	18.9	57.2	55.0C	21.8	54.8	53.4D	24.1	53.0	52.2D	27.9	50.4	50.1D
		10	17.1	58.5	55.9C	20.7	55.4	54.0C	23.6	53.2	52.5D	25.9	51.6	51.2D	29.5	49.4	49.2D
		12	18.5	57.0	55.2C	22.1	54.1	53.3C	24.9	52.1	51.8D	27.2	50.7	50.5D	30.6	48.7	48.5D
		14	19.6	55.9	54.6C	23.2	53.2	52.7C	26.0	51.4	51.2D	28.2	50.1	49.9D	31.5	48.2	48.0D
43F	300 FPM	06	8.3	57.8	54.7E	10.0	54.5	52.7E	11.2	52.1	51.1E	12.2	50.4	49.8E	13.6	48.1	47.9E
		08	9.4	55.1	53.4E	11.0	52.1	51.3E	12.2	50.1	49.7E	13.1	48.7	48.5E	14.3	47.0	46.8E
		10	10.2	53.3	52.4E	11.8	50.7	50.3E	12.9	49.0	48.8E	13.7	47.9	47.7E	14.8	46.4	46.2E
		12	10.8	52.1	51.5D	12.3	49.8	49.6E	13.4	48.4	48.2E	14.1	47.3	47.1E	15.0	46.0	45.8E
		14	11.3	51.2	50.9D	12.7	49.2	49.0E	13.7	47.9	47.7E	14.4	46.9	46.7E	15.2	45.7	45.5E
	400 FPM	06	9.8	59.7	55.9E	11.8	56.4	54.1E	13.5	54.0	52.5E	14.8	52.1	51.2E	16.8	49.5	49.2E
		08	11.1	57.0	54.7D	13.2	54.0	52.8D	14.8	51.8	51.2E	16.1	50.2	49.9E	17.9	48.2	48.0E
		10	12.1	55.1	53.7D	14.2	52.4	51.8D	15.8	50.5	50.2D	17.0	49.2	49.0E	18.7	47.5	47.3E
		12	13.0	53.8	53.0D	15.0	51.3	51.0D	16.5	49.7	49.5D	17.6	48.6	48.4D	19.2	47.0	46.8E
		14	13.6	52.9	52.4D	15.6	50.6	50.4D	17.0	49.2	49.0D	18.1	48.1	47.9D	19.5	46.6	46.4E
	500 FPM	06	10.9	61.2	56.8D	13.3	57.9	55.1D	15.3	55.5	53.6D	17.1	53.5	52.3E	19.7	50.8	50.3E
		08	12.5	58.6	55.7C	15.1	55.4	53.8D	17.1	53.2	52.3D	18.7	51.5	51.0D	21.2	49.3	49.1D
		10	13.7	56.6	54.8C	16.3	53.7	52.9D	18.3	51.8	51.4D	19.9	50.3	50.1D	22.2	48.5	48.3D
		12	14.8	55.2	54.1C	17.3	52.6	52.1D	19.2	50.8	50.6D	20.7	49.6	49.4D	22.9	47.9	47.7D
		14	15.6	54.2	53.4C	18.0	51.8	51.6D	19.9	50.3	50.1D	21.4	49.1	48.9D	23.4	47.5	47.3D
	600 FPM	06	11.9	62.5	57.5C	14.7	59.2	55.9D	17.0	56.7	54.5D	19.0	54.7	53.2D	22.2	51.8	51.2D
		08	13.8	59.8	56.4C	16.6	56.6	54.7C	19.1	54.3	53.2D	21.0	52.6	52.0D	24.1	50.2	50.0D
		10	15.2	57.9	55.6C	18.1	54.9	53.8C	20.5	52.8	52.3D	22.5	51.3	51.0D	25.3	49.3	49.1D
		12	16.3	56.4	54.9C	19.3	53.7	53.1C	21.6	51.8	51.6D	23.5	50.5	50.3D	26.2	48.7	48.5D
		14	17.3	55.4	54.3C	20.3	52.8	52.4C	22.5	51.2	51.0D	24.3	50.0	49.8D	26.9	48.3	48.1D
	700 FPM	06	12.8	63.5	58.0B	15.9	60.2	56.5C	18.5	57.7	55.2C	20.8	55.7	54.0D	24.5	52.8	52.0D
		08	14.9	60.9	57.0B	18.1	57.7	55.4C	20.8	55.3	54.0C	23.1	53.6	52.7D	26.7	51.1	50.8D
		10	16.5	59.0	56.2B	19.8	55.9	54.5C	22.6	53.8	53.0C	24.8	52.2	51.8D	28.2	50.1	49.9D
		12	17.8	57.5	55.5B	21.1	54.7	53.8C	23.9	52.7	52.3D	26.0	51.3	51.1D	29.3	49.5	49.3D
		14	18.8	56.4	55.0B	22.2	53.8	53.2C	24.9	52.0	51.8D	27.0	50.8	50.6D	30.1	49.0	48.8D
44F	300 FPM	06	7.9	58.3	55.2E	9.5	55.1	53.3E	10.7	52.8	51.8E	11.6	51.1	50.5E	13.0	48.9	48.7E
		08	9.0	55.7	53.9D	10.5	52.8	52.0E	11.6	50.9	50.5E	12.5	49.6	49.4E	13.7	47.9	47.7E
		10	9.7	53.9	52.9D	11.2	51.4	51.0D	12.3	49.8	49.6E	13.1	48.8	48.6E	14.1	47.3	47.1E
		12	10.3	52.7	52.2D	11.8	50.5	50.3D	12.8	49.2	49.0D	13.5	48.2	48.0E	14.4	46.9	46.7E
		14	10.8	51.9	51.6D	12.2	50.0	49.8D	13.1	48.7	48.5D	13.8	47.8	47.6E	14.6	46.7	46.5E
	400 FPM	06	9.3	60.2	56.3D	11.2	57.0	54.6D	12.8	54.6	53.1D	14.1	52.8	51.9E	16.1	50.3	50.0E
		08	10.6	57.6	55.2D	12.6	54.6	53.3D	14.1	52.5	51.8D	15.4	51.0	50.6D	17.1	49.1	48.9E
		10	11.6	55.7	54.3D	13.6	53.0	52.4D	15.1	51.2	50.9D	16.2	50.0	49.8D	17.8	48.4	48.2E
		12	12.4	54.4	53.5D	14.3	52.0	51.7D	15.8	50.5	50.3D	16.8	49.4	49.2D	18.3	47.9	47.7E
		14	13.0	53.5	52.9D	14.9	51.3	51.1D	16.3	50.0	49.8D	17.3	48.9	48.7D	18.6	47.5	47.3E
	500 FPM	06	10.4	61.7	57.2C	12.7	58.4	55.5D	14.6	56.0	54.1D	16.3	54.1	52.9D	18.8	51.5	51.0D
		08	11.9	59.1	56.1C	14.3	56.0	54.4D	16.3	53.8	52.9D	17.9	52.2	51.7D	20.2	50.1	49.9D
		10	13.1	57.2	55.2C	15.6	54.4	53.5D	17.5	52.4	52.0D	19.0	51.0	50.8D	21.2	49.3	49.1D
		12	14.1	55.8	54.5C	16.5	53.2	52.7D	18.4	51.5	51.3D	19.8	50.4	50.2D	21.8	48.7	48.5D
		14	14.9	54.8	54.0C	17.3	52.4	52.2D	19.0	51.0	50.8D	20.4	49.9	49.7D	22.3	48.3	48.1D
	600 FPM	06	11.4	62.9	57.8B	14.0	59.7	56.3C	16.3	57.2	54.9D	18.1	55.3	53.8D	21.2	52.5	51.8D
		08	13.2	60.3	56.7B	15.9	57.2	55.1C	18.2	54.9	53.7C	20.1	53.2	52.6D	23.0	50.9	50.7D
		10	14.6	58.4	55.9B	17.3	55.5	54.3C	19.6	53.5	52.9D	21.4	52.0	51.7D	24.2	50.1	49.9D
		12	15.7	57.0	55.3B	18.4	54.3	53.6C	20.7	52.5	52.2D	22.4	51.2	51.0D	25.1	49.5	49.3D
		14	16.6	55.9	54.7C	19.3	53.5	53.0C	21.5	51.9	51.7D	23.2	50.7	50.5D	25.7	49.1	48.9D
	700 FPM	06	12.4	63.8	58.2A	15.2	60.7	56.8B	17.7	58.2	55.6C	19.9	56.3	54.5C	23.4	53.4	52.6D
		08	14.3	61.3	57.3A	17.3	58.2	55.8C	19.9	55.9	54.5C	22.1	54.2	53.3C	25.5	51.7	51.4D
		10	15.9	59.5	56.5B	19.0	56.5	54.9C	21.5	54.4	53.6C	23.7	52.8	52.4D	27.0	50.8	50.6D
		12	17.1	58.1	55.9B	20.2	55.2	54.3C	22.8	53.3	52.9C	24.9	52.0	51.8D	28.0	50.2	50.0D
		14	18.1	56.9	55.4B	21.2	54.3	53.7C	23.8	52.6	52.4C	25.8	51.5	51.3D	28.8	49.8	49.6D
45F	300 FPM	06	7.5	58.8	55.6D	9.0	55.7	53.9E	10.2	53.5	52.4E	11.1	51.9	51.3E	12.4	49.8	49.5E
		08	8.5	56.3	54.4D	10.0	53.5	52.6D	11.1	51.7	51.2D	11.9	50.3	50.1E	13.1	48.8	48.6E
		10	9.3	54.6	53.5D	10.7	52.1	51.7D	11.7	50.6	50.4D	12.5	49.6	49.4D	13.5	48.2	48.0E
		12	9.8	53.4	52.8D	11.2	51.3	51.1D	12.2	50.0	49.8D	12.8	49.1	48.9D	13.7	47.9	47.7E
		14	10.3	52.6	52.3D	11.6	50.8	50.6D	12.5	49.6	49.4D	13.1	48.7	48.5D	13.9	47.6	47.4E
	400 FPM	06	8.8	60.7	56.7C	10.7	57.5	55.1D	12.2	55.2	53.7D	13.4	53.5	52.5D	15.3	51.1	50.7D
		08	10.1	58.1	55.6C	12.0	55.2	53.9D	13.5	53.2	52.5D	14.6	51.7	51.4D	16.3	49.9	49.7D
		10	11.0	56.3	54.8C	12.9	53.7	53.0D	14.4	51.9	51.6D	15.4	50.8	50.6D	17.0	49.2	49.0D
		12	11.8	55.0	54.1C	13.6	52.7	52.3D	15.0	51.2	51.0D	16.0	50.2	50.0D	17.4	48.8	48.6D
		14	12.4	54.1	53.5C	14.2	52.0	51.8D	15.5	50.7	50.5D	16.5	49.8	49.6D	17.8	48.4	48.2D
	500 FPM	06	9.9	62.2	57.5B	12.1	59.0	56.0C	14.0	56.6	54.6D	15.5	54.7	53.5D	17.9	52.2	51.7D
		08	11.4	59.6	56.4B	13.7	56.6	54.9C	15.5	54.4	53.5D	17.0	52.9	52.4D	19.2	50.8	50.6D
		10	12.6	57.8	55.6B	14.8	55.0	54.0C	16.7	53.1	52.6D	18.1	51.7	51.5D	20.2	50.1	49.9D
		12	13.5	56.4	55.0C	15.7	53.9	53.3C	17.5	52.2	52.0D	18.9	51.1	50.9D	20.8	49.6	49.4D
		14	14.2	55.4	54.4C	16.4	53.1	52.8C	18.1	51.7	51.5D	19.4	50.7	50.5D	21.3	49.2	49.0D
	600 FPM	06	11.0	63.2	58.0A	13.4	60.2	56.6B	15.5	57.8	55.4C	17.3	55.8	54.3C	20.2	53.2	52.5D
		08	12.7	60.7	57.0A	15.2	57.7	55.6C	17.3	55.5	54.3C	19.2	53.9	53.2C	21.9	51.6	51.4D
		10	14.0	59.0	56.3B	16.6	56.1	54.7C	18.7	54.1	53.5C	20.4	52.7	52.3D	23.0	50.9	50.7D
		12	15.1	57.6	55.6B	17.6	54.9	54.1C	19.7	53.1	52.8C	21.4	51.9	51.7D	23.9	50.3	50.1D
		14	15.9	56.5	55.1B	18.5	54.0	53.6C	20.5	52.5	52.3C	22.1	51.4	51.2D	24.5	49.9	49.7D
	700 FPM	06	12.0	64.3	58.4A	14.6	60.9	57.1B	16.9	58.8	56.0C	18.9	56.9	55.0C	22.3	54.1	53.2D
		08	13.9	61.9	57.5A	16.6	58.8	56.1B	19.0	56.5	54.9C	21.0	54.8	53.9C	24.3	52.4	52.1D
		10	15.3	60.0	56.8A	18.2	57.0	55.3B	20.6	55.0	54.1C	22.6	53.4	53.0C	25.7	51.5	51.3D
		12	16.5	58.4	56.2A	19.4	55.8	54.7B	21.7	53.9	53.5C	23.7	52.6	52.4C	26.7	51.0	50.8D
		14	17.4	57.5	55.7B	20.3	54.9	54.2C	22.7	53.2	53.0C	24.6	52.1	51.9C	27.4	50.5	50.3D

TOT. (MBH) is per square foot of face area

FIG. 10-64 McQuay Inc., Evaporator Catalog

FIGURE NO. 3   **AIR PRESSURE DROP**

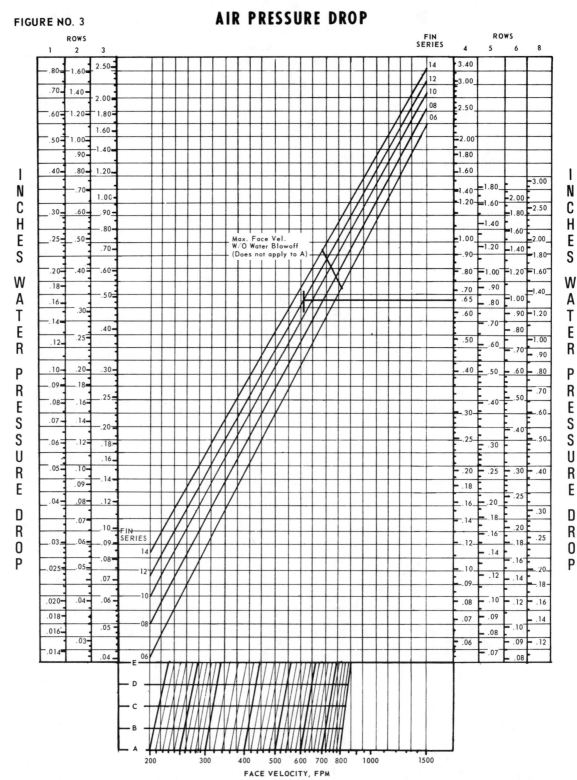

NOTE: The letters A, B, C, D, or E following the Lvg. WB temperature indicates the degree of wetness at which the coil would be operating. Dry coils are shown by the letter A, wet coils by the letter E. In between conditions are shown by the letters B, C, and D.

FIG. 10-65 McQuay-Perfex Inc., Coil Catalog

HEAT TRANSFER VALUES — R_f

WATER HEATING COILS

FIGURE NO. 6

R_{f_1} FOR ALL COILS

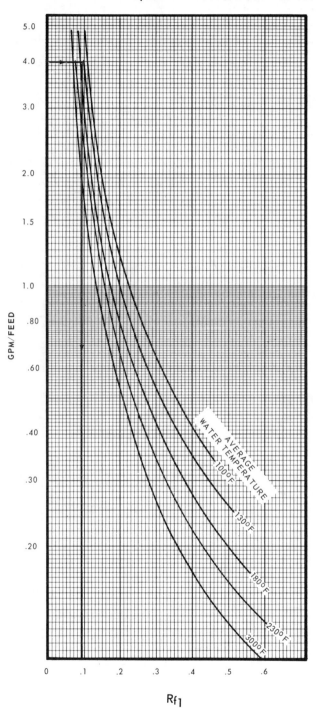

FIGURE NO. 7

R_{f_2} FOR ALL COILS

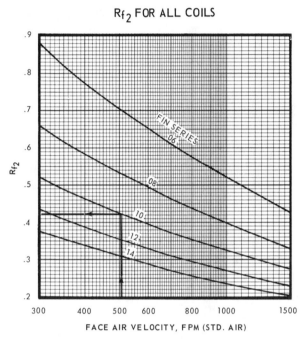

$$R_{f_t} = R_{f_1} + R_{f_2}$$

FIG. 10-65 *(Continued)*

HEAT TRANSFER VALUES — M$_t$

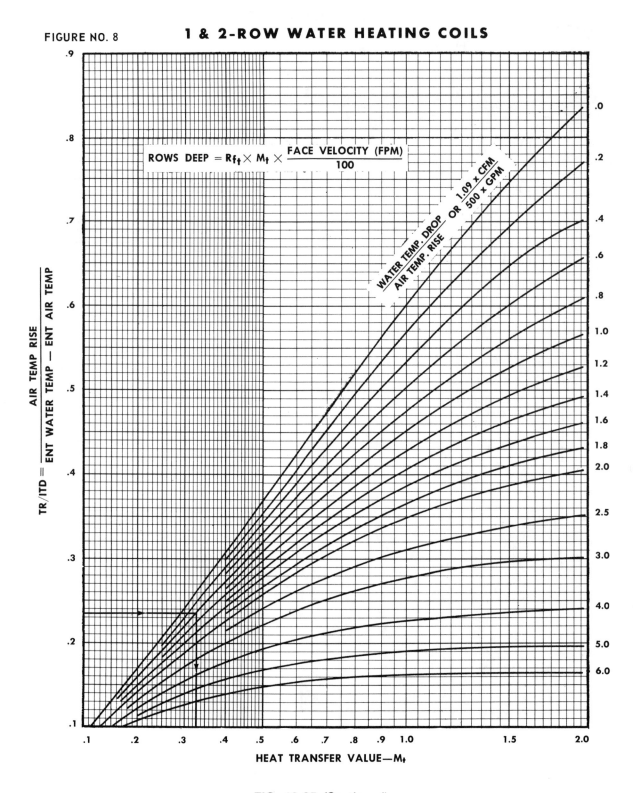

FIGURE NO. 8   **1 & 2-ROW WATER HEATING COILS**

FIG. 10-65 *(Continued)*

Model MSL

15" FC

DRAW-THRU

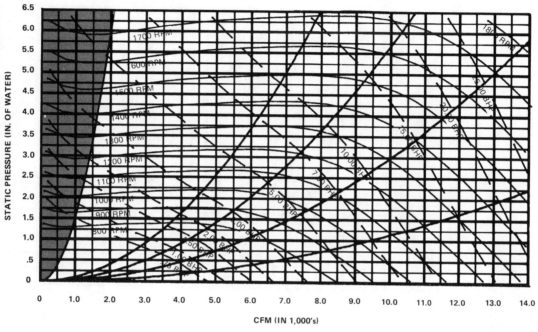

Maximum fan RPM: 1910

CFM STD. AIR	FAN OUTLET VEL FPM	COIL FACE VEL FPM			0.25		0.50		0.75		1.00		1.25		1.50		1.75		2.00		2.25	
		EFA	LFA	SFA	RPM	BHP	RPM	BHP	RPM	BHP	RPM	BHP	RPM	BHP	RPM	BHP	RPM	BHP	RPM	BHP	RPM	BHP
3300	857	258	283	357	384	0.32	483	0.49	581	0.70	671	0.92	753	1.16	827	1.41	894	1.66	958	1.92	1017	2.18
3500	1000	301	330	416	419	0.45	501	0.62	587	0.84	671	1.08	750	1.35	823	1.62	891	1.90	954	2.16	1013	2.48
4000	1142	344	377	476	456	0.61	528	0.80	602	1.02	677	1.27	751	1.55	822	1.85	888	2.15	951	2.47	1010	2.79
4500	1285	387	424	535	496	0.82	560	1.02	625	1.25	690	1.50	757	1.79	824	2.10	887	2.43	949	2.77	1007	3.12
5000	1428	431	471	595	536	1.07	596	1.30	653	1.53	711	1.79	771	2.08	831	2.40	891	2.74	949	3.09	1006	3.47
5500	1571	474	518	654	578	1.37	633	1.62	686	1.87	738	2.14	791	2.43	845	2.75	900	3.10	955	3.46	1009	3.85
6000	1714	517	566	714	621	1.73	672	2.00	721	2.28	769	2.56	817	2.85	866	3.18	915	3.52	966	3.89	1016	4.29
7000	2000	603	660	833	707	2.63	753	2.95	796	3.27	838	3.59	879	3.91	920	4.25	961	4.60	1003	4.98	1045	5.37
8000	2285	689	754	952	795	3.80	836	4.18	876	4.55	913	4.91	949	5.27	985	5.64	1021	6.01	1057	6.41	1094	6.82
9000	2571	775	849	1071	884	5.30	922	5.73	958	6.15	992	6.56	1025	6.96	1058	7.37	1090	7.78	1121	8.20	1153	8.63

CFM STD. AIR	FAN OUTLET VEL FPM	COIL FACE VEL FPM			2.50		2.75		3.00		3.25		3.50		3.75		4.00		4.25		4.50	
		EFA	LFA	SFA	RPM	BHP	RPM	BHP	RPM	BHP	RPM	BHP	RPM	BHP	RPM	BHP	RPM	BHP	RPM	BHP	RPM	BHP
3300	857	258	283	357	1074	2.45	1129	2.73	1180	3.02	1230	3.31	1278	3.61	1325	3.92	1370	4.24	1413	4.55	1457	4.88
3500	1000	301	330	416	1069	2.77	1123	3.08	1174	3.39	1224	3.70	1271	4.02	1318	4.35	1363	4.69	1407	5.03	1449	5.37
4000	1142	344	377	476	1066	3.12	1120	3.45	1170	3.78	1220	4.12	1267	4.47	1313	4.82	1356	5.17	1400	5.54	1442	5.90
4500	1285	387	424	535	1063	3.48	1117	3.83	1167	4.20	1216	4.56	1263	4.94	1309	5.32	1352	5.69	1396	6.08	1437	6.47
5000	1428	431	471	595	1062	3.85	1114	4.24	1165	4.63	1213	5.02	1260	5.43	1305	5.83	1349	6.24	1392	6.65	1433	7.07
5500	1571	474	518	654	1062	4.25	1113	4.66	1163	5.08	1210	5.50	1257	5.93	1302	6.36	1346	6.80	1388	7.24	1430	7.69
6000	1714	517	566	714	1065	4.70	1114	5.12	1163	5.57	1210	6.01	1256	6.47	1300	6.92	1343	7.39	1386	7.87	1427	8.34
7000	2000	603	660	833	1088	5.79	1131	6.23	1174	6.69	1217	7.16	1259	7.64	1301	8.14	1342	8.65	1383	9.17	1424	9.71
8000	2285	689	754	952	1130	7.24	1167	7.68	1204	8.14	1241	8.62	1279	9.12	1317	9.63	1355	10.17	1391	10.69	1428	11.25
9000	2571	775	849	1071	1185	9.07	1217	9.52	1250	9.99	1283	10.48	1316	10.98	1348	11.49	1381	12.02	1415	12.57	1449	13.14

CFM STD. AIR	FAN OUTLET VEL FPM	COIL FACE VEL FPM			4.75		5.00		5.25		5.50		5.75		6.00		6.25		6.50	
		EFA	LFA	SFA	RPM	BHP	RPM	BHP	RPM	BHP	RPM	BHP	RPM	BHP	RPM	BHP	RPM	BHP	RPM	BHP
3300	857	258	283	357	1498	5.22	1538	5.56	1578	5.90	1616	6.25	1654	6.61	1690	6.96	1728	7.34	1762	7.70
3500	1000	301	330	416	1490	5.73	1530	6.08	1569	6.44	1607	6.81	1645	7.19	1682	7.57	1718	7.95	1753	8.35
4000	1142	344	377	476	1483	6.28	1523	6.65	1561	7.03	1599	7.42	1637	7.81	1674	8.22	1708	8.61	1743	9.01
4500	1285	387	424	535	1477	6.86	1517	7.26	1556	7.67	1593	8.07	1630	8.49	1666	8.90	1701	9.33	1737	9.76
5000	1428	431	471	595	1473	7.49	1512	7.91	1551	8.34	1588	8.77	1625	9.21	1661	9.65	1696	10.09	1730	10.55
5500	1571	474	516	654	1470	8.14	1509	8.60	1547	9.04	1584	9.51	1620	9.97	1657	10.44	1691	10.90	1727	11.39
6000	1714	517	566	714	1466	8.81	1506	9.30	1543	9.79	1581	10.27	1617	10.76	1654	11.26	1687	11.74	1722	12.25
7000	2000	603	660	833	1462	10.23	1501	10.77	1538	11.31	1576	11.87	1611	12.41	1647	12.95	1681	13.51	1715	14.06
8000	2285	689	754	952	1466	11.83	1503	12.41	1539	12.99	1575	13.59	1609	14.17	1644	14.78	1679	15.40	1712	16.01
9000	2571	775	849	1071	1482	13.72	1515	14.31	1549	14.92	1582	15.55	1615	16.16	1648	16.81	1679	17.44	1713	18.11

FIG. 10-66 McQuay Air Conditioning, Fan Catalog

Models MMM & LML

 14⁹/₁₆" AF

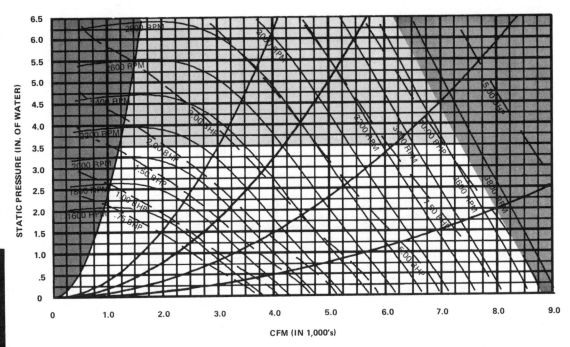

Maximum fan RPM: 3706

BLOW-THRU

CFM STD. AIR	FAN OUT-LET VEL FPM	COIL FACE VEL FPM			0.25		0.50		0.75		1.00		1.25		1.50		1.75		2.00		2.25	
		EFA	LFA	SFA	RPM	BHP	RPM	BHP	RPM	BHP	RPM	BHP	RPM	BHP	RPM	BHP	RPM	BHP	RPM	BHP	RPM	BHP
2200	785	258	297	372	1076	0.26	1195	0.38	1295	0.49	1384	0.61	1468	0.73	1548	0.86	1626	0.99	1701	1.13	1774	1.28
2600	928	305	351	440	1225	0.38	1336	0.52	1431	0.65	1514	0.79	1592	0.92	1666	1.06	1736	1.21	1805	1.36	1872	1.51
3000	1071	352	405	508	1378	0.53	1482	0.69	1571	0.85	1651	1.00	1725	1.16	1794	1.31	1860	1.47	1923	1.63	1985	1.80
3400	1214	400	459	576	1533	0.72	1629	0.90	1715	1.09	1792	1.26	1862	1.44	1928	1.61	1991	1.79	2050	1.96	2108	2.14
3800	1357	447	513	644	1691	0.96	1780	1.16	1861	1.36	1934	1.57	2002	1.76	2066	1.96	2126	2.15	2184	2.35	2239	2.54
4200	1500	494	567	711	1850	1.26	1932	1.47	2009	1.69	2080	1.92	2145	2.14	2207	2.36	2265	2.57	2320	2.79	2373	3.00
4600	1642	541	621	779	2011	1.61	2087	1.83	2159	2.07	2227	2.32	2290	2.57	2350	2.81	2407	3.05	2460	3.29	2512	3.52
5000	1785	588	675	847	2173	2.03	2243	2.26	2311	2.52	2376	2.78	2437	3.06	2495	3.32	2550	3.59	2602	3.85	2652	4.10
5400	1928	635	729	915	2336	2.52	2401	2.76	2465	3.03	2527	3.31	2586	3.61	2641	3.90	2695	4.19	2745	4.47	2795	4.76
6200	2214	729	837	1050	2664	3.74	2721	3.99	2778	4.28	2833	4.60	2887	4.93	2939	5.26	2989	5.59	3037	5.93	3084	6.26

CFM STD. AIR	FAN OUT-LET VEL FPM	COIL FACE VEL FPM			2.50		2.75		3.00		3.25		3.50		3.75		4.00		4.25		4.50	
		EFA	LFA	SFA	RPM	BHP	RPM	BHP	RPM	BHP	RPM	BHP	RPM	BHP	RPM	BHP	RPM	BHP	RPM	BHP	RPM	BHP
2200	785	258	297	372	1846	1.43	1915	1.59	1984	1.75	2051	1.92	2116	2.08	2181	2.26	2245	2.44	2307	2.62	2368	2.80
2600	928	305	351	440	1936	1.67	1999	1.84	2061	2.01	2123	2.19	2184	2.37	2242	2.55	2301	2.74	2359	2.93	2414	3.13
3000	1071	352	405	508	2044	1.97	2103	2.15	2160	2.32	2216	2.51	2271	2.70	2326	2.89	2381	3.10	2434	3.30	2485	3.50
3400	1214	400	459	576	2164	2.32	2219	2.51	2272	2.70	2326	2.90	2377	3.10	2427	3.30	2477	3.51	2527	3.72	2575	3.93
3800	1357	447	513	644	2292	2.74	2345	2.94	2395	3.15	2444	3.35	2493	3.56	2540	3.77	2588	3.99	2634	4.21	2680	4.44
4200	1500	494	567	711	2425	3.22	2475	3.44	2523	3.65	2570	3.87	2616	4.09	2662	4.32	2707	4.55	2751	4.78	2794	5.02
4600	1642	541	621	779	2561	3.76	2609	3.99	2656	4.23	2702	4.47	2747	4.71	2790	4.94	2832	5.19	2874	5.43	2916	5.68
5000	1785	588	675	847	2701	4.37	2747	4.62	2792	4.87	2836	5.13	2879	5.38	2922	5.64	2963	5.90	3003	6.16	3044	6.42
5400	1928	635	729	915	2842	5.04	2887	5.31	2931	5.59	2974	5.86	3016	6.14	3057	6.42	3097	6.69	3137	6.97	3175	7.25
6200	2214	729	837	1050	3128	6.59	3171	6.91	3214	7.24	3255	7.56	3295	7.88	3334	8.20	3372	8.52	3410	8.84	3447	9.15

CFM STD. AIR	FAN OUT-LET VEL FPM	COIL FACE VEL FPM			4.75		5.00		5.25		5.50		5.75		6.00		6.25		6.50	
		EFA	LFA	SFA	RPM	BHP	RPM	BHP	RPM	BHP	RPM	BHP	RPM	BHP	RPM	BHP	RPM	BHP	RPM	BHP
2200	785	258	297	372	2428	2.99	2486	3.18	2544	3.38	2603	3.58	2658	3.78	2712	3.98	2767	4.19	2820	4.40
2600	928	305	351	440	2471	3.33	2525	3.53	2581	3.74	2636	3.95	2687	4.16	2740	4.38	2792	4.60	2842	4.82
3000	1071	352	405	508	2537	3.71	2589	3.93	2640	4.14	2689	4.36	2739	4.59	2789	4.82	2838	5.05	2886	5.28
3400	1214	400	459	576	2624	4.15	2671	4.38	2719	4.60	2766	4.84	2812	5.07	2858	5.31	2903	5.55	2949	5.80
3800	1357	447	513	644	2725	4.67	2770	4.90	2814	5.14	2858	5.38	2901	5.62	2944	5.87	2987	6.12	3029	6.37
4200	1500	494	567	711	2837	5.26	2880	5.50	2921	5.75	2962	6.00	3003	6.25	3043	6.50	3085	6.77	3123	7.02
4600	1642	541	621	779	2957	5.93	2997	6.19	3037	6.44	3077	6.71	3115	6.96	3154	7.23	3192	7.50	3230	7.77
5000	1785	588	675	847	3083	6.69	3121	6.95	3160	7.23	3197	7.49	3235	7.77	3272	8.05	3308	8.33	3345	8.61
5400	1928	635	729	915	3213	7.53	3250	7.81	3287	8.09	3323	8.37	3359	8.66	3394	8.95	3430	9.24	3465	9.54
6200	2214	729	837	1050	3482	9.46	3518	9.78	3552	10.10	3586	10.41	3621	10.74	3654	11.05	3688	11.38	****	*****

FIG. 10-66 (Continued)

ENTERING WATER 42F

5-ROW COILS

WATER COOLING COILS CAPACITY DATA

ENTERING AIR { 80F db / 64F wb

COIL TYPE			COIL FINNED LENGTH (inches)																	
5WH–½ SERP. 5WL–¾ SERP. 5WS–1 SERP. 5WM–1½ SERP.			12 18 24 36			18 27 36 54			24 36 48 72			36 54 72 108			48 72 96 144			72 108 144 —		
FACE VEL.	WATER TEMP. RISE	FIN SERIES	TOT. (MBH)	LVG. D.B.	LVG. W.B.	TOT. (MBH)	LVG. D.B.	LVG. W.B.	TOT. (MBH)	LVG. D.B.	LVG. W.B.	TOT. (MBH)	LVG. D.B.	LVG. W.B.	TOT. (MBH)	LVG. D.B.	LVG. W.B.	TOT. (MBH)	LVG. D.B.	LVG. W.B.
300 FPM	10	06	–	–	–	*9.2	55.2	53.7C	9.9	54.2	52.8C	10.7	53.0	51.8D	11.1	52.3	51.2D	11.6	51.6	50.5D
		08	–	–	–	*10.2	53.0	52.3C	11.1	51.8	51.2D	12.0	50.5	50.1D	12.5	49.8	49.4D	13.1	49.0	48.6D
		10	*9.6	53.6	53.1C	*11.1	51.5	51.2C	11.9	50.3	50.1D	12.9	49.0	48.8D	13.5	48.2	48.0D	14.1	47.4	47.2E
		12	*10.1	52.7	52.4C	11.7	50.6	50.4C	12.6	49.4	49.2D	13.6	48.0	47.8D	14.2	47.2	47.0D	14.8	46.4	46.2E
		14	*10.6	52.1	51.9C	12.2	50.0	49.8C	13.1	48.7	48.5D	14.2	47.3	47.1D	14.7	46.4	46.2D	15.3	45.6	45.4E
	12	06	–	–	–	*8.3	56.6	54.7C	*9.1	55.4	53.8C	9.9	54.1	52.7C	10.5	53.3	52.0D	11.0	52.5	51.3D
		08	–	–	–	*9.3	54.4	53.5C	10.2	53.1	52.4C	11.2	51.7	51.1D	11.8	50.9	50.4D	12.4	50.0	49.5D
		10	–	–	–	*10.1	53.0	52.5C	11.0	51.6	51.3C	12.1	50.1	49.9D	12.7	49.3	49.1D	13.4	48.3	48.1D
		12	–	–	–	*10.7	52.0	51.8C	11.7	50.6	50.4C	12.8	49.2	49.0D	13.4	48.3	48.1D	14.1	47.3	47.1D
		14	–	–	–	*11.1	51.4	51.2C	12.2	50.0	49.8C	13.3	48.4	48.2D	14.0	47.5	47.3D	14.7	46.5	46.3D
	14	06	–	–	–	–	–	–	*8.3	56.6	54.7C	9.2	55.2	53.6C	9.8	54.3	52.9C	10.4	53.5	52.1C
		08	–	–	–	–	–	–	*9.4	54.3	53.4C	10.4	52.8	52.1C	11.1	51.9	51.3C	11.7	50.9	50.4D
		10	–	–	–	*9.2	54.3	53.6B	*10.2	52.9	52.4C	11.3	51.2	50.9C	12.0	50.3	50.1C	12.7	49.3	49.1D
		12	–	–	–	*9.7	53.3	53.0C	*10.7	51.9	51.7C	12.0	50.3	50.1C	12.7	49.3	49.1D	13.5	48.2	48.0D
		14	–	–	–	*10.1	52.7	52.5C	*11.2	51.3	51.1C	12.5	49.5	49.3C	13.2	48.6	48.4D	14.0	47.4	47.2D
	16	06	–	–	–	–	–	–	*7.7	57.8	55.5B	8.6	56.2	54.4C	9.2	55.3	53.7C	9.8	54.4	52.9C
		08	–	–	–	–	–	–	*8.6	55.6	54.3B	9.7	53.9	53.0C	10.3	52.9	52.2C	11.1	51.9	51.2C
		10	–	–	–	–	–	–	*9.3	54.1	53.4B	10.5	52.4	51.9C	11.2	51.3	51.0C	12.0	50.2	50.0C
		12	–	–	–	–	–	–	*9.9	53.1	52.8C	11.2	51.3	51.1C	11.9	50.3	50.1C	12.8	49.2	49.0D
		14	–	–	–	–	–	–	*10.3	52.4	52.2C	11.7	50.7	50.5C	12.5	49.6	49.4C	13.3	48.4	48.2D
400 FPM	10	06	*9.7	58.7	56.0B	11.1	56.8	54.7C	12.0	55.7	53.9C	13.0	54.6	52.9D	13.7	53.9	52.3D	14.3	53.2	51.7D
		08	*10.9	56.4	54.9B	12.6	54.4	53.3C	13.7	53.2	52.3C	14.9	51.9	51.1D	15.5	51.2	50.5D	16.3	50.4	49.8D
		10	*11.9	54.8	54.0B	13.7	52.8	52.3C	14.9	51.5	51.1C	16.2	50.2	49.9D	16.9	49.4	49.1D	17.7	48.6	48.3D
		12	*12.6	53.7	53.3C	14.6	51.7	51.4C	15.8	50.4	50.2D	17.2	49.0	48.8D	18.0	48.2	48.0D	18.8	47.3	47.1D
		14	*13.1	53.0	52.8C	15.3	50.9	50.7C	16.5	49.7	49.5D	18.0	48.2	48.0D	18.8	47.4	47.2D	19.6	46.5	46.3D
	12	06	–	–	–	*10.1	58.1	55.6B	11.0	56.9	54.8C	12.1	55.7	53.8C	12.8	54.9	53.2C	13.5	54.1	52.5D
		08	–	–	–	*11.4	55.8	54.4C	12.5	54.5	53.4C	13.8	53.1	52.2C	14.6	52.2	51.4D	15.4	51.3	50.6D
		10	–	–	–	*12.5	54.2	53.4C	13.7	52.8	52.3C	15.1	51.3	50.9C	15.9	50.5	50.1D	16.8	49.5	49.2D
		12	*11.2	55.4	54.6B	13.2	53.1	52.7C	14.6	51.7	51.4C	16.1	50.1	49.9D	16.9	49.3	49.1D	17.9	48.3	48.1D
		14	*11.7	54.6	54.1B	13.9	52.3	52.1C	15.3	50.9	50.7C	16.9	49.3	49.1D	17.7	48.5	48.3D	18.7	47.4	47.2D
	14	06	–	–	–	*9.3	59.4	56.3B	*10.1	58.1	55.6B	11.3	56.7	54.5C	11.9	55.9	53.9C	12.7	55.0	53.2C
		08	–	–	–	*10.4	57.1	55.3B	11.5	55.7	54.3C	12.9	54.2	53.1C	13.6	53.3	52.3C	14.5	52.3	51.5C
		10	–	–	–	*11.3	55.5	54.5B	12.6	54.1	53.3C	14.1	52.5	51.9C	14.9	51.5	51.1C	15.9	50.4	50.1D
		12	–	–	–	*12.0	54.4	53.8B	13.4	53.0	52.6C	15.0	51.3	51.0C	16.0	50.3	50.1C	17.0	49.2	49.0D
		14	–	–	–	*12.6	53.6	53.3B	14.0	52.1	51.9C	15.7	50.5	50.3C	16.7	49.5	49.3C	17.8	48.4	48.2D
	16	06	–	–	–	–	–	–	*9.3	59.4	56.3B	10.5	57.8	55.3B	11.2	56.9	54.6C	12.0	55.9	53.9C
		08	–	–	–	–	–	–	*10.6	57.0	55.2B	12.0	55.2	53.9C	12.8	54.3	53.2C	13.7	53.2	52.3C
		10	–	–	–	*10.3	57.0	55.4B	*11.6	55.3	54.3B	13.1	53.6	52.9C	14.0	52.6	52.0C	15.0	51.4	51.0C
		12	–	–	–	*11.0	55.8	54.8B	*12.3	54.2	53.6B	13.9	52.4	52.0C	14.9	51.3	51.1C	16.1	50.2	50.0C
		14	–	–	–	*11.5	55.0	54.4B	12.9	53.4	53.0B	14.6	51.6	51.4C	15.7	50.5	50.3C	16.9	49.3	49.1C
500 FPM	10	06	*11.2	59.9	56.6B	12.9	58.0	55.4C	13.9	57.0	54.7C	15.1	55.9	53.8C	15.9	55.2	53.2D	16.7	54.5	52.6D
		08	*12.7	57.5	55.5B	14.7	55.5	54.1C	16.0	54.4	53.1C	17.4	53.1	52.1D	18.2	52.4	51.4D	19.1	51.6	50.7D
		10	*13.9	55.9	54.7B	16.1	53.8	53.0C	17.5	52.6	51.9C	19.1	51.2	50.7D	20.0	50.5	50.0D	21.0	49.6	49.2D
		12	*14.9	54.7	54.0B	17.2	52.6	52.2C	18.8	51.3	51.0C	20.4	49.9	49.7D	21.4	49.1	48.9D	22.5	48.2	48.0D
		14	*15.6	53.8	53.4B	18.1	51.7	51.5C	19.7	50.5	50.3C	21.5	49.0	48.8D	22.5	48.2	48.0D	23.6	47.3	47.1D
	12	06	–	–	–	*11.7	59.4	56.3B	12.8	58.2	55.5C	14.1	56.9	54.6C	14.8	56.2	54.0C	15.7	55.4	53.4C
		08	*11.4	59.3	56.5A	13.4	56.9	55.1B	14.7	55.6	54.1C	16.2	54.2	53.0C	17.1	53.4	52.3C	18.1	52.5	51.5D
		10	*12.4	57.6	55.8B	14.6	55.2	54.1B	16.1	53.8	53.1C	17.8	52.3	51.7C	18.8	51.5	50.9C	19.9	50.6	50.1D
		12	*13.2	56.4	55.2B	15.6	54.0	53.4C	17.2	52.6	52.2C	19.1	51.0	50.8C	20.1	50.2	49.9D	21.3	49.2	49.0D
		14	*13.9	55.5	54.7B	16.4	53.1	52.8C	18.1	51.7	51.5C	20.1	50.1	49.9C	21.2	49.3	49.1D	22.4	48.2	48.0D
	14	06	–	–	–	*10.7	60.7	57.0B	11.8	59.4	56.2B	13.1	58.0	55.3C	13.9	57.2	54.7C	14.8	56.3	54.0C
		08	–	–	–	*12.2	58.3	55.9B	13.5	56.8	55.0B	15.0	55.3	53.8C	16.0	54.4	53.1C	17.1	53.5	52.3C
		10	–	–	–	*13.4	56.6	55.1B	14.8	55.1	54.0B	16.6	53.5	52.7C	17.6	52.5	51.9C	18.8	51.5	50.9C
		12	–	–	–	*14.2	55.4	54.5B	15.8	53.9	53.3C	17.8	52.2	51.8C	18.9	51.2	50.9C	20.2	50.1	49.8D
		14	–	–	–	*14.9	54.5	53.9B	16.6	53.0	52.6C	18.7	51.2	51.0C	20.0	50.2	50.0C	21.3	49.2	49.0D
	16	06	–	–	–	–	–	–	*10.8	60.6	56.9B	12.2	59.0	55.9B	13.0	58.1	55.3C	13.9	57.2	54.7C
		08	–	–	–	*11.1	59.8	56.7A	*12.4	58.1	55.8B	14.0	56.4	54.6B	15.0	55.4	53.9C	16.1	54.4	53.1C
		10	–	–	–	*12.1	58.1	56.0B	13.6	56.4	54.9B	15.4	54.6	53.6C	16.5	53.6	52.8C	17.8	52.4	51.8C
		12	–	–	–	*12.9	56.8	55.4B	14.5	55.1	54.2B	16.5	53.3	52.7C	17.7	52.2	51.8C	19.1	51.1	50.8C
		14	–	–	–	*13.6	55.9	54.9B	15.3	54.2	53.7B	17.4	52.3	52.1C	18.7	51.3	51.1C	20.2	50.1	49.9C
600 FPM	10	06	*12.6	61.0	57.1B	14.5	59.1	56.0B	15.7	58.1	55.3C	17.0	57.0	54.5C	17.9	56.3	53.9C	18.8	55.7	53.4D
		08	*14.4	58.5	56.0B	16.7	56.5	54.7C	18.1	55.3	53.8C	19.7	54.1	52.8C	20.7	53.4	52.1D	21.8	52.6	51.5D
		10	*15.8	56.8	55.2B	18.4	54.6	53.6C	20.0	53.4	52.6C	21.8	52.1	51.4D	22.9	51.4	50.8D	24.1	50.5	50.0D
		12	17.0	55.5	54.5B	19.7	53.4	52.8C	21.5	52.1	51.7C	23.5	50.7	50.3D	24.6	49.9	49.6D	*25.9	49.0	48.8D
		14	17.9	54.5	53.9B	20.8	52.4	52.1C	22.7	51.1	50.9C	24.8	49.7	49.5D	26.0	48.9	48.7D	*27.3	48.0	47.8D
	12	06	*11.4	62.7	57.8A	13.2	60.4	56.8B	14.4	59.2	56.1B	15.8	58.0	55.2C	16.7	57.3	54.7C	17.7	56.5	54.1C
		08	*12.9	60.4	56.9A	15.1	57.9	55.6B	16.6	56.6	54.7C	18.3	55.2	53.7C	19.4	54.4	53.0C	20.6	53.5	52.3C
		10	*14.1	58.6	56.2A	16.7	56.1	54.7B	18.3	54.7	53.7C	20.3	53.2	52.4C	21.5	52.4	51.7C	22.8	51.4	50.8D
		12	*15.1	57.3	55.6B	17.9	54.8	54.0B	19.7	53.4	52.8C	21.9	51.8	51.4C	23.1	51.0	50.6C	24.5	50.0	49.7D
		14	*15.8	56.3	55.2B	18.8	53.8	53.3C	20.8	52.4	52.1C	23.1	50.8	50.6C	24.4	49.9	49.7D	25.9	48.9	48.7D
	14	06	–	–	–	*12.0	61.8	57.4A	13.2	60.4	56.7B	14.7	59.0	55.8B	15.6	58.2	55.3C	16.7	57.4	54.7C
		08	–	–	–	*13.8	59.3	56.4B	15.3	57.8	55.5B	17.1	56.2	54.4C	18.1	55.4	53.8C	19.4	54.4	53.0C
		10	–	–	–	15.1	57.5	55.6B	16.8	56.0	54.6B	18.9	54.3	53.3C	20.1	53.4	52.5C	21.5	52.4	51.7C
		12	*13.4	59.5	56.6A	16.2	56.2	55.0B	18.1	54.6	53.8B	20.3	53.0	52.4C	21.7	52.0	51.5C	23.2	50.9	50.5C
		14	*14.1	58.5	56.2A	17.1	55.2	54.4B	19.1	53.7	53.2C	21.5	51.9	51.7C	23.0	50.9	50.7C	24.6	49.8	49.6C
	16	06	–	–	–	*11.0	63.3	58.0A	*12.2	61.6	57.3B	13.7	60.0	56.4B	14.6	59.2	55.9B	15.7	58.2	55.3C
		08	–	–	–	*12.6	60.9	57.1A	14.0	59.1	56.3B	15.9	57.3	55.2B	17.0	56.4	54.5C	18.3	55.3	53.7C
		10	–	–	–	*13.8	59.1	56.4A	15.5	57.3	55.4B	17.5	55.4	54.1B	18.8	54.4	53.4C	20.3	53.3	52.4C
		12	–	–	–	*14.7	57.8	55.8A	16.6	55.9	54.7B	18.9	54.0	53.3C	20.3	53.0	52.4C	21.9	51.9	51.4C
		14	–	–	–	*15.5	56.7	55.4B	17.5	54.9	54.2B	20.0	53.0	52.6C	21.5	52.0	51.7C	23.2	50.8	50.5C
700 FPM	10	06	*14.0	61.9	57.5A	16.0	60.0	56.4B	17.3	59.0	55.8C	18.8	57.9	55.0C	19.7	57.3	54.5C	20.7	56.6	54.0D
		08	*16.0	59.4	56.4B	18.5	57.3	55.2B	20.1	56.2	54.4C	21.9	55.0	53.4C	23.0	54.3	52.8C	24.2	53.5	52.1D
		10	17.6	57.5	55.6B	20.4	55.4	54.2C	22.3	54.2	53.2C	24.4	52.9	52.1C	25.5	52.2	51.4D	*26.9	51.3	50.6D
		12	18.9	56.2	55.0B	22.0	54.0	53.3C	24.0	52.8	52.2C	26.3	51.4	51.0C	27.6	50.6	50.3D	*29.0	49.8	49.4D
		14	20.0	55.2	54.4B	23.3	53.0	52.6C	25.5	51.7	51.4C	27.9	50.3	50.1D	29.2	49.5	49.3D	*30.8	48.6	48.4D
	12	06	*12.6	63.6	58.1A	14.5	61.3	57.2B	15.9	60.1	56.5B	17.5	58.9	55.7C	18.4	58.2	55.2C	19.5	57.5	54.6C
		08	*14.3	61.3	57.3A	16.8	58.7	56.0B	18.4	57.4	55.2B	20.4	56.0	54.2C	21.5	55.3	53.6C	22.8	54.4	52.9C
		10	*15.7	59.5	56.6A	18.6	56.8	55.1B	20.4	55.5	54.2C	22.6	54.0	53.0C	24.0	53.2	52.3C	25.5	52.2	51.4C
		12	*16.8	58.1	56.0A	20.0	55.5	54.4B	22.0	54.1	53.3C	24.5	52.5	52.0C	26.0	51.6	51.2C	27.6	50.7	50.3D
		14	*17.7	57.1	55.6A	21.1	54.5	53.8B	23.3	53.0	52.6C	26.0	51.5	51.2C	27.5	50.6	50.3C	29.2	49.5	49.3D
	14	06	–	–	–	*13.3	62.7	57.8A	14.6	61.3	57.1B	16.3	59.9	56.3B	17.3	59.1	55.8C	18.4	58.3	55.2C
		08	–	–	–	15.3	60.2	56.8A	16.9	58.6	56.0B	18.9	57.1	54.9C	20.1	56.2	54.3C	21.5	55.3	53.6C
		10	*14.0	61.7	57.4A	16.9	58.3	56.0B	18.7	56.7	55.0B	21.1	55.1	53.8C	22.4	54.2	53.1C	24.0	53.2	52.3C
		12	*15.0	60.4	57.0A	18.1	56.9	55.4B	20.2	55.3	54.3B	22.8	53.6	52.9C	24.3	52.7	52.1C	26.0	51.6	51.1C
		14	*15.8	59.4	56.5A	19.2	55.9	54.8B	21.4	54.3	53.7B	24.2	52.6	52.2C	25.8	51.6	51.3C	27.7	50.5	50.2C
	16	06	–	–	–	*12.1	64.2	58.4A	13.5	62.5	57.7A	15.1	60.9	56.9B	16.2	60.0	56.4B	17.3	59.1	55.8C
		08	–	–	–	*13.9	61.8	57.5A	15.6	59.9	56.7B	17.6	58.2	55.6B	18.9	57.2	55.0B	20.3	56.2	54.2C
		10	–	–	–	*15.4	60.0	56.8A	17.2	58.0	55.8B	19.6	56.2	54.6B	21.0	55.2	53.9C	22.7	54.1	53.0C
		12	–	–	–	*16.5	58.5	56.2A	18.5	56.6	55.2B	21.2	54.7	53.8B	22.7	53.7	52.9C	24.6	52.6	52.0C
		14	–	–	–	17.4	57.5	55.7A	19.6	55.6	54.6B	22.4	53.7	53.1C	24.2	52.6	52.2C	26.1	51.4	51.1C

* These ratings are not Certified Ratings since they fall outside of the Range of Certified Rating Conditions specified in ARI Standard 410-64.

FIG. 10-67 McQuay Inc., Coil Catalog

Water pressure drop

HI-F5 & E-F5 – 5W coils

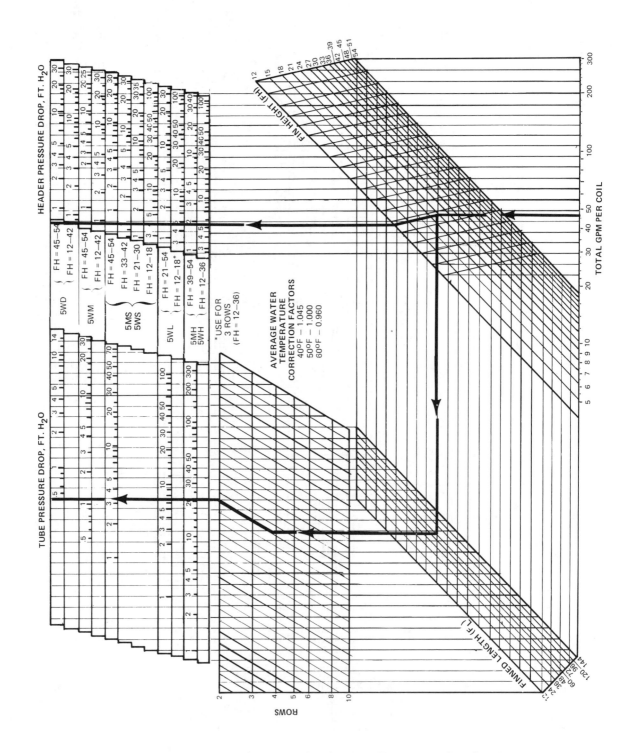

FIG. 10-68 McQuay Inc., Water Cooling and Evaporator Coil Catalog

E-F5 coils air pressure drop

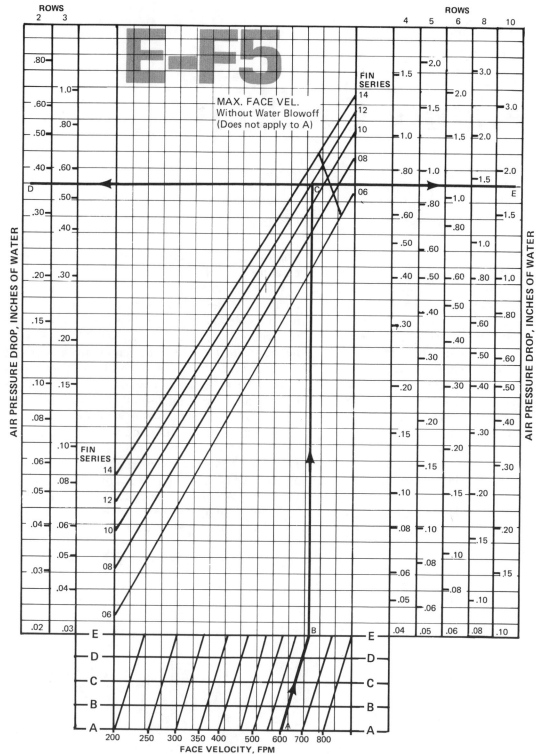

NOTE: The letters A, B, C, D or E following the Lvg. WB temperature indicate the degree of wetness at which the coil would be operating. Dry coils are shown by the letter A, wet coils by the letter E. In between conditions are shown by the letters B, C and D.

FIG. 10-68 *(Continued)*

CHAPTER 11

Temperature Controls

BY NOW I am sure you have noticed that Chapters 1 to 9 built up to Chapter 10. In an effort to keep your thoughts moving smoothly toward Chapter 10, I left out three important factors in the design of air conditioning systems: temperature controls, moisture in construction, and sound. While these are important they do not affect the design directly. For example, temperature controls are generally described in the specifications. Moisture in construction is a factor that you discuss with the architect. You tell him where to put vapor barriers, etc. Then we have sound. You must alert the architect to possible trouble spots and advise him how to handle them. For example, you will warn him that a classroom next to a boiler room could cause some hearing problems in the classroom. You must tell him how to treat the boiler room and what the noise level will be in the boiler room so he can erect a suitable barrier between the boiler room and

classroom. So let's get on with it and start with temperature controls.

The sales engineer from the temperature control company is the design engineer's best friend. While it is possible to learn all the "ins and outs" of temperature controls, few of us rarely do. Although it is not necessary to be able to design control systems, you should be able to tell the control specialist what you want. The designer of environmental systems must be able to describe intelligently a control sequence.

The early heating systems were simple to control. If your office became too hot, you closed the radiator valve. When it became too cold, you opened the valve. If it didn't warm up, you banged on the radiator to awaken the fireman who then shoveled some coal into the boiler. When it got too hot, you opened a window. You must admit that these were simple controls. If any of you conservatives long for these good old days,

FIG. 11—1 Primitive Control System

let me tell you one of my experiences. I opened a window in my office because I was warm. Someone called me and, as I turned to answer, my completed drawing flew out the window and down Main Street. The drawing represented about 100 hours of work. I never saw it again. We soon put in air conditioning and controls.

Well, on with temperature controls. Temperature controls systems have three parts: a sensing device, signal transmission system, and the control device. When the temperature is controlled manually, as shown in Fig. 11–1, a man feels cold (sensing device), he walks to the radiator (signal transmission), and closes the valve (control). When automatic controls are used, we replace the man with a thermostat which senses temperature change. It sends a signal electrically through wires to a motor-operated valve which opens or closes as needed. The signal can also be a change in air pressure, which is transmitted through air piping to an air-operated valve. Temperature control systems can be electric, electronic, pneumatic (air), or a combination of the three. Which is best? I don't know. The design engineer should write the sequence of operation he desires and let the temperature control specialist decide which is best for the particular job. The sensing device can send a signal to a motorized valve which controls the flow of the heating fluid, water or steam. It can also control the operation of motorized dampers to control the flow of air. The simplest control system is one where you simply turn the heating or cooling on or off. The on or off control may be the starting or stopping of a fan or unit heater, furnace or fancoil unit. It may also be the opening or closing of a motorized shut-off valve on a hot-water, chilled-water, or steam line. On and off control tends, alternately, to overheat and underheat the space. If you made a graph of room temperature against time, the resulting curve would look like a roller coaster.

The high and low points on the curve can be smoothed out in several ways. When steam is the heating medium, a modulating valve may be used. A modulating valve varies the amount of fluid that enters the heat emission device. So, if a room is too cold, the valve opens a little and a small amount of steam passes. If this satisfies the thermostat, the valve opens no further. If the room is still too cold, it opens some more. The reverse happens if it is too warm.

When hot water is used, a modulating valve works little better than a two-position valve. If the flow of water is cut in half, the heat output is only reduced 10%. Reducing the flow by 75% only reduces heat output 30%. So, you can see that modulating flow

isn't much better than on-off. In hot-water systems it is better to vary water temperature in accordance with outdoor temperature and then use either on-off or modulating control valves on the room heat emitter.

Air systems have two points of control: the temperature of the heating fluid and the fan. Turning a fan on and off gives a "roller coaster" type curve, just like on-off controls. The air temperature should be controlled and the fan permitted to run continuously. If steam is the heating medium, a modulating valve should be used. With hot water, again we need outdoor reset control and a valve. If the unit is direct-fired, a modulating gas valve or, at least, a high-low valve should be used. Another method of controlling air temperature is with face and bypass dampers. Dampers are installed on the coil and in a bypass around the coil (Fig. 11–2). The room thermostat controls the dampers and in this way the air temperature. The only problem encountered here is "coil wiping." Sometimes the bypass air comes in contact with the coil before it reaches the bypass and is heated slightly. The air "kinda" bounces off the coil or face dampers and then goes through the bypass. This can cause overheating. It is advisable to install an end switch on the damper motor which will close a valve on the heating line (steam or hot water) when the bypass damper is fully open and the face (coil) damper fully closed. Another method of reducing the "wiping action" is to use outdoor reset control on the hot-water system. In this way, the coil temperature is only as warm as needed. This reduces the amount of heat wiped off and tends to keep air flowing through the coil.

When ventilation is incorporated into a system, additional controls are required. Generally, a motorized damper is installed to operate the return-air and outside-air dampers. These dampers can be controlled

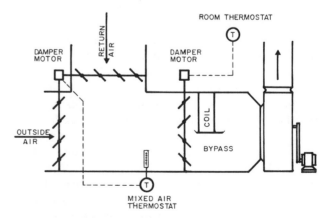

FIG. 11–2 Face And Bypass Control

so that they open to a set position any time the fan runs. Another system is to use a thermostat to sense mixed air temperature. This thermostat positions the damper motor or motors to maintain the desired temperature. Still another method is for the room thermostat to position these dampers. The thermostat, on a rise in temperature, would first close the coil control valve or face damper and then modulate open the outside air and close the return air. The reverse happens on a drop in temperature.

Controlling water temperature can be done through the use of a three-way mixing valve or by controlling the boiler directly. The problem that can arise when controlling boiler water temperature is condensation of the flue gases when low water temperatures are being used. This can cause severe corrosion, because condensed flue gases can be acid. The three-way mixing valve (Fig. 11–3) should always be used. As you can see, return water is blended with boiler water to get desired temperature. An outdoor thermostat senses outdoor temperature. This thermostat sends a signal to a thermostat in the mixed-water temperature and operates the three-way valve to get the desired water temperature. For instance, a schedule may be set up which provides 170° water at 0° outdoors, 100° at 70°, and a directly proportional temperature in between, such as 135° at 35° outdoors. This is a 1-to-1 ratio; that is, for every 1° change in outdoor temperature there is a 1° change in water temperature. The ratio can be any desired, such as 1.5-to-1, 2-to-1, or 0.5-to-1. Two types of instruments are available. The least expensive is the fixed-ratio type which gives only the ratio specified. You can change the control point but not the ratio. For instance, if you were set at 100° water at 70° outdoors and 170° at 0°, you could raise the set point so that you got 120° at 70° outdoors and 190° at 0° outdoors. With an adjustable-ratio instrument, however, the ratio can be changed. For instance, the set point can remain at 100° water at 70° outdoors, but at 0°, the water temperature is 205°. You can see that an adjustable-ratio instrument is quite superior to a fixed-ratio instrument.

When a steam-to-water heat exchanger or convertor is used instead of a boiler, water temperature can be reset in two ways. A three-way mixing valve may be used in the same way as with a boiler. The steam is uncontrolled, that is, full steam pressure is on the exchanger at all times. I prefer the method used in Fig. 11–4. Notice that we have no three-way mixing valve, but we do have a modulating valve on the steam main. The outdoor thermostat and a thermostat in the leaving water position the steam valve to give the desired water temperature.

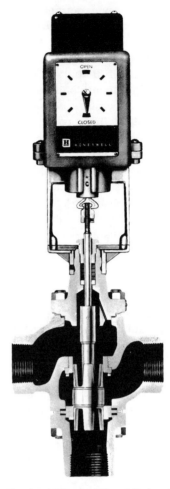

FIG. 11–3 Typical Three-Way Mixing Valve (Courtesy: Honeywell)

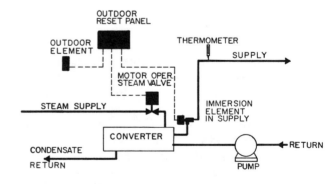

FIG. 11–4 Outdoor Reset Control Of Converter

The control of chilled-water systems is similar to the control of hot water. However, chilled-water temperature should not be varied with outdoor temperature. If the water temperature is raised, insufficient dehumidification takes place. The water chiller should be controlled with self-contained controls installed by the manufacturer. The chiller may be started manually or automatically from an outdoor thermostat or time clock. Reversing relays must be installed when the room thermostat controls a valve on a combination heating and cooling coil, fan, or a face and bypass damper. Remember when on a heating cycle, the valve closes or fan stops or bypass damper opens on a rise in temperature. But when cooling, a rise in temperature requires that the valve open, fan start, or bypass damper close. A reversing relay is used to reverse the action of the thermostat. This reversing relay can be controlled by a summer-winter switch, a thermostat in the supply water, or a strap-on aquastat. The only time a relay is not needed is when the thermostat is controlling a device that has a pool of hot and cold water or hot and cold air to draw from. In a dual duct system the thermostat positions dampers on a hot duct and cold duct to get desired mixed-air temperature. The same is true in a multizone system, except here the dampers are right at the unit. A three-pipe system is similar, because here a three-way valve mixes hot and cold water in response to signals from a thermostat.

If a fancoil unit is used for cooling, the fan must run all the time. If the fan stops while chilled water is running through the coil, condensation can occur on the enclosure cover and cause a nuisance or damage to the room. Either valve control or face and bypass control can be used, but face and bypass control gives superior humidity control. There are available multiported valves (Fig. 11–5) which serve multi-circuited coils. These give control equal to face and bypass. The valve may have four ports. The coil is divided into four parts or, in effect, four coils. Each coil is then connected to a port in the valve. The thermostat closes a port at a time.

When direct expansion coils are used in a central air conditioning unit, face and bypass control gives the best results. If face and bypass control is not used, the room thermostat (or cold duct thermostat in a multizone) can start and stop the compressor. A better method is to have the room thermostat open and close a solenoid valve (shut-off valve) on the liquid refrigerant line. The condensing unit then operates from its own pressure control. In the multizone, or if face and bypass is used on a single-zone unit, no solenoid valve is used. The condensing unit is merely

energized by clock, switch, or outside temperature and cycled by its own pressure controls. The use of the condensing unit pressure controls is recommended for two reasons. First, it is difficult to sense duct temperature accurately in a multizone cold deck. Secondly, duct temperature control can cause short cycling of the condensing unit, which can be harmful.

Another item to be controlled is the day-night or occupied-unoccupied period. Not only may you want to reduce the space temperature during unoccupied periods, but you should close the outside air dampers. The occupied-unoccupied periods can be set by a switch or a seven-day-programming clock with override switches. The clock is really quite superior to the switch, particularly if the temperature is reduced during unoccupied periods. Once a temperature is reduced, it takes time to get back to the desired occupied temperature. On a cold day, three hours are necessary for a five-degree pickup in temperature, while twelve or more are needed for a thirty-degree pickup. Always remember that you not only have to warm

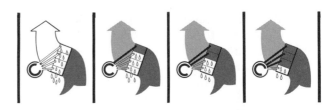

FIG. 11–5 Multiported Valve (Courtesy: ITT Nesbitt)

up the air, but also the building and furniture, and this takes time. If a switch is used, someone must get to the building three hours before it is to be used to provide for warmup. But a clock does this for you. While during occupied cycles we operate fans continuously to achieve maximum comfort, fans can be cycled during unoccupied periods. When unit ventilators or fancoil units are used, one unoccupied thermostat will cycle several units. If the unit ventilators to be controlled are wired from the same lighting panel, a relay, electric, or pneumatic electric is wired in series with the fuses or circuit breakers. The unoccupied thermostat then opens or closes the relay as required. When the relay closes, the circuit is completed and the fans run. Buildings are usually zoned for the unoccupied periods. Natural zoning generally occurs. In the case of a school, you would generally put the administration area on a zone, the library on a zone, gymnasium and locker rooms on a zone, and zone the classrooms by wings or floors.

When a multizone or dual duct system is used the unit fan is cycled by an unoccupied thermostat. The room thermostats will still position their dampers, so that some individual room control will be possible even during unoccupied periods. Remember that the unoccupied thermostat can only maintain the set temperature in the room where it is located. All other rooms will also vary, which isn't too important because they are unoccupied. The room thermostats will, however, tend to even off the extremes.

Radiation systems can best be controlled by reducing the water temperature during unoccupied periods. You could start and stop the pump instead, but this can create air binding problems. Again, the individual room controls will even out the extremes.

The following are some typical temperature control sequences of operation. Use these as a guide, but don't be sterile. Add to them, improve them, refine them as your knowledge of the field increases.

A. *System Water Control*

    1. An adjustable-ratio outdoor reset controller shall reset water temperature by modulating a three-way mixing valve. Water temperature to be____at____outdoors and____at 70° outdoors. Controller to shut off boiler at 65° outdoors, and pump at 70°.

    2. An adjustable-ratio outdoor reset controller shall modulate the steam valve on the convertor to maintain desired water temperature. Water temperature to be____at____outdoors and____at 70° outdoors. Pump to stop at 70° outdoors.

B. *Unit Heater Control*

The space thermostat shall cycle the unit heater fan to maintain desired temperature. Thermostat shall include an override switch to provide for continuous fan operation, if desired.

C. *Direct Radiation Control*

Where indicated, furnish and install a heat-anticipating room thermostat and modulating valve to control finned radiation and convectors.

D. *Cabinet Unit Heater Control*

Furnish and install an electric thermostat to start and stop the fan as required. Install an aquastat to prevent fan from operating if hot water is not available.

E. *Classroom Unit Ventilators* (Sampling Chamber Control)

Classroom unit ventilators are to be controlled by a thermostat in a sampling chamber which will operate face and bypass dampers, and outside-air and return-air dampers, to maintain desired space temperature. The sampling chamber stat will operate through a resetting sensing element to reset discharge temperature. A one degree change in temperature in the sampling chamber sensing element shall change the discharge temperature 20°. During unoccupied periods, groups of unit ventilators will be controlled by a zone-unoccupied thermostat which will start and stop the units to maintain desired temperature. (Note: a wall thermostat may be used instead of a sampling chamber; however, sampling chambers sense a more representative air sample).

Classroom Unit Ventilators (Valve Type)

Classroom unit ventilators are to be controlled by a room thermostat and 50° low-limit thermostat which will operate a modulating valve on the heating coil and the outside-air/return-air dampers in sequence. On a rise in room temperature, the thermostat will first close the control valve and then modulate the outside-air damper open (return damper closed) from the minimum position. On a drop in temperature, the action shall reverse, close outside-air damper to its minimum setting and then modulate control valve open. During unoccupied periods, groups of unit ventilators will be controlled by a zone-unoccupied thermostat which will start and stop the units to maintain desired temperature. Outside-air damper to be closed during unoccupied periods. During warm-up periods, outside-air damper to stay closed until within 2° of thermostat setting.

F. *Single-Zone Heating and Ventilating Units* (Face and Bypass Control)

Room thermostat to modulate face and bypass dampers and outside- and return-air dampers to maintain desired temperature. Room thermostat to operate through a 50° low-limit thermostat. On a rise in temperature, face damper will modulate closed and bypass damper open. If the temperature continues to rise after bypass damper is fully open, then outside-air damper shall modulate open from its minimum setting. The reverse shall occur on a drop in temperature. Provide a warm-up set point to keep outside-air damper closed until room is warmed up. (If desired, a control valve on the heating coil can take the place of the face and bypass dampers.)

G. *Occupied-Unoccupied Control* (Typical School)

The unit ventilators and heating and ventilating units will be controlled from zone-unoccupied thermostats, as shown on the drawings. These zone thermostats will cycle the fans of the various unit ventilators and heating and ventilating units. The building shall be automatically switched from occupied to unoccupied, and vice-versa, by seven-day-programming clocks. Provide one clock for the gymnasium section, one clock for the shop section, and one clock for the remainder of the building. Provide a manual reset switch for each zone to override the clock. Relays for the occupied-unoccupied control will be mounted in a cabinet adjacent to the lighting panel that serves the units in the particular zone. All zone-unoccupied switches and the time clocks shall be located where shown on the drawings. The outside-air dampers shall be closed during night operation, but shall open (if needed) whenever the zone switch is put on day position.

H. *Exhaust Fan Control*

Certain exhaust fans are to be started and stopped by the day-night temperature control system. Install electric relays or pneumatic electric relays, as required, to start and stop these fans.

I. *Smoke Detectors*

Install Fireye smoke detectors in the return air duct of all heating and ventilating units. When smoke is detected, the fan is to stop, outside-air damper to open and return-air damper to close.

J. *Multizone Heating and Ventilating Units*

Mixed-air thermostat to maintain 50° mixed-air temperature during occupied cycle. Room thermostats to modulate zone hot and cold duct dampers to maintain desired temperature. During unoccupied periods, an unoccupied thermostat will start and stop the fan to maintain desired reduced temperature. Outside-air damper to close.

K. *Single-Zone Air Conditioning Unit* Winter-Occupied Cycle

Room thermostat to operate face and bypass dampers (modulate control valve) and outside-air return dampers to maintain desired temperature. On a rise in temperature, bypass damper to modulate to open position and face damper to closed (control valve to modulate closed), and outside-air damper to modulate open from minimum position, and return-air damper to modulate closed. On a drop in temperature, the reverse to take place.

Winter-Unoccupied Cycle

Outside-air damper to be closed. Unoccupied thermostat to cycle unit fan to maintain desired temperature. When unit is switched to occupied cycle, outside-air damper to remain closed unitl room temperature is within 2° of thermostat setting.

Summer-Occupied Cycle

Outside-air damper to be in minimum position. When summer-winter switch is in summer position, thermostat action to be reversed, so that a rise in temperature causes face damper to open and bypass damper to close. Reverse to take place on a drop in temperature.

Summer-Unoccupied Cycle

Unit to be off and outside-air damper to be closed.

Note: If a control valve is used on the heating coil, then a valve must be used on the chilled-water valve. If direct expansion is used, then the room thermostat will operate a refrigerant solenoid valve.

L. *Multizone Air Conditioning Unit* Winter-Occupied Cycle

Mixed-air thermostat to maintain 50° mixed air by modulating outside-air and return-air dampers. Zone thermostats to modulate hot and cold deck dampers to maintain desired temperature.

Winter-Unoccupied Cycle

Outside-air damper to be closed. Unoccupied thermostat to cycle unit fan to maintain desired temperature.

Summer-Occupied Cycle

Outside air damper to be at minimum setting. Room thermostat to modulate hot and cold deck dampers to maintain desired temperature.

Summer-Unoccupied Cycle

Unit to be off and outside-air damper closed.

M. *Dual Duct Mixing Boxes*

Room thermostat to modulate hot and cold duct dampers on mixing box to maintain desired temperature.

N. *Fancoil Units* (Face and Bypass Type)
Occupied Cycle

Room thermostat to modulate face and by-pass dampers to maintain desired temperature. Unit fan to run continuously. Action of thermostat to be reversed by summer-winter switch. (Action of thermostat to be reversed by a strap on aquastat sensing water temperature). Outside-air damper to be set manually. (Motorized damper to open outside-air damper when fan operates, and close when fan stops.)

Unoccupied Cycle

During winter season, room thermostat to cycle a group of fancoil units. During summer cycle, units to be off.

O. *Fancoil Units* (Valve Type)
Occupied Cycle

Room thermostat to modulate a control valve on the combination heating and cooling coil while the unit fan runs continuously. Action of thermostat to be reversed by summer-winter switch. (Action of thermostat to be reversed by a strap on aquastat sensing water temperature). Outside-air damper to be set manually. (Motorized damper to open outside-air damper when fan operates, and close when fan stops.)

Unoccupied Cycle

During winter season, room thermostat to cycle a group of fancoil units. During summer cycle, units to be off.

P. *Fancoil Units* (Multiport Type)

Use same description as valve type, except substitute multiport valve for control valve.

Q. *Summer-Winter Control* (Combination Hot-Water/Chilled-Water System)

Furnish and install a summer-winter switch to index system to summer or winter. The individual unit summer-winter cycle is specified under the unit control sequence. When switch is placed in summer position, boiler to be de-energized and water chiller energized. Switch-over valve shall not open to chiller circuit and pump shall circulate water until return-water temperature is at 90°.

R. *Summer-Winter Control* (Single-Zone Air Conditioning Units with Direct Expansion Unit)

Furnish and install a summer-winter switch to index system to summer or winter. The individual unit summer-winter cycle is specified under the unit control sequence. When the switch is placed in the summer position, the condensing unit is to be energized, and boiler to be de-energized. (If system contains radiation or heating units, change the last sentence to: When the switch is placed in summer position, a two-position valve on the air conditioning unit heating coil to close, the condensing unit to be energized, and the boiler to continue to operate until 65°.)

S. *Summer-Winter Control* (Multizone Unit)

Furnish and install a summer-winter switch to index system to summer or winter. The individual unit summer-winter cycle is specified under the unit control sequence. When the unit is placed on the summer position, condensing unit (chiller) to be energized. Boiler to continue to operate until 65° outdoors.

Note: If desired, the action of the summer-winter switch can be accomplished by an outdoor thermostat which switches the system at the desired temperature. Be sure to specify all required inter-connecting wiring for the condensing unit or water chiller.

CHAPTER 12

Moisture in Construction

MOISTURE IS A necessary part of comfort and health. But, moisture does introduce problems. Moisture causes visible surface condensation, which is annoying. Worse yet, it can cause hidden condensation in walls and roofs and produce serious damage. The damage can take the form of rusting, dry rot, or spalling.

It also causes mildew, with all its problems. I remember two schools that had a cast-in-place insulating roof deck. The mildew here was so bad, and the odor so fierce, that the parents had two sets of clothes for the children—school clothes and home clothes. When the children arrived home they undressed on the back porch and changed into their home clothes. The school clothes were aired overnight. Many theories were advanced about the cause and about who was to blame. The true fact is that hidden condensation was the villain.

An incident took place in my own home a few years ago which taught me my lesson. I replaced some glass with a homemade sandwich panel. The panel consisted of two inches of rigid insulation between two pieces of plywood. Within the first year, a musty odor began to develop. By the time I removed the panel two years later, most of the sash had rotted away. Again, hidden condensation.

How can this happen? It happens because all material is porous to a degree. Materials that are waterproof prevent the passage of liquid water, but may permit the vapor phase of water to diffuse through them. The movement of vapor through a material is similar to heat transfer. Heat flows from the higher energy level, or hot side, to the low energy level or cold side of material. You might say it flows down hill. In the same fashion, vapor flows from a higher vapor pressure to the low pressure. The vapor pressure forces the vapor through a material.

Where does all this problem-causing moisture come from? People give off lots of moisture, as do the activities that people engage in, such as cooking, bathing, laundering, etc. (See Table 12–3.) An obvious source is a large open surface of water, such as a swimming pool. The amount of vapor entering the air from a swimming pool varies with the water and air temperature and, of course, the room relative humidity. If the water temperature is 75° and the room conditions are 75° and 50% RH, 0.004216 lb of water per sq ft of water surface will enter the room.

Swimming pools are expected to be a problem because we can see the large amount of water. But people are surprised to find that they, themselves, can cause a problem. The amount of moisture given off by people varies with their size and degree of activity.

*TABLE 12–1
Moisture From People in Pounds-Per-Hour

Activity	Room Dry-Bulb Temperature		
	80°F	75°F	70°F
Seated at rest in theatre or grade school	0.155	0.12	0.09
Office work	0.250	0.205	0.17
Shopping	0.30	0.244	0.21
Light factory work	0.53	0.45	0.28
Dancing	0.605	0.525	0.45
Bowling	0.98	0.92	0.85

These values are for an adult. To convert to values for children multiply by 75%.

*Extrapolated from Table 48, Carrier System Design Manual, published by Carrier Corporation, Syracuse, N.Y.

Let's take a theater with 100 people. The theater has a volume of 20,000 cu ft and the room temperature is 75°F. The amount of moisture entering the theater per hr is $100 \times 0.12 = 12$ lbs per hr. If the room started perfectly dry or 0% relative humidity, the room would have at the end of the first hour:

$$\frac{12 \text{ lbs}}{20,000 \text{ cu ft} \times 0.075 \text{ lb/cu ft}} = \frac{0.008 \text{ lb}}{\text{lb of air}}$$

Consulting a psychrometric chart, we will find that the room relative humidity is now 42%. At the end of the second hour, the relative humidity will be 80%. These values assume, of course, that no dilution took place because of infiltration or ventilation. So you can see that it is no trick to get excess moisture. The trick is what to do about it.

Let's check a school room with a population of 26 pupils and a teacher. The room starts at 70° and 5% RH. The occupants give off 1.85 lb of moisture per hr ($26 \times 0.09 \times 75\% + 1 \times .09$). Assume these rooms are not mechanically ventilated, so we must depend on infiltration to remove the moisture. At 5-mph wind velocity, 600 cu ft/hr, or 45 lb/hr of outside air enter the room. If this air is at 0°F and 90% RH, its moisture content will be 0.0007 lb/lb dry air.

TABLE 12–2 Permeance And Permeability Of Metals To Water Vapor

Material	Perme-ance Perm	% RH_1–RH_2	Method†	Ref.‡	Material	Perme-ance Perm	% RH_1–RH_2	Method†	Ref.‡
AIR (still)	120*	92–73	b	4	Plywood (Interior type 3 ply D.F.), ¼ in.	1.86	50–	4	13
INSULATION			d						
Cellular glass	0.0*		d		MASONRY				
Corkboard	2.1–2.6*	75–0	d	5	Concrete (1:2:4 Mix)	3.2*	100–45	w	11
Corkboard	9.5*	100–45	w	11	Concrete (8 in. cored block wall, limestone agrgt.)	2.4	79–68	t	4
Structural Insulating Board (vegetable, uncoated)	20–50*	40–x	t	9	Brick wall—with mortar—4 in.	0.8	50–x	t	10
Mineral Wool (unprotected)	116*	100–30	w	8	Tile wall—with mortar—4 in.	0.12	50–x	t	10
INTERIOR FINISH					FILMS				
Plaster on wood lath	11	100–30	w	8	Aluminum foil, 1 mil	0.0	100–0	d	21
Plaster on metal lath—¾ in.	15	40–x	t	9	Aluminum foil, 0.35 mil	0.05	100–0	d	21
Plaster on plain gypsum lath (with studs)	20	40–85	t	4	Polyethylene, 2 mil	0.16	50–0	d	18
Gypsum wall board—plain—⅜ in.	50	50–20	v	15	Polyethylene, 4 mil	0.08	50–0	d	18
Insulating wall board (un-coated)—½ in.	50–90	40–x	t	9	Polyester, 1 mil	0.72	50–0	d	20
Hardboard, ⅛ in.	11			19	Cellulose acetate, 10 mil	4.1	50–0	d	20
Hardboard, tempered, ⅛ in.	5			19					

Material	Perme-ance Perm	% RH_1–RH_2	Method†	Ref.‡	Material	Lb per 500 sq ft	Permeance-Perms dry cup	Permeance-Perms wet cup	Ref.‡
**PAINT—2 coats					**BUILDING PAPERS, FELTS				
Asphaltic paint on plywood	0.4	100–30	w	8	Duplex sheet, asphalt lami-nae, *aluminum foil* one side	43	0.002	0.176	14
Aluminum in varnish on wood	0.3–0.5	95–0	d	12	Saturated and *coated* felt heavy roll roofing	326	0.05	0.24	14
Enamels, brushed on smooth plaster	0.5–1.5	92–0	b	4	Kraft and *asphalt laminae*, Reinforced 30-120-30	34	0.3	1.8	14
Primers or *Sealers* on insulat-ing wall board	0.9–2.1	40–x	t	9	Insulation back up, asphalt-sat., one side glossy	31	0.4	0.6–4.2	14
Various *Primers*+1 coat flat paint on plaster	1.6–3.	40–x	t	9	Asphalt-saturated and coated sheathing paper	43	0.3	0.6	14
Flat paint (alone) on insulat-ing wall board	4	40–x	t	9	Asphalt-saturated sheathing paper	22	3.3	20.2	14
Water Emulsion on insulat-ing wall board	30.–85.	40–x	t	9	15-pound asphalt felt	70	1.0	5.6	14
**PAINT—Exterior, 3 coats					15-pound tar felt	70	4.0	18.2	14
White *lead & oil* prepared paint on wood siding	0.3–1.0	50–0	d	15	Single sheet Kraft, double infused	16	30.8	41.9	14
White lead-zinc oxide & lin-seed oil on wood	0.9	95–0	d	12					
WOOD									
Sugar Pine	0.4–5.4*	various	tv	4					
Plywood (Exterior type 3 ply D.F.), ¼ in.	0.72	50–	4	13					

* These boldface values are permeability in perm-inches.
** Description is a guide only, and does not insure permeance.
† Methods: d—dry cup; w—wet cup; t—two temperatures; b—special cell; v—air velocity both sides; 4—average of four methods.
‡ References. No. 9 also includes *Bulletins* 22 and 25 of the Engineering Experiment Station, University of Minnesota. No. 15 includes values from unpublished tests at the Pennsylvania State University, Engineering Experiment Department.

Reprinted by permission from ASHRAE Guide and Data Book, 1965.

To determine the moisture content at the end of six hours use the following:

$$\log \frac{G + Q\,(Z - c)}{G + Q\,(Z - c_0)} = -\frac{Q\,t}{2.3P}$$

G = rate generation of moisture lb/hr = 1.85 lb/hr
P = pounds of air in room = 600 lb
t = time in hrs = 6 hrs
c = room moisture concentration at end of time t in lb/lb dry air
c_0 = original room moisture concentration in lb/lb dry air = 0.0007 lb/lb
Q = pounds of air introduced per hr = 45 lb/hr
Z = pounds of moisture in the infiltrating air (Q) = 0.0007 lb/lb

$$-\frac{Q\,t}{2.3P} = \frac{45\,(6)}{2.3\,(600)} = -0.19$$

Let's say that $\dfrac{G + Q\,(Z - c)}{G + Q\,(Z - c_0)} = X$

Then $\text{Log } X = -\dfrac{Qt}{2.3P}$

$$\text{Log } X = -\frac{45\,(6)}{2.3\,(600)}$$

$$\text{Log } X = -0.19$$
$$X = 0.645$$

If $X = \dfrac{G + Q\,(Z - c)}{G + Q\,(Z - c_0)}$

then

$$\frac{G + Q\,(Z - c)}{G + Q\,(Z - c_0)} = 0.645$$

$$\frac{1.85 + 45\,(0.0007 - c)}{1.85 + 45\,(0.0007 - 0.0007)} = 0.645$$

c = 0.0155 lb/lb dry air

As you can see on a psychrometric chart, this would give us 70° and 98% RH, a very bad condition.

How do you know you are going to have trouble? How do you analyze a structure? First, determine the temperature gradient in the structure. Then, through the use of Fick's Law, determine the amount of moisture flowing into the structure and the amount that flows out. If more flows in than flows out, then condensation will take place.

The temperature gradient is determined by using:

$$\frac{R_x}{R_a} = \frac{t_i - t_x}{t_i - t_0}$$

Where R_x = resistance to heat transfer from the room to any point in the structure at which the temperature is to be determined

R_a = overall resistance from the room to the outdoors
t_i = inside temperature
t_0 = outside temperature
t_x = temperature to be determined

The flow of vapor is computed through the use of:

$$W = A\,\emptyset\,M\,\frac{\Delta P}{L}$$

Where W = total vapor transmitted in grains (7000 grains = one pound)

$\overline{M}$ = average permeability in perm-inches or the grains per (hour) (square foot of surface) (inch of mercury vapor pressure difference per inch thickness.)

Where $\emptyset$ = time in hours
P = pressure difference in inches of mercury (in Hg)
L = thickness of material in inches.

Many times the permeability is given for the overall thickness of the material, in which case we use:

$$W = MA\emptyset\Delta P$$

Where M = perms or grains per hour, per square foot of surface, per inch of mercury vapor pressure difference.

The permeabilities of various materials in perms or perm inches are listed in Table 12–2. Notice that some values are determined by the dry-cup method and some by the wet-cup. In some cases both dry-cup and wet-cup values are given. The difference in values is caused by the methods themselves. In the dry-cup, the material is placed over a cup containing a desiccant. The wet-cup method uses a cup of water instead of a desiccant. The cups are kept in an atmosphere held at 50% RH. The difference in vapor pressures is about the same because vapor pressure varies directly with relative humidity if the temperature is constant. But, in the case of the dry cup, the room side is at 50% RH. At 70° temperature, the vapor pressure in the room is 0.37 in Hg. In the dry cup, the 0% RH gives 0 vapor pressure. So, with the use of a dry cup, we have 0.37 in the room and 0 in the cup, while in the wet cup, the room is at 0.37 and the cup is at 100% RH, or 0.74 in Hg vapor pressure.

In each case, the difference is 0.37, but the results are quite different. Wet-cup methods give higher values than dry-cup; as much as six times. Which value do you use? If only one value is given, use the value given. But, when both values are given, use dry-cup if you are concerned with winter vapor transmissions from 70°, 50% rooms to low outdoor values. If you are concerned with rooms at high relative humidities, say 80°, 75%, transmitting to atmospheres at normal values, such as 70°, 30%, then use the wet-cup values.

To determine the permeability of several different materials in series:

$$M_t = \frac{1}{\dfrac{1}{M_1} + \dfrac{1}{M_2} + \dfrac{L}{M_3} + \cdots \dfrac{1}{M_n}}$$

TABLE 12–3
Moisture Production by Various Domestic Operations

Operation			Pounds of Moisture
Floor mopping (per sq. ft.)............................			0.03
Clothes drying (per week when dried indoors) .			26.40
Clothes washing (per week).........................			4.33
Cooking (per meal)	From Food	From Gas	
Breakfast	0.34	0.56	0.90
Lunch	0.51	0.66	1.17
Dinner	1.17	1.52	2.69
Dish washing (per meal)			
Breakfast			0.20
Lunch..			0.15
Dinner			0.65
Bathing (per shower).................................			0.50
(per tub)			0.12
Human contribution (family of four, per hour)..			0.46
Gas refrigeration (per hour)			0.12
House plants (each, per hour)			0.04
Humidifier (per hour).................................			2.00

S. C. Hite and J. L. Bray, Research in Home Humidity Control, Purdue University Engineering Experiment Station Bulletin. Research Series No. 106, 1948, p. 24.

Saturated vapor pressures are found in Table 7–2 in Chapter 7. In the same chapter, the following formula is given:

$$P_v = RH (P_{v\ sat})$$

So, to determine the vapor pressure at a specific temperature and relative humidity, first find the saturated vapor pressure and multiply it by the relative humidity.

Let's do a sample problem. Assume a room is at 70°DB and 50% RH, while the outdoor conditions are 0° and 54% RH. The outside wall is constructed of $^1/_8$-in. tempered hardboard paneling on $^3/_8$-in. gypsum wall board, 2 × 4 studs, $^1/_4$-in. plywood sheathing, 15-lb asphalt felt building paper, and 4-in. brick. The overall resistance is 3.05.

Point in Wall	R_x	t_x	$P_{v\ sat}$
Inside surface of	0.68	54°	0.42
Gypsum wall board	0.86	50°	0.362
Air space side of gypsum wall board	1.18	43°	0.278
Air space side of sheathing	2.12	21°	0.105
Building paper	2.43	14°	0.077
Brick	2.49	13°	0.073
Outside surfaces of brick	2.88	4°	0.046
Outdoors		0°	0.038

The t_x temperatures were determined by the following procedure, repeated for each wall element:

$$\frac{R_x}{R_a} = \frac{t_1 - t_x}{t_1 - t_o}$$

$$\frac{0.68}{3.05} = \frac{70 - t_x}{70 - 0}$$

$$t_x = 54°$$

The next step is to use the principle of continuity to determine the actual vapor pressures. All "continuity" means is that what comes in, must go out. If you determine the vapor transmission from the room to the air space, then this must equal what goes from the air space outdoors. First, determine the vapor transmission through the complete wall.

$W = M_t (P_i - P_o)$
$P_i = 0.50 (0.74) = 0.37$ (room vapor pressure)
$P_o = 0.54 (0.037) = 0.02$ (outdoor vapor pressure)

$$M_t = \frac{1}{\frac{1}{M_1} + \frac{1}{M_2} + \frac{L}{M_3} + \frac{1}{M_4} + \frac{1}{M_5} + \frac{1}{M_6}}$$

Material	Permeability	Vapor	Resistance
$^1/_8$-in. Wallboard	5	$\frac{1}{5}$	0.20
$^3/_8$-in. Gypsum	50	$\frac{1}{50}$	0.02
Air space	120	$\frac{3.625}{120}$	0.03
Plywood	0.72	$\frac{1}{0.72}$	1.4
Asphalt felt	1	$\frac{1}{1}$	1.0
Brick	0.8	$\frac{1}{0.8}$	1.25

$$M_t = \frac{1}{0.20 + 0.02 + 0.03 + 1.4 + 1.0 + 1.25}$$

$$M_t = \frac{1}{3.90}$$

$$M_t = 0.256$$

$$W = 0.256 (0.37 - 0.02)$$

$$W = 0.09 \text{ grains/sq ft hr}$$

If 0.09 grains per sq ft are transmitted per hr, through the wall, then at any point in the wall the vapor transmission must be 0.09 grains/sq ft hr.

$W = M_1 (P_i - P_1)$
$0.09 = 5 (0.37 - P_1)$
$P_1 = 0.352$
$W = M_2 (P_1 - P_2)$
$0.09 = 50 (0.352 - P_2)$
$P_2 = 0.349$

Continuing this process, we find $P_3 = 0.347$, $P_4 = 0.222$, $P_5 = 0.132$.

The values of the vapor pressure gradient and saturated vapor pressure gradient are all plotted in Fig. 12–1. Notice that the saturated vapor pressure

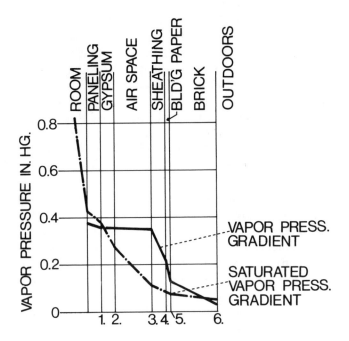

FIG. 12–1 Vapor Pressure Gradient Through Wall

drops below the vapor pressure gradient about $\frac{1}{3}$ of the way through the gypsum wall board.

Let's check a roof deck. A favorite building material at present is a pre-cast roof deck consisting of wood fibers with a cement binder (Fig. 12–2). Let's assume −10° and 80% outdoors, and 70° and 50% indoors. Our first step, of course, is to determine the temperature at various points. Using again

$$\frac{R_x}{R_1} = \frac{t_i - t_x}{t_i - t_o}$$

we find, as shown in Fig. 12–3, the temperatures in the deck. The saturation vapor pressures and actual vapor pressures would be:

	t	$P_{v\,sat}$	$P_v\,actual$
Room	70	0.74	0.37
Surface	63	0.58	0.37
$\frac{3}{4}$ in.	46.5	0.317	0.37
1-$\frac{1}{2}$ in.	30	0.164	0.369
2-$\frac{1}{4}$ in.	13.5	0.074	0.369
3 in.	−3	0.032	0.0368
Outdoors	−10	0.022	0.017

As you can see, the two lines would cross at about 50° or $\frac{1}{4}$ through the deck. From here on we would have condensation, and at 32° we would get freezing. Is this freezing harmful? Not to the material, because this material is treated with silicones to prevent moisture damage. But, this is a problem to the heating

system. This material is sponge like; $\frac{2}{3}$ of the volume is wood and $\frac{1}{3}$ air. Water cannot penetrate the silicone treated wood, so it must lodge in the air portion. Enough water will condense and freeze to fill $\frac{3}{4}$ of the air layer. This will change our U factor from 0.21 to 0.29—an increase of 38%. Condensation will increase the heat transmission coefficient of all materials.

O.K! We now know all the troubles that moisture can cause. What do we do about it? First of all, we can take steps to reduce the room moisture content by ventilation. For example: In the example of the wall, if the room RH is reduced to 30% by bringing in outside air, the vapor pressure stays below the saturation line. In the deck above, we would have to drop the RH to almost 0 to prevent condensation. But, if we reduce the RH to 20%, the condensation will take place as ice. This will not stain as liquid water will. When the outdoor temperature rises, the ice will go back to the vapor state. However, it still increases our U factor, which is not desirable.

Another solution is to use a vapor barrier on the room side to reduce the flow of vapor into the wall to what can leave the wall. This is a good solution, but what do you do when you can't use a vapor barrier such as on our preformed deck? Here, a solution would be to add insulation and a vapor barrier to the top of the deck. If we wish to maintain 70°, 50% RH, we will need 3 in. of insulation. This isn't a very practical solution, is it, to buy a preformed, insulated roof deck and then add 3 in. of insulation?

Another solution would be to reduce the room relative humidity and add insulation to the top of the

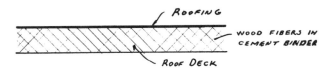

FIG. 12–2

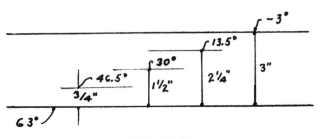

FIG. 12–3

deck. Now, while 3 in. of insulation is out of the question, 1 in. is within reason. The temperature at the room side of the insulation would be:

Resistance of deck	6.65
Resistance of 1 in. of insulation	4.00
	10.65

$$\frac{6.00}{10.65} = \frac{70 - t_x}{70 - (-10)}$$

$$t_x = 25°$$

The saturated vapor pressure at 25° = 0.13.

For all practical purposes the room vapor pressure must be 0.13.

So the room conditions must be

$$RH = \frac{0.13}{0.74}$$

$$RH = 17.5\%$$

So, if enough outside air is introduced to reduce the room relative humidity to 17.5% at −10°F, 1 in. of insulation on top of the deck will be sufficient. At 20° outdoors, the room relative humidity can be 28%.

What else can you do? In searching through the literature on moisture migration, I came across an interesting bit of information. The whole problem of hidden condensation came to light with the use of blown-in rock wool insulation. The early researchers found that the sloppy insulators DID NOT HAVE CONDENSATION PROBLEMS on their jobs. The sloppy insulators did not plug the holes they drilled in the stud spaces. It appears that this gives the vapor an avenue of escape. None of the researchers had an answer. They could only state that it worked. All I can figure is that the higher vapor pressure is sufficient to drive the vapor through before it can condense. The vapor appears as a fog as it leaves the holes. The same principle could be applied to our preformed roof deck. Some of the roofing could be removed and the deck protected from the elements by a suitable cap. But, how much roofing must be removed? Actually, at this time, no one really knows. My own guess would be about 1%.

In conclusion, I would recommend that the following steps be taken to prevent hidden condensation:

1. Reduce the room moisture content by ventilation.
2. Apply vapor barriers to the warm side of structures.
3. Add insulation to top of the roof deck to raise the temperature at the vapor barrier.
4. Pierce the roofing and protect it with a cap and so give the moisture a path.

Just remember, place the best vapor barrier you can use on the warm side of a space and make the exterior as permeable to vapor transmission as possible.

CHAPTER 13

Sound

WHEN YOU THINK of energy, what comes to mind? Light, heat, electricity. Anything else? How about sound? Yes, sound is energy, just like the others. It is a vibrating energy. It shows itself as variations in pressure and density of the atmosphere. Sound needs a transmitting medium. This can be air or any gas, a liquid, or solid. Sound will not be transmitted through a vacuum. The classic example in physics is to place a ringing alarm clock under a bell jar. A vacuum pump is connected to the jar; as the pressure is reduced, the sound grows fainter, disappearing when vacuum conditions are reached. Conversely, sound is trans-mitted more readily in a liquid or solid. Next time you are swimming, notice how much louder an outboard motor sounds under water than out of water. Or, if you are a brave one, place your ear on a railroad track and notice how long you can hear an approaching train through the track before you can hear it through the air.

The air itself does not move. It's like water when a pebble is dropped in it. Waves are generated, rising and falling, but the water does not move. A leaf floating in the water will rise and fall, but does not move. An illustrative example can be performed by lining up a number of billiard balls so they are touching. Now cause the cue ball to strike the first ball squarely and watch the last ball shoot off toward the pocket.

This is how sound is transmitted through air. When sound is transmitted, this relative motion results in pressure fluctuations above and below atmospheric pressure. These fluctuations are known as compressions (increased pressure), and rarefactions (decreased pressure). This appears as a sine wave as shown in Fig. 13-1.

These sine waves are not a simple single wave as shown in Fig. 13-1, unless we have a pure tone. Sound waves more often are a complex mixture of sound waves similar to Fig. 13-2.

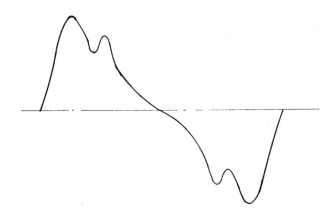

FIG. 13–1 Sound Wave For Pure Tone   **FIG. 13–2** Sound Wave For Two Pure Tones

The speed of sound, 'c' (a constant), is equal to the wave length, 'λ', times the frequency, 'f', where frequency is equal to the number of waves per second.

$$c = f\lambda$$

So, an increase in wave length results in a decrease in frequency, and vice-versa. The lower limit of sound is 20 cycles per second (cps) and the upper limit is 20,000 cps. At age 10, we begin to lose acuity at the upper end. By age 50, our acuity has dropped from 20,000 cps to about 10,000 cps. The bulk of speech lies between 100 to 6,000 cps. Frequency is what musicians call pitch. Middle 'C' is 264 cps.

Sound may consist of a single pure tone (a simple sine wave), or a complex combination of such tones. Sound is rarely a pure tone. Noise can be defined as unwanted sound. Noise may be narrow band, broad band, steady state, or highly pitched.

   a. Narrow band noise is one in which acoustic energy is confined to a group of closely spaced frequencies.
   b. Broad band noise is one in which acoustic energy is present throughout most of the audible frequency range.
   c. Pitched noise is one in which acoustic energy is confined to one or a few discrete frequencies.
   d. Steady state noise has little sudden change in level. Few or no variations in intensity appear when switching a sound meter from fast to slow response. Duct noise is steady state noise.

In sound work, you will hear the terms "sound pressure," "sound power," and "sound intensity." What are they and how are they related?

SOUND PRESSURE

What is it that you feel when you hear sound? You feel the change in pressure. When you hit a tuning fork, the tines vibrate. The harder you hit it, the further the tines travel back and forth. The farther the tines travel, the larger the pressure wave. The average of this wave is called the sound pressure. The unit of measure is the microbar. One bar equals one barometric pressure. A microbar is one millionth of a bar.

SOUND POWER

The energy required to produce a sound is called the sound power. You can't measure sound power but you can convert sound pressure into sound power. Sound power is expressed in watts, microwatts, or kilowatts. A voice speaking normally radiates 10 microwatts (0.000,01 watts), while a two-engine airplane radiates 1000 watts.

SOUND INTENSITY

Sound intensity is the sound power per unit area. Remember our illustration of the pebble dropped into the water. The circles got bigger and bigger as the waves traveled out from the pebble and, finally, disappeared. So you can see that the intensity reduced as we got further from the center. Why? Because the force had to be spread over a larger area. So it is in sound, except that now we are concerned, not with a flat circle, but a sphere. If sound emanates from a point in space, the waves go out in all directions. As the radius doubles, the area generated quadruples. This is the inverse square law. Every time you double the radius, the sound intensity is reduced by four. The sound power is the same at all distances, but it must spread itself over a larger area. The inverse square law only applies to free-field conditions where there are no boundaries or other objects to interfere with a true spherical propagation of the sound wave. The sound intensity can be determined from the sound pressure by the following equation:

I $= (1.52p)^2$
I   = intensity in microwatts/sq ft
p   = sound pressure in microbars
1.52 = a constant

A word of caution. This equation will not apply when very close to the sound source, or if the sound is reflected from adjacent surfaces.

DECIBEL

When someone wishes to express the intensity of sound you will hear him speak of decibels. What is a decibel? The decibel developed because of the desirability of compressing the wide range of sound intensities. Notice in Table 13–1 that intensity varies from 1 watt/sq ft to 0.000,000,000,000,1. This is quite a range. Decibels, which are logarithmic, vary from 130 to 0 for the same range of intensities. There are 10 decibels in a bel. The bel is a logarithmic unit. It is the exponent of a base 10.

$1 \times 10^0 = 1$	0 bel
$1 \times 10^1 = 10$	1 bel
$1 \times 10^2 = 100$	2 bel
$1 \times 10^3 = 1000$	3 bel

0 bel was picked as the threshold of hearing. You can see that 3 bel is 1000 times the intensity of the threshold of hearing.

The bel is a large cumbersome unit, so the decibel which equals 10 bels is used. Decibels cannot be added directly. They are exponental and must be

TABLE 13–1
Comparison of Sound Measurement Scales

Intensity (watts/sq ft)	Intensity Level (db)	Sound Pressure Level (db)	R.M.S. Sound Pressure (ubars)
1.0	130	130	
0.1	120	120	200
0.01	110	110	
0.001	100	100	20
0.000,1	90	90	
0.000,01	80	80	2
0.000,001	70	70	
0.000,000,1	60	60	0.2
0.000,000,01	50	50	
0.000,000,001	40	40	0.02
0.000,000,000,1	30	30	
0.000,000,000,01	20	20	0.002
0.000,000,000,001	10	10	
0.000,000,000,000,1	0	00	0.000,2

converted back to exponents and then added. To add the intensities of two equal sounds at 50 db:

$$50 \text{ db} = 5 \text{ bel} = 10^5$$
$$10^5 + 10^5 = 2\,(10^5)$$
$$2\,(10^5) = 200,000$$
$$\text{Log } 200,000 = 5.3$$
$$2\,(10^5) = 10^{5.3}$$
$$10^{5.3} = 53 \text{ db}$$

The relationship of decibels to everyday events is as follows:

40 db = quiet room
60 db = normal speech
70 db = loud voice
90 db = upper limit of speech
120 db = physical discomfort

At about 90 db we begin to worry about damage to the ear. How do you tell if you are in a 90 db environment? A good rule to observe is, if you can't hear a shout, then you are in the trouble range. Only 10 or 15 second exposure to 120 db sound can cause permanent damage to the ear.

SOUND PRESSURE LEVEL

The decibel is, as we have seen, the unit for expressing intensity levels. It is also the unit for expressing the sound pressure level.

Sound pressure level = $20 \log_{10} (p/p_0)$
p = root mean square pressure in microbars
p_0 = the reference pressure 0.0002 microbar

We stated earlier, in discussing sound intensity, that the equation $I = (1.52p)^2$ applies only if we are not too close to the source or reflecting objects. Under these same conditions, we can equate sound pressure level to intensity level. Both p_0 and I_0 are reference levels equal to the threshold of hearing.

Intensity level = $10 \log_{10} (I/I_0)$
$I = (1.52p)^2$
$I_0 = (1.52\,p_0)^2$
$$\frac{I}{I_0} = \frac{(1.52p)^2}{(1.52p_0)^2}$$

Intensity level = $10 \log_{10} (p/p_0)^2$
Intensity level = $20 \log_{10} (p/p_0)$
Sound pressure level = $20 \log_{10} (p/p_0)$
Intensity level = Sound pressure level

Again, this only applies precisely if we are not too close to the sound source or reflecting objects. However, in practice, they are used interchangeably, because the difference is generally small.

SOUND POWER LEVELS

Sound power level cannot be measured directly, but is derived from some other measurable quantity. For free field conditions:

Sound Power Level (PWL) = $10 \log_{10} W_1/W_0$
W_1 = acoustic power in watts
W_0 = reference value = 10^{-12} watts

ANECHOIC CHAMBER

An anechoic chamber is a "dead room." The walls, ceilings, and floors are treated to prevent any reverberation. An anechoic chamber is close to being a true free field. It has 0.5% reverberation.

REVERBERENT TEST ROOM

A reverberent room is one that is constructed completely of hard surfaces which give a maximum reverberation.

FREQUENCY

Our ears do not respond equally to all frequencies. For example: At 30 cps, a sound pressure level (SPL) of 62 db is barely audible; at 100 cps, 38 db; at 1000 cps, 0 db. But, at 1000 cps, 60 db is quite loud. So, here we have two 60 db sound levels; one can barely be heard, the other is annoying. So, you can see that a statement that a sound level is X db is meaningless unless the frequency is specified. In an attempt to improve this situation, equal loudness contours were developed which are plotted on a scale of SPL against frequency. These are called Fletcher-Munsen curves, N-C (noise criteria) curves, or sones. (See Fig. 13–3.) Low frequency sounds have a rumbling sound. A subway train is an example of a low frequency sound. High frequency sounds hiss like the sound of compressed air. Low frequency sounds are much harder to dissipate than high frequency sounds.

SOUND PRESSURE LEVEL SPECTRUM

Sound pressure level spectrum of a noise is the sound pressure level described as a function of individual frequencies. A spectrum may also be de-

termined by analyzing the noise in terms of sound pressure level existing in each of eight frequency bands of one octave width. Octave is the interval between two sounds having a frequency ratio of 2-to-1. The octave bands are as follows:

Band	*Midpoint*
1 = 20–75 cps	53 cps
2 = 75–150 cps	106 cps
3 = 150–300 cps	212 cps
4 = 300–600 cps	425 cps
5 = 600–1200 cps	850 cps
6 = 1200–2400 cps	1700 cps
7 = 2400–4800 cps	3400 cps
8 = 4800–10,000 cps	6900 cps

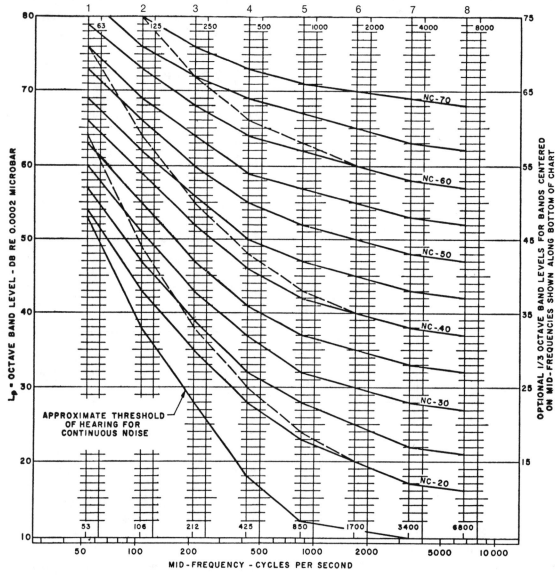

FIG. 13–3 Noise Criterion Curves For Specifying The Design Level In Terms Of The Maximum Permissible Sound Pressure Level For Each Frequency Band (Reprinted by permission from *ASHRAE Guide and Data Book*, 1965)

REVERBERATION AND ABSORPTION

Whenever a sound wave comes in contact with a surface, a reflected wave is produced. The intensity of this reflected wave depends on the hardness of the surface. If the surface is hard and non-porous, it will be practically equal to the incident wave. When a sound is enclosed in a room with hard surfaces, the sound will be reflected back and forth from surface to surface many times. The sound is said to reverberate. The action of the sound wave is similar to the action of a handball in a closed handball court. The handball is very lively and comes off a wall with nearly as much speed as it hits. In fact, as it bounces from wall to wall, the beginner handball player is convinced that more than one ball is being used. On the other hand, if a handball is thrown against a padded wall, it falls limply off the wall. In sound work, a similar action takes place when sound waves hit a soft, porous material. Very little sound is reflected from a soft surface and you might say that the sound waves fall limply from the wall.

If the sound source is stopped, the continuation of the sound is called reverberation. The time that it takes for the sound to drop 60 db after the sound stops is called the reverberation time. Reverberation time varies with the hardness of the walls, ceiling, and floor. In a hard room, the reverberation time could be six seconds, while in a soft room, the reverberation time may be as little as 0.33 seconds.

In a free field, sound intensity decays as we move away from the source. If we start with a sound intensity of 100 db, the direct sound intensity drops to 80 db at 10 ft from the source, 68 db at 40 ft, and 60 db at 100 ft. But in a hard room, the sound pressure level *STARTS* at 103 db because of reverberation and never drops below 100 db, even at 100 ft from the source. This build-up sound pressure is called the reflected sound, while the sound from the source alone is called direct sound. It is important that you realize that sound absorption materials do not and cannot reduce the sound power. All they can do is try to simulate free field conditions.

The amount of sound a surface absorbs varies with the material. A concrete wall, smoothly troweled, will reflect almost 99% of the sound falling on it. On the other hand, an acoustic tile ceiling will absorb almost 90%. The sound absorption coefficient is called α. Normally absorption coefficients are given for six frequencies; 125, 250, 500, 1000, 2000, and 4000 cps. This is important because the coefficient will vary considerably over these ranges. Let's say you are faced with a source whose sound power is at a maximum at 2000 cps. If you select an absorption material that is poor at 2000 cps, you will not do a very good job.

You will hear the term "Sabin" used. A "Sabin" is one square foot of perfect absorbing material and the symbol is "a." For one surface a $= \alpha s$ where "s" is the area of the surface. For several surfaces a $= \alpha_1 s_1 + \alpha_2 s_2 \cdots + \alpha_n s_n$.

The average coefficient of absorption "$\bar{\alpha}$" for a room can be determined by:

$$\bar{\alpha} = \frac{\alpha_1 s_1 + \alpha_2 s_2 + \alpha_3 s_3 \cdots + \alpha_n s_n}{s_1 + s_2 + s_3 \cdots + s_n}$$

Example #1

If you have a room 10 ft $\times$ 10 ft $\times$ 10 ft, with a terrazo floor $\alpha_1 = 0.02$, plaster walls $\alpha_2 = 0.04$, and plaster ceiling $\alpha_3 = 0.04$, find $\bar{\alpha}$ and a:

$$\bar{\alpha} = \frac{0.02\,(10 \times 10) + 0.04\,(4)\,(10 \times 10) + 0.04\,(10 \times 10)}{10 \times 10 + 4\,(10 \times 10) + 10 \times 10}$$

$$\bar{\alpha} = \frac{0.02\,(100) + 0.04\,(400) + 0.04\,(100)}{600}$$

$\bar{\alpha} = 0.0366$
$a = \alpha_1 s_1 + \alpha_2 s_2 + \alpha_3 s_3$
$a = 0.02\,(100) + 0.04\,(400) + 0.04\,(100)$
$a = 22$ sabins

To determine the amount of reflected sound, we use the formula $L_r = L_w - 10 \log a + 16.4$ where $L_w =$ the sound power in db.

Example #2

If the sound power in example #1 is 100 db, what is the reflected sound?

$$L_r = L_w - 10 \log a + 16.4$$
$$L_r = 100 - 10 \log 22 + 16.4$$
$$L_r = 103 \text{ db}$$

Now, if we increased the surface treatment to 500 sabins, what would be reflected sound?

$$L_r = L_w - 10 \log a + 16.4$$
$$L_r = 100 - 10\,(\log 500) + 16.4$$
$$L_r = 100 - 27 + 16.4$$
$$L_r = 86.6 \text{ db}$$

The direct sound at various distances from the source can be determined by the formula: $L_d = L_w - 20 \log r - 0.6$ where r is the distance from the source. To determine the total sound level, you add (logarithmically) the direct and reflected sound. Table 13-2 can be used to add the two sound intensities.

TABLE 13–2
Increments for Combining Intensity Levels

Excess of Stronger Component (db)	Add to the Stronger to Get Combined Level (db)
0	3.0
1	2.5
2	2.0
3	1.8
4	1.5
5	1.2
6	1.0
7	0.8
8	0.7
9	0.6
10	0.5
12	0.3
14	0.2
16	0.1

Example #3

What is the combined direct and reflected sound level, 2 ft from the source, if the room absorption is 22 sabins and the sound power level is 100 db?

$$L_d = L_w - 20 \log r - 0.6$$
$$L_d = 100 - 20 (\log 2) - 0.6$$
$$L_d = 100 - 6 - 0.6$$
$$L_d = 93.4$$
$$L_r = L_w - 10 \log a + 16.4$$
$$L_r = 100 + 3$$
$$L_r = 103$$

From Table 13–2 we find that for a difference of 9.6 (103 − 93.4) we add to the larger value 0.55 db.

$$L_d + L_r = 103.55$$

At 50 ft:

$$L_d = 100 - 17.5$$
$$L_d = 82.5$$
$$L_r \text{ still equals } 103$$
$$L_d + L_r = 103 + 82.5$$
$$103 - 82.5 = 20.5$$

From Table 13–2 we find that for a difference of 16 or more we add 0.1

$$L_d + L_r = 103.1$$

So, you can see that with low sound absorption the sound level actually is higher than the sound power.

As we stated earlier, sound absorbing material does not stop sound transmission. In fact, the best sound absorber, an open window, permits all of the sound to pass through it. The mass of an enclosure determines how effective it is in stopping sound transmission. As the sound waves try to get through the enclosure, they change to heat and are dissipated.

The effectiveness of a sound barrier varies with the frequency. The reduction in sound is called the trans-mission loss (TL). The TL for a 12-in. common brick wall averages 53 db and varies as follows:

cps	125	250	500	1000	4000
TL	45	44	53	59	61

Be careful not to rely on the average rating. If you had a machine that had the bulk of its noise in the 250 frequency range, the TL would be 9 below average.

Lead is an excellent material to use for stopping sound transmission, because of its heavy mass. It is also a fine material to use when concerned with low frequency sounds. Its compliance gives it an excellent low frequency TL. When faced with a noisy piece of equipment to enclose, don't overlook the importance of sound absorbing material on the equipment side of the enclosure. Example #4 points up the importance and necessity of both a sound barrier and a sound absorber.

Example #4

A machine has a sound power of 100 db at 1000 cps and it is necessary that the noise level in an adjacent room be 30 db. The machine is in the center of a 20-ft by 10-ft-high room. Assume the room has a room absorption of 50 sabins.

$$L_d = L_w - 20 \log r - 0.6$$
$$r = 10 \text{ ft}$$
$$L_d = 100 - 20 \log 10 - 0.6$$
$$L_d = 100 - 20 - 0.6$$
$$L_d = 79.4$$
$$L_r = L_w - 10 \log a + 16.4$$
$$L_r = 100 - 16.9 + 16.4$$
$$L_r = 99.5 \text{ db}$$
$$L_d + L_r = 99.5 \text{ db}$$

TABLE 13–3
Sound Absorption Coefficients of Building Materials

Material	125 cps	250 cps	500 cps	1000 cps	2000 cps	4000 cps
Brick wall	0.02	0.02	0.03	0.04	0.05	0.06
Concrete	0.01	0.01	0.02	0.02	0.02	0.03
Wood floor	0.01	0.01	0.01	0.02	0.02	0.02
Linoleum	0.04	0.03	0.04	0.04	0.03	0.02
Gypsum plaster	0.02	0.02	0.02	0.03	0.04	0.04
½-in. acoustical plaster	0.30	0.23	0.45	0.74	0.80	0.78
Typical acoustic tile on furring	0.08	0.52	0.52	0.69	0.71	0.68

Acoustical tile values are given for illustrative purposes only. Consult manufacturer's ratings for actual values.

Extracted from *Acoustics* by L. L. Beranek. Copyright, 1954, McGraw-Hill, Inc. Used with permission of McGraw-Hill Book Company.

The difference between L_d and L_r is over 15 so use higher value.
TL = 59 db at 1000 cps for a 12-in. brick wall.
Sound level in next room = 99.5 − 59 = 40 db.

If the room absorption were 1000 sabins:

L_d = 79.4 db
L_r = 100 − 10 log 1000 + 16.4
L_r = 100 − 14.6

L_r = 85.4
The difference between L_d and L_r is 6 db so from Table 13–2 we find that we must add 1 to the higher value 85.4.
$L_d + L_r$ = 86.4 db
Sound level in next room = 86.4 − 59 = 27.4 db.

Table 13–4 gives the transmission loss for various enclosures.

TABLE 13–4*   Transmission Loss of Various Materials

Material	125 cps	175 cps	250 cps	350 cps	500 cps	700 cps	1000 cps	2000 cps	4000 cps
1 12-in. common brick, 4-in. cinder block	45	49	44	52	53	54	59	60	61
2 plaster on both sides	36	37	37	41	44	47	51	55	62
3 8-in. concrete block + 4-in. concrete block	47	49	43	43	46	50	53	54	56
4 12-in. lightweight block	13	17	16	20	22	19	20	25	30
5 2 × 4 studs, gypsum plaster on lathe, both sides	31	26	34	32	28	44	43	45	61
6 staggered 2 × 4 studs, gypsum plaster on lathe, both sides	44	44	47	48	47	50	50	52	53
7 2 × 4 studs, gypsum plaster on lathe, using spring clips	47	50	48	51	52	55	54	51	61
8 wood floor and sub floor in 2 × 8, with ½-in. plaster on fiberboard ceiling	23	28	34	44	47	52	55	54	69
9 same as 8 with second ceiling on 1 × 3 furring	31	28	32	43	45	49	48	54	79
10 same as 8 with second floor on furring	30	30	37	47	50	52	57	65	79
11 4-in. concrete slab	37	33	36	44	45	50	52	60	67
12 2½-in. thick door, rubber gasket sides and top, drop felt at sill	30	30	30	29	24	25	26	37	36
13 1¾-in. hollow core door wall fitted and sealed	14	21	27	24	25	25	26	29	31
14 20-in. × 40-in. double strength window	15		26		27		31	33	29
15 20-in. × 40-in., ¼-in. plate glass	25		33		31		34	34	32

*Extracted with permission from *Handbook of Noise Control*, Edited by Cyril M. Harris, Ph.D., McGraw-Hill Book Co. Inc. (1952).

Often a heavy wall is built around a noisy piece of equipment, and the results are not as good as predicted. When this happens, look for holes. A hole in a wall is just like a hole in a balloon. A 150-sq ft wall's TL will be reduced from 40 db to 38 db by one, 1-in. dia. hole. The gap between a door and a floor can defeat the installation of the door. A ½-in. space under a door will pass more sound than a 2-in.-thick door stops. Watch, when piercing a double wall with a pipe or duct, that you don't give sound a perfect path. Pierce the one wall, offset in the air space, and then pierce the second wall.

In example #4 we assumed that the machine was a point source of sound. Most air conditioning equipment is too large to be a point source. But all equipment acts like a point source at a distance two to three times its largest dimension. We must also consider that air conditioning equipment either sits on a floor, is suspended from a ceiling, or is on a wall. The directivity factor Q is used to correct for these reflecting surfaces. If the sound source is freely suspended in space, permitting true spherical radiation, Q = 1. A diffuser propagates a hemispherical pattern because of the single reflecting surface, and Q = 2. If the source is near two surfaces, such as a sidewall grille near the ceiling, Q = 4. The same would be true of a floor register or a fancoil unit. For a single sound source, the direct plus reflected sound will equal:

$$(L_d + L_r) = L_w - 10 \log \left[\frac{1}{\frac{Q}{4\pi r^2} + \frac{4}{R}} \right] + 10.5$$

$$r = \text{distance from sound source in ft}$$
$$R = \text{room constant}$$

$$R = \frac{s\bar{\alpha}}{1 - \bar{\alpha}}$$

When dealing with more than one source of sound, the above formula becomes:

$$(L_d + L_r) = L_w - 10 \log \left[\frac{1}{\frac{Q}{4\pi n}\left(\frac{1}{r_1^2} + \frac{1}{r_2^2} \cdots + \frac{1}{r_n^2}\right) + \frac{4}{R}} \right] + 10.5$$

$$n = \text{number of sound sources of equal sound power.}$$

The noise from a piece of equipment can be isolated by a suitable enclosure, as we have shown. But in air conditioning, we must transmit fluids from machinery, and sound can ride along a duct or pipe. A duct is a natural speaking tube and readily transmits sound. Our first problem is to either make sure that the sound does not leave the equipment room, or to attenuate the sound as the air moves down the duct.

Table 13–5 gives the attenuation of sheet metal ducts at various octave band centers. For example, a

TABLE 13–5
Approximate Natural Attenuation in Bare Rectangular Sheet Metal Ducts[a]

Duct	Size, in.	Octave Band Center, cps			
		53 63	106 125	212 250	Above 250
		Attenuation, db/ft			
Small	6 × 6	0.2	0.2	0.15	0.1
Medium	24 × 24	0.2	0.2	0.1	0.05
Large	72 × 72	0.1	0.1	0.05	0.01

[a] If a duct is covered with thermal insulating material, attenuation will be approximately twice the listed values.

Reprinted by permission from *ASHRAE Guide and Data Book,* 1965.

24 × 24 duct will attenuate 0.05 db/ft at frequencies above 250 cps. So if our sound power at 500 cps is 75 db, we will attenuate 5 db in a 100-ft duct. As we will show later, lining the duct with 1-in. sound insulation will increase the attenuation from 5 db to 50 db. Often forgotten is that a duct is the same as a room. While we are attenuating the noise in the duct we do not stop the sound transfer through the sides of the duct. If this noise will cause a problem, then the duct sides must have mass added to them to increase the TL. Usually, gypsum board is used on the outside of the duct.

Duct fittings, such as elbows and divided-flow fittings, will provide some attenuation. The use of turning vanes may provide regenerated noise. Table 13–6 gives the attenuation provided by radius elbows, both round and square cross section. Table 13–7 gives the attenuation of square elbows without turning vanes. The sound power generated by turning vanes is given

TABLE 13–6
Approximate Attenuation of Round Elbows

Diameter or dimensions in.	Octave Band Center, cps						
	106 125	212 250	425 500	850 1000	1700 2000	3400 4000	6900 8000
	Attenuation, db						
5–10	0	0	0	1	2	3	3
11–20	0	0	1	2	3	3	3
21–40	0	1	2	3	3	3	3
41–80	1	2	3	3	3	3	3

Reprinted by permission from *ASHRAE Guide and Data Book,* 1965.

in Table 13-8 for 2000 fpm duct velocity. Use Table 13-9 to adjust these values for other velocities. In Table 13-10, the reduction in sound power is given for branch takeoffs.

How do you use all these tables? Let's do a simple sample problem.

Example #5

A room requires a sound level which does not exceed NC 30. Using the duct layout shown in Fig. 13-4, determine first what the sound level is and what can be done to reduce it if necessary. The room is 15 ft × 15 ft × 10-ft high. The room has an $\overline{\alpha}$ of 0.2 which classifies it as an average room (Table 13-11).

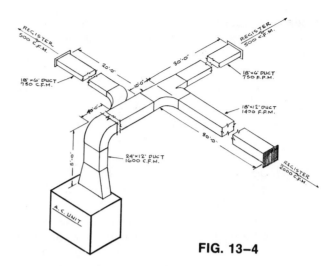

FIG. 13-4

TABLE 13-7   Attenuation of Square Elbows without Turning Vanes,[a] in Decibels

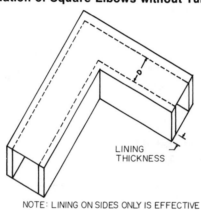

NOTE: LINING ON SIDES ONLY IS EFFECTIVE FOR ELBOW ATTENUATION

	Octave Band Center, cps															
	53	106	212	425	850	1700	3400	6900	63	125	250	500	1000	2000	4000	8000
(A) No Lining																
5″ Duct width (D)	..	..	..	..	4	8	5	3	..	..	..	1	5	7	5	3
10″ Duct width	..	..	..	4	8	5	3	3	..	..	1	5	7	5	3	3
20″ Duct width	..	..	4	8	5	3	3	3	..	1	5	7	5	3	3	3
40″ Duct width	..	4	8	5	3	3	3	3	1	5	7	5	3	3	3	3
(B) Lining* Ahead of Elbow																
5″ Duct width	..	..	..	..	4	8	6	7	..	..	..	1	5	8	6	8
10″ Duct width	..	..	..	4	8	6	7	11	..	..	1	5	8	6	8	11
20″ Duct width	..	..	4	8	6	7	11	11	..	1	5	8	6	8	11	11
40″ Duct width	..	4	8	6	7	11	11	11	1	5	8	6	8	11	11	11
(C) Lining* After Elbow																
5″ Duct width	..	..	..	..	4	11	10	10	..	..	..	1	6	11	10	10
10″ Duct width	..	..	..	4	11	10	10	10	..	..	1	6	11	10	10	10
20″ Duct width	..	..	4	11	10	10	10	10	..	1	6	11	10	10	10	10
40″ Duct width	..	4	11	10	10	10	10	10	1	6	11	10	10	10	10	10
(D) Lining* Ahead of and After Bend																
5″ Duct width	..	..	..	..	4	11	13	15	..	..	..	1	6	12	14	16
10″ Duct width	..	..	..	4	11	13	15	18	..	..	1	6	12	14	16	18
20″ Duct width	..	..	4	11	13	15	18	18	..	1	6	12	14	16	18	18
40″ Duct width	..	4	11	13	15	18	18	18	1	6	12	14	16	18	18	18

*Based on lining extending for a distance of at least two duct widths "D" and lining thickness of 10% of duct width "D." For thinner lining, lined length must be proportionally longer.
[a] For square elbows with short turning vanes, use average between Tables 13-6 and 13-7.
Reprinted by permission from *ASHRAE Guide and Data Book*, 1965.

Before we can select a diffuser we must determine what the maximum sound power level of the diffuser may be at NC 30.

Octave Band Midpoint	106	212	425	850	1700	3400
NC30	51	43	37	32	30	28
Room Effect	4	5	6	6	6	6
	55	48	43	38	36	34

So, be sure your diffuser does not exceed the above values. The manufacturer either shows this in his catalogue, or makes it available on request.

Now that we know what the maximum sound level is at the diffuser, it stands to reason that we should be at least 2 db lower in the duct so that the combined levels will not exceed the maximum allowable.

From the fan manufacturer, we get the fan's sound spectrum.

Octave Band Midpoint	106	212	425	850	1700	3400
Fan	82	82	77	72	67	62

TABLE 13–8

Approximate Levels of Sound Power Generated by Air Flowing Through Elbows[a] With Turning Vanes at 2000 fpm

Duct Size Sq Ft	Octave Band Center, cps						
	106	212	425	850	1700	3400	6900
0.25	52	52	51	48	44	38	31
0.5	57	56	55	51	47	41	34
1	62	60	58	54	50	44	37
2	73	66	62	57	53	47	40
	125	250	500	1000	2000	4000	8000
0.25	52	52	50	47	43	36	29
0.5	57	56	54	50	45	39	32
1	62	60	57	53	49	42	35
2	71	65	61	56	52	45	38

[a]Valid for 30 to 90 deg elbow turn angles.
Reprinted by permission from *ASHRAE Guide and Data Book*, 1965.

Next, we calculate the attenuation of the bare duct to the first branch (40 ft), using Table 13-5, and the elbow, Table 13-6.

Octave Band Mid Point	106	212	425	850	1700	3400
Duct	8	8	4	2	2	2
Elbow	0	2	5	4	3	3
Total	8	10	9	6	5	5

TABLE 13–10

Power Level Division at Branch Take-Offs

Area of Continuing Duct in Percent of the Total Area of all Ducts after Branch Take-Off	5	10	15	20	30	40	50	80
Decibels to be subtracted from power level before take-off in order to get power level in continuing duct	13	10	8	7	5	4	3	1

Reprinted by permission from *ASHRAE Guide and Data Book*, 1965.

TABLE 13–11

Typical Room Surface Absorption Ratings

Type of Room	
Radio & TV Studios Theaters, Lecture Halls	Soft $\bar{\alpha} = 0.4$
Concert Halls, Stores, Restaurants, Offices, Conference Rooms, Hotel Rooms, School Rooms, Hospitals, Private Homes, Libraries, Business Machine Rooms, Churches (Protestant)	Average $\bar{\alpha} = 0.2$
Large Churches (Catholic) Gymnasiums Factories	Hard $\bar{\alpha} = 0.1$

Extracted by permission from *ASHRAE Guide and Data Book*, 1965.

TABLE 13–9

Effect of Air Flow Velocity on Noise Generated by Air Flow in Duct Fittings[a]

Velocity, fpm	800	1000	1200	1500	2000	2500	3000	4000
Decibels to be added to noise generated at 2000 fpm	−24	−18	−13	−7.5	0	+6	+10	+18

[a]To interpolate between velocity values listed, add 1 db for each 4 percent increase in velocity.
Reprinted by permission from *ASHRAE Guide and Data Book*, 1965.

Subtract these values from the fan sound power and:

82	82	77	72	67	62
−8	−10	−9	−6	−5	−5
74	72	68	66	62	57

The sound power generated by the turning vanes corrected to 1600 fpm is:

63	56	52	47	43	37

The difference in sound power between the fan and the turning vanes is large enough so that the value to be added to the higher sound power is negligible.

At the first branch, we take off 500 cfm. The area of the main is 2 sq ft before and after the branch. The area of the branch is 0.75 sq ft. The total of the branch and the main is $2 + 0.75 = 2.75$ sq ft. The ratio of the branch to the branch plus main is $\frac{0.75}{2.75} \times 100 = 27\%$.

The attenuation from Table 13–10 is about 5 db.

PWL before Takeoff	74	72	68	66	62	57
Takeoff	−5	−5	−5	−5	−5	−5
PWL after Takeoff	69	67	63	61	57	52

Next, subtract the attenuation for 20 ft of bare duct.

69	67	63	61	57	52
−4	−2	−1	−1	−1	−1
65	65	62	60	56	51

The attenuation for the end reflection is taken from Fig. 13–5 for a directivity factor Q = 2.

	65	65	62	60	56	51
	−9	−5	−1	0	0	0
Actual PWL	56	60	61	60	56	51

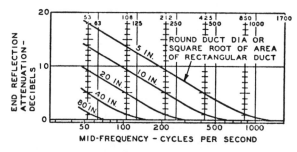

(a) Above curves based on directivity factor, Q = 2, when opening is in a room surface several feet from other room boundaries.
(b) If opening is at the junction of two room surfaces, Q = 4, end reflection attenuation will be reduced to that of a duct with twice the area. (1.4 times the radius or square root of area.)
(c) For other values of Q, multiply actual area by Q/2 for equivalent area to be used in this figure.

FIG. 13–5 Sound Reflection At Air Outlet (Reprinted by permission from *ASHRAE Guide and Data Book*, 1965)

Allowable PWL	53	46	41	36	34	32
Additional Attenuation Required	3	14	20	24	22	19

We can try lining the ductwork, including the square elbow, from the fan outlet to 15 ft past the elbow to see if we get enough attenuation. First, let's check the effect of lining the elbow.

Table 13–7 is based on a duct lining equal to 10% of duct width. If we use 1-in.-thick lining we must reduce Table 13–7 values by $1/2.4 = 0.416$. Then, average this reduced elbow value with the value from Table 13–6.

Lined Elbow	0	4	11	13	15	18
Multiplied by 0.416	0	1.66	4.4	5.2	6	7.2
Averaged with Table 13–6	0	1.33	3.2	3.6	4.5	5.1

To get the value of lining the duct we must use Fig. 13–6, Table 13–13, and the formula:

$$\text{Attenuation} = 12.6\,(\text{L})\,\frac{(P)}{(A)}\,\alpha^{(1.4)}$$

L = length of duct = 40 ft
P = perimeter of duct (inside) in inches = 72 in.
A = clear inside area in square inches = 288 sq in.

$$\text{Attenuation} = 12.6\,(40)\,\frac{(72)}{(288)}\,\alpha^{(1.4)}$$

$$\text{Attenuation} = 126\,\alpha^{(1.4)}$$

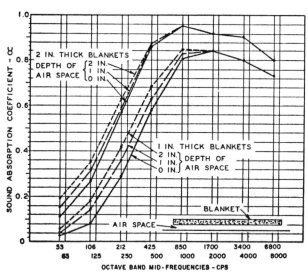

FIG. 13–6 Approximate Sound Absorbing Coefficients (Reprinted by permission from *ASHRAE Guide and Data Book*, 1965)

TABLE 13–12

Range of Design Goals for Air Conditioning System Sound Control

Type of Area	A-Sound Level Decibels			NC Level Decibels		
	Low	Average	High	Low	Average	High
RESIDENCES						
Private homes (rural and suburban)	25	30	35	20	25	30
Private homes (urban)	30	35	40	25	30	35
Apartment houses, 2- and 3-family units	35	40	45	30	35	40
HOTELS						
Individual rooms or suites	35	40	45	30	35	40
Ball rooms, Banquet rooms	35	40	45	30	35	40
Halls and corridors, Lobbies	40	45	50	35	40	45
Garages	45	50	55	40	45	50
Kitchens and laundries	45	50	55	40	45	50
HOSPITALS AND CLINICS						
Private rooms	30	35	40	25	30	35
Operating rooms, Wards	35	40	45	30	35	40
Laboratories, Halls and corridors / Lobbies and waiting rooms	40	45	50	35	40	45
Washrooms and toilets	45	50	55	40	45	50
OFFICES						
Board room	25	30	35	20	25	30
Conference rooms	30	35	40	25	30	35
Executive office	35	40	45	30	35	40
Supervisor Office, Reception room	35	40	50	30	35	45
General open offices, Drafting rooms	40	45	55	35	40	50
Halls and corridors	40	50	55	35	45	55
Tabulation and computation	45	55	65	40	50	60
AUDITORIUMS AND MUSIC HALLS						
Concert and opera halls / Studios for sound reproduction	25	30	35	20	22	25
Legitimate theaters, Multi-purpose halls	30	35	40	25	27	30
Movie theaters, TV audience studios / Semi-outdoor amphitheaters / Lecture halls, planetarium	35	40	45	30	32	35
Lobbies	40	45	50	35	40	45
CHURCHES AND SCHOOLS						
Sanctuaries	25	30	35	20	25	30
Libraries	35	40	45	30	35	40
Schools and classrooms	35	40	45	30	35	40
Laboratories	40	45	50	35	40	45
Recreation halls	40	45	55	35	40	50
Corridors and halls	40	50	55	35	45	50
Kitchens	45	50	55	40	45	50
PUBLIC BUILDINGS						
Public libraries, Museums, court rooms	35	40	45	30	35	40
Post offices, General banking areas, Lobbies	40	45	50	35	40	45
Washrooms and toilets	45	50	55	40	45	50
RESTAURANTS, CAFETERIAS, LOUNGES						
Restaurants	40	45	50	35	40	45
Cocktail lounges	40	50	55	35	45	50
Night clubs	40	45	50	35	40	45
Cafeterias	45	50	55	40	45	50

TABLE 13–12
Range of Design Goals for Air Conditioning System Sound Control *(Continued)*

Type of Area	A-Sound Level Decibels			NC Level Decibels		
	Low	Average	High	Low	Average	High
STORES RETAIL						
Clothing stores	40	45	50	35	40	45
Department stores (upper floors)						
Department stores (main floor)	45	50	55	40	45	50
Small retail stores						
Supermarkets	45	50	55	40	45	50
SPORTS ACTIVITIES INDOOR						
Coliseums	35	40	45	30	35	40
Bowling alleys, gymnasiums	40	45	50	35	40	45
Swimming pools	45	55	60	40	50	55
TRANSPORTATION (RAIL, BUS, PLANE)						
Ticket sales offices	35	40	45	30	35	40
Lounges and Waiting rooms	40	50	55	35	45	50
MANUFACTURING AREAS						
Foreman's office	45	50	55	40	45	50
Assembly lines, Light machinery	50	60	70	45	60	70
Foundries, heavy machinery	60	70	80	55	65	75

Reprinted by permission from *ASHRAE Guide and Data Book,* 1965.

From Fig. 13–6, we find that 1-in. insulation with no air space has the following values of α:

Midpoint cps	106	212	425	850	1700	3400
	0.08	0.25	0.58	0.8	0.84	0.80

We can get $\alpha^{1.4}$ from Table 13–13, and multiplying these values by 126, we get the following attenuation:

$$\text{Attenuation} = 126\alpha^{(1.4)}$$

For example, at 106 cps:

$$\text{Attenuation} = 126\,(0.08)^{1.4}$$
$$\text{Attenuation} = 5\text{ db}$$

Repeating for each band we get:

Midpoint cps	106	212	425	850	1700	3400
Attenuation	5	17	22	50	50	50
Elbow	0	1	3	4	6	7
	5	18	25	54	56	57
Required Attenuation	3	14	16	21	19	14

As you can see, we ended up with too much attenuation in the higher frequencies to get enough in the 106 cps and 212 cps mid-band point.

PACKAGE SOUND ATTENUATORS

Package sound attenuators, as shown in Fig. 13–8, will provide good attenuation in a short distance. As a practical matter, it is my feeling that the sound power level should be quite close to the acceptable level by the time the ducts leave the equipment room. I would advise either sufficient duct lining, or the use of a package attenuator on the fan or unit discharge and suction connections.

It is important in selecting a sound attenuator to be certain that the regenerated noise level of the attenuator does not offset the expected attenuation. Be certain that the manufacturer furnishes you with charts

TABLE 13–13
Values of $(\alpha)^{1.4}$

Absorption Coefficient α	0.10	0.15	0.20	0.25	0.30	0.35	0.40	0.45	0.5	0.6	0.7	0.8	0.9	1.0
$(\alpha)^{1.4}$	0.04	0.07	0.11	0.14	0.19	0.23	0.28	0.33	0.38	0.49	0.61	0.73	0.86	1.0

Extracted by permission from *ASHRAE Guide and Data Book,* 1965.

that give the attenuation WITH AIR FLOWING and the regenerated noise. Tables 13–14 and 13–15, furnished by the Industrial Acoustics Company, give the attenuation and noise regeneration, respectively.

We require the following attenuation:

Octave Band	2	3	4	5	6	7
Midpoint cps	106	212	425	850	1700	3400
Required Attenuation	3	14	16	21	19	14

From Table 13–14, for model 3L 1600 fpm velocity, interpolating between 0 and 2000 fpm we get, to the nearest whole number:

Octave Band	2	3	4	5	6	7
Attenuation	5	7	10	18	24	19

This provides sufficient attenuation; but let's check the regenerated noise from Table 13–15A, again interpolating:

Octave Band	2	3	4	5	6	7
Regenerated Noise	51	44	45	45	47	48

Subtract the adjustment factor from Table 13–15B:

51	44	45	45	47	48
−3	−3	−3	−3	−3	−3
48	41	42	42	44	45

The allowable PWL is:

53	46	41	36	34	32

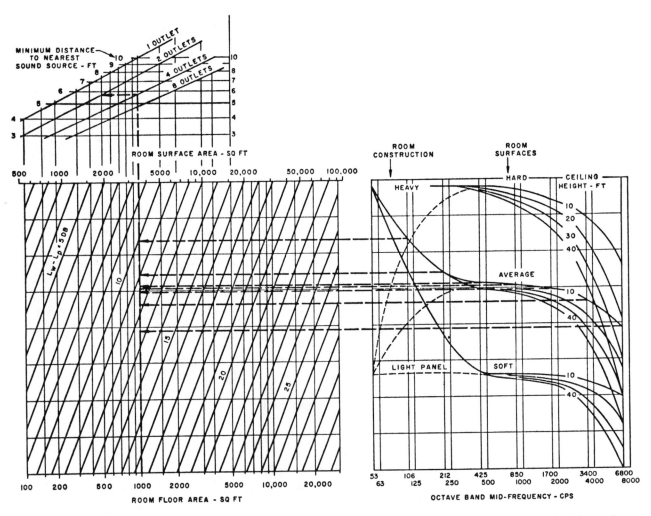

FIG. 13–7 Chart for Estimating L_w-L_p (Reprinted by permission from *ASHRAE Guide and Data Book,* 1965)

**Quiet-DUCT® silencer
for rectangular applications**

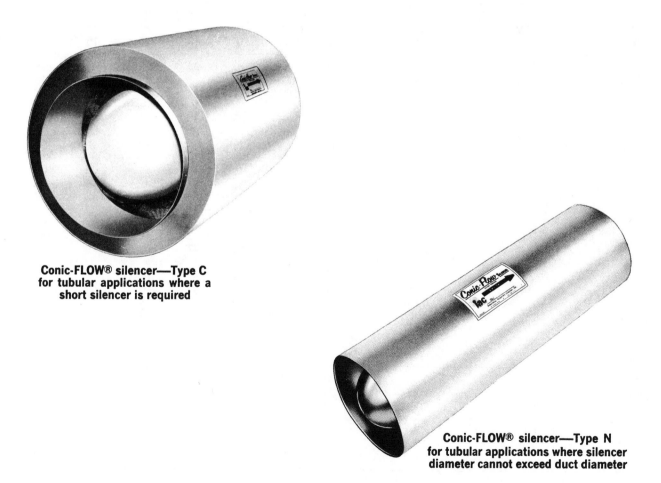

**Conic-FLOW® silencer—Type C
for tubular applications where a
short silencer is required**

**Conic-FLOW® silencer—Type N
for tubular applications where silencer
diameter cannot exceed duct diameter**

FIG. 13–8   Typical Package Sound Attenuator (Courtesy: Industrial Acoustics Co.)

TABLE 13–14

Dynamic Insertion Loss (D.I.L.) Ratings Quiet-DUCT and Conic-FLOW Silencers

Config-uration	IAC SILENCER MODEL	Octave Band Mid-Frequencies, CPS	106	212	425	850	1700	3400	6900
		Silencer Face Velocity, FPM	Dynamic Insertion Loss in Decibels						
RECTANGULAR Quiet-DUCT SILENCERS	3L	0	7	9	12	20	25	20	18
		2000	5	7	10	18	24	19	17
		5000	4	6	9	17	23	18	16
	5L	0	10	14	22	31	39	30	23
		2000	7	13	18	29	37	29	21
		5000	6	11	17	26	36	28	20
	7L	0	17	20	30	43	48	37	29
		2000	13	18	28	41	46	35	26
		5000	9	15	24	37	44	33	25
	3M	0	10	12	14	16	16	13	13
		2000	8	10	12	14	15	12	12
		4000	7	9	11	13	14	11	11
	5M	0	15	20	24	24	23	17	17
		2000	10	15	17	20	21	16	16
		4000	8	14	16	19	20	15	15
	7M	0	19	25	29	31	29	21	19
		2000	13	20	23	26	28	20	18
		4000	11	19	22	25	27	19	17
	3S	0	14	18	23	30	37	35	30
		1000	13	16	21	27	35	33	28
		2000	10	15	20	26	33	32	26
	5S	0	20	29	40	50	52	48	37
		1000	18	24	35	44	48	44	34
		2000	17	21	32	41	47	42	31
	7S	0	24	37	50	54	55	52	42
		1000	21	35	45	50	52	50	40
		2000	18	32	43	47	51	49	39
TUBULAR Conic-FLOW SILENCERS	CL	0	10	14	19	26	32	27	23
		2000	8	12	16	24	30	24	20
		8000	6	10	14	22	28	22	18
	CS	0	14	18	27	36	38	34	27
		2000	12	16	23	34	35	28	22
		6000	10	14	22	32	32	26	23
	NL	0	10	12	17	18	15	12	11
		2000	9	11	15	16	14	11	10
		6000	8	10	13	14	13	10	9
	NS	0	12	15	22	26	23	16	16
		2000	10	14	18	24	21	12	12
		6000	9	13	16	23	20	10	10

(Courtesy: Industrial Acoustics Co. Inc.)

TABLE 13–15a

Self-Noise Power Levels, db re 10⁻¹² Watts, Quiet-DUCT and Conic-FLOW Silencers

Configuration	IAC SILENCER MODEL	Octave Band Mid-Frequency, CPS	106	212	425	850	1700	3400	6900
		Silencer Face Velocity, FPM	Self-Noise Power Levels in Decibels						
RECTANGULAR Quiet-DUCT SILENCERS	3L / 5L / 7L	500	25	*	*	*	20	21	*
		750	34	25	25	28	30	30	26
		1000	41	32	33	35	36	37	33
		1500	50	43	44	44	46	47	42
		2000	57	50	51	51	53	54	49
		3000	66	61	62	61	61	64	58
		4000	73	68	69	68	69	70	66
		6000	82	79	80	77	79	80	76
	3M / 5M / 7M	500	29	*	20	22	23	25	*
		750	38	29	31	32	33	35	29
		1000	44	37	38	38	40	42	36
		1500	53	47	49	48	50	51	45
		2000	60	55	56	54	57	58	52
		3000	69	65	67	64	66	68	62
		4000	75	72	74	72	73	75	69
	3S / 5S / 7S	500	46	35	30	30	31	30	29
		750	54	45	40	40	41	40	39
		1000	62	53	48	47	47	47	45
		1500	70	63	58	56	57	56	55
		2000	77	70	65	63	63	63	62
TUBULAR Conic-FLOW SILENCERS	CL	500	28	22	*	*	*	*	*
		750	36	31	27	26	23	*	*
		1000	41	36	33	33	30	23	*
		1500	49	45	42	42	40	34	28
		2000	54	50	48	49	48	42	37
		3000	62	58	57	58	58	54	50
		4000	68	64	64	65	65	62	58
		6000	76	72	73	74	76	74	70
		8000	81	78	79	81	83	82	79
	CS / NS	500	33	28	24	24	20	*	*
		750	41	36	33	33	31	25	*
		1000	47	42	40	40	38	34	25
		1500	54	50	49	49	48	45	38
		2000	60	56	55	56	56	53	46
		3000	68	64	64	65	66	64	59
		4000	73	70	70	72	74	72	68
		6000	81	78	80	81	84	84	80
		8000	86	83	86	88	91	92	89
	NL	500	30	24	20	20	*	*	*
		750	38	32	30	30	25	*	*
		1000	43	38	36	36	33	27	20
		1500	51	46	45	45	43	38	32
		2000	56	52	51	52	50	46	41
		3000	64	60	60	61	61	58	54
		4000	70	66	67	68	68	66	62
		6000	78	74	76	77	78	77	75
		8000	83	80	82	84	85	85	83

*Less than 20 db.

(Courtesy: Industrial Acoustics Co. Inc.)

TABLE 13–15b

Face Area Adjustment Factors

To be added to or subtracted from TABLE 15a PWL values, as indicated below.

Quiet-DUCT Face Area—Sq. Ft.‡	.5	1	2	4	8	16	32	64	128
Conic-FLOW Face Area—Sq. Ft.	–	.75	1.5	3	6	12	24	–	–
PWL ADJUST-MENT FACTOR db	–9	–6	–3	0	+3	+6	+9	+12	+15

‡For intermediate face areas, interpolate to nearest whole number.
(Courtesy: Industrial Accoustics Co. Inc.)

As you can see, we exceed the allowable values in the 4, 5, 6, and 7 bands. If we use a unit with a lower face velocity, such as 750 fpm (4 sq ft face area), we will have:

Octave Band	2	3	4	5	6	7
Regenerated Noise	34	25	25	28	30	30

The face adjustment value for 4 sq ft is 0. But, we are still not 5 db lower in the seventh band. However, when we are this close, we can go ahead and depend on the ductwork system to pick up the small amount required. Remember that the elbow and the bare duct will pick up 5 db in the seventh band.

With these last words on sound we come to the end of the book. I hope I have given you the foundation you need to start your education in the design of comfort control systems. What next? Contact every manufacturer you know and get all the literature he has. Get his catalogues and any training aids he has. Read, study, devour, and you will learn. Subscribe to some of the many journals available in our field and read, read, read. Good luck.

CHAPTER 14

Air Cleaners

THERE WAS A TIME when one merely opened the window to purge the room with clean, fresh, outside air. There are still places in our world where outside air is clean, fresh air but these are not the places where buildings are built.

Outside air and inside air contain a variety of contaminants. These contaminants are a complex mixture of smokes, mists, fumes, dusts and fibers.

   a. Smokes are a suspension of small particles that result from a combustion process. These particles are mostly carbon but many other particles can be present.

   b. Mists are a dispersion of liquid particles which result from the shattering of a liquid. Fogs are mists that are the result of a meteorological phenomenon.

   c. Fumes are oxides of metals formed by the condensation of metal vapors.

   d. Dusts are dry granular particles that are larger than colloidal and capable of temporary suspension in air or other gases. Dust particles are irregular in shape and are not spherical. They result from the attrition of a larger mass. If one cubic centimeter of quartz is crushed into particles one cubic micron in size, we will have 10^{12} (1,000,000,000,000) particles with a total surface area of 6 square meters.

I have just used a unit of measurement called the micron. A micron is equal to 0.001 millimeter or 0.000001 meter. So, as you can see, this is darn small. To give a better idea of how small a micron is—a human hair has a diameter of 100 microns. The abbreviation for micron is the Greek letter μ (mu).

The size of contaminants in the air will range in size from $0.01\,\mu$ to the size of leaves and insects. This makes it difficult, if not impossible, to design a single air filter that will stop all contaminants. A typical sample of atmospheric air will contain not only smokes, mists, fumes and dust but also lint, plant fibers, bacteria, spores and pollens. Obviously, outside air is not necessarily clean air and filters are necessary.

Normally, when you look across a room, you can see no contamination. But have you ever noticed what a ray of sunshine brings out? When a sunbeam enters a room you see thousands of particles dancing in the air. And yet, there are many more that you cannot see.

The following will give you an idea of the sizes of typical contaminants:

1.	Human hair	$40\,\mu$ to $300\,\mu$
2.	Oil smokes	$0.03\,\mu$ to $1\,\mu$
3.	Fertilizer	$10\,\mu$ to $1,000\,\mu$
4.	Tobacco smoke	$0.01\,\mu$ to $1\,\mu$
5.	Coal dust	$1\,\mu$ to $100\,\mu$
6.	Beach sand	$100\,\mu$ to $2,000\,\mu$
7.	Plant spores	$10\,\mu$ to $30\,\mu$
8.	Pollens	$10\,\mu$ to $1,000\,\mu$
9.	Typical atmospheric dust	$0.001\,\mu$ to $30\,\mu$

Filters collect contaminants in a number of ways: impingement or inertial impaction, straining, diffusional effects and electrostatic effects.

Impingement or inertial impaction filters make use of inertia. The air stream is made to change direction quickly. The air stream changes direction as it comes in contact with the fibers in a filter but the heavier contaminant particles do not. The contaminant impinges on the filter fibers and is held on the fiber by a viscous coating on the fiber. This type of filter is most effective on large particles but does collect some small particles.

TABLE 14-1

Size Distribution of a Typical Atmospheric Dust Sample

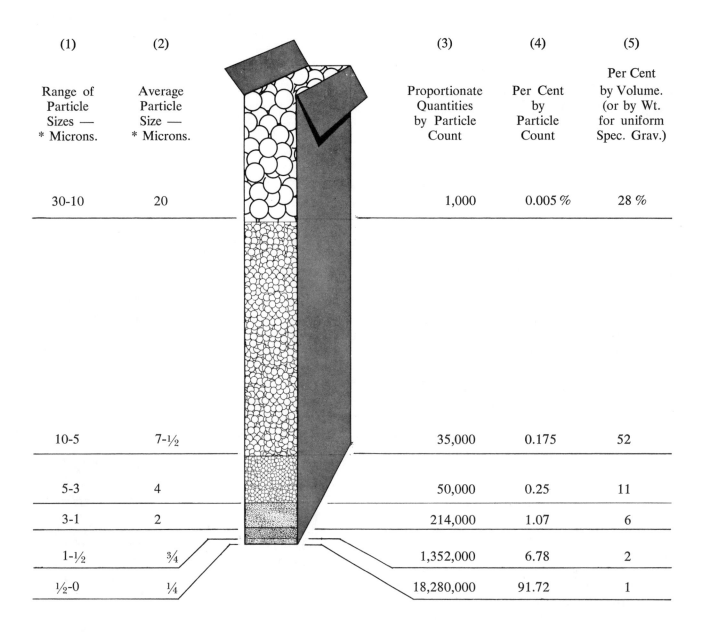

(1) Range of Particle Sizes — * Microns.	(2) Average Particle Size — * Microns.	(3) Proportionate Quantities by Particle Count	(4) Per Cent by Particle Count	(5) Per Cent by Volume. (or by Wt. for uniform Spec. Grav.)
30-10	20	1,000	0.005 %	28 %
10-5	7-½	35,000	0.175	52
5-3	4	50,000	0.25	11
3-1	2	214,000	1.07	6
1-½	¾	1,352,000	6.78	2
½-0	¼	18,280,000	91.72	1

It should be pointed out that atmospheric dust varies considerably in particle size as well as constituents. In the above sample there were very few particles noted which were in excess of 30 microns in average diameter. With this as an upper limit, the particles were divided into six size ranges as indicated, with Column (2) indicating the average particle size for each group. For example the largest group consisted of those particles ranging between 30 and 10 microns — or an average of 20. In this particular size range it will be noted that the number of particles present is indicated as 1000 as shown in Column 3. This represents the proportionate quantities by count and simply indicates the relative number of particles in each size range based upon 1000 particles for the average 20 micron size.

Straining works as long as the strainer is smaller than the particle. A one-inch mesh screen is a very good strainer for golf balls but not too effective for flies. A typical fly screen keeps out insects but permits pollen, dirt, etc., to pass. It is possible to provide a strainer small enough to remove 0.01 μ particles but at the cost of a large pressure drop.

Diffusional effects are caused by the action of the random motion of air particles. Air molecules continuously bombard contaminant particles and move them toward the filter fibers where they are captured. This is only successful with very small particles.

Electrostatic effects are used in electronic air cleaners. The particles are ionized as they pass through an electric field and all given the same charge. Collector plates are maintained at an opposite charge and, because opposites attract, the particles are collected on the plates.

The biggest problem in filter selection and evaluation has been the confusion arising from the variety of test methods used. A particular filter can have an efficiency ranging from 20% to 80% depending on which test is used. This confusion results from the complex nature of the makeup of atmospheric dust. As you can see in Table 14-1, a typical sample will, if analyzed by weight, consist mostly of large size particles. On the other hand, if we consider the number of particles, 99% of the particles are less then 3 μ in size. So, if you have a filter which is 95% efficient in collecting particles 10 μ or larger, then its efficiency by weight will be 80% or better. However, it has permitted 99% of the particles to pass through. And herein lies the problem. Most dirt streaking, which is attributed to heating and air conditioning systems, is caused by these small particles.

ASHRAE, in 1968, integrated the various test methods in use into the ASHRAE *Test Standard for Air Cleaning Devices.* There are some eight tests in this standard. However, we will only discuss the three of most interest to comfort air conditioning: weight arrestance, dust spot efficiency and DOP penetration test.

The weight arrestance test states efficiency in terms of the weight of standard dust captured by the filter. Standard dust consists of 72% by weight of standardized air cleaner dust fine; 23% by weight of Molocco Black; 5% by weight of No. 7 cotton linters.

The dust spot efficiency test states efficiency in terms of discoloration. White filter paper is placed upstream and downstream of the filter being tested. The amount of light that can be transmitted through the two filters is an indication of the filter efficiency. This is really a better indication than weight of how well a filter will prevent surface soiling.

The DOP test is the best test for filters that must remove extremely small particles. A smoke cloud of uniformly sized (0.3 μ) of DOP (Di-Octyl Phthalate) is generated and introduced into the test chamber. The concentration of DOP upstream and downstream of the test filter indicates its efficiency.

TYPES OF FILTERS

Figure 14-1 shows the filter most of us are familiar with. It consists of a cardboard and light metal frame which holds the filter media. The filter media is most often spun glass fiber although it can be hemp, felt, animal fibers, metal wool and other materials. The filter media is coated with oil or other viscous adhesive and so it falls in the impingement type class. The efficiency of this filter is only 10% for dust spot and 72% by weight. The filter is most often 24″ x 24″ x 1″ deep although it is available in ½″, 2″, 3″ and 4″ depth. Initial resistance is 0.1 in. wg. (water gauge).

Figure 14-2 is another throwaway type filter. It is constructed of glass fiber mats in depths of 9″ and 15″. The efficiency of this filter is 95% by weight, 30-35% dust spot and 15-20% DOP. The initial resistance is 0.15 in. wg for the 9″ and 0.30 in. wg for the 15″.

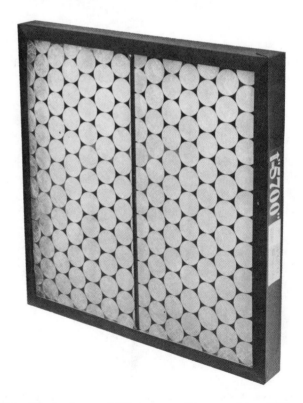

FIG. 14–1 Low Efficiency Disposable Filter (Courtesy: American Air Filter Company, Inc.)

Another dry type throwaway filter is shown in Figure 14-3. Its depth ranges from 23″ to 38″. The deeper the unit the more air it can handle. Depending on the density of the fibers, the efficiency varies from 96 to 99% by weight, 45 to 97% dust spot and 20 to 85%

DOP. The initial resistance varies from 0.15 in. wg to 0.60 in. wg depending on the filter media and filter depth.

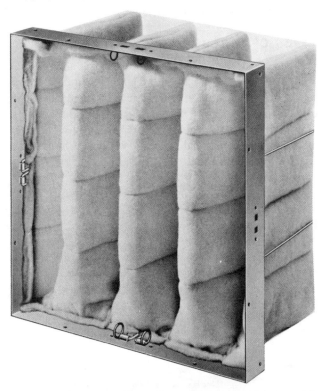

FIG. 14–2 30% Dust Spot Efficiency Disposable Filter (Courtesy: Cambridge Filter Corp.)

FIG. 14–4 HEPA Filter (Courtesy: Cambridge Filter Corp.)

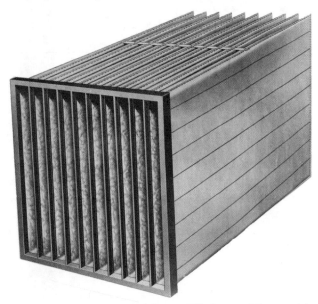

FIG. 14–3 45% Dust Spot Efficiency Disposable Filter (Courtesy: Cambridge Filter Corp.)

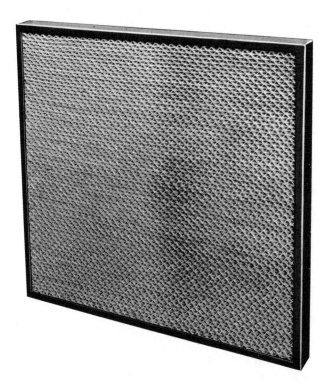

FIG. 14–5 Washable Filter (Courtesy: American Air Filter Company, Inc.)

Still classed as a throwaway filter is the filter shown in Figure 14-4. This filter is constructed of wet laid papers of mostly submicron glass and asbestos fibers. They are called HEPA (High Efficiency Particulate Air) filters and, by some, absolute filters. They were developed during WW II for the Atomic Energy Commission. They are virtually 100% efficient (99.99) by the DOP Test. The initial pressure drop is one inch wg.

Figure 14-5 shows a washable filter. This filter is constructed of crimped screen wire which is coated with an adhesive oil. The filter is cleaned by washing with steam or hot water and detergent. After cleaning, it must be coated with adhesive. The initial resistance is 0.10 in. wg. Efficiencies are generally quite low, generally less than the impingement type throwaway shown in Figure 14-1.

Renewable type filters are shown in Figures 14-6 and 14-7. The filter shown in Figure 14-6 is commonly called a roll type filter. It consists of a spool of clean media which is connected to an empty spool. When the media collects the maximum amount of contaminant it is wound on the empty spool. This is accomplished manually or by a pressure sensor or timer-actuated motor.

FIG. 14–7 Pleated Renewable Filter (Courtesy: American Air Filter Company, Inc.)

FIG. 14–6 Roll Filter (Courtesy: American Air Filter Company, Inc.)

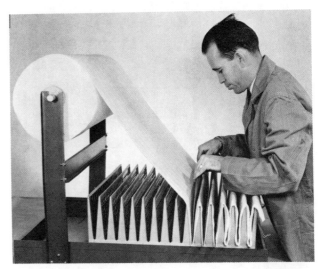

FIG. 14–8 Manual Replacement Of Filter Media In Pleated Renewable Filter (Courtesy: American Air Filter Company, Inc.)

The timer is most often used because it is by far the most foolproof. An efficiency of 80% by weight and 20% dust spot can be expected along with an initial resistance of 0.15 in. wg. The media is similar to the type used in the impingement type throwaway.

The paper used in the pleated filter is constructed of multiple layers of very thin cellulose material stitched together. Replacement of the media can be done by machine or manually as shown in Figure 14-8. The efficiency is 94% by weight and 52% dust spot. Initial pressure drop is 0.10 in. wg.

An electronic air cleaner is shown in Figure 14-9. High efficiencies and ease of maintenance are possible with this unit. The efficiency ranges from 85-95% dust spot while the initial resistance is 0.20 in. wg. Unlike other filters, the resistance remains virtually unchanged because the collected dust is continuously removed.

These filters are cleaned by an automatic or manual washing of the collector plates with a hot detergent

FIG. 14–10 Electronic Air Cleaner With Agglomerator (Courtesy: American Air Filter Company, Inc.)

FIG. 14–9 Electronic Air Cleaner (Courtesy: American Air Filter Company, Inc.)

FIG. 14–11 Activated Carbon Filter (Courtesy: Cambridge Filter Corp.)

solution. The plates must immediately be coated with an adhesive.

A prefilter, usually a throwaway type (Figure 14-1), must be used with an electronic air cleaner. Prefilters are needed to screen out large particles whose inertia would enable them to pass through the filter uncollected. The prefilters also even out the flow across the electronic filter.

In some cases several filters are used in combination. For example: in Figure 14-10 we see the filter in Figure 14-9 used with the filter in Figure 14-3. The collector plates are used without adhesive. This causes the collected contaminant to gather together into larger particles (agglomerate) which blow off and are collected by the downstream dry filter. This combination filter has a higher efficiency than either of the individual filters.

In general, high efficiency filters are used in combination with low efficiency filters to extend the life of the high efficiency filter.

We think of air cleaning only as particle removal but odor removal is also possible. Activated charcoal is often used for odor removal. Odor causing gases and vapors are adsorbed on the charcoal. It is necessary to use particle removal filters with the activated charcoal filters to extend the life of the charcoal filter. Any high efficiency filter will remove tobacco smoke from the air stream but it will not remove the irritating gases and vapors given off by tobacco. Figure 14-11 shows an activated carbon filter which can be used with a high efficiency filter.

FILTER SELECTION

The selection of a filter can only be made by considering the required or desired efficiency, first cost, replacement costs and maintenance costs. Through the courtesy of the American Air Filter Company, we present an owning and operating cost analysis for various filters (Table 14-2). This report assumes a 30,000 cfm installation operating 12 hours a day, 6 days per week. Investment costs were based on a 15-year amortization at 6% interest. Maintenance labor was assumed to average $6.75 per hour over the 15-year life.

TABLE 14-2
Filter Cost Comparison

FIG. NO.	FILTER TYPE	EFFICIENCY		MONTHS BETWEEN SERVICE PERIODS	HOURS OF SERVICE PER YR.	FIRST COST Dollars	YEARLY COSTS				COST PER 1% DUST SPOT EFF.
							Replacement		AMORTI-ZATION	TOTAL	Cents/Year/
		Weight	Dust Spot				Material	Labor			1000 CFM
14-1	Low Efficiency Throwaway	75%	20%	2	15	285.00	282.50	101.25	29.36	383.10	64
14-5	Washable	73%	18%	3	26⅔	518.50	7.25	180.00	53.40	240.65	44
14-6	Renewal Roll Type	82%	25%	12	1	967.00	106.00	6.75	99.60	212.35	28
14-9	Electronic	*	90%	½	26	4825.00	370.50	175.50	496.97	1042.97	38
14-3	High Efficiency Throwaway .	*	83%	13	2½	754.50	563.50	16.85	77.71	658.06	26
14-10	Electronic plus High Efficiency	*	97%	6†	3½	3755.00	279.00	23.60	386.76	689.36	24

* Not Applicable   † Agglomorator service only.

CHAPTER 15

Designing for Energy Conservation

ENERGY CONSERVATION has always been *good practice*. As a matter of fact, legislation has been enacted in many areas of the country to assure that energy conserving measures are implemented in new building designs. And architects, engineers, manufacturers and installers must all strive to do their part in this energy conservation movement.

The ASHRAE standard 90-80, for new buildings, and the 100 Series, for existing buildings, are preliminary steps towards formally recognizing energy conservation measures in building design. Both of those documents are actually step-by-step prescriptions for conserving energy. And some have criticized these documents for just that reason. Many of the critics don't want a prescriptive standard but prefer a performance standard.

At present, there is insufficient data available to determine exactly what constitutes reasonable yearly energy usage. Consequently, it's not feasible to construct a valid performance standard. At the same rate, the ASHRAE 90-80 can be used to determine what constitutes reasonable energy consumption for heating, cooling and lighting a building. In other words, a reasonable performance standard is constructed each time we use the 90-80. We are then free to make any architectural changes to the building as long as they do not increase the energy consumption over what the 90-80 prescribes.

One of the most prominent questions posed in the conservation controversy is: must comfort be sacrificed to conserve energy? The answer, no. We can design a system that will be comfortable and still conserve both money and energy by using the technology that's readily available to us. As a refresher, it might help to go back and re-read Chapter 2 so that you'll have comfort and its variables in mind before reading any further.

What can we do to conserve energy?
• Reduce space temperatures.
• Insulate.
• Reclaim waste heat.
• Examine new architectural concepts.
• Consider new heating systems.

In the average 1,500 ft² home, lowering the temperature to 68 °F in a 6,000 degree-day area will save 390 therms of gas at a dollar savings of approximately $175.50. A therm is 100,000 Btus; gas is estimated to cost $0.45 per therm. Lets's determine how much we can save if we take additional steps that will allow us to maintain a comfortable environment.

First of all, let's take a home that was built prior to 1975. Its perimeter was probably ⅓ glass, and its walls uninsulated — except for insulating sheathing. The attic insulation was probably two inches thick. As stated earlier, this home should save on the order of 390 therms — provided the temperature is maintained at 65 °F instead of 75 °F. If we insulate the house with 3½" of insulation in the walls and increase the attic insulation to 6", we'll save an additional 390 therms, and the temperature can then be kept at 75 °F. The installed cost of this insulation should only run about $500. This extra insulation should render a savings of about $175.50/year.

Next, let's look at a new house that will only have 15% glass and is built with 2" × 6" studs which enables us to use 5½" of insulation. In addition, we will use 1" styrofoam for sheathing. The attic will have 12" of insulation. This house will use 730 therms less than the poorly insulated house and will still have a 75 °F space temperature. Will this cost more? Probably not. The increased cost of 6" studs and the additional insulation will be offset by reduced costs for the heating and air conditioning systems.

Further savings can be generated by changing the architectural design. Do all parts of a house need to be above ground? We don't need large windows in sleeping quarters, so why not put the bedrooms below ground. This relocation can tentatively reduce gas consumption by 480 therms.

The use of earth berms or heavy masonry can also keep a home cooler in summer. Due to the mass of the wall, the heat from the sun never penetrates the interior living area. And by the time the temperature of the wall is high enough to add heat to the interior, the sun sets and the outdoor temperature drops. The wall will then reject heat. If we combine wall mass with shaded glass and forced ventilation of attics, the interior temperature should be less then 80 °F — even on a day when the outdoor temperature reaches 95 °F. But this can pose a problem in humid areas because of uncomfortably high humidity.

Our everyday activities create moisture that must be removed in order to maintain a comfortable relative humidity. For example, cooking for a family of four can add 4.86 pounds of moisture/day to a home. Washing dishes can add another 3.6 pounds. If each family member takes a shower, another 2 pounds can enter the home. These three activities alone can add 10.46 pounds of water/day. If the home started with a morning temperature of 70 °F and 50% relative humidity, by late afternoon the house could have a relative humidity of close to 100%. We have two alternatives to help prevent this: we can open windows, that will also admit heat, or we can add a dehumidifier. A dehumidifier is the best bet. It uses much less power than an air conditioner. Actually, these homes have such a small cooling load that a combination air conditioner/dehumidifier can be used with good results.

Let's consider windows for a moment. These are the greatest source of heat loss and heat gain in a building. In a well insulated home, with a perimeter that is 15% glass, the windows lose twice as much heat as the walls. In other words, 15% of the perimeter has twice the heat loss of the remaining 85%. Logically, we should examine the windows to determine what can be done to reduce their heat loss and heat gain. One solution is to use an insulated panel, similar to a sectional garage door, that can be kept in the attic and lowered to insulate the windows at night, on rainy days and/or when the house is empty. Another possibility is the use of insulated bifold doors that could be drawn across the window when desired. Even insulated draperies can help.

FIGURE 15-1

U_o Walls – Type "B" – Buildings – Heating

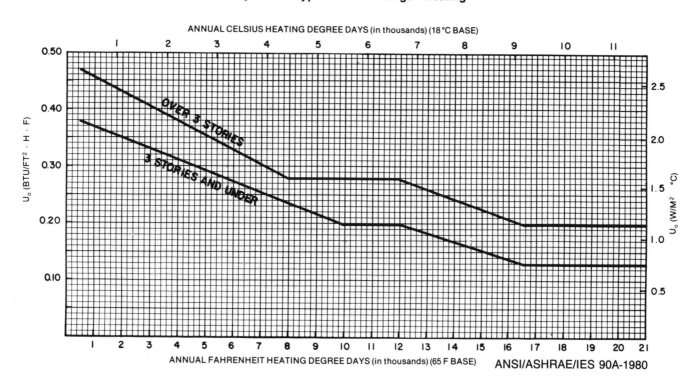

ANSI/ASHRAE/IES 90A-1980

Let's put together the ideal combination of glass, wall and insulation to reduce the heat loss and heat gain to a practical minimum. Selecting heat transmission coefficients is not as simple as it once was. Until energy became a primary consideration in building design, an architect merely selected a wall, window or roof that he felt was best suited for his building. More often than not, little consideration was given to the insulating value of a building element. ASHRAE standard 90-80 makes it simple to select the proper combination of wall, window and proper combination of wall, window and roof to achieve optimum energy savings. (Refer to Fig. 15-1 to select the proper overall U value for a wall window combination.)

Let's go back to the clinic used in Example #19 found in Chapter 3 to see how close it comes to complying with 90-80. If it doesn't comply, let's examine what can be done to assure compliance with 90-80. From Example#19, the glass area was determined to be 350 ft², the door area — 84 ft² and the net wall area 2,514 ft².

Pontiac, Michigan is not listed in our abbreviated degree-day table. We consult the weather bureau and find that Pontiac has 6591 degree-days in an average year. We note from Fig. 15-1 that with 6591 degree-days we must have a U_o of 0.265.

$$\frac{U_o}{A_o} = \frac{U_g\,A_g + U_w\,A_w + U_d\,A_d}{A_g + A_w + A_d}$$

U_g = glass U factor
U_w = wall U factor
U_o = door U factor
A_g = glass area
A_w = wall area
A_d = door area

For our building,

$$U_g = 1.10, \quad A_g = 350, \quad U_w = 0.14$$
$$A_w = 2514, \quad U_d = 0.49, \quad A_d = 84$$

$$\frac{U_o}{A_o} = \frac{1.10(350) + 0.14(2514) + 0.49(84)}{350 + 2514 + 84}$$

$$\frac{U_o}{A_o} = \frac{778}{2948} = 0.264$$

Next, let's consider the roof. The roof U factor is selected from Fig. 15-2. But there's an exception here that must be taken into account. If a cathedral ceiling is used, the U value shall be 0.05 for areas with less than 8000 degree-days and 0.04 for areas with more than 8000 degree-days. For our roof, we need a U factor of 0.07 or less. And if skylights are to be used, substitute skylights and net roof areas and U factors in place of glass and wall in equation 15-1.

FIGURE 15-2

U_o Roof/Ceilings – Type "B" Buildings

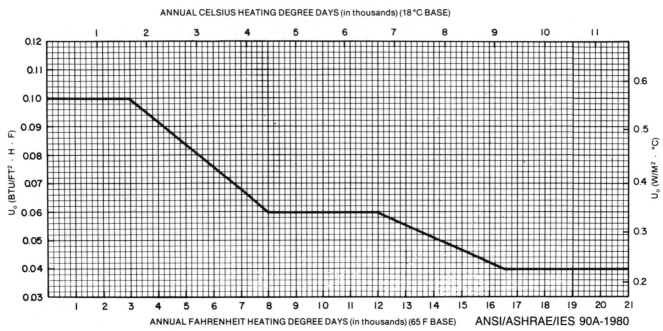

ANNUAL CELSIUS HEATING DEGREE DAYS (in thousands) (18 °C BASE)

ANNUAL FAHRENHEIT HEATING DEGREE DAYS (in thousands) (65 F BASE)   ANSI/ASHRAE/IES 90A-1980

FIGURE 15-3
Overall Thermal Transfer Values (OTTV$_w$)
Walls – Cooling Type "B" Buildings

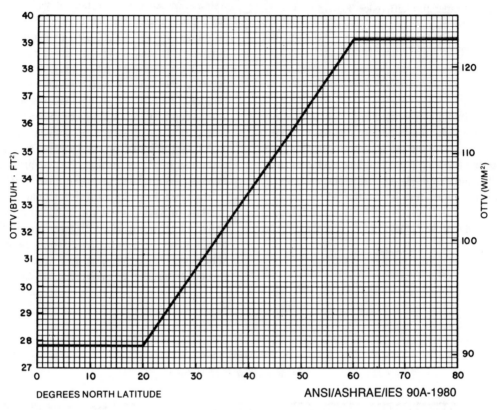

DEGREES NORTH LATITUDE

ANSI/ASHRAE/IES 90A-1980

We can indulge in some horse trading if we wish. Let's say that you want to keep the roof U factor at 0.15. First calculate the required average U for the building.

$$\text{Wall } 2948 \text{ ft}^2 \times 0.264 = 778.3$$
$$\text{Roof } 5383 \text{ ft}^2 \times 0.070 = 376.8$$
$$\text{Total } 8{,}331 \qquad\qquad 1{,}155.1$$

Now divide the total UA (1,155.1) by the total area (8,331) and get the overall envelope U of 0.139. Then solve to see what overall wall U is needed if a roof U of 0.15 is to be used.

$$8{,}331(0.139) = 5{,}383(0.15) + 2{,}948 \, U_o$$
$$U_o A = 0.119$$

With single glass (U = 1.10) and with the original wall (U = 0.14) we could never reach this value. However, if we switch to 2" of roof insulation (U = 0.087) and replace the single glass with insulating glass (U = 0.58) we will meet the required overall objective with a U of 0.139.

$$\frac{U_o}{A_o} = \frac{U_g \, A_g + U_w \, A_w + U_d \, A_d}{A_g + A_w + A_d}$$

$$\frac{U_o}{A_o} = \frac{(0.58)(350) + (0.14)(2514) + (0.49)(84)}{2{,}948}$$

$$\frac{U_o}{A_o} = 0.202$$

Now let's check the total building.

$$\text{Wall } 2948 \times 0.202 = 595.5$$
$$\text{Roof } 5383 \times 0.087 = 468.3$$
$$595.5 + 468.3 = 1063.8$$

$$\frac{1068.3}{2{,}948 + 5{,}383} = 0.128$$

That takes care of the skin load for heating but we can't forget cooling. In cooling, we use the OTTV which stands for overall thermal transfer value. It is calculated through the use of following equation.

$$\frac{[(U_w \times A_w \times TD_{cq}) + (A_g \times SF \times SC) + (U_g \times A_g \times \Delta T)]}{A_o}$$

where

OTTV = overall thermal transfer values (Fig. 15-3)
SF   = solar factor (Fig. 15-4)
TD_{eq} = equivalent temperature difference (Fig. 15-5)
SC   = shade coefficient
ΔT = temperature difference using the 2½% value for the outdoor condition and 70°F indoors.

Assume that our building is located in 42° north latitude. From Fig. 15-4 we find that 129.5 is the solar factor. Assume that we have changed the glass to insulating glass (U = 0.61) and that the wall U = 0.14. The wall mass/unit area is 77 lbs/ft² and from Fig. 15-5 the TD_{eq} = 23°. If the outside design temperature is 88° the ΔT = 10.

$$OTTV = \frac{[.14\ (2514)(23)\ +\ (350)(129)(.75)\ +\ (.61)(129)(10)]}{2,864}$$

$$OTTV = 0.27$$

By examining Fig. 15-3 we find that the OTTV should not exceed 34 Btu/hr/ft² and that we comply.

Now let's consider heating beginning with gas fueled systems. When gas is burned, it yields 80% efficiency at best. This means that for every therm (100,000 Btu) we purchase, we get 80,000 Btu in usable heat. Instead of burning this gas, if we opt use it to fuel a heat pump, we can get a minimum of 140,000 Btu. But how is this done? In the past, it's been tried with converted automotive engines. The problem — engines had short lives operating in that capacity. Running a combustion engine over a one year period for this type of application is the equivalent to 200,000 miles of on-road operation. A better solution is the use of a gas turbine. Gas turbines are just as efficient as gas engines, but they have much longer lives.

Engineers have known about heat pumps every since the days of Sadi Carnot, the father of refrigeration. Essentially, a heat pump is a refrigeration machine. Refrigeration machines transport heat from one place to another. When used for cooling, as in a window unit, heat is absorbed from the environment of the house and then

FIGURE 15-4
Solar Factor (SF) Values

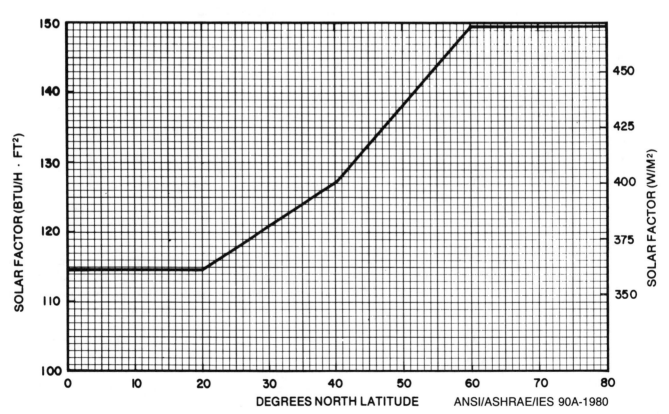

ANSI/ASHRAE/IES 90A-1980

FIGURE 15-5
For Temperature Difference (TD$_{EQ}$) Walls

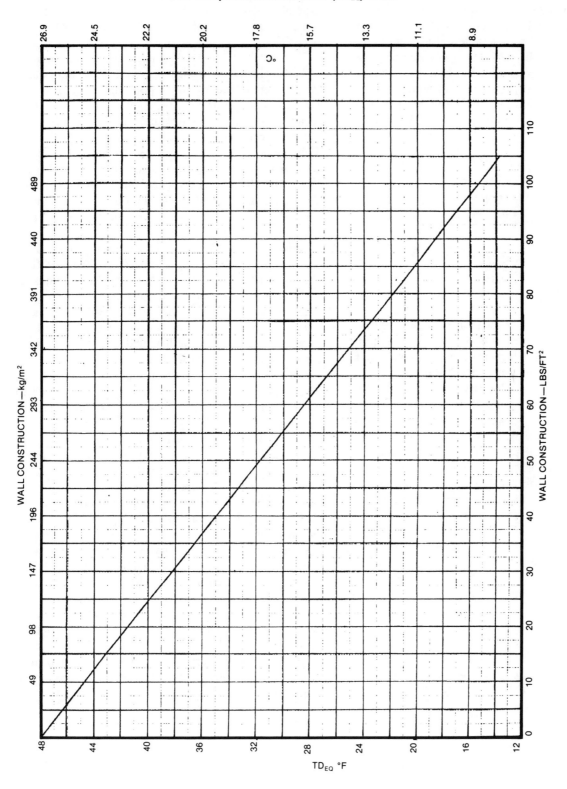

ANSI/ASHRAE/IES 90A-1980

discarded outside of the building. On the other hand, if we want to heat a house with a heat pump, the process is just reversed so that heat is deposited inside the house. This works exceptionally well at 40 °F outdoor temperature and above but with somewhat less efficiency at temperatures less than 40 °F. Table 15-1 gives the energy input and output for a typical air source heat pump. Notice the COP (coefficient of performance) column. COP represents the ratio of heat output to heat input. At 0 °F we get 2.3 kW of heat (7,854 Btu) for every kW of energy input.

Table 15-1

Outdoor Temp°F	kW/hr 1,000 Btu	COP
0	0.154	1.9
20	0.113	2.6
40	0.083	3.5
50	0.073	4.0

This is certainly a lot better than resistance heating where we get 1 kW of heat for every kW of energy purchased. Notice how much higher the COP is at 40 °F than at 0 °F. This leads us to search for 40 °F or higher heat sources. If the building was located near a lake, flowing stream or a well, we could utilize these as energy sources. If we had a 50 °F source, we would use only ¼ the energy in comparison to what we'd use with resistance heating. And don't count out the possibility of using exhaust air or waste water as a source.

Another way to utilize a heat pump is to make ice that can be stored and used in the summer for cooling. For every pound of ice we produce, 144 Btu must be extracted. The COP with this system would be 3. If we use this system in our super-insulated 1,500 ft² house, we would make 5787 ft³ of ice (or an equivalent of 181 tons). This would fill a storage cell 24' × 24' × 10' deep. In Michigan, that's three times as much ice as we can use in the summer. One solution would be to use solar heating to melt the ice so less ice would have to be stored. By utilizing solar heat, our storage area can be reduced to 1,000 ft³ that translates into a smaller 10' × 10' × 10' deep cell. If additional ice is required during the summer, it can be produced at night and stored for use during the day. In fact, during the winter, ice can be produced at night and the hot water resulting from the reduction can be stored for daytime use — provided the power company has an off-peak rate that can be utilized.

Another combination of heat pump and solar heating utilizes a eutectic salt that freezes at 80 °F. Just as with mixing rock salt with ice to depress the freezing, or rather

the melting point of ice, to yet colder temperatures, we can use certain salts to raise the freezing point. With this type of system we would have a COP of 4, and it'd require a larger storage pit of approximately 15' × 10' × 10' deep.

Everyday each household wastes considerable amounts of heat by literally pouring it down the drain. Bath water, dishwasher and laundry wastes can range from temperatures from 100° to 140 °F. Every gallon of hot water used means that a gallon of cold water, that ranges from temperatures between 50° to 70°, must be heated to 140 °F. Why not heat this water with the waste water? Launderettes that I designed years ago discharged laundry wastes into a pit. The pit was equipped a lint screen and a bundle of pipes. The cold makeup water was circulated through these pipes and was consequently preheated by the waste water.

In new homes, the discharge from sanitary facilities, such as toilets, could be piped to keep it separate from the hot wastes. The hot wastes could then be piped to a pit or heat exchanger to preheat cold water much in the same way I preheated the launderette makeup water. And vacuum breakers, check valves and double-walled heat exchangers could be installed to prevent the contamination of drinking water.

What about non-residential buildings? Can much energy be saved by reducing temperatures to 68 °F? In non-residential buildings, outside air is brought into the building for ventilating purposes. Reducing the heat required to warm this ventilation air may prove more economical than an reducing environment temperatures.

Remember from Chapter 5 that for non-industrial buildings the primary use for makeup air is dilution. A 20,000 ft² commercial building requires 5,000 cfm of outside air for contaminant control. A high efficiency filter, coupled with activated charcoal, will remove the particulate portion of contaminants along with the gases. This way, even cigarette smoke can be filtered out. Filtering can actually reduce the outside air requirement from 5,000 cfm to 1,000 cfm which will save some 3,700 therms in a 6,000 degree-day area. Consider the fact that temperature reduction alone in this building would only have saved a mere 500 therms in comparison. The figures in this example should serve to show you just how practical and dollar-wise energy conscious system designing can prove to be.

Index

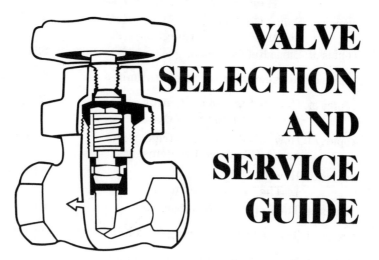

VALVE SELECTION AND SERVICE GUIDE

John T. Mead

This guide was written for contractors, servicemen and anyone who must size, select and install or repair valves on the job. The information in this book targets the air conditioning, heating and refrigeration industry, but the general principles and basic valve designs covered apply to a number of other fields as well.

Many manufacturers can help you select the right valve design required for your job, but the recommendations they make are based on the information you provide. If your information is sketchy or not thoroughly planned out, improper selections are likely to occur. The result, equipment failure, or even worse, personal injury and/or a lawsuit that can translate into considerable financial loss.

Get the Valve Selection and Service Guide so you can learn to select the proper valves for your applications and **prevent costly mistakes.** It may prove to be the best investment you'll ever make!

BNP Business News Publishing Company
P.O. Box 2600, Troy, MI 48007

Rush me the following books:

_____copies of: Smart Contracting Dollar & Sense Strategies	$12.95 each
_____copies of: Basic Refrigeration	$24.95 each
_____copies of: Refrigeration Licenses Unlimited	$29.95 each
_____copies of: Getting Started in Heating and Air Conditioning Service	$27.95 each
_____copies of: How to Solve Your Refrigeration and A/C Service Problems	$19.95 each
_____copies of: How To Make It In The Service Business	$19.95 each
_____copies of: The Reference Notebook Set	$15.95 each
_____copies of: The Schematic Wiring Book Set	$24.95 each
_____copies of: Hydronics, the art of cooling and heating with water	$9.95 each
_____copies of: Good Piping Practice	$5.95 each
_____copies of: The Valve Selection & Service Guide	$39.95 each
_____copies of: The Service Hot Line Handbook Vol. I	$12.95 each
_____copies of: The Service Hot Line Handbook Vol. II	$12.95 each
_____copies of: The Service Hot Line Handbook Vol. III	$12.95 each

Total number of books_____   **Total $_____**
(count set as one book)

All orders over $40.00 must be prepaid.*

☐Check ☐Money Order enclosed for $_____
☐Bill me and add shipping/handling charges.

Allow 3–6 weeks for delivery.
For express delivery, add extra 15% of order total.

Send to:
Name _____

Company_____

Address _____

City/State/Zip _____

Please check one: ☐New Customer ☐Previous Customer

*Only prepaid orders eligible for free shipping/handling.
(Prices subject to change without notice.)